# Intertidal Fishes

## LIFE IN TWO WORLDS

---

EDITED BY
**Michael H. Horn**
*Department of Biological Science*
*California State University*
*Fullerton, California*

**Karen L. M. Martin**
*Department of Biology*
*Pepperdine University*
*Malibu, California*

**Michael A. Chotkowski**
*California Department of Fish & Game*
*Stockton, California*

**ACADEMIC PRESS**
San Diego London Boston New York Sydney Tokyo Toronto

*Front cover photographs:* Inset photographs are (top) *Cebidichthys violaceus,* taken by Michael Horn; (middle) *Oligocottus maculosus,* taken by Stephen Norton; and (bottom) *Xererpes fucorum,* taken by Michael Horn. Background photograph is the intertidal zone at San Simeon, California, taken by Karen Martin.

This book is printed on acid-free paper. ∞



Academic Press
*a division of Harcourt Brace & Company*
525 B Street, Suite 1900, San Diego, California 92101-4495, USA
http://www.apnet.com

Academic Press
24-28 Oval Road, London NW1 7DX, UK
http://www.hbuk.co.uk/ap/

Library of Congress Catalog Card Number: 98-85619

International Standard Book Number: 0-12-356040-3

PRINTED IN THE UNITED STATES OF AMERICA
98 99 00 01 02 03 QW 9 8 7 6 5 4 3 2 1

# CONTENTS

**Contributors** xi

**Preface** xiii

## I. BACKGROUND, METHODS, AND BASIC PATTERNS

### 1. Introduction

I. Definition and Rationale 1
II. The Intertidal Zone as Fish Habitat 2
III. Tidal Ebb and Flow and the Rhythmic Behavior of Intertidal Fishes 2
IV. Early Knowledge and Use of Intertidal Fishes 3
V. Conservation of Intertidal Fishes 4
VI. Overall Goals of This Book 5
References 5

### 2. Methods for Studying Intertidal Fishes

I. Introduction 7
II. Describing the Environment of Intertidal Fishes 7
III. Observing Intertidal Fishes in the Field 8
IV. Capturing Intertidal Fishes 9
V. Estimating the Abundance of Intertidal Fishes 16
VI. Marking Intertidal Fishes 16
VII. Transporting Intertidal Fishes 20
VIII. Laboratory Experiments Involving Intertidal Fishes 20
IX. Standard Techniques Applied to the Study of Intertidal Fishes 21

X. Safety Considerations 21
References 21

## 3. Vertical Distribution Patterns

I. General Ecological Patterns 26
II. Vertical Distribution of Intertidal Fishes Is the Result of Inter- and Intraspecific Habitat Partitioning 32
III. Temporal Variations in Vertical Distribution 39
IV. Coping with a Dynamic Habitat 40
V. Future Work 48
References 48

# II. PHYSIOLOGICAL SPECIALIZATIONS

## 4. Respiration in Water and Air

I. Introduction 54
II. The Respiratory Environment of the Intertidal Zone 55
III. Emergence Behaviors 56
IV. Respiratory Organs of Intertidal Fishes 60
V. Respiratory Gas Exchange in Water and Air 63
VI. Ventilation 64
VII. Circulation 67
VIII. Acid–Base Balance 70
IX. Blood Oxygen Affinity 73
X. Future Work 73
References 74

## 5. Osmoregulation, Acid–Base Regulation, and Nitrogen Excretion

I. Introduction 79
II. Osmoregulation 79
III. Acid–Base Regulation 83
IV. Nitrogen Excretion 85
V. Summary and Future Directions 92
References 92

# III. BEHAVIOR AND THE SENSORY WORLD

## 6. Movement and Homing in Intertidal Fishes


I. Introduction 97
II. Spatial and Temporal Scales of Intertidal Movement 97
III. Intertidal Movements in Different Habitats 98
IV. Homing 112
V. Functions of Intertidal Movements and Homing 112
VI. Control Mechanisms 118
VII. Modeling Intertidal Movements 119
VIII. Conclusions 120
References 121


## 7. Sensory Systems


I. What Sensory Systems Are "Typical" in Intertidal Fish? 126
II. Approaches 127
III. Toward the Surface 127
IV. Small Size as a Beneficial(?) Constraint 133
V. The Chemosensory Anterior Dorsal Fin in Rocklings: A Case Study 134
VI. Clues from the Deep Sea? 136
VII. Conclusions: Does the Intertidal Produce Convergent Sensory Systems? 138
References 139


# IV. REPRODUCTION

## 8. Intertidal Spawning


I. Introduction 143
II. Taxonomic Survey 144
III. Reproductive Biology 145
IV. Environmental Influences 151
V. Ultimate Factors and Intertidal Spawning 154
VI. Summary and Conclusions 158
References 158

## 9. Parental Care in Intertidal Fishes


I. Introduction 165
II. Patterns 166
III. Process 174
IV. Conclusions and Recommendations 177
References 178


## 10. Recruitment of Intertidal Fishes


I. Introduction 181
II. The Timing of Recruitment 182
III. Factors Affecting the Success of Recruitment 184
IV. Patterns of Recruitment 188
V. Population Consequences of Recruitment Patterns 190
VI. Future Research 192
References 193


# V. TROPHIC RELATIONSHIPS AND COMMUNITY STRUCTURE

## 11. Herbivory


I. Introduction 197
II. Taxonomic Diversity and Latitudinal Patterns 198
III. Feeding Mechanisms and Food Habits 203
IV. Gut Morphology and Digestive Mechanisms 207
V. Case Studies of Herbivorous Taxa 211
VI. Ecological Impacts of Intertidal Herbivorous Fishes 215
VII. Recommendations for Future Studies 216
References 218


## 12. Predation by Fishes in the Intertidal


I. Introduction 223
II. Recent Studies of the Diets of Intertidal Fishes 227
III. Influence of Habitat on the Diets of Intertidal Fishes 228
IV. A Functional Approach to Understanding Fish Diets 239
V. Ontogenetic Changes in the Diets of Intertidal Fishes 245

VI. Other Factors Influencing Diet Choice in Fishes 250
VII. Impact of Fish Predation on Intertidal Communities 251
VIII. Summary 254
References 255

## 13. Intertidal Fish Communities

I. Introduction 264
II. Functional Components of Community Structure 265
III. Spatial Variation in Community Structure 270
IV. Temporal Variation in Community Structure 279
V. Interspecific Interactions 284
VI. Intraspecific Interactions 286
VII. Predation 287
VIII. Overview 288
References 290

# VI. SYSTEMATICS AND EVOLUTION

## 14. Systematics of Intertidal Fishes

I. Introduction 297
II. The Intertidal Ichthyofauna 298
III. Taxonomic Summary of Intertidal Fishes 323
IV. Phylogenetic Relationships among Intertidal Fishes 326
V. Conclusions 326
References 329

## 15. Biogeography of Rocky Intertidal Fishes

I. Introduction 332
II. Scope of Study and Methodology 333
III. Overview of Faunas 339
IV. Similarity Relationships among Regional Intertidal Ichthyofaunas 347
V. Latitudinal Patterns of Species Richness 351
VI. Conclusions 352
References 353

## 16. Convergent Evolution and Community Convergence: Research Potential Using Intertidal Fishes

I. Introduction 356
II. Description of Six Intertidal Fish Communities 358
III. Convergent Adaptation 366
IV. Community Convergence 370
V. Recommendations for Future Studies 371
References 371

## 17. The Fossil Record of the Intertidal Zone

I. Introduction 373
II. Criteria for Recognizing Fossil Shores 373
III. Freshwater–Marine Controversy 380
IV. Fossil Examples 381
V. The Fossil Record of Intertidal Fishes 387
References 388

**Index** **393**

# CONTRIBUTORS

*Numbers in parentheses indicate the pages on which the authors' contributions begin.*

**Christopher Robert Bridges** (54), Institut für Zoophysiologie–Stollwechselphysiologie, Heinrich Heine Universität, D-40225 Dusseldorf, Germany

**Donald G. Buth** (297, 332), Department of Biology, University of California, Los Angeles, California 90095

**Michael A. Chotkowski** (1, 297, 332), California Department of Fish and Game, Stockton, California 95205

**J. B. Claiborne** (79), Mt. Desert Island Biological Laboratory, Salsbury Cove, Maine 04672; and Department of Biology, Georgia Southern University, Statesboro, Georgia 30460

**Ronald M. Coleman** (165), Department of Integrative Biology, University of California, Berkeley, California 94720

**Amy E. Cook** (223), Department of Ecology and Evolutionary Biology, University of California, Irvine, California 92697

**Edward E. DeMartini** (143), National Marine Fisheries Service, Southwest Fisheries Science Center, Honolulu, Hawaii 96822

**David H. Evans** (79), Mt. Desert Island Biological Laboratory, Salsbury Cove, Maine 04672; and Department of Zoology, University of Florida, Gainesville, Florida 32611

**R. N. Gibson** (7, 97, 264), Centre for Coastal and Marine Sciences, Dunstaffnage Marine Laboratory, Oban, Argyll PA34 4AD, Scotland

**Michael H. Horn** (1, 197, 356), Department of Biological Science, California State University, Fullerton, California 92834

**Gregg A. Kormanik** (79), Mt. Desert Island Biological Laboratory, Salsbury Cove, Maine 04672; and Department of Biology, University of North Carolina, Asheville, North Carolina 28807

**Kurt Kotrschal** (126), Zoologisches Institut der Universität Wien and Konrad-Lorenz-Forschungsstelle, A-4645 Grünau 11, Austria

**Karen Martin** (1, 26, 54), Department of Biology, Pepperdine University, Malibu, California 90236

**Jürgen Nieder** (26), Botanisches Institut, Universität Bonn, D-53115 Bonn, Germany

**Stephen F. Norton** (223), Department of Biology, East Carolina University, Greenville, North Carolina 27858

**F. Patricio Ojeda** (197), Departamento de Ecologia, Facultad de Ciencias Biologicas, Pontificia Universidad Catolica de Chile, Casilla 114-D, Santiago, Chile

**Catherine A. Pfister** (181), Department of Ecology and Evolution, University of Chicago, Chicago, Illinois 60637

**Kim Prochazka** (297, 332), Zoology Department, University of Cape Town, Rondebosch 7700, South Africa

**Hans-Peter Schultze** (373), Institut für Paläontologie, Museum für Naturkunde der Humboldt-Universität, Berlin, Germany

**R. M. Yoshiyama** (264), Department of Wildlife, Fish and Conservation Biology, University of California, Davis, California 95616

**C. Dieter Zander** (26), Zoologisches Institut und Museum, Universität Hamburg, D-2000 Hamburg 13, Germany

# PREFACE

Intertidal fishes attract attention because many spend part of their lives out of water at the very edge of the ocean environment, yet they are small and secretive and thus have remained poorly known. These fishes are of wide biological interest for several reasons, and this book is designed to meet the needs of biologists in a number of fields.

Intertidal fishes live at the interface of land and sea where they are alternately submerged in water and exposed to the air, often under turbulent conditions. Different species have met these challenges to varying degrees, and thus they are of interest to ecologists, behaviorists, and physiologists.

The disjunct distribution of the rocky intertidal habitats poses some potential barriers to dispersal and gene flow. The fishes represented in the intertidal habitat presumably evolved from fully aquatic, deeper water relatives. The richness or absence of species in certain genera and families at particular localities may provide intriguing opportunities for phylogenetic and biogeographic analyses by systematic biologists.

These fishes may provide a present-day glimpse of some of the requirements for the colonization of land that took place in the early evolution of terrestrial vertebrates. As such, they are of interest to evolutionary biologists, vertebrate biologists, and paleontologists.

Intertidal fishes occupy a habitat readily subject to human alteration. Although they may be suited to life in a turbulent, fluctuating environment, intertidal fishes are nevertheless vulnerable to the effects of pollution, sedimentation, and other insults. They may serve as effective biomonitors of the health of the intertidal ecosystem. Their local absence may signal that the habitat has become excessively damaged and that the species themselves are in decline. In this regard, intertidal fishes are relevant to the interests of conservation biologists and coastal zone managers.

Impetus for this book arose from a symposium on intertidal fishes we convened at the annual meeting of the American Society of Ichthyologists and Herpetologists in Los Angeles in June 1994. This meeting gave us the opportunity to hear new information about intertidal fishes and learn of innovative approaches to studying them. Our discussions with the participants indicated that contributors were available to produce the multiauthored volume we envisioned. Charles Crumly at Academic Press showed an early interest in the project and sustained that interest through to the final product.

M.H.H. extends sincere thanks to Janine Kido for hours spent exercising her illustration skills; Chris Paulin for loaning specimens of intertidal fishes, providing information on various aspects of the New Zealand fauna, and giving permission to use some of his published work; Patricio Ojeda and Alejandro Muñoz for providing specimens of Chilean intertidal fishes; Kim Prochazka for sending specimens of South African intertidal fishes; Robin Gibson for searching out information on early knowledge and use of intertidal fishes; Ken Gobalet for sharing his knowledge of intertidal fishes found in early California middens; Steve Murray for information about marine life refuges; and Anna Gawlicka for appreciating intertidal fishes, reading manuscript material in progress, and offering needed inspiration throughout the writing process.

K.L.M. gratefully acknowledges Doug Martin, a Pisces, and Gregory and Alex Martin, two accomplished intertidal fishermen, for many hours of discussion and fun. Many thanks to Stuart Sumida for his distinguished counsel.

M.A.C. cannot adequately thank Cindy Dole for her boundless patience, enthusiastic use of the "tidepool trawl" on many occasions, and unwavering support. M.A.C. also thanks James Clegg, Director of the UC Davis Bodega Marine Laboratory, for the use of space and facilities during 1997–1998.

All three of us appreciate the professionalism, patience, and consistent enthusiasm for this project shown by editor Charles Crumly and editorial assistant Hazel Emery at Academic Press.

Mike Horn
Karen Martin
Mike Chotkowski

# 1

# Introduction

M. H. Horn

*Department of Biological Science, California State University, Fullerton, California*

K. L. M. Martin

*Department of Biology, Pepperdine University, Malibu, California*

M. A. Chotkowski

*California Department of Fish & Game, Stockton, California*

## I. Definition and Rationale

Fishes that occupy the narrow band of habitat between the tidemarks on marine shores have proven to be difficult to classify in terms of their use of this restricted living space. Some species spend most of their lives there, others descend to deeper waters as they grow larger, and still others enter this zone only during high tide periods. A spectrum of occupancy is apparent, and it is possible to imagine that virtually all inshore fish species may at some time or other be found in the intertidal zone. Recognition of a continuum only, however, fails to serve an operational purpose and blurs the real differences that exist among the fish species found in this habitat. This book follows the distinction that fishes can be classified as resident species if they occupy the habitat on a permanent basis or as transients or visitors if they spend varying periods of their life in the intertidal zone (see Gibson and Yoshiyama, Chapter 13, this volume). Although transient species are included in the accounts in most of the chapters, the primary focus is on the resident members of the intertidal zone. These latter species, usually small, and cryptic in habit and coloration, are the taxa generally considered to be specialized for intertidal life.

The impetus for this volume stems from the fact that intertidal fishes have often been overlooked in studies of intertidal biota. Seaweeds and invertebrates have traditionally received greater attention, perhaps in large part because they are more obvious, more abundant, and more easily studied. The small size, secretive nature, and apparent rarity of many intertidal fishes have caused them to be frequently ignored by biologists interested in intertidal organisms. However, a growing awareness that resident fishes are integral components of nearshore food webs, specialized to cope with the rigors of the intertidal environment, has led to increased research on these animals in recent decades. New methods and innovative approaches have contributed to this expanding knowledge. The purposes of this book are to capture the essence of the recent work and to provide a broad survey of what is known about intertidal fishes in the context of their relatives in other habitats.

## II. The Intertidal Zone as Fish Habitat

The turbulent, wave-swept rocky intertidal habitat is highly productive (Leigh *et al.,* 1987), rich in plant and invertebrate life, and highly variable in space and time. The shallow water of the coastal environment magnifies these variations and is also influenced by nearshore natural processes as well as by pollution runoff and other human uses of the land.

Variation in space occurs at different levels of scale. Rocky shores themselves are separated from one another, interspersed with sandy beaches, steep bluffs, mudflats, or some combination of these. Some rocky shores are near estuaries or other sources of freshwater inflow. Within the habitat, rocky shores encompass boulder flats, reef outcrops, and rock-pool microhabitats, and many rocky coasts include combinations of all of these, along with gravel, sand, and shell substrata. The boulders and substrata are subject to movement and remodeling by storms and winter waves, and, on a longer timescale, by erosion and tectonic activity. Intertidal fishes survive these disturbances by finding refuge in the complex physical structure of the rocky shore environment.

Variation in time includes the effects of tides, time of day, and season. The tides influence the presence or absence of water and the effects of waves on a predictable cycle, with tidal heights varying hourly and changing according to lunar rhythms. The combination of tidal cycle and time of day can lead to profound and rapid physical and chemical changes in isolated tidal pools. Temperature and salinity can fluctuate dramatically as a result of insolation and evaporation during daytime low tides. Oxygen levels rise during periods of photosynthesis but dip at night when relatively high amounts of respiration occur. As respiration continues, carbon dioxide accumulates and can lower the pH of an isolated pool (Davenport and Woolmington, 1981).

Fishes that emerge from the water completely and those that occur high in the intertidal zone are subject to greater stresses than those species that remain in the water or live lower on the shore (Zander *et al.,* Chapter 3; Martin and Bridges, Chapter 4, this volume). Low tides can make intertidal animals vulnerable to terrestrial predators, including birds, snakes, raccoons, feral cats, and, of course, ichthyologists. Seasonal cycles cause changes in the amount of freshwater runoff, in water and air temperatures, and in the amount of solar radiation and plant productivity. Some fishes move into the intertidal zone seasonally for reproduction, depositing eggs and occasionally even remaining to guard the nest (see DeMartini, Chapter 8, this volume and Coleman, Chapter 9, this volume). Thus, the community of fishes can change seasonally as well as tidally.

## III. Tidal Ebb and Flow and the Rhythmic Behavior of Intertidal Fishes

Because the daily progression of the tides is profoundly important and highly predictable, both resident and transient occupants of the intertidal zone generally have strongly entrained behavioral and physiological rhythms (Gibson, 1992; Burrows *et al.,* 1994). In many nearshore fishes, these rhythms are manifested in programmed endogenous variations in foraging activity, oxygen consumption rates, and use of habitat features. Although it is difficult to observe them directly in the field (except on the calmest of days), most resident intertidal fishes are probably diurnal and most active during high tides (Gibson, 1982, 1992). There also exist tidal rhythms with longer periods; fortnightly and monthly (lunar) cycles are especially important. The intensity of moonlight at night and timing of moonrise and moonset both vary with daily tidal excursion on a lunar-monthly cycle. This cycle creates a predictable schedule of tidal height and nighttime lighting that many fishes, especially those that breed in the high intertidal zone, exploit during spawning (see DeMartini, this volume, Chapter 8, Section III.B.3, for a review and examples). Entrained behavioral and physiological rhythms strongly affect the laboratory behavior of intertidal fishes and may persist or reassert themselves many days after removal of the fishes from the field (Gibson, Chapter 2, this volume; Northcott *et al.,* 1990).

## IV. Early Knowledge and Use of Intertidal Fishes

Most prehistoric peoples with access to coastal waters must have known about intertidal fishes, and many probably used them as food, yet documentation of this knowledge and use is sparse and unevenly distributed. Materials obtained from archaeological sites provide some evidence of the capture and use of shorefishes. For example, young saithe (*Pollachias virens*, Gadidae), which use shore pools in the temperate North Atlantic as nursery areas (Rangeley and Kramer, 1995), were caught in large numbers by hook and line in Neolithic times in northern Scotland where land-based foods were minimal (Wheeler and Jones, 1989). Also, the numerous pharyngeal bones of *Symphodus* (=*Crenilabrus*) *melops* recovered from a Neolithic tomb in the Orkney Islands of Scotland is proposed as evidence by Wheeler (1977) that this transient intertidal species of wrasse (Labridae) was more abundant at that time than at present in these islands. In New Zealand, the Maori used the intertidal zone extensively as a source of food in pre-European times, although the small bones of intertidal fishes are rarely preserved in middens (Paulin and Roberts, 1992). Wrasse bones (especially the vomer) are among the most common bones recorded in Maori middens, showing that intertidal and subtidal fishes were captured but suggesting that the bones of less well-ossified species had disintegrated (C. D. Paulin, personal communication). The advent of European fishing methods focused attention on demersal subtidal species because intertidal fishes were considered not suitable for consumption. Attempts by historians in the early twentieth century to document Maori fishing methods and their use of intertidal fishes came too late for by that time much of the information had been lost. This loss is illustrated by the fact that Maori names are now available for only 9% of rockpool fish species as compared to 63% of species in adjacent subtidal waters (Paulin and Roberts, 1992).

Perhaps the best available records of use of intertidal fishes in historic and prehistoric times come from the coast of western North America, especially California. The diversity of Native American peoples in what is now California and the extensive use of marine resources by many tribes (see Heizer, 1978) are factors contributing to the numerous middens rich in the remains of coastal fish species. Analysis of fish remains from 51 central California archaeological sites dated from 6200 B.C. to A.D. 1830 shows that more than half of the rocky coast material represented large inshore species, including the rockfishes *Sebastes* spp. (Scorpaenidae), the kelp greenling *Hexagrammos decagrammus* (Hexagrammidae), the cabezon *Scorpaenichthys marmoratus* (Cottidae), and the pricklebacks *Cebidichthys violaceus* and *Xiphister* spp. (Stichaeidae) (Gobalet and Jones, 1995). The stichaeids are resident intertidal species and at least some of the others are transients in the intertidal zone, especially as juveniles (see Chotkowski *et al.,* Chapter 14, this volume). Other studies of California middens provide similar lists of species. More than 80% of remains collected at Humboldt County sites on the northern California coast consisted of cabezon, greenling, and rock prickleback, suggesting extensive exploitation of rocky intertidal fishes by native peoples, probably using hand collection, spearing, and even poisoning (Gobalet, 1997). Trade or importation of intertidal species is implied in a study showing that intertidal fishes appear in inland archaeological sites at distances ranging to more than 80 km from the central California coast (Gobalet, 1992). Currently, several species of rockfishes that visit the intertidal zone are of sport and commercial importance on the west coast of North America (Leet *et al.,* 1992), and the cabezon is a sport fish and one of the few cottids of commercial importance (Eschmeyer *et al.,* 1983). The monkeyface and rock pricklebacks are subjects of a small sport fishery ("poke poling") on the central and northern California coasts (Eschmeyer *et al.,* 1983; personal observation).

## V. Conservation of Intertidal Fishes

Although intertidal fishes are mostly small and cryptic and, as already mentioned, often overlooked in scientific studies of intertidal communities, they are nevertheless vulnerable to overexploitation through indiscriminate collecting and to population reduction from the effects of pollution, trampling, and habitat loss. These fishes together occupy a variety of microhabitats, including tidepools, vegetation, and under-rock spaces, each of which is subject to somewhat different degradative forces. Pools not only concentrate fishes, making them easier to collect, but are also sinks for silt and other contaminants, even if on a temporary basis. Vegetative (seaweed or seagrass) habitat can be obliterated by trampling or overharvesting of the plants. Habitat space beneath fixed benchrock or movable boulders can also be disturbed by siltation and pollution. The fishes occupying this space, especially those found beneath small loose boulders, can be easily collected. Unfortunately, little is known about these impacts on intertidal fish populations. In addition to the removal and recolonization studies that have thus far been completed (see Gibson and Yoshiyama, Chapter 13, this volume), research is needed to assess the degree to which these populations change in response to short- and long-term disturbances.

Global climate change, especially ocean warming, is one of the disturbances that may exert short- and long-term impacts on intertidal fish distributions and other aspects of their biology. Major impacts have been predicted for marine ecosystems (e.g., Fields *et al.,* 1993) and evidence is accumulating that human-induced warming trends of the past few decades have caused shifts in the distribution and abundance of marine organisms. For example, during this period zooplankton biomass has declined by 80% in the California Current (Roemmich and McGowan, 1995), and anchovy and sardine abundances have changed dramatically in the Southern California Bight (Smith, 1995). Some intertidal invertebrates of southern affinities show increases in abundance on the cool-temperate central California coast, whereas other, more northerly species show declines (Barry *et al.,* 1995). Subtidal reef fishes at two sites in southern California waters have declined in diversity and abundance over the past two decades, apparently in response to lowered productivity associated with an abrupt temperature increase in 1976–1977 (Holbrook *et al.,* 1997). No studies, however, appear to have been completed as yet on the effects of ocean warming on intertidal fish populations; research is clearly needed in this area.

The growing realization of the variety of impacts facing marine organisms has resulted in increased efforts to design, establish, and manage marine refuges to protect coastal biodiversity (e.g., Cole *et al.,* 1990; Carr and Reed, 1992; Rowley, 1994; Allison *et al.,* 1998). These authors argue that such refuges can provide protection for critical areas, spatial escape for heavily exploited species, and a buffering effect against unforeseen or unusual conditions (perhaps such as global warming) and management miscalculations. Although reserves are being established at a rapid rate in coastal regions in different parts of the world (Rowley, 1994), design principles are still being formulated, and the need for increased scientific imput to ensure their effectiveness has been advocated (Allison *et al.,* 1998). Thus, far, intertidal fishes as a group have not been included explicitly in plans to protect coastal organisms.

Marine reserves have been shown to support higher densities and larger sizes of heavily fished species; less evidence is available to demonstrate a "spillover" of individuals across refuge borders to enhance local catches, and even less evidence has been garnered to show that larval export from reserves augments recruitment over larger regions (Rowley, 1994). Not surprisingly, marine refuges have been designed and established with a focus on subtidal fishes. These include species of economic importance, and this factor has been a principal driving force in refuge designation. Again, just as with efforts to study them, the small size, cryptic nature, and apparent rarity of many intertidal fish species has worked to their disadvantage in terms of recognized importance and increased protection. This situation prevails despite some evidence showing that certain coastal species occur in the intertidal habitat only as juveniles and thus use this habitat as a nursery area (see Gibson and Yoshiyama, Chapter 13, this volume). There now exist marine protected areas that are supposed to serve as the primary agent for protecting and maintaining intertidal ecosystems. For example, along the southern California coast, a heavily urbanized region, California Marine Life Refuges are the main form of marine protected area designed to protect intertidal populations (Murray, 1997). Although restrictions are placed on collecting most species of seaweeds and invertebrates, recreational fishing is allowed, and visitor access is not limited. No attention is given to protecting intertidal fishes. After a year-long study, Murray (1997) concluded that these California coastal refuges are ineffective in protecting coastal populations in areas of high visitor density. Plans to improve refuge effectiveness should recognize the existence of intertidal fishes and incorporate methods to ensure their protection.

## VI. Overall Goals of This Book

The major objective of this volume is to provide an up-to-date survey of what is known about intertidal fishes. Coverage includes aspects of their behavior, physiology, and ecology as well their systematic relationships, distributional patterns, and evolutionary histories. More than 700 species have been recorded in the intertidal zone, and many of these are specialized for an amphibious life in this habitat and are integral members of inshore food webs. A primary purpose of the book is to stimulate further study, greater appreciation, and stronger protection of this fascinating array of fishes.

## References

Allison, G. W., Lubchenco, J., and Carr, M. H. (1998). Marine reserves are necessary but not sufficient for marine conservation. *Ecol. Appl.* **8**(Suppl.), S79–S92.

Barry, J. P., Baxter, C. H., Sagarin, R. D., and Gilman, S. E. (1995). Climate-related, long-term faunal changes in a California rocky intertidal community. *Science* **267,** 672–675.

Burrows, M. T., Gibson, R. N., and MacLean, A. (1994). Effects of endogenous rhythms and light conditions on foraging and predator-avoidance in juvenile plaice. *J. Fish Biol.* **45**(Suppl. A), 171–180.

Carr, M. H., and Reed, D. C. (1992). Conceptual issues relevant to marine harvest refuges: Examples from temperate fishes. *Can. J. Fish. Aquat. Sci.* **50,** 2019–2028.

Cole, R. G., Ayling, T. M., and Creese, R. G. (1990). Effects of marine reserved protection at Goat Island, northern New Zealand. *New Zeal. J. Mar. Freshw. Res.* **24,** 197–210.

Davenport, J., and Woolmington, A. D. (1981). Behavioural responses of some rocky shore fish exposed to adverse environmental conditions. *Mar. Behav. Physiol.* **8,** 1–12.

Eschmeyer, W. N., Herald, E. S., and Hammann, H. (1983). "A Field Guide to Pacific Coast Fishes of North America." Houghton Mifflin, Boston.

Fields, P. A., Graham, J. B., Rosenblatt, R. H., and Somero, G. N. (1993). Effects of expected global climate change on marine faunas. *Trends Ecol. Evol.* **8,** 361–367.

Gibson, R. N. (1982). Recent studies on the biology of intertidal fishes. *Oceanogr. Mar. Biol. Annu. Rev.* **20,** 363–414.

Gibson, R. N. (1992). Tidally-synchronized behaviour in marine fishes. *In* "Rhythms in Fishes" (M. A. Ali, Ed.), pp. 63–81. Plenum Press, New York.

Gobalet, K. W. (1992). Inland utilization of marine fishes by Native Americans along the central California coast. *J. Calif. Great Basin Anthropol.* **14,** 72–84.

Gobalet, K. W. (1997). Fish remains from the early 19th century native Alaskan habitation at Fort Ross. *In* "The Archaeology and Ethnobotany of Fort Ross, California," Volume II, "The Native Alaskan Neighborhood Multiethnic Community at Colony Ross" (K. G. Lightfoot, A. M. Schiff, and T. A. Wake, Eds.), pp. 319–327. Contrib. Univ. Calif. Archaeol. Res. Facility, No. 55, Berkeley.

Gobalet, K. W., and Jones, T. L. (1995). Prehistoric Native American fisheries of the central California Coast. *Trans. Am. Fish. Soc.* **124,** 813–823.

Heizer, R. F. (Ed.). (1978). "California," Volume 8, "Handbook of North American Indians" (W. C. Sturtevant, Gen. Ed.). Smithsonian Institution, Washington.

Holbrook, S. J., Schmitt, R. J., and Stephens, J. S., Jr. (1997). Changes in an assemblage of temperate reef fishes associated with a climate shift. *Ecol. Appl.* **7,** 1299–1310.

Leet, W. S., Dewees, C. M., and Haugen, C. W. (1992). "California's Living Marine Resources and Their Utilization." California Sea Grant Extension Program Publication UCSGEP-92-12, Davis.

Leigh, E. G., Jr., Paine, R. T., Quinn, J. F., and Suchanek, T. H. (1987). Wave energy and intertidal productivity. *Proc. Natl. Acad. Sci. USA* **84,** 1314–1318.

Murray, S. N. (1997). Effectiveness of marine life refuges on southern California shores. *In* "California and the World Ocean '97," pp. 1453–1465. Proc. Conf. Am. Soc. Civil Engineers, San Diego.

Northcott, S. J., Gibson, R. N., and Morgan, E. (1990). The persistence and modulation of endogenous circatidal rhythmicity in *Lipophrys pholis* (Teleostei). *J. Mar. Biol. Assoc. U.K.* **70,** 815–827.

Paulin, C., and Roberts, C. (1992). "The Rockpool Fishes of New Zealand, Te ika aaria o Aotearoa." Museum of New Zealand, Te Papa Tongarewa, Wellington.

Rangeley, R. W., and Kramer, D. L. (1995). Use of rocky intertidal habitats by juvenile pollock *Pollachius virens. Mar. Ecol. Progr. Ser.* **126,** 9–17.

Roemmich, D., and McGowan, J. (1995). Climatic warming and the decline of zooplankton in the California Current. *Science* **267,** 1324–1326.

Rowley, R. J. (1994). Case studies and reviews: Marine reserves in fisheries management. *Aquat. Conserv.: Mar. Freshw. Ecosyst.* **4,** 233–254.

Smith, P. E. (1995). A warm decade in the Southern California Bight. *Calif. Coop. Oceanic Fish. Invest.* **36,** 120–126.

Wheeler, A. (1977). Fish bone. *In* "Excavations in King's Lynn 1963–1970" (H. Clarke and A. Carter, Eds.), pp. 46–468. Soc. Medieval Archaeol. Monogr. Ser. 7.

Wheeler, A., and Jones, A. K. G. (1989). "Fishes." Cambridge Manuals in Archaeology. Cambridge University Press, Cambridge.

# 2

# Methods for Studying Intertidal Fishes

R. N. Gibson

*Centre for Coastal and Marine Sciences, Dunstaffnage Marine Laboratory, Oban, Argyll, Scotland*

## I. Introduction

For the ichthyologist, the intertidal zone has many advantages as a location for studying fishes, including its ease of access, shallow water, and the fact that fishes are often concentrated in restricted areas at low tide. These advantages are counterbalanced, however, by the nature of the habitat and of the fishes themselves. Soft sediments, the rugosity of rocky shores, dense plant cover, or the turbulence caused by swell and breaking waves can all make capture and observation of fishes difficult. In addition, many intertidal species are small, well camouflaged, and cryptic in their behavior. Nevertheless, none of these problems is unique or insuperable and in most cases methods have been developed to overcome the difficulties, although the varied nature of intertidal habitats means that few methods are universally applicable.

This chapter concentrates on methods that have been specially devised for the study of intertidal fishes *in situ* but also includes more generally applicable methods where they have been applied to intertidal species. It deals principally with methods of observation, capture, and marking and summarizes representative studies in which particular methods are described or illustrated in detail. It is intended to be a practical guide to methodology and further sources of information rather than an exhaustive review of the literature.

## II. Describing the Environment of Intertidal Fishes

An important part of the study of intertidal fishes is the ability to relate their distribution, physiology, and behavior to the peculiar conditions in the environment. For this purpose it is necessary to be able to measure those variables that are considered appropriate. Numerous standard methods are available for measuring the more important physical (temperature, salinity, tidal level, substratum character, exposure) and biological (cover) variables in the intertidal zone (Raffaelli and Hawkins, 1996) and are not detailed here. Measurement of the characteristics of tidepools, however, is particularly relevant because pools constitute an important habitat for fishes. The volume of a tidepool is often a good predictor of the number of fish it will contain (Bennett and Griffiths, 1984; Prochazka and Griffiths, 1992; Mahon and Mahon, 1994) and several measurement methods have been described. The simplest method is to calculate approximate volumes from measurements of average length, width, and depth. A more accurate method is to add a known volume of dye to the water and disperse it thoroughly. A sample of the diluted dye can then be taken and its concentration determined photometrically; pool volume can then be calculated by proportion (Green, 1971a; Pfister, 1995). Volumes calculated using this technique can be very close to those obtained by direct measurement of pool volume (Green, 1971a). A more complicated method involves constructing detailed contour maps of the pools and multiplying planimeter contour areas by mean depth between contours (Weaver, 1970). Pool perimeters (Willis and Roberts, 1996) and rugosity or roughness can be measured using flexible measuring tape. A factor for rugosity can be then be calculated by comparing the distance covered by a tape that follows the pool contours with that covered when the tape is stretched taut between the same two points.

## III. Observing Intertidal Fishes in the Field

A great deal of valuable information on the behavior, ecology, and distribution of individual fishes can be obtained by observing fishes in the wild. Such information is free of the constraints imposed by experimentation in the laboratory and is not obtainable in any other way. Numerous methods for direct and indirect observation and recording of fishes in the wild have been developed and the actual techniques used depend on the objectives of the study in question.

The simplest method is direct observation by eye from above the surface of the water, but such a method is dependent on calm sea conditions so that the water surface is not disturbed by wind or turbulence. On calm days it is possible to observe the movements of fishes within the intertidal zone (Williams, 1957; Norris, 1963) or in rock pools (Green, 1971c; Marsh *et al.*, 1978). This task is made easier where the fish leave the water, as do the amphibious mudskippers and blennies. These forms can be watched and their behavior and distribution recorded directly (Columbini *et al.*, 1995) or with the use of telescopes (Ikebe and Oishi, 1996) or binoculars (Cancino and Castillo, 1988) much in the same way as birds or terrestrial animals. Direct observations can also be used as a form of censusing populations (Jones and Clare, 1977; Christensen and Winterbottom, 1981; Beckley, 1985) but are unlikely to give unbiased estimates of, for example, population size because of the cryptic nature of many species and the likelihood that many individuals are inactive, particularly at low tide. Nevertheless, direct observations are valuable in providing information that supplements other forms of data collection.

The problem of not being able to see through the water surface is best solved by snorkeling or diving but viewers with transparent windows can also be used (Gibson, 1980). Numerous studies have been done of the behavior of fish in the intertidal zone by diving, usually at high tide or in large tidepools where the depth of water is sufficient to allow it. These studies include the study of intertidal movements (Green, 1971b; Black and Miller, 1991; Rangeley and Kramer, 1995b), territoriality (Carlisle, 1961; Craig, 1996), distribution (Macpherson, 1994; Yamahira *et al.*, 1996), and feeding (Whoriskey, 1983; Fishelson *et al.*, 1987).

Direct observation either from above or below the surface is limited by the time the observer can spend making the observations. Also, the presence of the observer may affect the behavior of the fish; thus, in some cases indirect methods are more useful and appropriate. Two such methods, underwater television and telemetry, that are widely used in deeper water have also been applied to intertidal situations. Underwater television cameras have been surprisingly little used for the study of intertidal fishes, perhaps because of the difficulties of anchoring the equipment in turbulent water. The two studies that have employed this technique were both carried out on sandy beaches to study the intertidal movements of flatfishes (Tyler, 1971; Burrows *et al.,* 1994b). Cameras were deployed at or near the low water mark and fish entering and leaving the intertidal zone were directed beneath the camera by a series of nets. Use of infrared lights at night permits observations to be made in darkness provided the depth of the field of view is not too great (Burrows *et al.,* 1994b). Use of this technique on rocky shores may be of limited value because of the complex nature of the environment but it could provide a means of recording behavior of fish in pools at low tide or small fish with a limited range of movement. It is unlikely to be of value in turbid waters of estuaries or salt marshes. Under such conditions ultrasonic telemetry comes into its own. Tags emitting ultrasound pulses and attached externally or inserted into the alimentary canal or body cavity have been widely used for the study of large free-swimming species. The lower size limit of the tags is determined by battery volume and they have only recently become available for smaller individuals. The signal emitted from the tag is detected by fixed or portable hydrophones and used to determine the position of the tagged fish. Using this technique it has been possible to follow the intertidal migrations of flounders in estuaries (Wirjoatmodjo and Pitcher, 1980), the movements of eels (Helfman *et al.,* 1983) and summer flounders (Szedlmayer and Able, 1993) in salt marsh creeks, and the very restricted intertidal activities of the large *Cebidichthys violaceus* on a rocky shore (Ralston and Horn, 1986).

## IV. Capturing Intertidal Fishes

Observing fishes in the wild provides one source of information on their biology but in most investigations it usually becomes necessary at some point to capture them. Capture allows individuals to be identified, measured, marked, and returned; preserved for further investigation of their age and diet; or transported to the laboratory for experimental studies. A wide range of capture methods, which vary in their ingenuity and sophistication, have been devised and can be grouped into four main categories: manual methods, chemical methods, netting, and trapping. The method employed at any one site depends on the nature of the site and the purpose of the investigation but, in general, manual and chemical methods are used at low tide on rocky shores. Netting techniques can be used on any occasion but have been mainly used at high tide and on sediment shores.

### A. Manual Methods

Capture of fishes by hand is probably the oldest and simplest technique that can be used. It is most effective in areas such as boulder shores where other techniques are not possible or of limited value. Stones and boulders are simply lifted or moved to one side and any fishes sheltering beneath them picked up. Small hand nets are usually useful to pick up actively moving individuals or elongate species. To prevent damage to the environment all stones should be replaced after collection is complete. Fishes can also be collected by hand from tidepools but the larger number of refuges available to them rarely makes this method efficient. The main advantage of capture by hand is that specimens are obtained in good condition and are undamaged by entanglement in nets or subject to the aftereffects of chemical treatments. Where quantitative measures or large numbers of individuals of an unbiased size frequency distribution are required, manual collections are usually preceded by some form of chemical treatment.

Another manual method is to drain tidepools by bailing with buckets or using portable petrol-powered pumps. The fish can be captured as they are concentrated in the reduced volume of water or after they have been stranded on the bottom and sides of the pools. The ability of many intertidal fish species to withstand periods of aerial exposure makes this an effective technique for capturing the fish with a minimum of stress. It is necessary, of course, to refill the pools with water if they are likely to be emersed for long so that other tidepool organisms are not affected.

## B. Chemical Methods

On shores where fish cannot be readily captured by active or passive netting techniques the addition of an ichthyocide or anesthetic to the water greatly improves capture rates. Addition of such chemicals ensures an almost complete catch of all fish present provided that sufficient time is allowed for the chemical to act and the affected area is thoroughly searched. The use of chemicals is of particular value in complex habitats such as rocky shores or mangroves. In open bodies of water the area to be treated is usually surrounded by netting to prevent escape of affected fish. In restricted areas such precautions are not necessary and tidepools are ideally suited to the capture of fishes by this technique. Several compounds have been used for the purpose but the two most widely used are rotenone and quinaldine.

### 1. Rotenone

Rotenone is a toxic compound found in several genera of African, Asian, and American leguminous plants and has been used for centuries by local people of these regions as a means of capturing fish for food (Morrison, 1988). It acts as a vasoconstrictor and by narrowing the capillaries in the gills it reduces oxygen uptake from the water. It also causes damage to the gill epithelia and the first symptoms shown by affected fish are usually erratic swimming, gasping, and rising to the surface. The principal current use is in fishery management for removing unwanted fish populations from lakes and rivers (Holden, 1980) and consequently it can be obtained in several formulations, as emulsions under trade names such as Chem Fish and Pronoxfish or as a powder. Although particularly toxic to fish, rotenone also affects amphibians and crustaceans. When required, its toxic effects can be counteracted by the addition of potassium permanganate to treated areas.

Rotenone can be toxic at concentrations as low as 5 ppm but its effect varies according to species. Eels, for example, take longer to succumb than most other species (Mahon and Mahon, 1994) and wrasse (Labridae) seem particularly susceptible (Weaver, 1970). This differential susceptibility must be taken into account when collecting in the field.

For small areas such as tidepools, solutions or emulsions of rotenone are best dispensed from a flexible plastic bottle with a narrow aperture that allows the liquid to be squirted into holes and crevices and under boulders. After application, the water should be thoroughly mixed to ensure even dispersion. The first signs of its effect, often after only a few minutes, are usually the appearance of fish swimming erratically out from cover or floating to the surface; they should be removed immediately. Not all affected fish necessarily appear in this way and pools should also be searched visually and by hand to ensure as complete a catch as possible. In larger areas rotenone can be sprayed onto the water surface or mixed in after pouring from larger containers. The amount to be added is best determined from experience but as a general guide concentrations producing a slight milkiness to the water are generally sufficient. If too much is added, although the effect is faster, the water may become too opaque to allow affected fish to be detected.

The main disadvantage of rotenone for collecting fish is its toxic effect. This may not be an important consideration if fish are to be preserved for later examination after collection but rotenone is not generally recommended if fish are to be returned live or used for experimentation.

## 2. Quinaldine

Quinaldine (2-methyl quinoline) is an oily brown liquid with a pungent odor and was first described as an anesthetic for fishes by Muench (1958) and as an aid to collecting in the field by Gibson (1967a) and Moring (1970). It is readily available as the practical grade from suppliers of chemical reagents. It is insoluble in water and before use needs to be dissolved in a carrier such as acetone, ethyl alcohol, or isopropanol at a dilution of approximately 20%. This formulation ensures adequate dispersion when added to the area to be treated.

Quinaldine is a very potent fish anesthetic and is active at concentrations between 5 and 10 ppm. Anesthesia at these concentrations takes only a few minutes and recovery time is also rapid. Although it also appears to affect crustaceans, little is known of its effects on invertebrates. For adequate recovery of fish an *in situ* concentration of 10 ppm is recommended (Gibson, 1967a).

Methods for dispensing quinaldine are the same as those described for rotenone but immediate mixing is essential. When affected, fish dart out from cover and collapse onto the sides of pools, where they can be netted. Some care is necessary when netting the first fish to appear because they often respond with a further spasmodic escape reaction. This reaction is not seen after longer anesthesia. As with rotenone, species differ in their susceptibility (Gibson, 1967a; Yoshiyama, 1981) and allowance should be made for the differences before assuming that all fish have been collected from the treated area.

The main advantages of quinaldine are its potency, lack of toxicity at low concentrations, rapid recovery after anesthesia, and the fact that it does not discolor the water. During anesthesia, however, oxygen consumption is markedly reduced and osmoregulatory functions are affected (Milton and Dixon, 1980); these effects may persist after apparent recovery. Some postanesthesia effects on fishes have also been reported, including the impairment of a chemically mediated fright response (Losey and Hughie, 1994). Its disadvantages for the user are its persistent, unpleasant odor and its irritant effect when in contact with the eyes and mucous membranes. Although there is some evidence of a mutagenic effect in rats, there is no evidence that it has carcinogenic properties (BDH Hazard Data Sheet). For these reasons, solutions of quinaldine should be kept in well-sealed containers, contact with the face must be avoided, and rubber gloves should be worn during its use.

### 3. Other Chemicals

Although rotenone and quinaldine are the main chemicals used for collecting fishes, others have been employed in the past, including Antimycin (Reis and Dean, 1981), sulphuric ether (Arruda, 1979), hypochlorite (Ibáñez *et al.,* 1989), tertiary amyl alcohol (McFarland, 1959), methyl parafynol (McFarland, 1960), and sodium cyanide (Lee, 1980). None of these has been widely used.

### 4. General Precautions

Specific precautions that should be taken while using rotenone and quinaldine have been mentioned in the relevant sections but some general guidelines may also be valuable. In particular, it is necessary to ensure that relevant permission for the use of chemicals has been obtained from local or national authorities and that, if available, the manufacturer's instructions are closely followed.

## C. Netting

Chemical methods are suitable for use in restricted areas but where a wider area is to be fished or catches over a longer time period are required, some form of netting technique is usually employed. These techniques fall essentially into two categories: active and passive. Active techniques involve moving a net to catch fish whereas passive techniques are those in which the net is stationary. Table 1 summarizes these netting techniques and provides selected examples of studies in which they have been used. All netting techniques impose some form of mesh selection on the samples they capture.

### 1. Active Netting Techniques

#### *a. Seines*

Seine nets consist of a wall of netting weighted at the bottom and provided with floats at the top and can vary in length from >100 m to <10 m. The mesh size usually decreases from the ends toward the center, which may or may not be extended into a bag to retain the catch on hauling. Seine nets may be set parallel or at right angles to the shore and then hauled onto the beach surrounding fishes in their path. To standardize the area swept by each haul they have also been set around previously placed poles (Ross *et al.,* 1987). Their efficiency can be tested by measuring the proportion of marked fish recaptured after release into a fixed area. In tests of this sort efficiency has been found to vary greatly with fish length, behavior, and species, and with the nature of the bottom (Weinstein and Davis, 1980; Ross *et al.,* 1987; Pierce *et al.,* 1990). They are best used on shores where the bottom is smooth and the foot rope does not ride up over, or snag on, obstacles when the net is hauled.

#### *b. Trawls and Push Nets*

Trawls consist of bags of netting that are hauled over the bottom and are principally used in the capture of benthic and demersal species. Most studies of intertidal fishes using trawls have been made on sand or mud shores where the progress of the net is not impeded. They are normally towed from a boat and the mouth may be held open by a beam or by otter

**Table 1.** Methods for Capturing Intertidal Fishes Using Nets and Representative Examples of Studies Using the Methods in a Variety of Habitats and Locations

| Method | Habitat | Location | Reference |
|---|---|---|---|
| Seine nets | Rocky shore | Canada | Rangeley and Kramer, 1995a |
| | Sandy beach | USA | Ross *et al.*, 1987 |
| | Sandy beach | Japan | Senta and Kinoshita, 1985 |
| | Mangroves | Australia | Robertson and Duke, 1990 |
| | Sandy beach | Azores | Santos and Nash, 1995 |
| | Salt marsh | USA | Rountree and Able, 1992 |
| Trawls and push nets | Sand flats | Netherlands | Kuipers, 1973 |
| | Sand flats | Netherlands | Kuipers *et al.*, 1992 |
| | Sandy beach | UK | Gibson *et al.*, 1996 |
| | Estuary | USA | Horn and Allen, 1985 |
| | Sand flat | USA | Toole, 1980 |
| Drop nets and drop traps | Sandy beach | Sweden | Pihl and Rosenberg, 1982 |
| | Estuary | USA | Horn and Allen, 1985 |
| | Salt marsh | USA | Chamberlain and Barnhart, 1993 |
| | Mangroves | USA | Sheridan, 1992 |
| Lift nets and pop-up nets | Salt marsh | USA | Rozas, 1992 |
| | Salt marsh | USA | Crabtree and Dean, 1982 |
| | Rocky shore | Canada | Black and Miller, 1991 |
| | Seagrass beds | Australia | Connolly, 1994 |
| Fyke nets | Sand flats | Netherlands | van der Meer *et al.*, 1995 |
| | Mudflats | UK | Raffaelli *et al.*, 1990 |
| | Mudflats | Kuwait | Abou-Seedo, 1992 |
| | Salt marsh | USA | Varnell and Havens, 1995 |
| | Mangrove | Guadeloupe | Lasserre and Toffart, 1977 |
| | Salt marsh | Netherlands | Cattrijsse *et al.*, 1994 |
| Weirs, flume, and stake nets | Salt marsh | USA | McIvor and Odum, 1986 |
| | Salt marsh | USA | Rountree and Able, 1992 |
| | Mangrove | Australia | Morton, 1990 |
| | Salt marsh | USA | Shenker and Dean, 1979 |
| | Salt marsh | USA | Kneib, 1991 |
| Trammel nets | Rocky shore | Canada | Black and Miller, 1991 |
| Gill nets | Seagrass beds | USA | Sogard *et al.*, 1989 |
| | Rocky shore | Canada | Green, 1971b |
| | Mangroves | Australia | Robertson and Duke, 1990 |
| | Estuary | USA | Horn and Allen, 1985 |
| | Mudflats | Arabian Gulf | Ali and Hussain, 1990 |

boards that plane outwards when the net is in motion. In shallow water where a boat cannot operate, small trawls can be pulled by hand or modified so that they can be pushed in front of the operator (Holme and McIntyre, 1971). The efficiency of small trawls varies considerably with fish length and species and their use for sampling young flatfish populations in shallow water is discussed in detail by Kuipers *et al.* (1992).

### *c. Drop Nets*

Both seines and trawls have the disadvantage that they disturb fish in their path and so are rarely 100% efficient. To overcome this problem nets that are dropped from above the water surface and suddenly surround a set area were devised. They are usually relatively small and can be operated by hand or set in advance at low water and operated remotely when the tide is high. Drop traps are similar in concept but have solid rather than mesh walls. Fishes retained within the boundaries of the net are then removed by hand netting. Drop traps can be very efficient (>80%, Sheridan, 1992; >95%, Pihl and Rosenberg, 1982).

#### *d. Lift Nets*

Lift nets were devised for the same purpose as drop nets but work in the reverse direction. They are set at low tide and raised by hand when the area to be sampled is covered with water at high tide. Pop-up nets are similar but released remotely, either by pumping air into a hollow frame at the top (Black and Miller, 1991) or by suddenly removing the weights holding down a buoyant frame (Connolly, 1994). The net then rises quickly to the surface trapping fish within a netting enclosure held to the bottom by a weighted footrope. Measurement of their efficiency gave values of 26 to 75%, depending on the species in question (Black and Miller, 1991).

### 2. Passive Netting Techniques

#### *a. Fyke Nets*

Fyke nets consist essentially of tunnels of netting that are staked out on the bottom and into which fish are directed by a wall of netting, the "leader." They are often used commercially and can vary widely in size and design. For the study of intertidal fishes they are deployed between the tidemarks and by setting the net with the mouth in various orientations with respect to the shore line, it may be possible to estimate the numbers and direction of fishes entering and leaving the zone on each tide.

#### *b. Weirs, Flumes, and Stake Nets*

These types of nets are used in situations where it is possible to block off or completely surround an area to prevent fish escaping and they differ in this respect from fyke and gill nets. Weirs and flumes are frequently used in salt marshes where they are stretched across channels at high tide and collect all those fish leaving the channel on the ebb. Kneib (1991) gives a useful discussion of the various techniques for use under such conditions. The size and design of the net, which usually incorporates some form of bag to hold the catch, varies with the width and topography of the channel. Stake nets consist of walls of netting set at high tide around previously positioned stakes or poles. The enclosed fish are then trapped as the tide recedes or the area can be treated with an ichthyocide to aid complete capture and recovery of fish. Stake nets have been particularly used in mangroves. In all cases it is essential that the bottom of the net be fixed closely to the bottom. Estimates of efficiencies range from 58 to 75% for stake nets (Thayer *et al.,* 1987), 80 to 90% for weirs (Kneib, 1991), to 100% for channel nets (Bozeman and Dean, 1980).

#### *c. Gill and Trammel Nets*

Gill nets are rectangular panels of netting that can be suspended in the water or fixed to the bottom. Fishes are caught when they swim into the net and become entangled in the mesh. By using sections of different mesh size it is possible to use gill nets to sample a range of fish lengths. They can be used in all situations but may be ineffective where there is a large amount of floating debris. Trammel nets are similar to gill nets but consist of double panels of netting so that fishes become both entangled and enveloped in the net.

## D. Trapping

Traps used for catching intertidal fishes are usually based on the standard minnow trap design, that is, cylindrical or semicylindrical baskets of net or plastic mesh with one or two conical entrances. They may be set baited or unbaited at low tide and emptied and reset the following low tide. Traps have the advantage that they are easy to deploy and can be used in a wide variety of situations, although they have been mostly used on rocky shores and salt marshes. Addition of rotenone to pools after traps had been in position for 2 h suggested that a very high proportion (>85%) of *Oligocottus maculosus* and *Clinocottus globiceps* could be caught using this technique (Green, 1971a).

One other form of trapping, analogous to terrestrial pitfall traps, has been used on salt marshes. In this technique small containers are set into the marsh surface to simulate natural pools. Larval and juvenile fish are then collected from the containers at low tide (Kneib, 1984).

## E. Other Methods

Several other methods have also been used for capturing intertidal fishes, usually for specific purposes or where conventional techniques were ineffective. They include the use of spears for winter flounders (Wells *et al.,* 1973), angling (Beja, 1995), sets of baited hooks (Williams, 1957; Bennett *et al.,* 1983), and even air rifles for collecting mudskippers (Tytler and Vaughan, 1983).

## F. Choice of Capture Method

No one capture method is suitable for all species and all environments and so the method(s) to be used in any particular investigation depends primarily on its objective. A single method may be appropriate for the capture of individual species but where a measure of community structure is required, the use of more than one method is almost essential. The results of two studies can be used to illustrate this point. In a study of intertidal fishes on a rocky shore Beja (1995) found no correlation between the species composition of the catches made by trapping, angling, and hand netting. Trapping caught 14 species, 56% of which were gobies, angling caught 15 species, 81% of which were wrasse; and hand netting produced 7 species, 71% of which were blennies. A similar comparison between seine and trawl catches on a sandy beach (Gibson *et al.,* 1993) showed that the maximum similarity in species composition of the annual catches by the two gears was only 49% over the 4 years of the investigation. Clearly, reliance on the results of any single method for an adequate description of community structure in either of these investigations would be invalid. Where a range of environments is to be included within a study, several techniques may be required for adequate sampling of all species present (Horn and Allen, 1985).

A further consideration in the choice of method is the required condition of the specimens captured. Fish to be used for experimental purposes or returned to the wild need to be undamaged and so benign methods such as capture by hand or short hauls with a net are necessary. Passive techniques, such as gill netting, are rarely suitable for this purpose, although various forms of trapping with short emersion times may be adequate.

## V. Estimating the Abundance of Intertidal Fishes

Methods for measuring the abundance of intertidal fishes depend to a large extent on the nature of the species or community being investigated but all aim to produce as accurate an estimate as possible, usually in terms of numbers per unit area. The best estimates are probably obtained from techniques that surround or block off a predefined area; the drop and lift net techniques described above were designed for this purpose. Assuming 100% efficiency, catch per set is thus a direct measure of abundance and the variation between sets will determine the number of sets that need to be made for a given level of accuracy. Less efficient techniques, such as trawling or seining, require a knowledge of the area swept by the net and its efficiency. The catches of passive netting techniques are difficult to translate into meaningful measures of abundance and are most useful for obtaining relative measures of abundance in terms of catch per unit effort. Visual estimates of abundance (Jones and Clare, 1977; Christensen and Winterbottom, 1981) also fall into this category but are very susceptible to bias caused by the difference in behavior between species. Mark and recapture techniques do not seem to have been commonly used for estimating intertidal fish abundance (but see Moring, 1976; Pfister, 1996). As with many populations, however, we know little about how violating the assumptions underlying these models might affect estimates of survivorship and abundance. Indeed, an important assumption, that the marked individuals disperse randomly among the population, may not be met. Nevertheless, in a study of tidepool sculpin populations, Pfister (1996) found that estimates of survivorship based on an exponential decay function and mark recapture techniques were strongly positively correlated.

Attempts at estimating abundance on rocky shores illustrate two problems. First, it is highly probable that the distribution of fishes will be extremely patchy with most individuals in pools and relatively few elsewhere. Under these conditions, should abundances be expressed as numbers per unit pool area, volume, or some other measure? Alternatively, is abundance more meaningfully expressed as numbers per unit area of shore, including areas without pools? On shores without pools, the situation is simpler because standard quadrat and transect methods can be used and numbers expressed directly per unit area (Burgess, 1978; Koop and Gibson, 1991; Barber *et al.,* 1995). Sampling pools and using transects are mostly done at low tide although transect methods can also be used at high tide (Yamahira *et al.,* 1996). Low tide surveys only measure the abundance of resident species. To estimate the total intertidal community, that is, including those species present only at high tide, a variety of methods must be used, all with differing efficiencies.

## VI. Marking Intertidal Fishes

In many studies it is often necessary to recognize individuals or groups of fishes and numerous marking methods have been developed for this purpose. The main problems associated with marking intertidal fishes are their small size (often <10 cm), dense cryptic coloration, and for rocky shore species, the complex nature of their habitat where the loss of externally attached tags may be high. All of these problems can be overcome by the choice of a suitable marking technique. Marking techniques fall broadly into three categories: fin clipping, attachment of internal or external tags, and marking with colored dyes or paints. Anesthesia may be necessary to improve marking efficiency; MS222 (tricaine methanesulfonate) and benzocaine (ethyl *p*-aminobenzoate) are suitable for this purpose. Table 2 lists examples of studies where each of these techniques has been used.

## A. Fin Clipping

Removal of parts of the fins of a fish is the simplest of all marking techniques and requires little apparatus apart from a pair of fine scissors. One possible disadvantage of this method is that removal of the whole or parts of the fins may impair swimming ability and reduce foraging efficiency and escape from predators. A further disadvantage is that regeneration of clipped fins may be rapid in juvenile fishes and in warm waters so that clipped individuals can no longer be recognized. Nevertheless, fin clipping is very useful for short-term studies and, by removing combinations of individual rays from different fins, large numbers of fish can be given a unique mark. The actual number of useful combinations that are possible will be determined by the size and number of the fin rays.

## B. Tagging

### 1. External Tags

Attaching small tags to the exterior of fish is also a useful way of recognizing individuals. Such tags are usually attached to the dorsal musculature by means of nylon monofilament line. Numbered tags are commercially available but may not be small enough for use on juveniles or small species. Embroidery beads (Green, 1971b) or small pieces of plastic tubing (Koop and Gibson, 1991) can be used for such individuals. The advantages of external tags are that they are inexpensive, easy to attach, and usually readily visible. Their disadvantages are that they require puncture of the muscles, providing a possible source of infection, can be easily shed, and may make tagged fish more vulnerable to predation.

### 2. Internal Tags

Many of the disadvantages of external tags can be overcome by the use of tags that are inserted into the body musculature and do not protrude from the body surface. These tags can be very small and hence would be suitable for most intertidal fishes. The most suitable types are likely to be coded wire tags and visible implant tags. The former contain an intrinsic binary code and can be used as unique markers. They are, however, not visible externally and need to be detected magnetically, removed surgically, and the code read under a microscope. Visible implant tags are small plastic tags that are inserted beneath translucent tissue and are either numbered or fluoresce when viewed under ultraviolet light. Both types of tags have been tested on fish between 15 and 30 mm in length and been found to have high retention rates. In addition, tagged fish have no higher mortality than untagged fish in laboratory trials (Beukers *et al.*, 1995).

## C. Marking

A third method that has been found valuable for marking small fishes is the subcutaneous injection of colored compounds. When injected, the color shows through the skin and identifies the fish. In this technique it is clearly advantageous to inject the color into areas where

the skin is pale or translucent, which in most cases is the ventral surface. In this position the mark also has the advantage that it is not visible to predators from above. Whenever possible, the marks should be made posterior to the body cavity because accidental injection of paints into the body cavity may result in morbidity. Injection can be carried out with small-diameter hypodermic needles or by pressure spraying. Substances that have been injected include dyes, acrylic paints, and latex. Some paints may harden over time so excessive quantities should not be injected. In general, the method is cheap, rapid, simple, and suitable for batch marking fish from one tidepool, for example, or from one area of a beach. Small numbers of fish can be given individual marks by the judicious use of different colors and mark positions. Injected paint marks have been reported in some cases to persist for as long as 1 to 2 years (Yoshiyama *et al.*, 1992).

## VII. Transporting Intertidal Fishes

It is unlikely that it will be necessary to transport large numbers of intertidal fishes at any one time so special arrangements are rarely necessary. In most cases small insulated containers are adequate to prevent large temperature changes during transport. In temperate climates even insulation is usually unnecessary in winter. The use of small, portable, battery-powered aerators to maintain oxygen concentrations in the surrounding water can be beneficial, especially if the duration of transport is likely to be long. If fish are to be sent long distances by mail, they can be enclosed in double polythene bags inflated with pure oxygen and the bags surrounded with insulating material such as polystyrene chips. Many resident rocky shore species such as blennies and pricklebacks, which have the ability to withstand exposure to air, may not require the presence of liquid water at all and may be packed in quantities of moist algae or other materials (moist paper towels, sponges) for transport.

## VIII. Laboratory Experiments Involving Intertidal Fishes

Generally speaking, intertidal fishes make excellent subjects for laboratory studies because of their ability to withstand handling and transport and their tolerance of a wide range of environmental conditions. Their small size also means that large holding and experimental facilities are not required. No special techniques are necessary for behavioral or physiological experiments except for the general maxim that the more appropriate the conditions, the more realistic and relevant the response. For most species this means the provision of adequate and appropriate cover.

One feature of behavior, however, must always be borne in mind when experimenting with intertidal species and that is the likelihood that many aspects of their behavior and physiology will be strongly rhythmic, even under nontidal laboratory conditions (Gibson, 1982). Consequently, interpretation of the results of any sort, whether behavioral or physiological, must take into account both the time of day and the tidal phase at which the observations were made. Values of activity, feeding, aggression, or oxygen consumption, for example, are all likely to be strongly time dependent. Even components of apparently similar behavior, such as swimming on and off the bottom, may differ in the effect that endogenous rhythmicity has upon their expression (Burrows *et al.*, 1994a). Such precautions are particularly necessary with freshly collected individuals but also at other times because, even though overt rhythmicity may initially appear to fade, in some species, such as *Lipophrys pholis,* it can spontaneously reappear several days later (Northcott *et al.,* 1990).

## IX. Standard Techniques Applied to the Study of Intertidal Fishes

No special techniques are required for the study of conventional aspects of the biology of intertidal fishes, such as age determination, growth, or reproduction. The small size of many species may cause some difficulty in weighing and measuring whole or parts of individuals and the determination of diet composition may present a particular problem due to the small size of the constituent items. A useful discussion of the latter problem is given by Berg (1979).

## X. Safety Considerations

The intertidal zone is a fascinating but dangerous place and safety should be a prime concern when working on the shore. Many institutes now have their own codes for safe conduct during fieldwork and these should be followed because they often place particular emphasis on local conditions. Most safety precautions are common sense and depend on an awareness of the conditions likely to be met. Generally, fieldwork should not be done alone or, if this is essential, notification of the site and expected time of return should be left at base. Appropriate clothing and footwear is essential, together with an appreciation of the dangers of hypo- and hyperthermia. Head torches are a necessary accessory if working at night. Finally, work in the intertidal zone requires a knowledge of the times of the tides so that adequate time is allowed for vacating the site as the tide rises. A useful maxim in this respect, particularly when close to the water line, and one that also minimizes the danger of the rogue wave, is "never work with your back to the sea."

## References

Abou-Seedo, F. S. (1992). Abundance of fish caught by stake traps (hadra) in the intertidal zone in Doha, Kuwait Bay. *J. Univ. Kuwait (Science)* **19,** 91–99.

Ali, T. S., and Hussain, N. A. (1990). Composition and seasonal fluctuation of intertidal fish assemblage in Kohr al-Zubair, Northwestern Arabian Gulf. *J. Appl. Ichthyol.* **6,** 24–36.

Arruda, L. M. (1979). Specific composition and relative abundance of intertidal fish at two places on the Portuguese coast (Sesimbra and Magoito, 1977–78). *Arq. Mus. Bocage* **6,** 325–342.

Barber, W. E., McDonald, L. L., Erickson, W. P., and Vallarino, M. (1995). Effect of the *Exxon Valdez* oil spill on intertidal fish: A field study. *Trans. Am. Fish. Soc.* **124,** 461–475.

Beckley, L. E. (1985). Tide-pool fishes: Recolonisation after experimental elimination. *J. Exp. Mar. Biol. Ecol.* **85,** 287–295.

Beja, P. R. (1995). Structure and seasonal fluctuations of rocky littoral fish assemblages in south-western Portugal: Implications for otter prey availability. *J. Mar. Biol. Assoc. U.K.* **75,** 833–847.

Bennett, B., Griffiths, C. L., and Penrith, M-L. (1983). The diets of littoral fish from the Cape Peninsula. *S. Afr. J. Zool.* **18,** 343–352.

Bennett, B. A., and Griffiths, C. L. (1984). Factors affecting the distribution, abundance and diversity of rock pool fishes on the Cape Peninsula, South Africa. *S. Afr. J. Zool.* **19,** 97–104.

Berg, J. (1979). Discussion of methods of investigating the food of fishes, with reference to a preliminary study of the prey of *Gobiusculus flavescens* (Gobiidae). *Mar. Biol.* **50,** 263–273.

Berger, A., and Mayr, M. (1992). Ecological studies on two intertidal New Zealand fishes, *Acanthoclinus fuscus* and *Forsterygion nigripenne robustum. N. Z. J. Mar. Freshw. Res.* **26,** 359–370.

Beukers, J. S., Jones, G. P., and Buckley, R. M. (1995). Use of implant microtags for studies on populations of small reef fish. *Mar. Ecol. Prog. Ser.* **125,** 45–60.

Black, R., and Miller, R. J. (1991). Use of the intertidal zone by fish in Nova Scotia. *Environ. Biol. Fish.* **31,** 109–121.

Bozeman, E. L., Jr., and Dean, J. M. (1980). The abundance of larval and juvenile fish in a South Carolina intertidal creek. *Estuaries* **3,** 89–97.

Burgess, T. J. (1978). The comparative ecology of two sympatric polychromatic populations of *Xererpes fucorum* Jordan & Gilbert (Pisces: Pholididae) from the rocky intertidal zone of central California. *J. Exp. Mar. Biol. Ecol.* **35,** 43–58.

Burrows, M. T., Gibson, R. N., and MacLean, A. (1994a). Effects of endogenous rhythms and light conditions on foraging and predator-avoidance in juvenile plaice. *J. Fish Biol.* **45** (Suppl. A), 171–180.

Burrows, M. T., Gibson, R. N., Robb, L., and Comely C. (1994b). Temporal patterns of movement in juvenile flatfishes and their predators: Underwater television observations. *J. Exp. Mar. Biol. Ecol.* **177,** 251–268.

Cancino, J. M., and Castillo, J. C. (1988). Emersion behavior and foraging ecology of the common Chilean clingfish *Sicyases sanguineus* (Pisces: Gobiesocidae) *J. Nat. Hist.* **22,** 249–261.

Carlisle, D. B. (1961). Intertidal territory in fish. *Anim. Behav.* **9,** 106–107.

Cattrijsse, A., Makwaia, E. S., Dankwa, H. R., Hamerlynck, O., and Hemminga, M. A. (1994). Nekton communities of an intertidal creek of a European estuarine brackish marsh. *Mar. Ecol. Prog. Ser.* **109,** 195–208.

Chamberlain, R. H., and Barnhart, R. A. (1993). Early use by fish of a mitigation salt marsh, Humboldt Bay, California. *Estuaries* **16,** 769–783.

Christensen, M. S., and Winterbottom, R. (1981). A correction factor for, and its application to, visual censuses of littoral fish. *S. Afr. J. Zool.* **16,** 73–79.

Colombini, I., Berti, R., Ercolini. A., Nocita, A., and Chelazzi, L. (1995). Environmental factors influencing the zonation and activity patterns of a population of *Periophthalmus sobrinus* Eggert in a Kenyan mangrove. *J. Exp. Mar. Biol. Ecol.* **190,** 135–149.

Connolly, R. M. (1994). Comparison of fish catches from a buoyant pop net and a beach seine net in a shallow seagrass habitat. *Mar. Ecol. Prog. Ser.* **109,** 305–309.

Crabtree, R. E., and Dean, J. M. (1982). The structure of two South Carolina estuarine tide pool fish assemblages. *Estuaries* **5,** 2–9.

Craig, P. (1996). Intertidal territoriality and time-budget of the surgeonfish, *Acanthurus lineatus,* in American Samoa. *Env. Biol. Fish.* **46,** 27–36.

Fishelson, L., Montgomery, L. W., and Myrberg, A. H. (1987). Biology of surgeonfish *Acanthurus nigrofuscus* with emphasis on changeover in diet and annual gonadal cycles. *Mar. Ecol. Prog. Ser.* **39,** 37–47.

Gibson, R. N. (1967a). The use of the anaesthetic quinaldine in fish ecology. *J. Anim. Ecol.* **36,** 295–301.

Gibson, R. N. (1967b). Studies of the movements of littoral fish. *J. Anim. Ecol.* **36,** 215–234.

Gibson, R. N. (1980). A quantitative description of the behavior of wild juvenile plaice (*Pleuronectes platessa*). *Anim. Behav.* **28,** 1202–1216.

Gibson, R. N. (1982). Recent studies on the biology of intertidal fishes. *Oceanogr. Mar. Biol. Annu. Rev.* **20,** 363–414.

Gibson, R. N., Ansell, A. D., and Robb, L. (1993). Seasonal and annual variations in abundance and species composition of fish and macrocrustacean communities on a Scottish sandy beach. *Mar. Ecol. Prog. Ser.* **98,** 89–105.

Gibson, R. N., Robb, L., Burrows, M. T., and Ansell, A. D. (1996). Tidal, diel and longer term changes in the distribution of fishes on a Scottish sandy beach. *Mar. Ecol. Prog. Ser.* **130,** 1–17.

Green, J. M. (1971a). Local distribution of *Oligocottus maculosus* Girard and other tidepool cottids of the west coast of Vancouver Island, British Columbia. *Can. J. Zool.* **49,** 1111–1128.

Green, J. M. (1971b). High tide movements and homing behavior of the tidepool sculpin *Oligocottus maculosus. J. Fish. Res. Bd. Can.* **28,** 383–389.

Green, J. M. (1971c). Field and laboratory activity patterns of the tidepool cottid *Oligocottus maculosus* Girard. *Can. J. Zool.* **49,** 255–264.

Hart, P. J. B., and Pitcher, T. J. (1969). Field trials of fish marking using a jet inoculator. *J. Fish Biol.* **1,** 383–385.

Helfman, G. S., Stoneburner, D. L., Bozeman, E. L., Christian, P. A., and Whalen, R. (1983). Ultrasonic telemetry of American eel movements in a tidal creek. *Trans. Am. Fish. Soc.* **112,** 105–110.

Hill, J., and Grossman, G. D. (1987). Home range estimates for three north American stream fishes. *Copeia* **1987,** 376–380.

Holden, A. V. (1980). Chemical methods. *In* "Guidelines for Sampling Fish in Inland Waters" (T. Backiel and R. L. Welcomme, Eds.), EIFAC Technical Papers, No. 33, pp. 97–113. Rome.

Holme, N. A., and McIntyre, A. D. (1971). "Methods for the Study of Marine Benthos." International Biological Programme Handbook, No. 16. Blackwell, Oxford.

Horn, M. H., and Allen, L. G. (1985). Fish community ecology in southern California bays and estuaries. *In* "Fish Community Ecology in Estuaries and Coastal Lagoons: Towards an Ecosystem Integration" (A. Yáñez-Arancibia, Ed.), pp. 169–190. DR (R) Unam Press, Mexico.

Ibáñez, M., Miguel, I., San Millán, D., and Ripa, I. (1989). Intertidal ichthyofauna of the Spanish Atlantic coast. *Sci. Mar.* **53,** 451–455.

Ikebe, Y., and Oishi, T. (1996). Correlation between environmental parameters and behavior during high tides in *Periophthalmus modestus. J. Fish Biol.* **49,** 139–147.

Jones, D., and Clare, J. (1977). Annual and long-term fluctuations in the abundance of fish species inhabiting an intertidal mussel bed in Morcambe Bay, Lancashire. *Zool. J. Linn. Soc.* **60,** 117–172.

Kneib, R. T. (1984). Patterns in the utilization of the intertidal saltmarsh by larvae and juveniles of *Fundulus heteroclitus* (Linnaeus) and *Fundulus luciae. J. Exp. Mar. Biol. Ecol.* **83,** 41–51.

Kneib, R. T. (1991). Flume weir for quantitative collection of nekton from vegetated intertidal habitats. *Mar. Ecol. Prog. Ser.* **75,** 29–38.

Koop, J. H., and Gibson, R. N. (1991). Distribution and movements of intertidal butterfish *Pholis gunnellus. J. Mar. Biol. Assoc. U.K.* **71,** 127–136.

Kuipers, B. (1973). On the tidal migration of young plaice (*Pleuronectes platessa*) in the Wadden Sea. *Neth. J. Sea Res.* **6,** 376–388.

Kuipers, B. R., MacCurrin, B., Miller, J. M., van der Veer, H. W., and Witte, J. IJ. (1992). Small trawls in juvenile flatfish research: Their development and efficiency. *Neth. J. Sea Res.* **29,** 109–117.

Lasserre, G., and Toffart, J. L. (1977). Échantillonnage et structure des populations ichthyologiques des mangroves de Guadeloupe en Septembre 1975. *Cybium* **2,** 115–127.

Lee, S-C. (1980). Intertidal fishes of the rocky pools at Lanyu (Botel Tobago), Taiwan. *Bull. Inst. Zool. Acad. Sin.* **19,** 1–13.

Losey, G. S., and Hughie, D. M. (1994). Prior anesthesia impairs a chemically mediated fright response in a gobiid fish. *J. Chem. Ecol.* **20,** 1877–1883.

Lotrich, V. A., and Meredith, W. H. (1974). A technique and the effectiveness of various acrylic colors for subcutaneous marking of fish. *Trans. Am. Fish. Soc.* **1,** 140–142.

Macpherson, E. (1994). Substrate utilisation in a Mediterranean littoral fish community. *Mar. Ecol. Prog. Ser.* **114,** 211–218.

Mahon, R., and Mahon, S. D. (1994). Structure and resilience of a tidepool fish assemblage at Barbados. *Environ. Biol. Fish.* **41,** 171–190.

Marsh, B., Crowe, T. M., and Siegfried, W. R. (1978). Species richness and abundance of clinid fish (Teleostei: Clinidae) in intertidal rock pools. *Zool. Afr.* **13,** 283–291.

McFarland, W. N. (1959). A study of the effects of anaesthetics on the behavior and physiology of fishes. *Pub. Inst. Mar. Sci. Univ. Tex.* **6,** 23–35.

McFarland, W. N. (1960). The use of anaesthetics for the handling and transportation of fish. *Calif. Fish Game* **46,** 407–432.

McIvor, C. C., and Odum, W. E. (1986). The flume net: A quantitative method for sampling fishes and macrocrustaceans on tidal marsh surfaces. *Estuaries* **9,** 219–224.

Milton, P., and Dixon, R. N. (1980). Further studies of the effects of the anaesthetic quinaldine on the physiology of the intertidal teleost *Blennius pholis. J. Mar. Biol. Assoc. U.K.* **60,** 1043–1051.

Moring, J. R. (1970). Use of the anesthetic quinaldine for handling Pacific coast intertidal fishes. *Trans. Am. Fish. Soc.* **1970,** 802–805.

Moring, J. R. (1976). Estimates of population size for tidepool sculpins, *Oligocottus maculosus,* and other intertidal fishes, Trinidad Bay, Humboldt County, California. *Calif. Fish Game* **62,** 65–72.

Morrison, B. R. S. (1988). "The Use of Rotenone in Fisheries Management." Scottish Fisheries Information Pamphlet, No. 15, pp. 1–17.

Morton, R. N. (1990). Community structure, density and standing crop of fishes in a subtropical Australian mangrove. *Mar. Biol.* **105,** 385–394.

Muench, B. (1958). Quinaldine, a new anaesthetic for fish. *Prog. Fish Cult.* **20,** 42–44.

Norris, K. S. (1963). The functions of temperature in the ecology of the percoid fish *Girella nigricans* (Ayres). *Ecol. Monog.* **33,** 23–62.

Northcott, S. J., Gibson, R. N., and Morgan, E. (1990). The persistence and modulation of endogenous circatidal rhythmicity in *Lipophrys pholis* (Teleostei). *J. Mar. Biol. Assoc. U.K.* **70,** 815–827.

Pfister, C. A. (1995). Estimating competition coefficients from census data: A test with field manipulations of tidepool fishes. *Am. Nat.* **146,** 271–291.

Pfister, C. A. (1996). The role and importance of recruitment variability to a guild of tide pool fishes. *Ecology* **77,** 1928–1941.

Pierce, C. L., Rasmussen, J. B., and Leggett, W. C. (1990). Sampling littoral fish with a seine: Corrections for variable capture efficiency. *Can. J. Fish. Aquat. Sci.* **47,** 1004–1010.

Pihl, L., and Rosenberg, R. (1982). Production, abundance, and biomass of mobile epibenthic marine fauna in shallow waters, western Sweden. *J. Exp. Mar. Biol. Ecol.* **57,** 273–301.

Prochazka, K., and Griffiths, C. L. (1992). The intertidal fish fauna of the west coast of South Africa—Species, community and biogeographic patterns. *S. Afr. J. Zool.* **27,** 115–120.

Raffaelli, D., and Hawkins, S. (1996). "Intertidal Ecology." Chapman and Hall, London.

Raffaelli, D., Richner, H., Summers, R., and Northcott, S. (1990). Tidal migrations in the flounder (*Platichthys flesus*). *Mar. Behav. Physiol.* **16,** 249–260.

Ralston, S. L., and Horn, M. H. (1986). High tide movements of the temperate-zone herbivorous fish *Cebidichthys violaceus* (Girard) as determined by ultrasonic telemetry. *J. Exp. Mar. Biol. Ecol.* **98,** 35–50.

Rangeley, R. W., and Kramer, D. L. (1995a). Use of rocky intertidal habitats by juvenile pollock *Pollachius virens. Mar. Ecol. Prog. Ser.* **126,** 9–17.

Rangeley, R. W., and Kramer, D. L. (1995b). Tidal effects on habitat selection and aggregation by juvenile pollock *Pollachius virens* in the rocky intertidal zone. *Mar. Ecol. Prog. Ser.* **126,** 19–29.

Reis, R. R., and Dean, J. M. (1981). Temporal variation in the utilization of an intertidal creek by the bay anchovy (*Anchoa mitchilli*). *Estuaries* **4,** 16–23.

Richkus, W. A. (1978). A quantitative study of intertidepool movement of the wooly sculpin *Clinocottus analis. Mar. Biol.* **49,** 277–284.

Riley, J. D. (1966). Liquid latex marking techniques for small fish. *J. Cons. Perm. Int. Explor. Mer.* **30,** 354–357.

Riley, J. D. (1973). Movements of 0-group plaice *Pleuronectes platessa* L. as shown by latex tagging. *J. Fish Biol.* **5,** 323–343.

Robertson, A. I., and Duke, N. C. (1990). Mangrove fish communities in tropical Queensland, Australia: Spatial and temporal patterns in densities, biomass and community structure. *Mar. Biol.* **104,** 369–379.

Ross, S. T., McMichael, R. H., Jr., and Ruple, D. L. (1987). Seasonal and diel variations in the standing crop of fishes and macroinvertebrates from a Gulf of Mexico surf zone. *Estuar. Coast. Shelf Sci.* **25,** 391–412.

Rountree, R. A., and Able, K. W. (1992). Fauna of polyhaline subtidal marsh creeks in southern New Jersey: Composition, abundance and biomass. *Estuaries* **15,** 171–185.

Rozas, L. P. (1992). Bottomless lift net for quantitatively sampling nekton on intertidal marshes. *Mar. Ecol. Prog. Ser.* **89,** 287–292.

Santos, R. S., and Nash, R. D. M. (1995). Seasonal changes in a sandy beach fish assemblage at Porto Pim, Faial, Azores. *Estuar. Coast. Shelf Sci.* **41,** 579–591.

Saucerman, S. E., and Deegan, L. A. (1991). Lateral and cross-channel movement of young-of-the-year winter flounder (*Pseudopleuronectes americanus*) in Waquoit Bay, Massachusetts. *Estuaries* **14,** 440–446.

Senta, T., and Kinoshita, I. (1985). Larval and juvenile fishes occurring in surf zones of western Japan. *Trans. Am. Fish. Soc.* **114,** 609–618.

Shenker, J. M., and Dean, J. M. (1979). The utilization of an intertidal salt marsh creek by larval and juvenile fishes: Abundance, diversity and temporal variation. *Estuaries* **2,** 154–163.

Sheridan, P. F. (1992). Comparative habitat utilization by estuarine macrofauna within the mangrove ecosystem of Rookery Bay, Florida. *Bull. Mar. Sci.* **50,** 21–39.

Sogard, S. M., Powell, G. V. N., and Holmquist, J. G. (1989). Utilization by fishes of shallow, seagrass-covered banks in Florida Bay. 2. Diel and tidal patterns. *Environ. Biol. Fish.* **24,** 81–92.

Szedlmayer, S. T., and Able, K. W. (1993). Ultrasonic telemetry of age-0 summer flounder, *Paralichthys dentatus,* movements in a southern New Jersey estuary. *Copeia* **1993,** 728–736.

Thayer, G. W., Colby, D. R., and Hettler, W. F., Jr. (1987). Utilization of the red mangrove prop root habitat by fishes in south Florida. *Mar. Ecol. Prog. Ser.* **35,** 25–38.

Toole, C. L. (1980). Intertidal recruitment and feeding in relation to optimal utilization of nursery areas by juvenile English sole *(Parophrys vetulus:* Pleuronectidae). *Environ. Biol. Fish.* **5,** 383–390.

Tyler, A. V. (1971). Surges of winter flounder, *Pseudopleuronectes americanus,* into the intertidal zone. *J. Fish. Res. Bd. Can.* **28,** 1727–1732.

Tytler, P., and Vaughan, P. (1983). Thermal ecology of the mudskipper, *Periophthalmus koelreuteri* (Pallas) and *Boleophthalmus boddarti* (Pallas) of Kuwait Bay. *J. Fish Biol.* **23,** 327–337.

van der Meer, J., Witte, J. IJ., and van der Veer, H. W. (1995). The suitability of a single intertidal fish trap for the assessment of long-term trends in fish and epibenthic invertebrate populations. *Environ. Monitor. Assess.* **36,** 139–148.

Varnell, L. M., and Havens, K. J. (1995). A comparison of dimension-adjusted catch data methods for assessment of fish and crab abundance in intertidal salt marshes. *Estuaries* **18,** 319–325.

Weaver, P. L. (1970). Species diversity and ecology of tidepool fishes in three Pacific coastal areas of Costa Rica. *Rev. Biol. Trop.* **17,** 165–185.

Weinstein, M. P., and Davis, R. W. (1980). Collection efficiency of seine and rotenone samples from tidal creeks, Cape Fear River, North Carolina. *Estuaries* **3,** 95–105.

Wells, B., Steele, D. H., and Tyler, A. V. (1973). Intertidal feeding of winter flounders (*Pseudopleuronectes americanus*) in the Bay of Fundy. *J. Fish. Res. Bd Can.* **30,** 1374–1378.

Whoriskey, F. C., Jr. (1983). Intertidal feeding and refuging by cunners, *Tautogolabrus adspersus* (Labridae). *Fish. Bull.* **81,** 426–428.

Williams, G. C. (1957). Homing behavior of California rocky shore fishes. *Univ. Calif. Pub. Zool.* **59,** 249–284.

Willis, T. J., and Roberts, C. D. (1996). Recolonisation and recruitment of fishes to intertidal rockpools at Wellington, New Zealand. *Environ. Biol. Fish.* **47,** 329–343.

Wirjoatmodjo, S., and Pitcher, T. J. (1980). Flounders follow the tides to feed: Evidence from ultrasonic tracking in an estuary. *Estuar. Coast. Shelf Sci.* **19,** 231–241.

Yamahira, K., Kikuchi, T., and Nojima, S. (1996). Age specific food utilization and spatial distribution of the puffer, *Takifugu niphobles,* over an intertidal sandflat. *Environ. Biol. Fish.* **45,** 311–318.

Yoshiyama, R. M. (1981). Distribution and abundance patterns of rocky intertidal fishes in central California. *Environ. Biol. Fish.* **6,** 315–332.

Yoshiyama, R. M., Gaylord, K. B., Philippart, M. T., Moore, T. R., Jordan, J. R., Coon, C. C., Schalk, L. L., Valpey, C. J., and Tosques, I. (1992). Homing behavior and site fidelity in intertidal sculpins (Pisces: Cottidae). *J. Exp. Mar. Biol. Ecol.* **160,** 115–130.

# 3

# Vertical Distribution Patterns

Claus Dieter Zander

*Zoologisches Institut und Museum, Universität Hamburg, Hamburg, Germany*

Jürgen Nieder

*Botanisches Institut, Universität Bonn, Bonn, Germany*

Karen Martin

*Department of Biology, Pepperdine University, Malibu, California*

## I. General Ecological Patterns

The vertical distribution of sessile organisms in the intertidal zone is conspicuous and has received considerable attention in marine ecological literature (overview in Russel, 1991). Highly mobile animals like fishes are often disregarded, presumably because the dynamic character of their distribution in general, and in the intertidal zone in particular, makes the study of this aspect of their ecology difficult.

Two groups of fishes can be found in the intertidal zone: resident species and transient species (Gibson, 1982). These groups diverge in body shape and behavior, due to their different dependence on the benthic environment. Resident intertidal fishes are genuinely benthic or demersal species that dwell on the bottom and cannot swim rapidly for long periods. They are better suited than suprabenthic and pelagic fishes to the extreme conditions of the intertidal or even supralittoral zones, characterized by strong wave action and desiccation. Benthic species are relatively small; in most cases body length does not exceed 10 cm, and the largest species measure 25–30 cm. Their body shapes are either short and somewhat dorso-ventrally flat or elongated and eel-like (Abel, 1961, 1962). Another characteristic is the absence or reduction of the swim bladder in the adult, which is only rarely reactivated, as in the gobiid *Pomatoschistus* (Hesthagen, 1977). Several species of the short and somewhat flattened shape type, common among Cottidae and Gobiesocidae, have dermal calcifications that provide greater density and help the fish maintain contact with the substratum. In addition, many benthic fishes have a clinging organ to withstand wave force, for example, a sucker, claw-like tips of fin rays, or a specially structured underside of the head or belly that functions as an adhesive organ (see Section IV).

Transient species regularly visit the intertidal zone during high tides (Black and Miller, 1991). These fishes actually find their optimal conditions in the sublittoral zone. Although there are several benthic species, many of these transient species have a suprabenthic appearance and swim a certain distance above the bottom. They have a compressed body shape and a well-functioning swim-bladder, with effective fins for movement in the water column (Abel, 1962). Although these species do not show morphological adaptations similar to those of benthic species, they have strong connections to the substratum because they search for food, spawn, or hide on the substratum. Some, especially juveniles of small size, may be able to survive intertidally during low tide in water-filled grooves or rock pools.

Some fishes exhibit an amphibious lifestyle and are able to leave the water for at least short periods of time. For fishes that are not, vertical distribution in the intertidal habitat is correlated with temporal restriction of access to food. The higher the tidal height of a pool within the intertidal zone, the less time intertidal fishes have for feeding outside that tide pool, assuming that fishes in tidepools do not restrict foraging to their home pool. Stomach contents of herbivorous fishes clearly show that they feed during high tide (Horn *et al.*, 1986; Ralston and Horn, 1986; John and Lawson, 1991). Intensive growth of algae in the intertidal zone is exploited by a number of herbivorous fishes with marked preference for the upper intertidal zone (Horn, 1989; see Horn and Ojeda, Chapter 11, this volume).

In the following, examples of characteristic habitat types of the rocky intertidal zone and their fish communities are described.

## A. Rock Flats in Temperate and Tropical Seas

Rock flats that are characteristic of the temperate sea are found off the island of Helgoland (German Bight), where 15 fish species may be found at low tide: 10 benthic, 3 suprabenthic, and 2 pelagic (Zander, personal observations, Table 1). Only *Pholis gunellus* (benthic) and 3 small-sized gobies (2 benthic *Pomatoschistus* spp. and the suprabenthic *Gobiusculus flavescens*) as well as *Gasterosteus aculeatus* (suprabenthic) are common as adults; 6 benthic species and the suprabenthic *Ctenolabrus rupestris* are young-of-the-year.

**Table 1.** Compilation of Intertidal Fishes off Helgoland, North Sea, and Sarso Island, Red Sea, According to Body Shape and Main Feeding Habits

| | Benthic herbivores | Benthic carnivores | Plankton feeder |
|---|---|---|---|
| Helgoland | | | |
| Eel-like shape | | *Pholis gunellus* | |
| Flattened shape | | *Pomatoschistus microps*<br>*Pomatoschistus pictus* | |
| Suprabenthic | | | *Gobiusculus flavescens*<br>*Gasterosteus aculeatus* |
| Sarso Island | | | |
| Eel-like shape | *Istiblennius edentulus*<br>*Antennablennius hypenetes*<br>*Halmablennius flaviumbrinus* | | |
| Flattened shape | | *Acentrogobius ornatus*<br>*Eviota meteori* | |
| Suprabenthic | | | *Aphanius dispar*<br>*Terapon jarhua* |

Two pelagic clupeids appear only as larvae and the benthic *Agonus cataphractus* lay eggs between *Laminaria* rhizoids at the low tide level. The common goby, *Pomatoschistus microps,* is the most abundant species that can survive low tides even under water levels of only 5 mm. In higher abundance are *Pholis gunellus,* hidden among wet seaweed such as *Fucus,* as are young *Zoarces viviparus,* young *Taurulus bubalis,* and young *Ctenolabrus rupestris* that withdraw into the grooves. The egg clumps of *Agonus cataphractus* may be found on seaweed as well, laid by fertilized females that migrate to the rock flat while the males remain in the deeper sublittoral, the typical habitat of this species (Kruess, 1989). Only 3 of these 15 species are able to survive low tides with no or only partial water cover in this boreal habitat.

A similarly structured rock flat in the tropics was studied by Klausewitz (1967) and Zander (1967) off the isle of Sarso, Farasan Archipelago, Red Sea, where 7 residential species (5 benthic and 2 suprabenthic), and 6 transient species (all suprabenthic) are found in the intertidal zone (Table 1). The blennies (Blenniidae) *Istiblennius edentulus, Antennablennius hypenetes,* and *Halmablennius flaviumbrinus* as well as the gobies (Gobiidae) *Acentrogobius ornatus* and *Eviota meteori* retreat into shaded, wet hollows or depressions during low tide (Zander, 1967). The cyprinodont *Aphanius dispar* and the teraponid *Terapon jarhua* remain in crevices. Additionally, 6 species from 6 different fish families are present during high tide, 3 by day and 3 by night. *Istiblennius edentulus* undertakes frequent terrestrial sojourns at night when temperature is lower and humidity is higher. Toward the open sea the intertidal zone is fringed by a knobby rock flat that is dry only at spring tides but covered at high tides with 70 to 80 cm of water. Here 10 benthic species, including 7 Gobiidae, 1 Blennidae, 1 Mullidae, and 1 Bothidae, and 8 suprabenthic fish are present, 6 of which are also found in the intertidal zone: 2 Gobiidae, 1 Atherinidae, 1 Lutianidae, 1 Labridae, and 1 Acanthuridae.

In the tropics, the exchange between intertidal and subtidal zones does not appear to be as intensive as in colder regions. This may be due to higher temperatures and heating up of the intertidal refuge waters at lower latitudes. The main difference between these two regions may lie in the presence of several herbivorous families in the tropical habitat that are missing in the temperate.

## B. Rock Pools

Rock pools are characteristic habitats of richly structured inclined rocky shores. During low tide, water remains in holes and depressions that function as refuge habitats for many intertidal organisms. Insolation leads to rising temperature and salinity, and rainfall may lower salinity considerably. In winter in some temperate habitats, water temperature in tidepools may reach near freezing, and only then are the pools deserted by intertidal species (Kotrschal and Reynolds, 1982). In rock pools, only resident fishes and the transient species that are able to tolerate drastic changes in the physical environment can survive. Vertical zonation of sessile organisms is more pronounced on emergent substrata than in tidepools (Metaxas and Scheibling, 1993). Distribution of intertidal fishes is to some extent "azonal," in addition, because the physical conditions of tide pools depend on two factors: vertical position and specific qualities such as the size, depth, and exposure of the pool.

On the coast of Brittany near Roscoff, France, 9 fish species occur in several sheltered pools, and 11 species in exposed pools of intertidal rock walls. Of these, 7 species are common to both sites (Gibson, 1972), and 11 species are benthic (represented by Blenniidae, Gobiidae, Gobiesocidae, and Gadidae), with only 2 (*Nerophis lumbriciformis* and

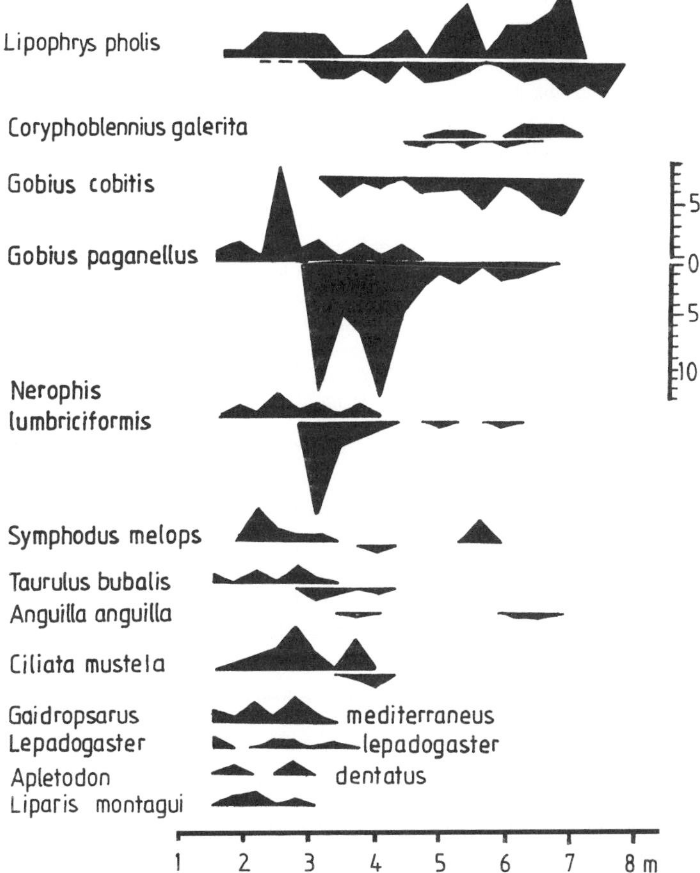


**Figure 1.** The vertical distribution of benthic fish in pools along a cliff off Roscoff, France. *X*-axis, distance above low-water level; *Y*-axis, numbers of specimens in each respective level; each distribution includes the occurrence in exposed (above) and sheltered (below zero line) pools. From Gibson (1972). (*Lipophrys* = *Blennius pholis.*)

*Symphodus melops*) suprabenthic (Fig. 1; Table 2). Apparently, the Gobiesocidae and Cyclopteridae with strong suckers are only present in the exposed habitat, whereas eel-like fishes and the large goby *Gobius cobitis* are missing. The most abundant species are *Gobius paganellus, Lipophrys pholis,* and *Gobius cobitis* at the sheltered sites, and *L. pholis, G. paganellus,* and *Ciliata mustela* at the exposed sites. Clear differences are obvious when examining the distribution above the low-tide level. Only *L. pholis, Coryphoblennius galerita,* and *G. cobitis* are more abundant in higher zones than at mid-tide level. All other species have their maximal abundance below the mid-tide level, which is covered by water during neap tides. This situation differs from that in the rock flats in the relation of benthic and suprabenthic fishes. This difference may be due to the behavior of transient species, which are able to escape faster from the rock walls than from the flat rocks during the time of receding water. During high tides, Gibson (1972) found 6 additional suprabenthic species and the rockpool-inhabiting *S. melops* in greater abundances at the rock walls.

**Table 2.** Compilation of Tidepool-Inhabiting Fishes off Roscoff, Northern Atlantic, and Marshall Islands, Tropic Pacific, According to Body Shape and Main Feeding Habits

| | Benthic herbivores | Benthic carnivores | Plankton feeder |
|---|---|---|---|
| Roscoff | | | |
| Eel-like shape | *Coryphoblennius galerita* | *Lipophrys pholis*<br>*Ciliata mustela*<br>*Gaidopsarus mediterraneus*<br>*Anguilla anguilla* | |
| Flattened shape | | *Taurulus bubalis*<br>*Lepadogaster lepadogaster*<br>*Apletodon denticulatus*<br>*Liparis montagui*<br>*Gobius paganellus*<br>*Gobius cobitis* | |
| Suprabenthic | | *Symphodus melops* | *Nerophis lumbriciformis* |
| Marshall Islands | | | |
| Eel-like shape | *Istiblennius edentulus*<br>*Istiblennius lineatus*<br>*Istiblennius paulus* | *Gymnothorax pictus*<br>*Runula snowi* | |
| Flattened shape | | *Bathygobius fuscus* | |
| Suprabenthic | *Abudefduf glaucus*<br>*Abudefduf sordidus*<br>*Acanthurus triostegus* | *Plesiops nigrans* | |

The study of Hiatt and Strasburg (1960) at the Marshall Islands, south Pacific, illustrates the importance of temperature in the tropics and in the supralittoral zone as an additional habitat. There were 10 fish species in an intertidal pool, similar to the species numbers in France and higher than in two supratidal pools. These included 4 Blenniidae, 1 Gobiidae, and 1 Muraenidae, all benthic; and 2 Pomacentridae, 1 Acanthuridae, and 1 Pseudochromidae species, all suprabenthic (Table 2). Two of the blennies and the pseudochromid were absent from the supratidal ponds, and only juvenile acanthurids were present in both habitats. The pools heated to 41°C in both supratidal and intertidal pools, whereas air temperatures at most reached 31°C. Even so, the only specimens that died were 1 *Istiblennius edentulus* and 2 *Acanthurus triostegus,* when the temperature reached 37°C in the intertidal pool.

The existence of more herbivorous species in tropical than in temperate regions may be a general trend, which was previously observed in rock flats (see Section I.A.).

## C. Supratidal Habitats

In addition to pools, fishes use damp areas above the water line as habitats in the supratidal zone. In boreal regions no examples are known, but in subtropical and tropical regions several fish species are amphibious and found out of water. The rocky supralittoral has been conquered only by fishes of the families Blenniidae, Clinidae (these two are closely related), Labrisomidae, and Gobiesocidae. The widely distributed *Periophthalmus* spp. live in the mangrove girdle in the intertidal zone, where they climb on the roots, stems, and branches of these trees or catch small crabs on the mud flat (Stebbins and Kalk, 1961;

Brillet, 1976; Magnus, 1981; Gordon *et al.,* 1985). Vertical zonation and different feeding behaviors allow the coexistence of several species of sympatric mudskippers (Nursall, 1981; Clayton and Vaughan, 1982). For these, as for other species, the supralittoral zone is an advantageous habitat with rich prey resources and little competition. In the rocky supratidal zone, sessile organisms such as barnacles, limpets, or algae are eaten by fishes. The tropical Pacific goby, *Gobionellus sagittula,* occupies a supratidal niche, leaving pools on mud flats for terrestrial grazing of algae (Todd, 1976). Amphibious behavior can be observed as a reaction to disturbance. Some blennies can move overland from one pool to another (Breder, 1948).

In some instances, a terrestrial habitat may mean increased predation pressure, as in the case of lizards preying on juvenile mudskippers in the intertidal zone (Fricke, 1970). The presence of *Scorpaena* sp. in shallow water at night in the same habitat suggests that terrestrial sojourns of blennies may also be a result of predator avoidance (Nieder and Zander, 1994). The amphibious labrisomid, *Mnierpes macrocephalus,* from the tropical Pacific Ocean remains just above the water line, making these fishes practically invisible and able to avoid predators (Graham, 1970, 1973).

In the Mediterranean Sea tidal forces are negligible and differences of water levels between tides are at most 10–20 cm. The blenny *Coryphoblennius galerita* on warm days often remains close to the water line, kept wet by an occasional wave (Zander, 1983). Its prey are mostly barnacles (Gibson, 1972), which are captured when these crustaceans are covered by waves and open their operculae (Soljan, 1932). *C. galerita* remains above the water line at night (Zander, 1972c; Heymer, 1982), up to 0.5 m above the water level on the rocks (Zander, 1983). The blenny *Lipophrys trigloides* (syntopic with *C. galerita*) emerges from the water during the night, preying on mussels and amphipods (Zander, 1983; Nieder, 1992; Nieder and Zander, 1994).

The rocky supratidal zone of the tropical Indian Ocean is inhabited by several blenniid species of the tribe Salariini that feed on small algae, scraping them off the rock with their tiny teeth. The habitat of *Alticus kirkii* is off Sarso Island, Farasan-Archipelago, Red Sea (Zander, 1967), also in the rocky supralittoral. *A. kirkii* is very abundant in the supralittoral of exposed rocks and climbs wet rocks up to 1 m above high-tide level. It is rarely found in the intertidal zone, where *Istiblennius edentulus* and the small *Antennablennius hypenetes* are the only resident fishes. The latter two species are also present on the rock flat (see Section 4.I.A). Abel (1973) observed a population of the related *A. saliens* in Ceylon, of about 20 specimens on an area of 4–6 square meters on steep rocky cliffs. The spawning hole of a male lies 40 cm above the mean water level, arising toward the inner rock, and is only covered when the tide is extremely high. Several places nearby at different tidal heights are used for feeding and for rolling around to keep the skin wet. Courtship behavior is performed out of water at the spawning hole.

The Chilean clingfish, *Sicyases sanguineus* (Gobiesocidae), emerges from the water on surfaces of large rocks, keeping in the splash zone ahead of the rising tide (Ebeling *et al.,* 1970). Its disc-like pelvic fins permit the fish an amphibious behavior in a most unlikely fish habitat, vertical rock walls above the water. It feeds on a wide variety of algae and animals from the intertidal zone and appears to digest its meals while exposed in the splash zone (Paine and Palmer, 1978), representing a unique trophic and habitat niche with no analog anywhere else in the world.

The rocky supralittoral, due to its terrestrial conditions, is an extreme habitat for intertidal fishes. Only in warmer regions have a few fish species successfully developed the necessary adaptations for surviving in this environment, where piscine competitors are rare.

Apart from the mudskipper species mentioned above, the blennies *Andamia* spp. from the Andaman Archipelago (Rao and Hora, 1938), *Scartichthys* sp. from Brazil (Gerlach, 1958), *Mnierpes macrocephalus* (Graham, 1970) from western Central America, and *Dialommus fuscus* from the Galapagos Islands (Munk, 1969) are known to emerge above the water level.

## II. Vertical Distribution of Intertidal Fishes Is the Result of Inter- and Intraspecific Habitat Partitioning

### A. Vertical Zonation in Intertidal Fish Assemblages

The following cases of fish zonation from the Atlantic, the Pacific west coast, and Japan have been selected because they represent the better known examples of the general patterns of vertical distribution. There are studies on intertidal fishes done in other areas, for example by Vivien (1973) in Madagascar; by Rao and Hora (1938) and Raju (1971) in India; by Chang *et al.* (1969, 1973, 1983), Lee (1980), Mok and Wen (1985), Yang and Chung (1978) in Taiwan; by Berger and Mayr (1992) in New Zealand; by Wagner *et al.* (1976) in Brazil; by Eger (1971) in the Gulf of California; and by others, but these do not yield a substantial amount of ecological information.

#### 1. From Northern Europe to Southern Africa

On the west coast of Ireland (Fives, 1980) and the southern coast of England (Milton, 1983), vertical zonation of intertidal fishes is evident.The ranges of small sized *Lipophrys pholis* and *Coryphoblennius galerita* overlap, but larger *L. pholis* occur at a lower tidal level than *C. galerita.* A third blenny, *Parablennius gattorugine,* occurs in the sublittoral and lower eulittoral zone, well separated from the other two species. The goby, *Gobius paganellus,* frequents tidepools only on sheltered shores.

On the Atlantic coast of France, a similar stratification of fish species abundance can be found in tidepools on different shore levels (Gibson, 1972; Fig. 1). The vertical range of one of the two species of Blenniidae (*Coryphoblennius galerita*) again overlaps the range of the other (*Lipophrys pholis*). The goby *Gobius cobitis* is numerous in higher tidepools and not present in lower ones, where *G. paganellus* can be found. *Lipophrys pholis* in tidepools high up the shore in some cases cannot leave their retreat for days on end because of the tides, and consequently their nutritional condition is reportedly poor (Wyttenbach and Senn, 1993). A trade-off between low food intake, competition, and predation pressure may lead to a particularly high intertidal level in this species.

Farther south, on the Portuguese coast, the intertidal fish community is more species rich, but analogous differences in the vertical distribution of 7 fish species present in the intertidal zone are apparent (Arruda, 1979a; Milton, 1983). On the southern coast of Portugal, juvenile specimens of the blenny *Coryphoblennius galerita* can be found according to tide pool size: on the same vertical level, smaller fish occupy significantly smaller tidepools, possibly as a result of interspecific competition (Nieder, 1993). An additional blenniid species, *Lipophrys trigloides,* occurs at the lower half of the intertidal zone. *C. galerita* is concentrated in the upper tidal zone, and *L. pholis* at the mid-tidal level (Arruda, 1990). Carvalho (1982) states that a clear vertical separation accounts for the niche separation in the different blenny species.

In the Azores, *L. pholis* is rather rare (Patzner *et al.*, 1992) and occurs predominantly at mid-tidal level, where *Gobius paganellus* and *L. trigloides* are also frequent. Again, *C. galerita* is more frequent in the upper zone. The two gobies (*G. cobitis* and *G. paganellus*) occupy markedly different strata, upper and lower pools, respectively (Arruda, 1979b). Overlap in community composition among intertidal pools is common (Santos *et al.*, 1994).

In conclusion, the vertical distribution of intertidal fish species on the rocky shores of the eastern Atlantic has a common structure. Despite different abundances and species compositions, species occupy roughly specific shore levels that vary in their absolute vertical extension, but not in position relative to each other.

The Mediterranean Sea offers the opportunity to consider vertical distribution of rocky shore fishes directly, without reference to tidepool position as a surrogate of actual depth of sojourn and activity. This is because tidal amplitudes are low, less than 0.5 m (Miller, 1983), and an intertidal zone in the strict sense does not exist. The small vertical extent of the intertidal zone in the Mediterranean leads to limited, but distinguishable, vertical distributions among fish species, similar to its effect on sessile intertidal organisms such as

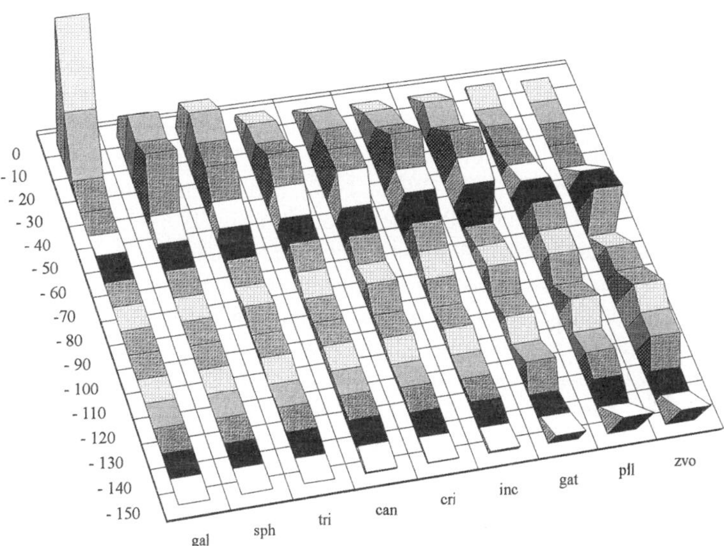


**Figure 2.** Vertical distribution of Mediterranean blennies at Playa de la Mora, Catalan coast (NW Spain). Bar height indicates relative proportion of sightings for each species in a particular depth category, based on underwater observations in January, April, July, September, and December 1990 (data from Nieder, 1992). Note marked preference for very shallow depth of sojourn in *C. galerita, A. sphinx,* and *L. trigloides,* a slightly lower layer of preferred depths in *L. canevai, S. cristata,* and *P. incognitus,* and relatively deep littoral strata in *P. gattorugine, P. pilicornis,* and *P. zvonimiri.* gal, *Coryphoblennius galerita;* sph, *Aidablennius sphinx;* tri, *Lipophrys trigloides;* can, *Lipophrys canevai;* cri, *Scartella cristata;* inc, *Parablennius incognitus;* gat, *Parablennius gattorugine;* pil, *Parablennius pilicornis;* zvo, *Parablennius zvonimiri.*

algae and barnacles. Of the littoral fishes, the blennies are particularly well documented (Zander, 1972a,b, 1980; Patzner, 1984, 1985; Nieder, 1988, 1992), and their vertical distributions show interspecific differences (Fig. 2).

Littoral blenniid fish assemblages in the Mediterranean tend to be relatively species rich. For example, there are 11 syntopic species in the northern Adriatic (Patzner, 1985). Consequently, habitat partitioning is to be expected among the 20 species of Blenniidae of the Mediterranean (Zander, 1986). Vertical habitat segregation is evident for several species present at one site, but overlap occurs between species with different food habits (Zander and Bartsch, 1972; Nieder, 1992). Depth distribution may be a convenient indicator of interspecific microhabitat differences, as in sympatric *Lipophrys* species in the Adriatic (Zander, 1980). Data on preferred depths reported by different authors are very similar for most Mediterranean blennies, as for *Parablennius incognitus* (around 0.8 m) by Zander (1972c) and Koppel (1988). For species also occurring in the Atlantic, *C. galerita, P. pilicornis, S. cristata,* and others, vertical ranges are analogous in the Atlantic and the Mediterranean.

*Aidablennius sphynx, Lipophrys canevai,* and *Parablennius sanguinolentus* are predominantly herbivorous Mediterranean blennies that are abundant on extremely shallow, horizontal rocks covered with green algae (Abel, 1962; Gibson, 1968; Zander, 1972c, 1979, 1980). The usual depth of sojourn for *A. sphynx* is less than 0.5 m (Nieder, 1992), reflecting its grazing habit.

The intertidal fish fauna of the west coast of Africa and adjacent islands, though poorly studied in its tropical sections, is a good example of considerable species turnover and at the same time relatively constant depth distribution of the widely distributed species (Table 3). On the Canary Islands, 6 species of blenniids, 1 clinid, and 1 tripterygiid can be found in the pools of the intertidal zone (Brito and Lozano, 1980). In Senegal, 5 blenniid species share the intertidal zone (Wirtz, 1980), and in Ghana only 4 of these species make up the resident blenniid fauna (Bauchot, 1966; Sanusi, 1980; John and Lawson, 1991), with *C. galerita* apparently missing. Little is known about intertidal fishes of other families. At the ichthyologically impoverished Ghanian coast, *Scartella cristata* and *Ophioblennius atlanticus* (Blenniidae) form the nucleus of an intertidal fish assemblage that toward the north is completed by species of northern origin (*Coryphoblennius galerita, Lipophrys trigloides, Parablennius parvicornis*), and toward the south, by endemic southern species (*Parablennius cornutus* and 1 Clinidae). On the coasts of northern Namibia, 6 intertidal fish species occur, 5 Blenniidae and 1 Clinidae (Penrith, 1970; Penrith and Penrith, 1972).

**Table 3.** Approximate Vertical Zonation of Blenniid Fishes in the Intertidal Zone of Selected Sites of the Western Coast of Africa and the Canary Islands[a]

| | Canary Islands | Senegal | Ghana | Namibia |
|---|---|---|---|---|
| High | *C. galerita* | *E. cadenati* | *S. cristata* | *S. cristata* |
| │ | *P. parvicornis* | *S. cristata* | | *L. velifer* |
| ↓ | *S. cristata* | *L. velifer* | | *P. cornutus* |
| | *L. trigloides* | *O. atlanticus* | | *P. pilicornis* |
| Low | *O. atlanticus* | *P. pilicornis* | *O. atlanticus* | *O. atlanticus* |

[a]Note the relative positions of *Scartella cristata* and *Omobranchus atlanticus,* which occur at all four sites. Based on the data of Brito and Lozano (1980), Wirz (1980), John and Lawson (1991), and Penrith and Penrith (1972). *C. galerita, Coryphoblennius galerita; P. parvicornis, Parablennius parvicornis; S. cristata, Scartella cristata; L. trigloides, Lipophrys trigloides; O. atlanticus, Omobranchus atlanticus; E. cadenati, Entomacrodus cadenati; L. velifer, Lipophrys velifer; P. pilicornis, Parablennius pilicornis; P. cornutus, Parablennius cornutus.*

*Scartella cristata* and *Ophioblennius atlanticus* occur at all four sites, stretching more than 5000 km, and their relative vertical distributions remain constant, with *S. cristata* occurring higher up the shore than *O. atlanticus.*

Within the species-rich intertidal fish communities of South Africa, the clinids are remarkable because they show a complicated intrageneric vertical zonation. The more strictly intertidal species, 14 of the total of 30 South African species, have smaller ranges of occurrence than the species that occur subtidally as well (Penrith, 1970). There is, however, little separation of the habitat between most species, as Prochazka and Griffiths (1992) demonstrated for the west coast of South Africa, with the exception of *Clinus acuminatus,* found high on the shore and not subtidally. Bennett and Griffiths (1984) distinguished a zonation pattern for three groups of species: high-level species, species occurring throughout the intertidal zone, and low-level species with subtidal preference. But even this coarse pattern is superimposed by intraspecific differences in size distribution (see Section II.B). Juvenile *Chorisochismus dentex* (Gobiesocidae) occur only subtidally, intermediate sized specimens occur throughout the intertidal zone, and very large individuals are again only found below the mean low-tide mark. Medium sized *Clinus superciliosus* occurred throughout the intertidal, but larger individuals preferred the lower tidal zone. In *Clinus acuminatus* and *Clinus heterodon* this is reversed: larger specimens occur near the upper distribution limits of the species (Bennett and Griffiths, 1984).

Interestingly, *Clinus cottoides* seems to occupy the food niche of the northeastern Atlantic blenny *Coryphoblennius galerita* (see above): it feeds on barnacle cirri and is present only where its food source is abundant (Penrith, 1970).

Temperature is probably a major factor influencing the distribution of clinids in South Africa, in particular, and intertidal fishes, in general. The intertidal fish fauna of the African west coast is an example of an inverted latitudinal diversity gradient, with the tropical coasts being poor in intertidal fish species and the temperate coast being much richer. Consequently, vertical distribution of intertidal fishes is more finely structured in cooler regions, presumably because temperature extremes do not exclude so many species from this habitat.

## 2. Japan

There is a detailed, though qualitative, description of the habitat preferences of blenniid species in the intertidal and shallow subtidal zone in Central Japan by Fukao (1985). This assemblage of intertidal blenniid fishes is unusually species rich, comprising 3 abundant species (*Istiblennius edentulus, Entomacrodus stellifer,* and *Scartella cristata*) and 4 common species (*Omobranchus loxozonus, Istiblennius lineatus, Praealticus amboinensis,* and *Plagiotremus tapeinosoma*). An additional 5 species are rarely found in the intertidal zone (*Istiblennius bilitonensis, Omobranchus fasciolatoceps, Plagiotremus margaritarius, Rhabdoblennius ellipes,* and *Salarias luctuosus*).

Habitat partitioning in the intertidal zone can be exemplified by the abundant and common species and follows along three dimensions: vertical position of habitat within the intertidal zone, exposure of habitat to wave force, and habitat type, whether tidepool or rock flat (Fig. 3). Small shallow pools of the upper intertidal zone are inhabited by *Omobranchus loxonus* in sheltered sites and *Istiblennius lineatus* in exposed sites. Between tidepools on the same vertical level *Entomacrodus stellifer* is found in moderately exposed sites and *Scartella cristata* in very exposed sites. On a lower level, *Parablennius yatabei* is common in sheltered sites and large, deep tide pools, whereas *Istiblennius edentulus* and *Praealticus amboinensis* occur in tide pools of exposed sites.

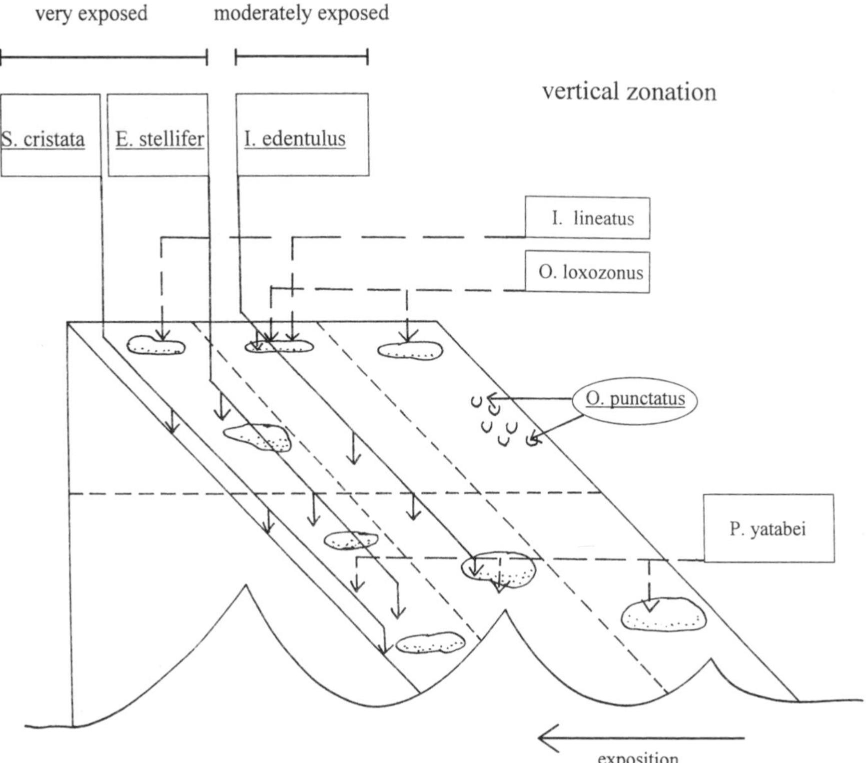


**Figure 3.** Zonation of intertidal blennies of Shirahama, Japan (from Fukao, 1985). Only "abundant" (names underlined) and "common" species (according to Fukao, 1985) are considered. Zonation of different species can be observed along a vertical gradient (upper or lower eulittoral, divided by a broken line in the scheme) and a horizontal gradient (exposition of rocky shore, with sheltered areas along the right edge of the block scheme and exposed areas on the left edge). Microhabitat differences distinguish *S. cristata* and *E. stellifer* (the latter not in tidepools) and *O. punctatus* (seeking shelter only in empty bivalve shells). *S. cristata, Scartella cristata; E. stellifer, Entomacrodus stellifer; I. edentulus, Istiblennius edentulus; I. lineatus, Istiblennius lineatus; O. loxozonus, Omobranchus loxozonus; O. punctatus, Omobranchus punctatus; P. yatabei, Parablennius yatabei.*

Two common species of gobies (*Chasmichthys dolichognathus* and *C. gulosus*) share the tidepools of the Japanese coast (Sasaki and Hattori, 1969). The two gobies have different food preferences and different vertical distributions: *C. dolichognathus* feeds mainly on amphipods and occurs in the lower tidepools, whereas the diet of *C. gulosus* contains a wide range of food items and this species is common in tidepools high up the shore.

## 3. North American Pacific Coast

The fishes of the intertidal zone of the North American Pacific coast are comparatively well studied. For an overview of tidepool fishes in California see Fitch and Lavenberg (1975).

The most important families are the Cottidae, Pholididae, and Stichaeidae (Cross, 1981). Their vertical distributions are consistent over large geographical areas, similar to those of the Atlantic blennies and gobies.

The Cottidae are particularly species rich. In coastal Washington there are 42 species, and in mid-intertidal pools 3 to 6 species coexist (Pfister, 1992), providing good examples of microhabitat differentiation. Within the species of *Oligocottus, O. maculosus* prefers upper level tidepools and *O. snyderi* lower level tidepools near Vancouver (Green, 1971; Nakamura, 1976) and in central California (Yoshiyama, 1981). Habitat separation between the species *Clinocottus globiceps* (lower intertidal) and *Oligocottus maculosus* (upper intertidal) is predominantly vertical in British Columbia (Mgaya, 1992), but the two species separate horizontally, too, in central California: *C. globiceps* is most abundant in high offshore pools and *O. maculosus* in high nearshore tidepools (Yoshiyama, 1981; Yoshiyama *et al.,* 1986). The latter species is particularly sensitive to wave action and avoids increased exposure (Green, 1971).

Obviously, elevation of tidepools is one important factor in the distribution of species; also exposure, substratum, and vegetation cover are relevant. In 6 species of intertidal fishes of the Pacific coast near Vancouver, larvae of intertidal species prefer the substratum of their adults' habitat for settling (Marliave, 1977). There is a persistent vertical and horizontal pattern of abundance for different species in the intertidal zone, as shown in Fig. 4 (Yoshiyama, 1981). Interspecific differences are particularly obvious in the case of the two *Oligocottus* and the two *Clinocottus* species. The out-of-water habitat is occupied by 3 species of stichaeids that seek cover under rocks and algae during low tide (Horn and Riegle, 1981).

Vertical distributions of five amphibious stichaeoid fishes in the rocky intertidal of central California overlap, but are clearly distinguishable (Table 4; Horn and Riegle, 1981; Barton, 1982). In these species, intertidal distribution patterns can be attributed to body size, body surface area, and tolerance to water loss, as determined by the ability to withstand the desiccation stress of low tides. Setran and Behrens (1993) showed interspecific

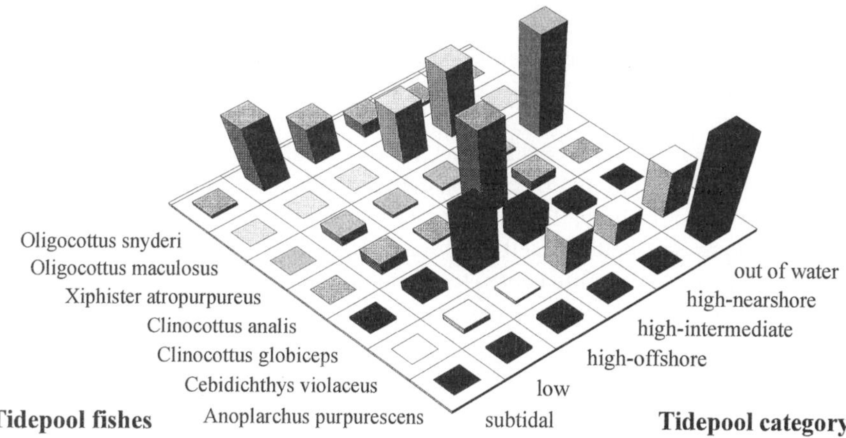


**Figure 4.** Distribution of intertidal fishes over tidepool categories in central California. Height of columns indicates relative proportion of captures for each species (after Yoshiyama, 1981). In *A. purpurescens, C. analis,* and *O. snyderi* only smaller specimens (<91 mm, <61 mm, and <61 mm, respectively) are considered.

**Table 4.** Size Range, Vertical Distribution of Greatest Abundance, and Maximum Vertical Height of Five Stichaeoid Fish Species in the Rocky Intertidal Zone Near Piedras Blancas, California[a]

| Species | Size range (cm SL, this study) | Vertical height range of greatest abundance (m) | Maximum vertical height (m) |
|---|---|---|---|
| *Cebidicthys violaceus* | 4.5–23.5 | +0.4 to +0.6 | +1.0 |
| *Anoplarchus purpurescens* | 6.0–10.0 | 0.0 to +0.2 | +0.8 |
| *Xiphister mucosus* | 7.0–20.0 | +0.2 to +0.4 | +0.8 |
| *Xiphister atropurpureus* | 6.5–20.0 | −0.2 to 0.0 | +0.2 |
| *Xererpes fucorum* | 5.5–13.5 | +0.2 to +0.5 | +1.0 (in seaweeds) |

[a] All species occur down to the lowest level of exposed intertidal zone. From Horn and Riegle (1981).

differences in vertical zonation of early juveniles of two sympatric stichaeids, with *Xiphister mucosus* consistently occurring at higher elevations than *Cebidichthys violaceus.*

## B. Intraspecific Vertical Zonation

Intraspecific abundance differences between size classes of intertidal fish species occur frequently. Differential size distribution has been observed on sandy shores (Gibson, 1973, 1993) and is common in the rocky intertidal zone as well. Smaller specimens are usually more abundant on a higher intertidal level; this has been observed in Gobiesocidae (Dunne, 1977; Stepien, 1990; Prochazka and Griffiths, 1992), Cottidae (Nakamura, 1976; Mgaya, 1992), and Blenniidae (Arruda, 1979b; Milton, 1983; Nieder, 1992). Surprisingly, among the Clinidae and Gobiesocidae of the Cape region, South Africa, size distributions do not always follow this trend (Bennett and Griffiths, 1984): juvenile *Chorisochismus dentex* (Gobiesocidae) occur only subtidally, intermediate sized specimens occur throughout the intertidal zone, and very large individuals are found only below the mean low-tide mark. Medium sized *Clinus superciliosus* occur throughout the intertidal, but larger individuals prefer the lower tidal zone. In *Clinus acuminatus* and *Clinus heterodon* this is reversed, as larger specimens occur near the upper distribution limits of the species.

Young specimens of sublittoral fishes, for example *Parablennius sanguinolentus* (Arruda, 1979a), can be found in midlittoral pools, and a preference for shallow water is commonplace among juvenile near-shore fishes. Hawaiian surge pools serve as "incubators" for juvenile fishes (Gosline, 1965); those on the Marshall Islands have been called "rearing ponds" (Hiatt and Strasburg, 1960). This phenomenon is common not only on rocky shores, but also on soft-bottom intertidal areas (Ruiz *et al.,* 1993). The juveniles of intertidal fishes with amphibious life habits, for example, stichaeids and pholids, that may potentially be subjected to dangerous desiccation because of small body size, still prefer positions high in the intertidal, relative to their species-specific vertical zonation (Horn and Riegle, 1981). This size-class distribution may result from intraspecific competition for shelter, because the outcome of intraspecific contests usually depends on body size (Shulman, 1985; Mayr and Berger, 1992). Graham *et al.* (1985) report aggressive behavior and movement within the intertidal zone of small sized *Entomacrodus nigricans* (Blenniidae). Smaller conspecifics are usually distributed in the upper tidal habitat of the Caribbean island of San Andres, whereas larger *E. nigricans* are found in the lower intertidal zone and show less intraspecific aggression. In the South African clinid *Clinus supercilio-*

*sus,* aggressive behavior of larger fishes forces juveniles to occupy poorly suited tide pools (Marsh *et al.,* 1978).

Predator avoidance may be another reason for the increased presence of juvenile fishes in the uppermost intertidal zone, despite the hazards of increased physiological stress. For other intertidal organisms, for example, snails of the Red Sea, predation by fishes affects vertical distribution (Ayal and Safriel, 1982). On the other hand, terrestrial predators such as birds may affect the abundance of fish in the upper intertidal (Skead, 1966).

An intriguing phenomenon is the differential vertical distribution of sexes among several species of clinid kelpfish in California. In the case of *Gibbonsia elegans,* lower intertidal tidepools hold approximately 90% females, but lower, to a depth of 30 m, the two sexes are equally abundant (Williams, 1954). Stepien (1986, 1987) and Stepien and Rosenblatt (1991) found sex-biased samples while collecting *G. elegans, G. metzi,* and *G. montereyensis,* as well as *Heterostichus rostratus,* with mostly males in the subtidal and mostly females and juveniles in the intertidal samples. These workers suggest that females migrate subtidally to mate and lay eggs and that males guard the eggs until they hatch. The same pattern is seen in the South American kelpfish, *Myoxedes viridis* (Stepien, 1990). Morphological and color differences seen between fish collected at these two vertical heights are suggested to be sexually dimorphic or related to the algal diet in each habitat.

## III. Temporal Variations in Vertical Distribution

### A. Seasonal Changes

The intertidal fish community outside tropical shores is subject to marked seasonal fluctuations. Abundances of some species of intertidal fishes decline in winter in France, Ireland, Portugal, and Italy (Gibson, 1968; Dunne, 1977; Arruda, 1979b; Patzner, 1985), California and North Carolina (Burgess, 1978; Chandler and Lindquist, 1981), and Taiwan (Chang *et al.,* 1969), presumably because of migration of fish to deeper waters. An exception may be *Coryphoblennius galerita,* which rests during the cold winter months in crevices of rock pools (Fives, 1980). How do seasonal changes in the intertidal fish community affect the vertical distribution of different species and what factors are responsible for them?

Temperature seems to be responsible for seasonal changes in vertical distribution. In the northern Adriatic, tidepools are deserted by intertidal fishes during winter, presumably because of near-freezing temperatures (Kotrschal and Reynolds, 1982). In central California, the mean vertical distribution of *Xererpes fucorum* (Pholididae) was higher in summer than in winter (Burgess, 1978). A seasonal shift in vertical distribution has been observed in *Pholis gunnellus* (Sawyer, 1967) and is suspected for *Lipophrys pholis* (Gibson, 1967), probably because of low winter temperatures.

Where fish assemblages are composed of temperate and tropical fish species, the latter are subject to severe winter kills. The abundance of tropical species is clearly correlated to sea temperatures, probably attributable to summer recruitment of juveniles. Species diversity and evenness are lowest in winter in the northern Gulf of California (Thomson and Lehner, 1976). Recolonization of artificially defaunated tidepools on the coast of South Africa was slower in winter than in summer (Beckley, 1985a), an indication of the importance of temperature as a limiting factor in vertical migration.

The influx of transient species, as both adults and juveniles, to the shallow intertidal zone is generally responsible for increased species richness in summer (Bennett, 1987). A "nursery function" of the intertidal zone in South Africa, of more (Beckley, 1985b) or less

(Bennett, 1987) importance, is at least part of the reason, as it is in northern California (Moring, 1986). On the central Chilean coast, juvenile Girellidae abound seasonally in tidepools, whereas they are absent from the subtidal habitat (Varas and Ojeda, 1990). Size class distribution of the blenny *Lipophrys pholis* in tide pools of the west coast of Portugal indicates their use as nursery habitats (Faria *et al.,* 1996).

An unusual exception to the rule of decreased winter abundance and species richness is an area of cold water upwelling off Baja California. There, the number of species and individuals increases during winter months (Stepien *et al.,* 1991). This may occur because the conditions are cooler and more suitable for fish species of northern origin that are not as sensitive to decreases in temperature.

Intertidal habitat suitability may be influenced by turbulence due to wave activity. Abundance of two species of stichaeids in California declines during winter, most likely as a result of heavy wave activity (Setran and Behrens, 1993). On the west coast of Vancouver Island, cottid abundance declines considerably in tidepools during winter. Horizontal differences in vertical distribution of *Oligocottus maculosus* corresponding to wave exposure suggest that the main environmental factor that determines seasonal occurrence is wave action, which tends to be much stronger during winter (Green, 1971); however, this may not be completely separated from the effects of lower temperatures. A highly positive correlation between low abundance of intertidal fishes and high wave action is often observed (Grossman, 1982).

In a South Carolina intertidal creek, the number of individuals does not decline in winter, though species diversity does (Cain and Dean, 1976). It seems that away from the rocky intertidal, not all transient species are affected by low winter temperatures to the same degree, possibly because there is little influence of wave action in an estuary.

### B. Diurnal Changes

Vertical distribution of some intertidal fishes can be subject to diurnal changes, but few observations and data exist. The Mediterranean blennies *Coryphoblennius galerita* and *Lipophrys trigloides* remain close to the surface or even emerge from the water during the night; *C. galerita* may emerge in order to "sleep" (Zander, 1972c, 1983; Heymer, 1982; Louisy, 1987). This fish has been observed to enter holes above the water level at night in Portugal (Almada *et al.,* 1983). In the Mediterranean, *L. trigloides* is active during the night at the level of the water line (Nieder and Zander, 1994). Both species probably seek to avoid nocturnal predators, particularly the *Scorpaena* sp. that is known to feed on blennies in shallow water (Kotrschal and Reynolds, 1982) and is present in the shallow sublittoral at night (Nieder, personal observation). On the coast of Israel, *Lipophrys pavo* is reported to move from shelters occupied during daytime to the surface of flat rocks barely covered with water at night (Fishelson, 1963).

## IV. Coping with a Dynamic Habitat

The boundary between water and air is characterized by extreme changes in physical factors. Waves and surf can develop enormous power and may drive organisms away, especially from steep rocky habitats. Living in the intertidal zone may be advantageous because predators may be rare or food supply may be rich. But these activites are not possible without special adaptations that guarantee the fish will remain in their characteristic microhabitats, by withstanding the power of waves to keep the fish on the substratum.

The higher fish live in the upper littoral or in the supralittoral above the water level, the greater the influence of the physical properties of air. Regular contact with water is necessary for aquatic organisms, including fish, to perform physiological functions such as breathing and osmoregulation and to prevent desiccation. Light rays have a different refractive index in air than in water, sound waves are slower in air than in water, neuromasts do not function in air, the danger of drying up of skin and gills is high, and importantly, air does not have the buoyancy of water. Therefore, dramatic adaptations in morphology and physiology are seen in intertidal and supratidal fishes.

## A. Clinging and Moving

In spite of the enormous power that waves may develop at rocky habitats, several species of fishes are able to withstand these extreme conditions. This is possible because, first, many of these species possess depressed or eel-like shaped bodies and the absence or reduction of swim bladders, which helps them stay near the bottom. Second, they have acquired behaviors to retire in narrow holes and clefts. Third, they display special organs like suckers or strong hooks that help them cling to the substratum. Fourth, their body size is relatively small. Fifth, parts of the body and fins that contact the substratum are protected against its roughness by a thick cuticular layer. Locomotion of fishes out of water is complicated because tail and fins, the primary moving organs of fishes, do not function well terrestrially, where animals generally move with the aid of lever-like pectoral fins. While none of these features is unique to intertidal fishes, in combination they provide excellent resilience to the intertidal conditions.

Blennies (Blenniidae) are slender with long dorsal and anal fins. Mediterranean species of the surf zone have acquired more strengthened tips on the ends of pectoral, pelvic, and anal fin rays than species of the zones of turbulences or currents (Zander, 1972c, 1973). The pelvic fins of *Coryphoblennius galerita, Lipophrys trigloides,* and especially *Aidablennius sphynx* are extremely short and can operate like tweezers to cling to the smallest overhangs of the substratum. *C. galerita* often rests as an inverted U or as an S in shallow depressions, where it clings with the rays of the anal fin wide apart, directing the anterior rays to the front and the posterior rays to the rear. The clinging is supported by four strong lower rays of the pectoral fin. *L. trigloides,* possessing five strong hooks in this fin, behaves similarly.

*C. galerita* and *Scartella cristata,* inhabitants of the surf zone, are characterized by a ventrally broadened head that is also suited to adhere to the substratum. Movements above the water surface amount to slight jumps that are performed with the aid of the laterally rotated tail. Terrestrially active salariin blennies from the Red Sea show similar behaviors and morphological adaptations (Zander, 1967, 1972a,b). The members of this blenniid tribe are characterized by ventrally placed mouths because they are grazers of algae. *Alticus kirkii,* a grazer on supratidal rocks, possesses two swellings at the angles of the mouth that may function together with the mouth, the ventral portion of the head, and the belly as a sucker, and thus facilitate adhesion even to steep rocks, as Abel (1973) observed in the related *A. saliens* from the Indo-Pacific (Fig. 5). In addition, these species have tweezer-like pelvic fins and strong hooks on the pectoral and anal fins. The four radialia of the pectoral fins of *A. kirkii* are freely movable at their distal ends. Their bases, arched like socket joints, originate at rounded projections of the cleithrum. Therefore, the pectoral fins can be twisted and squeezed flat onto the substrate to include an air bubble, supporting the adhesion (Zander, 1972b).

*A. kirkii* moves on steep walls by stemming steps in which the rotated tail is alternatively

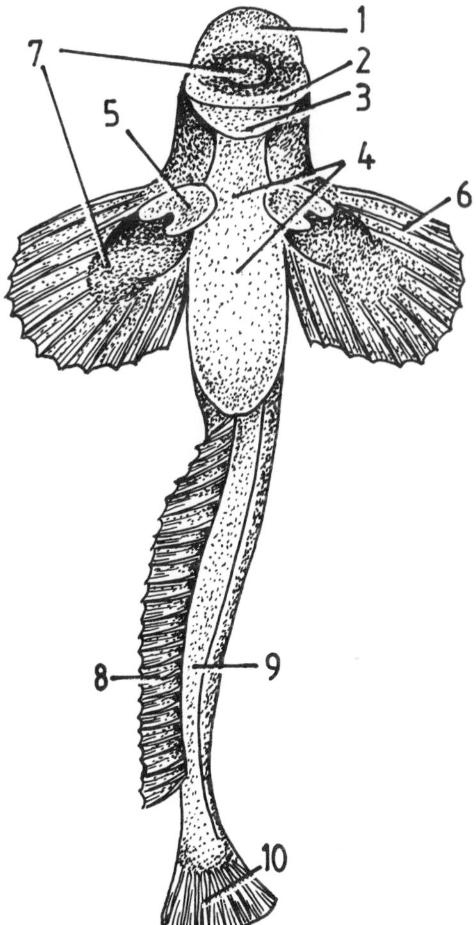


**Figure 5.** View of the ventral side of the blenniid *Alticus saliens* showing means of adhesion. 1–3 = margin of the mouth; 4 = throat; 5 = bases of pelvic fin; 6 = pectoral fin; 7 = air bubbles; 8 = anal fin; 9 = rotated tail; 10 = caudal fin. After Abel (1973).

flexed to the right and the left. By this movement the body is driven higher on the substrate. Rao and Hora (1938) also observed the same kind of locomotion in the terrestrially active salariin *Andamia* species from the Indo-Pacific. The rotated tail enables *A. kirkii* to perform short jumps on the rocks as well as on the water surface. On slanting substrata the tail is not rotated completely, only the anal fin. The hooks at the end of the single rays are independently movable and are used when the fish is climbing upward or downward. The intertidal *Istiblennius* species move on rock flats only by jumping. Males of *A. saliens* perform courtship behavior by short jumps above the water line in order to attract spawning females, but the males have difficulty in keeping their balance (Abel, 1973).

Representatives of other terrestrially active fishes display similar means of locomotion.

In the Pacific rockskipper *Mnierpes macrocephalus* (Clinidae), the anal fin rays are deeply incised and the pectorals take part in climbing (Graham, 1970). Gobies are generally shorter in length than blennies and clinids (although some are elongate) and their bodies are round or slightly compressed. The intertidal *Gobionellus sagittula* from Panama climbs upward by stemming steps using the tail and anal fin (Todd, 1976). This clearly differs from the locomotion in the related mudskipper *Periophthalmus,* which uses "crutching," in which the pectoral fins are two levers, the ventral fins lift the body when the pectorals swing forward, and the tail is used as a third leg (Harris, 1960). The pelvic fins of many gobies are fused to form a sucker that can help to withstand streaming water or even turbulence. While this is not unique to intertidal gobies, it provides a means for clinging to substrata in turbulent habitats. Some subtidal *Gobius* species and also most species of the supratidal *Periophthalmus* have separated pelvic fins.

Clingfishes (Gobiesocidae) have a dorso-ventrally flattened body shape. They possess a single dorsal fin on the hind body and ventral suckers built from the pelvic fins, which enable several species to live in the surf zone. *Eckloniaichthys scylliorhiniceps* from South Africa changes sites by moving its sucker along the substrate, thus holding contact with it (Smith, 1950). The Chilean clingfish *Sicyases sanguineus* clings above the water line on vertical rock walls for hours, so well developed is its ventral disc (Gordon *et al.,* 1970).

The Cottidae are demersal fishes that are somewhat rounded or triangular in cross section and negatively buoyant. Some intertidal cottids are found emerged during low tides in caves (Wright and Raymond, 1978) or protected under rocks and in crevices (Cross, 1981; Martin, 1996). Intertidal cottids actively emerge from hypoxic water (Martin, 1991, 1996; Yoshiyama *et al.,* 1995) by the stemming step described above for blennies or by pushing off the substrate using the tail. They maintain an upright posture in air with a tripod stance, using the large pectoral fins and the tail for balance.

Elongated intertidal fishes, including stichaeids such as *Xiphister* and *Anoplarchus,* are able to wriggle about thrashing rapidly, probably as a means of predator avoidance (Horn and Riegle, 1981), and their shape assists them to escape capture by allowing them to move into small crevices and between rocks. *Xererpes fucorum,* another elongate fish, weaves itself between fronds of intertidal seaweeds such as *Pelvetia* (Burgess, 1978), where it remains cryptic, emerged in air high above the water line and above the ground until the tide returns (Martin, 1993). Although these elongate fishes are capable of rapid thrashing about and directional sinusoidal locomotion, it is unlikely that they normally move very much during emergence at low tide (Ralston and Horn, 1986; Horn and Gibson, 1988).

The amphibious labrisomid *Mnierpes macrocephalus* has scales that form pockets that can hold water, possibly enabling capillary adhesion of the body's ventral surface (Graham, 1970). Thus it can move easily by stemming steps on horizontal or vertical substrata.

## B. Resistance to Desiccation

Fishes that emerge from water would be expected to face an immediate threat of desiccation. However, present knowledge does not seem to indicate any great physiological or anatomical resistance to desiccation among intertidal fishes; the challenge is typically addressed by behavioral means. In many cases, intertidal fishes emerge under conditions that minimize the danger of dessication, for example, at night (Wright and Raymond, 1978; Congleton, 1980; Laming *et al.,* 1982; Louisy, 1987; Martin, 1993), when hypoxic stresses are likely to be greatest in tidepools, or under cover such as in crevices, under boulders, or within algal mats (Horn and Riegle, 1981; Cross, 1981; Martin, 1995, 1996). These fishes may be consistently found out of water at each low tide (Cross, 1981; Martin, 1995); some of these fishes even spawn intertidally and remain guarding the eggs during tidal emergences (Marliave, 1981; Marliave and DeMartini, 1977; Coleman, 1992).

Some fishes, particularly the supralittoral rockskippers and mudskippers, are highly active terrestrially and fully emerged from water exposed on rocks or mudflats, without cover. The behavior of *Alticus kirkii* of falling or rolling over on each side in shallow puddles (Magnus, 1966; Zander, 1967) probably serves to reduce evaporative water loss of the body. *Entomacrodus,* the pearl blenny, is active diurnally in the supralittoral of Panama, but remains emerged only very briefly before returning to water or being washed by a wave (Graham *et al.,* 1985). *Mnierpes macrocephalus* is rarely in direct sunlight more than 5 min or out of water for more than 20 min at night (Graham, 1970). *Periophthalmus* mudskippers also tend to return to water frequently, not for full immersion but to fill the mouth cavity or to irrigate the gills with water (Gordon *et al.,* 1968). *Xiphister atropurpureus,* although relatively inactive terrestrially and emerged under boulders or other cover, also shows the behavior of periodically falling over on its side in shallow pools (Daxboeck and Heming, 1982).

Many studies have shown that if desiccation is prevented, some intertidal fishes can remain emerged in air for many hours (e.g., Gordon *et al.,* 1969, 1970; Horn and Riegle, 1981; Laming *et al.,* 1982; Martin, 1991). Mass loss over time does occur, probably as evaporative water loss. Horn and Riegle (1981) showed that of five species across a vertical tidal gradient, the stichaeid *Cebidichthys violaceus,* with the highest distribution, had the greatest water content and tolerance to desiccation. They found that two mid-intertidal species, *Xiphister atropurpureus* and *Xererpes fucorum,* continued to lose water even after reimmersion following desiccation in air, underscoring the importance of behavioral methods of avoiding desiccation in the first place. Five species of cottids occurring along a vertical tidal gradient show no difference in rate of water loss when corrected for differences in body mass (Martin, 1996).

## C. Sense Organs

The problem for fishes to see as well above the water level as below it is the change in refractive index of light between air and water. In air, the index is 1.00, whereas in water, it is 1.33. Light rays are not broken by the cornea, as it has the same refractive index as water, but are first broken by the lens itself, with a refractive index of 1.42 (Graham and Rosenblatt, 1970). The lens is round in fishes, in contrast to terrestrial vertebrates, in which it is elliptical. Accommodation for near sight also differs according to the respective habitat. Fish eyes retract the lens toward the retina, while higher vertebrates change the lens shape to a more flattened form (Zander, 1974).

The characteristic method of accommodation by the eye remains unchanged in fish during the adaptation to aerial vision (Graham, 1971). However, other parts of the eyes may undergo drastic changes. The lens of the mudskipper *Periophthalmus* is flattened (Karsten, 1923). In the nonamphibious four-eyed fish, *Anableps* (Anablepidae), the lens is egg-shaped for viewing above and below the water. Another adaptation is the flattening of the cornea, found in rockskippers climbing on land, including *Mnierpes macrocephalus* from the tropical east Pacific and *Dialommus fuscus* from the Galápagos archipelago. This allows the light rays to go straight to the lens without being bent in air (Munk, 1969; Graham and Rosenblatt, 1970). The eyes of these two fishes are divided by a transverse pigmented intersection, creating two different windows for seeing in the forward and lateral directions. The rock blenny *Entomacrodus nigricans* (Blenniidae) from eastern Panama, although less adapted to an amphibous way of life, has the same structure of divided and flattened eyes but with less pigmentation (Graham, 1970).

Other possibilities that permit fishes to have aerial vision are revealed in rock blennies of the Red Sea. In comparing the eye structures of the sublittoral *Salarias fasciatus,* the amphibious *Istiblennius edentulus,* and the supratidal *Alticus kirkii,* two different means of adaptation become apparent (Zander, 1974; Fig. 6). *I. edentulus* achieves focused vision by retracting the lens into a central depression of the retina that is without function at this region, casting sharp pictures on bordering parts of the retina. *A. kirkii* is able to retract the cornea propria, a structure that is separate from the anterior part of the cornea and connected at a central point with the lens. The separation of the parts of the cornea creates an additional eye chamber that can be extended by retracting the lens; thus it displaces the focal point according to whether the fish is in air or in water. Muscles found in the ventral limbal region may be able to flatten the cornea, as found in *Mnierpes* or *Dialommus.* Structures similar to those in *A. kirkii* are found in the amphibious *Coryphoblennius galerita* of the Mediterranean, where only the connection of cornea with lens is lacking (Ehlers, 1978).

Lateral lines are sense organs that function only under water. They are present in all aquatic vertebrates, including larval amphibians, but absent in terrestrial animals, including most adult amphibians. A comparison of the supratidal *Alticus kirkii* with 4 other blenniid species from the Red Sea reveals a greatly reduced lateral line system (Zander, 1972a). Along the vertical gradient, the subtidal *Salarias fasciatus* possesses a lateral line with pores around the pectoral fins, but in the intertidal *Istiblennius edentulus, I. flaviumbrinus,* and *Antennablennius hypenetes,* pores are absent, and the latter two species have shortened lateral lines also. On the other hand, a comparison of 12 species of Mediterranean blennies shows no clear correlation of vertical zonation and development of lateral lines (Zander, 1972c). Some species inhabiting the surf zone, like *Scartella cristata* or *Lipophrys trigloides,* possess the best developed organs, comparable to those in the sublittoral *Parablennius gattorugine* and *P. pilicornis,* although *L. trigloides* climbs above the water line at night as does *Coryphoblennius galerita* (Zander, 1983). The very short lateral line of *C. galerita* has only a few pores (Fig. 7), and the reduction is even more pronounced in *A. kirkii.*

## D. Aerial Respiration

Terrestrially active marine fishes have a dilemma. They have gills, which typically function in water, and they must protect their breathing organs against desiccation in air. A way out of this problem is achieved in the mudskipper *Periophthalmus* by reducing the number and length of gill filaments, and they may swallow water regularly during terrestrial sojourns and seal up the gill chambers (Magnus, 1981). In some amphibious or terrestrial blennies, no reduction of gill filaments is observed, but they can close their gill chambers when above the water line. In addition, these fishes have developed large blood vessels near the skin surface to breathe in air (Zander, 1972a,c, 1983). *Coryphoblennius galerita* from the Mediterranean, a fish that climbs regularly to or even above the water line, has concentrated these enlarged cutaneous vessels behind the eyes. *Lipophrys trigloides,* which stays above the surf zone only at night, lacks these enlarged surface vessels (Zander, 1972c, 1983). The terrestrially active *Alticus kirkii* from the Red Sea possesses many cutaneous blood vessels

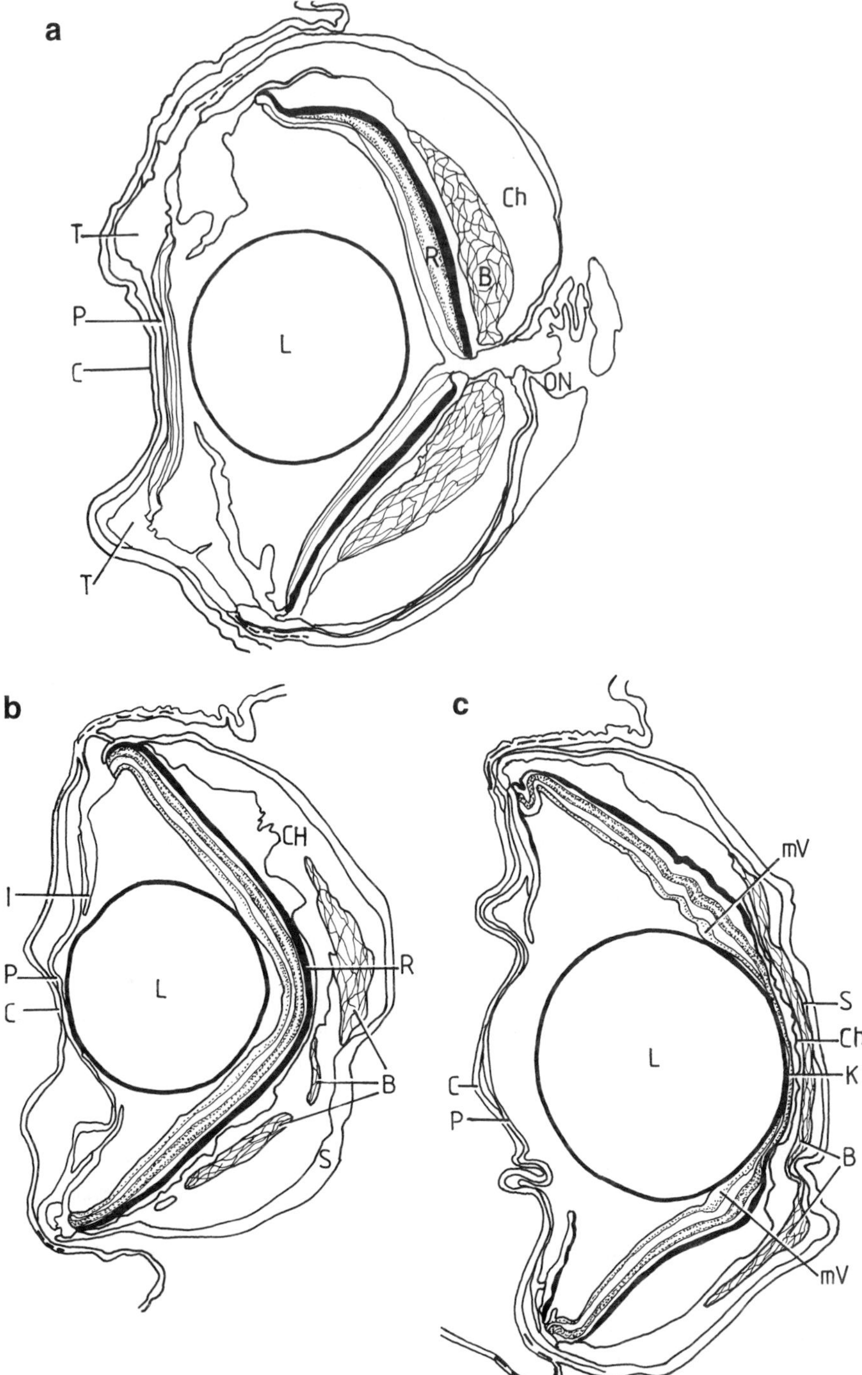


**Figure 6.** Cross sections of the eyes of an amphibious clinid and two amphibious blenniid fishes. (a) *Dialommus fuscus,* (b) *Alticus kirkii,* (c) *Istiblennius edentulus.* B, blood vessel; C, cornea conjunctiva; Ch, choroidea; I, iris; K, retinal depression; L, lens; mV, medial swelling; ON, optic nerve; P, cornea propria; R, retina; S, sclera; T, thickened part of the outer sclera. (a) after Munk (1969); (b, c) after Zander (1974).

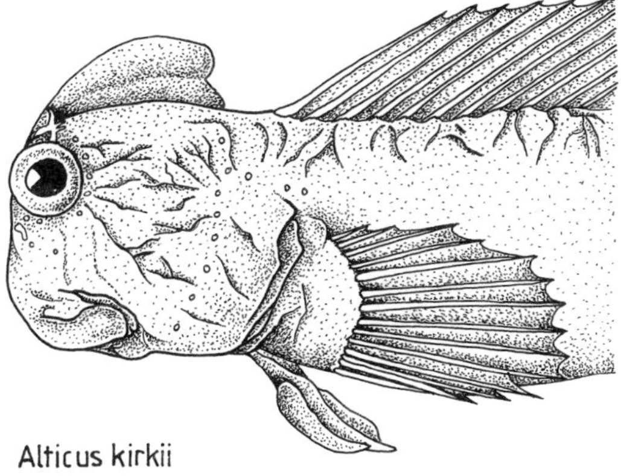


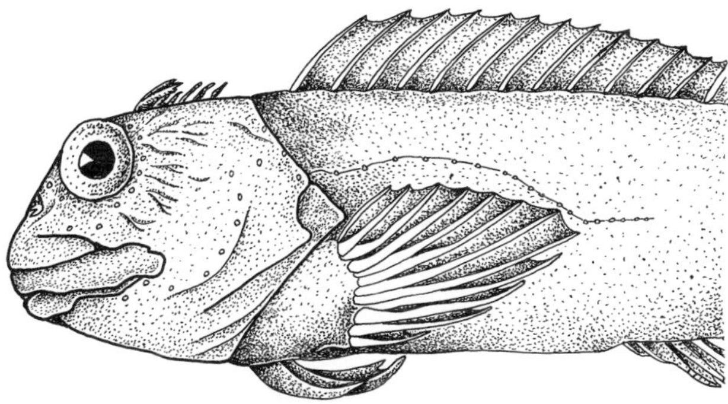


**Figure 7.** Head and front body of two amphibious blenniid fishes. The dermal blood vessels are visible on the head and, only in *Alticus kirkii,* below the dorsal fin. The lateral line is lost in *A. kirkii* and reduced in *Coryphoblennius galerita.* (After Zander 1972a,c.)

on the head and along the dorsal part of the body. It also has cartilagenous skeletal elements in the gill filaments for resisting collapse in air (Brown *et al.,* 1992). The development of cutaneous blood vessels, combined with the behavior of rolling over on the longitudinal axis on wet rocks, prevents drying out of these body regions (Magnus, 1966; Fig. 7). *Istiblennius edentulus* and *I. flaviumbrinus* from the mid-intertidal zone show fewer of these accessory breathing structures (Zander, 1972a).

Cottids occurring higher in the intertidal zone have behavioral and physiological adaptations for air-breathing that are not present in cottids occurring in the subtidal or deeper waters of Puget Sound (Martin, 1996). In contrast to cottids of the high intertidal zone, subtidal and lower intertidal cottids do not actively emerge from water and are not able to survive for long if taken from water.

Air-breathing of intertidal fishes is discussed at length by Martin and Bridges (Chapter 4, this volume).

## V. Future Work

It may be useful to add types of behavioral emergence to the concept of vertical zonation for fishes. Martin (1995) suggested the terminology of "skippers" for frequent emergers that are active terrestrially (and usually supralittoral), "remainers" for those that are passively emerged under cover of rocks or algae by a receding tide (usually mid to lower intertidal), and "tidepool emergers" for tidepool fishes that are typically aquatic but can deliberately emerge under certain conditions. Because of the complexity of habitat usage by mobile fishes, it is difficult to make comparisons between species that are resident and those that are transient in the intertidal zone, yet these comparisons could yield valuable new insights into niche partitioning and resource use.

Additional work on fishes in some geographical areas, particularly in tropical and high latitude (above 60°) regions, is needed for comparisons with the better studied assemblages of the temperate zone. The most prominent feature of fishes with high intertidal distributions is their hardiness, including an ability to survive prolonged aerial emergence and highly variable environmental conditions. Many aspects of these abilities remain to be studied, including the effects on temperature regulation, nitrogenous waste excretion, nutrition, relative predation risk, intraspecific communication, and aggregation or territoriality. The number of intertidal fishes known to be able to emerge has increased dramatically in the past decade, and study of the morphological or physiological adaptations in these fishes is just beginning.

## Acknowledgments

The authors thank Monika Haenel for drawing Figures 1, 5, 6, and 7, and Michael Horn and Robert Lea for many helpful discussions and suggestions.

## References

Abel, E. F. (1961). Über die Beziehungen mariner Fische zu Hartbodenstrukturen. *Sitzungsber. Österr. Akad. Wiss. Math. (Nat. Kl., Abt. I)* **170,** 223–263.

Abel, E. F. (1962). Freiwasserbeobachtungen an Fischen im Golf von Neapel als Beitrag zur Kenntnis ihrer Ökologie und ihres Verhaltens. *Int. Rev. Ges. Hydrobiol.* **47,** 219–290.

Abel, E. F. (1973). Zur Öko-Ethologie des amphibisch lebenden Fisches *Alticus saliens* (Förster) und von *Entomacrodus vermiculatus* (Val.) Blennioidea, Salariidae, unter besonderer Berücksichtigung des Fortpflanzungsverhaltens. *Sitzungsber. Österr. Akad. Wiss. Math. (Nat. Kl., Abt. I)* **191,** 137–153.

Almada, V., Dores, J., Pinheiro, A., Pinheiro, M., and Serrao Santos, R. (1983). "Contribuicao para o estudo do comportamento de *Coryphoblennius galerita* (L.) (Pisces: Blenniidae)," Vol. 2, pp. 1–163. Memorias do Museu do Mar, Serie Zoologica 2, No. 24, Cascais, Portugal.

Arruda, L. M. (1979a). Specific composition and relative abundance of intertidal fish at two places on the Portuguese coast (Sesimbra and Magoito, 1977–78). *Arq. Mus. Bocage, 2.a Serie* **6,** 325–342.

Arruda, L. M. (1979b). On the study of a sample of fish captured in the tidal range at Azores. *Bolm. Soc. Port. Ciênc. Nat.* **19,** 5–36.

Arruda, L. M. (1990). Population structures of fish in the intertidal ranges of the Portuguese coasts. *Vie Milieu* **40,** 319–323.

Ayal, Y., and Safriel, U. N. (1982). Role of competition and predation in determining habitat occupancy of Cerithiidae (Gastropoda: Prosobranchia) on the rocky, intertidal, Red Sea coast of Sinai. *Mar. Biol.* **70,** 305–316.

Barton, M. G. (1982). Intertidal vertical distribution and diets of five species of central California stichaeoid fishes. *Calif. Fish Game* **68,** 174–182.

Bauchot, M. L. (1966). Poissons marins de L'Est Atlantique tropical. Téléostéens Perciformes. V. Blennioidei. *Atlantide Rep.* **9,** 63–91.

Beckley, L. E. (1985a). Tide-pool fishes: Recolonization after experimental elimination. *J. Exp. Mar. Biol. Ecol.* **85,** 287–295.

Beckley, L. E. (1985b). The fish community of East Cape tidal pools and an assessment of the nursery function of this habitat. *S. Afr. J. Zool.* **20,** 21–27.

Bennett, B. A. (1987). The rock-pool fish community of Koppie Alleen and an assessment of the importance of Cape rock-pools as nurseries for juvenile fish. *S. Afr. J. Zool.* **22,** 25–32.

Bennett, B. A., and Griffiths, C. L. (1984). Factors affecting the distribution, abundance and diversity of rock-pool fishes on the Cape Peninsula, South Africa. *S. Afr. J. Zool.* **19,** 97.

Berger, A., and Mayr, M. (1992). Ecological studies on two intertidal New Zealand fishes, *Acanthoclinus fuscus* and *Forsterygion nigripenne robustum. N. Z. J. Mar. Freshw. Res.* **26,** 359–370.

Black, R., and Miller, R. J. (1991). Use of the intertidal zone by fish in Nova Scotia. *Environ. Biol. Fish.* **31,** 109–121.

Breder, C. M. (1948). Observations on coloration in reference to behavior in tide-pool and other marine shore fishes. *Bull. Am. Mus. Nat. Hist.* **92,** 285–311.

Bridges, C. R. (1988). Respiratory adaptations in intertidal fishes. *Am. Zool.* **28,** 79–96.

Brillet, C. (1976). Structure de terrier, reproduction et comportement des jeunes chez le poisson amphibie *Periophthalmus sobrinus* Eggert. *Terre Vie* **30,** 465–483.

Brito, A., and Lozano, G. (1980). El suborden Blennioidei (Pisces, Percomorphi) en las islas Canarias. *Bol. Inst. Espa. Oceanogr.* **6,** 7–17.

Brown, C. R., Gordon, M. S., and Martin, K. L. M. (1992). Aerial and aquatic oxygen uptake in the amphibious Red Sea rockskipper fish, *Alticus kirki* (Family Blenniidae). *Copeia* **1992,** 1007–1013.

Burgess, T. J. (1978). The comparative ecology of two sympatric polychromatic populations of *Xererpes fucorum* Jordan & Gilbert (Pisces: Pholididae) from the rocky intertidal zone of Central California. *J. Exp. Mar. Biol. Ecol.* **35,** 43–58.

Cain, R. L., and Dean, J. M. (1976). Annual occurrence, abundance and diversity of fish in South Carolina intertidal creek. *Mar. Biol.* **36,** 369–379.

Carvalho, F. P. (1982). Éthologie alimentaire de trois poissons Blenniidae de la côte Portugaise. *Bolm. Soc. Port. Ciênc. Nat.* **21,** 31–43.

Chandler, G. T., and Lindquist, D. G. (1981). The comparative behavioural ecology of two species of co-inhabiting tide-pool blennies. *Environ. Biol. Fish.* **6,** 126.

Chang, K.-H., Jan, R.-Q., and Shao, K.-T. (1983). Community ecology of the marine fishes on Lutao Island, Taiwan. *Bull. Inst. Zool., Acad. Sin.* **22,** 141–155.

Chang, K.-H., Lee, S.-C., and Wang, T.-S. (1969). A preliminary report of ecological study on some intertidal fishes of Taiwan. *Bull. Inst. Zool., Acad. Sin.* **8,** 59–70.

Chang, K. H., Lee, S. C., Lee, J. C., and Chen, C. P. (1973). Ecological study on some intertidal fishes of Taiwan. *Bull. Inst. Zool., Acad. Sin.* **12,** 45–50.

Clayton, D. A., and Vaughan, T. C. (1982). Pentagonal territories of the mudskipper *Boleophthalmus boddarti* (Pisces: Gobiidae). *Copeia* **1982,** 232–234.

Coleman, R. (1992). Reproductive biology and female parental care in the cockscomb prickleback, *Anoplarchus purpurescens* (Pisces: Stichaeidae). *Environ. Biol. Fish.* **41,** 177–186.

Congleton, J. L. (1980). Observations on the responses of some southern California tidepool fishes to nocturnal hypoxic stress. *Comp. Biochem. Physiol. A* **66,** 719–722.

Cross, J. N. (1981). "Structure of a Rocky Intertidal Fish Assemblage." Ph.D. dissertation, University of Washington, Seattle.

Daxboeck, C., and Heming, T. A. (1982). Bimodal respiration in the intertidal fish, *Xiphister atropurpureus* (Kittlitz). *Mar. Behav. Physiol.* **9,** 23–33.

Dunne, J. (1977). Littoral and benthic investigations on the west coast of Ireland. VII. (Section A: Faunistic and ecological studies). The biology of the shanny, *Blennius pholis* L. (Pisces) at Carna, Connemara. *Proc. R. Ir. Acad.* **77,** 207–226.

Ebeling, A. W., Bernal, P., and Zuleta, A. (1970). Emersion of the amphibious Chilean clingfish, *Sicyases sanguineus. Biol. Bull.* **139,** 115–137.

Eger, W. H. (1971). "Ecological and Physiological Adaptations of Intertidal Clingfishes (Teleostei: Gobeisocidae) in the Northern Gulf of California." Ph.D. dissertation, University of Arizona.

Ehlers, C. (1978). "Zur Morphologie des halbamphibischen Schleimfisches *Coryphoblennius galerita.*" Diploma thesis, Universität Hamburg.

Faria, C., Almada, V. C., and Goncalves, E. J. (1996). Juvenile recruitment, growth and maturation of *Lipophrys pholis* (Pisces: Blenniidae), from the west coast of Portugal. *J. Fish Biol.* **49,** 727–730.

Fishelson, L. (1963). Observations on littoral fishes of Israel. I. Behaviour of *Blennius pavo* Risso (Teleostei, Blenniida). *Isr. J. Zool.* **12,** 67–80.

Fitch, J. E., and Lavenberg, R. J. (1975). "Tidepool and Nearshore Fishes of California." University of California Press, Berkeley.

Fives, J. M. (1980). Littoral and benthic investigations on the west coast of Ireland. IX. The biology of Montagu's blenny, *Coryphoblennius galerita* L. (Pisces), on the Connemara coast. *Proc. R. Ir. Acad.* **80B,** 61–77.

Fricke, H. W. (1970). Die ökologische Spezialisierung der Eidechse *Cryptoblepharus boutoni cognatus* (Boettger) auf das Leben in der Gezeitenzone (Reptilia, Skinkidae). *Oecologia* **5,** 380–391.

Fukao, R. (1985). An annotated list of blenniid fishes from Shirahama, Wakayama Prefecture, Japan. *Pub. Seto Mar. Biol. Lab.* **30,** 81–124.

Gerlach, S. A. (1958). Die Mangroveregion tropischer Kuesten als Lebensraum. *Z. Morph. Okol. Tiere* **46,** 636–730.

Gibson, R. N. (1967). Studies on the movements of littoral fish. *J. Anim. Ecol.* **36,** 215–234.

Gibson, R. N. (1968). The food and feeding relationships of littoral fish in the Banyuls region. *Vie Milieu, Ser. A* **19,** 447–456.

Gibson, R. N. (1972). The vertical distribution and feeding relationships of intertidal fish on the Atlantic coast of France. *J. Anim. Ecol.* **41,** 189–207.

Gibson, R. N. (1973). The intertidal movements and distribution of young fish on a sandy beach with special reference to the plaice (*Pleuronectes platessa* L.). *J. Exp. Mar. Biol. Ecol.* **12,** 79–102.

Gibson, R. N. (1982). Recent studies on the biology of intertidal fishes. *Oceanogr. Mar. Biol. Annual Rev.* **20,** 363–414.

Gibson, R. N. (1993). Intertidal teleosts: Life in a fluctuating environment. In "Behaviour of Teleost Fishes" (T. J. Pitcher, Ed.), 2nd ed., pp. 513–536. Chapman & Hall, London.

Gordon, M. S., Boetius, J., Evans, D. H., and Oglesby, L. C. (1968). Additional observations on the natural history of the mudskipper, *Periophthalmus sobrinus. Copeia* **1968,** 853–857.

Gordon, M. S., Boetius, I., Evans, D. H., McCarthy, R., and Oglesby, L. C. (1969). Aspects of the terrestrial life in amphibious fishes. I. The mudskipper, *Periophthalmus sobrinus. J. Exp. Biol.* **50,** 141–149.

Gordon, M. S., Fischer, S., and Tarifeno, E. (1970). Aspects of the physiology of terrestrial life in amphibious fishes. II. The Chilean clingfish, *Sicyases sanguineus. J. Exp. Biol.* **53,** 559–572.

Gordon, M. S., Gabaldon, D. J., and Yip, A. Y.-w. (1985). Exploratory observations on microhabitat selection within the intertidal zone by the Chinese mudskipper fish *Periophthalmus cantonensis. Mar. Biol.* **85,** 209–215.

Gosline, W. A. (1965). Vertical zonation of inshore fishes in the upper water layers of the Hawaiian islands. *Ecology* **46,** 823–831.

Graham, J. B. (1970). Preliminary studies on the biology of the amphibious clinid *Mnierpes macrocephalus. Mar. Biol.* **5,** 136–140.

Graham, J. B. (1971). Aerial vision in amphibious fishes. *Fauna* **1971,** 14–23.

Graham, J. B. (1973). Terrestrial life of the amphibious fish *Mnierpes macrocephalus. Mar. Biol.* **23,** 83–91.

Graham, J. B., Jones, C. B., and Rubinoff, I. (1985). Behavioural, physiological, and ecological aspects of the amphibious life of the pearl blenny *Entomacrodus nigricans* Gill. *J. Exp. Mar. Biol. Ecol.* **89,** 255–268.

Graham, J. B., and Rosenblatt, R. H. (1970). Aerial vision: Unique adaptation in an intertidal fish. *Science* **168,** 386–388.

Green, J. M. (1971). Local distribution of *Oligocottus maculosus* Girard and other tide pool cottids of the west coast of Vancouver Island, British Columbia. *Can. J. Zool.* **49,** 1111–1128.

Grossman, G. D. (1982). Dynamics and organization of a rocky intertidal fish assemblage: The persistence and resilience of taxocene structure. *Am. Nat.* **119,** 611–637.

Harris, V. A. (1960). On the locomotion of the mud-skipper *Periophthalmus koelreuteri* (Pallas): Gobiidae. *Proc. Zool. Soc. London* **134,** 107–135.

Hesthagen, I. H. (1977). Migration, breeding and growth in *Pomatoschistus minutus* (Pallas) (Pisces, Gobiidae) in Oslofjorden, Norway. *Sarsia* **63,** 17–26.

Heymer, A. (1982). Le comportement pseudo-amphibie de *Coryphoblennius galerita* et *Blennius trigloides. Rev. Francaise d'Aquariol.* **9,** 91–96.

Hiatt, R. W., and Strasburg, D. W. (1960). Ecological relationships of the fish fauna on coral reefs of the Marshall Islands. *Ecol. Monogr.* **30,** 65–127.

Horn, M. H. (1989). Biology of marine herbivorous fishes. *Oceanogr. Mar. Biol. Annu. Rev.* **27,** 167–272.

Horn, M. H., and Gibson, R. N. (1988). Intertidal fishes. *Sci. Am.* **256,** 64–70.

Horn, M. H., Neighbors, M. A., and Murray, S. N. (1986). Herbivore responses to a seasonally fluctuating food supply: Growth potential of two temperate intertidal fishes based on the protein and energy assimilated from their macroalgal diets. *J. Exp. Mar. Biol. Ecol.* **103,** 217–234.

Horn, M. H., and Riegle, K. C. (1981). Evaporative water loss and intertidal vertical distribution in relation to body size and morphology of stichaeoid fishes from California. *J. Exp. Mar. Biol. Ecol.* **50,** 273–288.

John, D. M., and Lawson, G. W. (1991). Littoral ecosystems of tropical western Africa. *In* "Intertidal and Littoral Ecosystems" (A. C. Mathieson and P. H. Nienhuis, Eds.), pp. 297–322. Elsevier, Amsterdam.

Karsten, H. (1923). Das Auge von *Periophthalmus koelreuteri. Jena Z. Naturwiss.* **59,** 115–154.

Klausewitz, W. (1967). Die physiographische Zonierung der Saumriffe von Sarso. *Meteor Forsch. Ergebn. Reihe D* **2,** 44–68.

Koppel, V. H. (1988). Habitat selection and space partitioning among two Mediterranean Blenniid species. *Pubblicazione della Stazione Zoologica di Napoli I. Mar. Ecol.* **9,** 329–346.

Kotrschal, K., and Reynolds, W. W. (1982). Behavioral ecology of northern Adriatic reef fishes in relation to seasonal temperature regimes. *Contrib. Mar. Sci.* **25,** 99–106.

Kruess, A. (1989). "Die benthische Fischfauna des Helgolaender Felssockels." Diploma thesis, Universität Karlsruhe.

Laming, P. R., Funston, C. W., Roberts, D., and Armstrong, M. J. (1982). Behavioural, physiological and morphological adaptations of the shanny (*Blennius pholis*) to the intertidal habitat. *J. Mar. Biol. Assoc. U.K.* **62,** 329–338.

Lee, S. C. (1980). Intertidal fishes of the rocky pools at Lanyu (Botel Tobago), Taiwan. *Bull. Inst. Zool., Acad. Sin.* **19,** 1–14.

Louisy, P. (1987). Observations sur l'émersion nocturne de deux blennies méditerranéennes: *Coryphoblennius galerita* et *Blennius trigloides* (Pisces, Perciformes). *Cybium* **11,** 55–73.

Magnus, D. E. B. (1966). Bewegungsweisen des amphibischen Schleimfisches *Lophalticus kirkii magnusi* Klausewitz (Pisces, Salariidae) im Biotop. *Verh. Dtsch. Zool. Ges. Jena* **1965,** 542–555.

Magnus, D. E. B. (1981). "Bewegungsweisen, Nahrungserwerb und Fortpflanzungsverhalten des Schlammspringers *Periophthalmus kalolo* (Freilandaufnahmen)." Publ. Wiss.Film, IWF Goettingen, Sekt. Biol., Ser. 14, No. 7/D 1282.

Marliave, J. B. (1977). Substratum preferences of settling larvae of marine fishes reared in the laboratory. *J. Exp. Mar. Biol. Ecol.* **27,** 47–60.

Marliave, J. B. (1981). High intertidal spawning under rockweed, *Fucus distichus,* by the sharpnose sculpin, *Clinocottus acuticeps. Can. J. Zool.* **59,** 1122–1125.

Marliave, J. B., and DeMartini, E. E. (1977). Parental behavior of intertidal fishes of the stichaeid genus *Xiphister. Can. J. Zool.* **55,** 60–63.

Marsh, B., Crowe, T. M., and Siegfried, W. R. (1978). Species richness and abundance of clinid fish (Teleostei; Clinidae) in intertidal rock pools. *Zool. Afr.* **13,** 283–291.

Martin, K. L. M. (1991). Facultative aerial respiration in an intertidal sculpin, *Clinocottus analis. Physiol. Zool.* **64,** 1341–1355.

Martin, K. L. M. (1993). Aerial release of $CO_2$ and respiratory exchange ratio in intertidal fishes out of water. *Environ. Biol. Fish.* **37,** 189–196.

Martin, K. L. M. (1995). Time and tide wait for no fish: Intertidal fishes out of water. *Environ. Biol. Fish.* **44,** 165–181.

Martin, K. L. M. (1996). An ecological gradient in air-breathing ability among marine cottid fishes. *Physiol. Zool.* **69,** 1096–1113.

Mayr, M., and Berger, A. (1992). Territoriality and microhabitat selection in two intertidal New Zealand fish. *J. Fish Biol.* **40,** 243–256.

Metaxas, A., and Scheibling, R. E. (1993). Community structure and organization of tidepools. *Mar. Ecol. Prog. Ser.* **98,** 187.

Mgaya, Y. N. (1992). Density and production of *Clinocottus globiceps* and *Oligocottus maculosus* (Cottidae) in tidepools at Helby Island, British Columbia. *Mar. Ecol. Prog. Ser.* **85,** 219–225.

Miller, A. R. (1983). The Mediterranean Sea. A. Physical aspects. *In:* "Estuaries and Enclosed Seas," (B. H. Ketchum, Ed.), pp. 219–237. New York: Elsevier Scientific Pub. Co.

Milton, P. (1983). Biology of littoral blenniid fishes on the coast of southwest England. *J. Mar. Biol. Assoc. U.K.* **63,** 223–237.

Mok, H. K., and Wen, P. Y. (1985). Intertidal fish community ecology on Lu-Tao Island (Green Island), Taiwan. *J. Taiwan Mus.* **38,** 81–118.

Moring, J. R. (1986). Seasonal presence of tidepool fishes in a rocky intertidal zone of northern California, USA. *Hydrobiologia* **134,** 21–27.

Munk, O. (1969). The eye of the "four-eyed" fish *Dialommus fuscus* (Pisces, Blennioidei, Clinidae). *Vidensk. Meddr. Dansk. Naturh. Foren.* **132,** 7–24.

Nakamura, R. (1976). Temperature and the vertical distribution of two tidepool fishes (*Oligocottus maculosus, O. snyderi*). *Copeia* **1976,** 143–152.

Nieder, J. (1988). Zum Vorkommen von *Scartella cristata* (L.) und *Parablennius pilicornis* (Cuv.) (Teleostei, Blenniidae) an der nordspanischen Mittelmeerküste. *Zool. Anz.* **220,** 144–150.

Nieder, J. (1992). "Zur differenzierten Nutzung von Lebensraum und Nahrungsangebot durch Schleimfische (Pisces: Blenniidae) im westlichen Mittelmeer und ihre Einordnung in das Konzept der 'ökologischen Nische'." Doctor thesis, Universität Bonn.

Nieder, J. (1993). Distribution of juvenile blennies (Pisces, Blenniidae) in small tide pools: Result of low tide lottery or strategic habitat selection? *Bonn. Zool. Beitr.* **44,** 133–140.

Nieder, J., and Zander, C. D. (1994). Nocturnal activity of a blenny, *Lipophrys trigloides* (Pisces, Blenniidae) at the Spanish Mediterranean coast. *Misc. Zool.* **17,** 189–197.

Nursall, J. R. (1981). Behavior and habitat affecting the distribution of five species of sympatric mudskippers in Queensland. *Bull. Mar. Sci.* **31,** 730–735.

Paine, R. T., and Palmer, A. R. (1978). *Sicyases sanguineus:* A unique trophic generalist from the Chilean intertidal zone. *Copeia* **1978,** 75–81.

Patzner, R. A. (1984). Die Blenniiden von Ibiza (Balearen) und ihre Verbreitung im West-Mittelmeer (Pisces: Teleostei: Blennioidea). *Senckenberg. Biol.* **65,** 179–203.

Patzner, R. A. (1985). The blennies (Pisces, Blennioidea) at the marine biological station of Aurisina (Gulf of Triest, Italy). *Nova Thalassia* **7,** 107–119.

Patzner, R. A., Santos, R. S., and Nash, R. D. M. (1992). Littoral fishes of the Azores: An annotated checklist of fishes observed during the "Expedition Azores 1989." *Arquipélago. Life Earth Sci.* **10,** 101–111.

Penrith, M. L. (1970). The distribution of the fishes of the family Clinidae in Southern Africa. *Ann. S. Afr. Mus.* 55, 135–150.

Penrith, M. J., and Penrith, M.-L. (1972). The Blenniidae of western southern Africa. *Cimbebasia* **2,** 65–90.

Pfister, C. A. (1992). Sculpin diversity in tidepools. *Northwest Environ.* **8,** 156–157.

Prochazka, K., and Griffiths, C. L. (1992). The intertidal fish fauna of the west coast of South Africa—Species, community and biogeographic patterns. *S. Afr. J. Zool.* **27,** 115–120.

Raju, N. S. (1971). Breeding habits, development and life-history of *Blennius steindachneri* Day from Waltair coast. *Proc. Indian Acad. Sci. B* **74,** 37–45.

Ralston, S. L., and Horn, M. H. (1986). High tide movements of the temperate-zone herbivorous fish *Cebidichthys violaceus* (Girard) as determined by ultrasonic telemetry. *J. Exp. Mar. Biol. and Ecol.* **98,** 35–50.

Rao, H. S., and Hora, S. L. (1938). On the ecology, bionomics and systematics of the Blenniid fishes of the genus *Andamia* Blyth. *Rec. Indian Mus.* **40:** 377–401.

Ruiz, G. M., Hines, A. H., and Posey, M. H. (1993). Shallow water as a refuge habitat for fish and crustaceans in non-vegetated estuaries: An example from Chesapeake Bay. *Mar. Ecol. Prog. Ser.* **99,** 1–16.

Russel, G. (1991). Vertical distribution. *In* "Intertidal and Littoral Ecosystems" (A. C. Mathieson and P. H. Nienhuis, Eds.), pp. 43–65. Elsevier, Amsterdam.

Santos, R. S., Nash, R. D. M., and Hawkins, S. J. (1994). Fish assemblages on intertidal shores of the island of Faial Azores. *Arquipélago. Life Mar. Sci.* **12A,** 87–100.

Sanusi, S. S. (1980). "A Study on Grazing as a Factor Influencing the Distribution of Benthic Littoral Algae." University of Ghana, Ghana.

Sasaki, T., and Hattori, J. (1969). Comparative ecology of two closely related sympatric gobiid fishes living in tide pools. *Japan. J. Ichthyol.* **15,** 143–155.

Sawyer, P. J. (1967). Intertidal life history of the rock gunnel, *Pholis gunnellus,* in the western Atlantic. *Copeia* **1967,** 55–61.

Setran, A. C., and Behrens, D. W. (1993). Transitional ecological requirements for early juveniles of two sympatric fishes, *Cebidichthys violaceus* and *Xiphister mucosus. Environ. Biol. Fish.* **37,** 381–395.

Shulman, M. J. (1985). Coral reef fish assemblages: Intraspecific and interspecific competition for shelter sites. *Environ. Biol. Fish.* **13,** 81–92.

Skead, D. M. (1966). Birds frequenting the intertidal zone of the Cape peninsula. *Ostrich* **37,** 10–16.

Smith, J. L. B. (1950). "The Sea Fishes of Southern Africa." Central News Agency, South Africa.

Soljan, T. (1932). *Blennius galerita* L., poisson amphibien des zones supralittorale et littorale exposees de l'Adriatique. *Acta Adriatica* **2,** 1–14.

Stebbins, R. C., and Kalk, M. (1961). Observations on the natural history of the mud-skipper, *Periophthalmus sobrinus. Copeia* **1961,** 18–27.

Stepien, C. A. (1986). Regulation of color morphic patterns in the giant kelpfish, *Heterostichus rostratus* Girard: Genetic versus environmental factors. *J. Exp. Mar. Biol. Ecol.* **100,** 181–208.

Stepien, C. A. (1987). Color pattern and habitat differences between male, female, and juvenile giant kelpfish. *Bull. Mar. Sci.* **41,** 45–58.

Stepien, C. A. (1990). Population structure, diets and biogeographic relationships of a rocky intertidal fish assemblage in central Chile: High levels of herbivory in a temperate system. *Bull. Mar. Sci.* **47,** 598–612.

Stepien, C. A., Phillips, H., Adler, J. A., and Mangold, P. J. (1991). Biogeographic relationships of a rocky intertidal fish assemblage in an area of cold water upwelling off Baja California, Mexico. *Pacific Sci.* **45,** 63–71.

Stepien, C. A., and Rosenblatt, R. H. (1991). Patterns of gene flow and genetic divergence in the Northeastern Pacific Clinidae (Teleostei: Blennioidei), based on allozyme and morphological data. *Copeia* **1991,** 873–896.

Thomson, D. A., and Lehner, C. E. (1976). Resilience of a rocky intertidal fish community in a physically unstable environment. *J. Exp. Mar. Biol. Ecol.* **22,** 1–29.

Todd, E. S. (1976). Terrestrial grazing by the eastern tropical Pacific goby *Gobionellus sagittula. Copeia* **1976,** 374–377.

Varas, E., and Ojeda, F. P. (1990). Intertidal fish assemblages of the central Chilean coast: Diversity, abundance and trophic patterns. *Rev. Biol. Mar., Valparaiso* **25,** 59–70.

Vivien, M. (1973). Ecology of the fishes of the inner coral reef flat in Tulear (Madagascar). *J. Mar. Biol. Assoc. India* **15,** 20–45.

Wagner, H. J., Menezes, N. A., and Ali, M. A. (1976). Retinal adaptations in some Brazilian tide pool fishes (Teleostei). *Zoomorphologie* **83,** 209–226.

Williams, G. C. (1954). Differential vertical distribution of the sexes in *Gibbonsia elegans* with remarks on two nominal subspecies of this fish. *Copeia* **1954,** 267–273.

Wirtz, P. (1980). A revision of the Eastern-Atlantic Tripterygiidae (Pisces, Blennioidei) and notes on some West African blennioid fish. *Cybium* **11,** 83–101.

Wright, W. G., and Raymond, J. A. (1978). Air-breathing in a California sculpin. *J. Exp. Zool.* **203,** 171–176.

Wyttenbach, A., and Senn, D. G. (1993). Intertidal habitat: Does the shore level affect the nutritional condition of the shanny (*Lipophrys pholis,* Teleostei, Blenniidae)? *Experientia* **49,** 725–728.

Yang, H. C., and Chung, C. H. (1978). Studies on the intertidal fishes and their geographical distribution in Liuchiu Island. *Annu. Rep. Sci., Taiwan Mus.* **21,** 197–225.

Yoshiyama, R. M. (1981). Distribution and abundance patterns of rocky intertidal fishes in central California. *Environ. Biol. Fish.* **6,** 315–332.

Yoshiyama, R. M., Sassaman, C., and Lea, R. N. (1986). Rocky intertidal fish communities of California: Temporal and spatial variation. *Environ. Biol. Fish.* **17,** 23–40.

Yoshiyama, R. M., Valey, C. J., Schalk, L. L., Oswald, N. M., Vaness, K. K., Lauritzen, D., and Limm, M. (1995). Differential propensities for aerial emergence in intertidal sculpins (Teleostei; Cottidae). *J. Exp. Mar. Biol. Ecol.* **191,** 195–207.

Zander, C. D. (1967). Beiträge zur Ökologie und Biologie litoralbewohnender Salariidae und Gobiidae (Pisces) aus dem Roten Meer. *Meteor Forsch. Ergebn. Reihe* **D2,** 69–84.

Zander, C. D. (1972a). Beziehungen zwischen Körperbau und Lebensweise bei Blenniidae (Pisces) aus dem Roten Meer. I. Äußere Morphologie. *Mar. Biol.* **13,** 238–246.

Zander, C. D. (1972b). Beziehungen zwischen Körperbau und Lebensweise bei Blenniidae (Pisces) aus dem Roten Meer. II. Bau der Flossen und ihrer Muskulatur. *Z. Morph. Tiere* **71,** 299–327.

Zander, C. D. (1972c). Beiträge zur Ökologie und Biologie von Blenniidae (Pisces) des Mittelmeeres. *Helgoländer Meeresunters.* **23,** 193–231.

Zander, C. D. (1973). Zur Morphologie der Flossen von Blenniidae (Pisces) des Mittelmeeres. *J. Ichthyol., Rome, C.I.E.S.M.* **1970,** 93–96.

Zander, C. D. (1974). Beziehungen zwischen Körperbau und Lebensweise bei Blenniidae (Pisces) aus dem Roten Meer. III. Morphologie des Auges. *Mar. Biol.* **28,** 61–71.

Zander, C. D. (1979). Morphologische und oekologisache Untersuchungen der Schleimfische *Parablennius sanguinolentus* (Pallas, 1811) und *P. parvicornis* (Valenciennes, 1836) (Perciformes, Blenniidae). *Mitt. Hamburg. Zool. Mus. Inst.* **76,** 469–474.

Zander, C. D. (1980). Morphological and ecological investigations on sympatric *Lipophrys* species (Blenniidae, Pisces). *Helgoländer Meeresunters.* **34,** 91–110.

Zander, C. D. (1983). Terrestrial sojourns of two mediterranean blennioid fish. *Senckenberg. Marit.* **15,** 19–26.

Zander, C. D. (1986). Blenniidae. *In* "Fishes of the North-eastern Atlantic and the Mediterranean" (P. J. P. Whitehead, M.-L. Bauchot, J.-C. Hureau, J. Nielsen, and E. Tortonese, eds.), pp. 1096–1112. Unesco, Paris.

Zander, C. D., and Bartsch, I. (1972). In situ Beziehungen zwischen Nahrungsangebot und aufgenommener Nahrung bei 5 *Blennius*-Arten (Pisces) des Mittelmeeres. *Mar. Biol.* **17,** 77–81.

# 4

# Respiration in Water and Air

Karen L. M. Martin

*Department of Biology, Pepperdine University, Malibu, California*

Christopher R. Bridges

*Institut für Zoophysiologie–Stollwechselphysiologie, Heinrich Heine Universität, Düsseldorf, Germany*

## I. Introduction

The regular and predictable departure of water from the rocky intertidal zone during ebb tides results in air exposure for sessile epifauna and epiflora. The ability of mussels, sea anemones, barnacles, seaweeds, and various gastropods to tolerate emergence from water for hours is well documented. However, more active animals such as crabs and fishes appear to be able to exert more behavioral choice in their habitat and refugia, and thus one could predict that these animals would be able to avoid accidental exposure or stranding in air. The emergence of many species of rocky intertidal fishes into air during low tide, including adults guarding nests of eggs, must be considered an appropriate and natural behavior rather than an error or an occasional atypical excursion. Many intertidal fishes of temperate coasts can breathe air (Bridges, 1993a; Martin, 1993; Yoshiyama and Cech, 1994), although in the past far more attention has been paid to the most terrestrially active amphibious marine fishes, the mudskippers and rockskippers.

Intertidal fishes that breathe air on occasion share certain morphological and behavioral characteristics with the more actively amphibious mudskippers and rockskippers. Remarkably, marine air-breathing fishes generally have no air-breathing organs (Graham, 1976), unlike their freshwater counterparts. Therefore respiratory gas exchange must take place across the same surfaces in air as it does in water: the gills, the skin, and perhaps the linings of the opercular and buccal cavities (Zander, 1972). Since they typically have no swim bladder, they are negatively buoyant and demersal, perching upright on the substrate when not swimming. This pose is also used during terrestrial emergence. Activity may occur out of water, though the behavior of these intertidal fishes in general, both aquatically and terrestrially, is calm and subdued (Horn and Gibson, 1988). Tidepool fishes are usually small, but some stichaeids can grow to more than half a meter in length. Many marine


*Intertidal Fishes: Life in Two Worlds*

intertidal fishes breathe air well, consuming oxygen at the same rate in air as in water (Bridges, 1988) and releasing carbon dioxide to air at the same rate as it is formed by metabolism (Martin, 1993). These fish fully emerge from the water when breathing air, although occasionally only the head and operculae appear as a fish rests on a sloped substrate before emerging completely.

This chapter examines the features of the habitat that may have led to the evolution of air-breathing among amphibious intertidal fishes, and then describes the behavior and physiology of the transition from water to air in air-breathing intertidal fishes.

## II. The Respiratory Environment of the Intertidal Zone

Ocean waters generally are normoxic and may even be hyperoxic in regions of high surf or active photosynthesis by algae (Graham *et al.,* 1978). Oxygen solubility decreases as salinity increases, and seawater therefore holds less oxygen per volume than freshwater (Dejours, 1994), but the partial pressure of oxygen is maintained in equilibrium with the atmosphere and this partial pressure is the driving force for diffusion of respiratory gases into the blood. Because hypoxia in freshwater systems appears to act as a selection pressure for air-breathing ability in fishes, in the past the oceans have not been considered likely areas for the evolution of air-breathing ability (Packard, 1974; Graham *et al.,* 1978). The areas of the ocean that are exceptions to this high-oxygen scenario are the marine habitats of the intertidal zone. For the purposes of this chapter, we concentrate on the tidepools of the rocky intertidal zone; however, some of the hypoxic characteristics of this habitat also occur in salt marshes and mud flats.

At high tide, the marine intertidal zone is inundated with oceanic water and has the physicochemical characteristics of near-shore water, typically including a seasonably stable temperature, stable salinity, oxygen saturation or even supersaturation, a well-buffered pH of 7.2 to 8.4, and relatively low dissolved carbon dioxide (Bridges, 1993a). Wave action on exposed shores causes mixing and rapid recovery from weather or biologically induced changes in the surrounding water.

By contrast, low-tide conditions in the water may vary considerably within a short period of time (Bridges, 1993a). As the tide recedes, water levels drop until the substrate itself is exposed. Crevices and depressions in a rocky substrate may hold pools of water that are aquatic refuges during the low tide. However, these pools are much smaller than the open ocean and much more vulnerable to physicochemical changes due to weather and biological activity. During a low tide, a tide pool may show a drastic change in temperature or salinity, as a result of insolation, freshwater runoff, or rainfall (Bridges, 1993a; Martin *et al.,* 1996). In addition, biological activity, particularly oxygen consumption and carbon dioxide release, can dramatically alter the amounts of these gases dissolved in the water. Tidepools are generally shallow and small, no more than a few meters in any dimension. During low tides they are isolated from wave action by being above the water line, so the water within is still, not mixed or refreshed by incoming ocean water until the tide returns.

Oxygen solubility is lower in seawater than in freshwater, and considerably lower per unit volume in water than in air (Dejours, 1994). The solubility declines with increasing temperature, exacerbating hypoxia during daytime low tides in shallow pools without plants. Oxygen consumption by tidepool residents can result in aquatic hypoxia in tidepools, in the absence of plants by day, or at night in pools with both animals and plants

(Truchot and Duhamel-Jouve, 1980; Congleton, 1980; Zander, 1983; Louisy, 1987). The tidepool habitat that already contains sessile organisms and some fishes may also become a refuge during low tides for small obligate aquatic organisms of the intertidal zone, such as shrimp and some echinoderms. It is likely that tidepools concentrate these organisms during low tides, adding their biological burden to that of the resident organisms already present. The respiration of all of these animals and plants increases the partial pressure of carbon dioxide in this habitat as well. In the open ocean this carbon dioxide increase is buffered by bicarbonates and recycled by plants carrying out photosynthesis. But in the small volume of a tidepool, and particularly at night when photosynthesis does not occur, bicarbonate buffering may be overwhelmed and the pH may decline by several pH units over the course of a few hours (Bridges, 1993a,b).

Within a tidepool, conditions are still aquatic. Even more drastic physical changes occur when a fish emerges into air. Air contains a high percentage of oxygen relative to volume (Dejours, 1994), and diffusion occurs much more quickly in air, so that boundary layers are much less of a problem for respiratory surfaces than in water (Feder and Burggren, 1985). But the open air is a desiccating environment, with wind and insolation effects, and it shows much more profound and rapid temperature variability than water (Bridges, 1993a). The effect of gravity is much more pronounced on land than in water. The loss of buoyancy and support for the body may cause the collapse of the gills, reduced respiratory surface area, and changes in internal fluid osmolarities. Thus emergence from hypoxic water into air may provide quick access to increased levels of oxygen, but there is a trade-off between this advantage and the disadvantages of desiccation and collapse of respiratory surfaces (Randall *et al.,* 1981). Unlike freshwater air-breathing fishes, most marine intertidal fishes have no specialized or enclosed air-breathing organs (Graham, 1997). Dependent upon gills and cutaneous surfaces for respiration while fully emerged into air, they must remain close to water in moist, sheltered habitats or return frequently to pools to maintain the integrity of respiratory surfaces.

## III. Emergence Behaviors

As the tide ebbs and the water line moves lower in the intertidal zone, sessile organisms have no choice but to remain in place. Many marine intertidal plants and invertebrates spend the time of a low tide completely emerged out of water. Some examples are chitons, seaweeds, sea anemones, barnacles, and mussels. During low tides, exposure to air may be inevitable because of the location of these animals and plants on the rocky substrate. Presumably their survival and adult distribution is related both to the settling position of larvae and to adult survival during the unavoidable time out of water in low tides. Animals and plants of these same species are frequently also found in tidepools, so air emergence appears to be tolerated, but does not seem to be required for survival. Locating high in the intertidal zone may provide advantages of reducing predation pressure from fully aquatic predators (Paine, 1974).

In addition to sessile plants and animals, crabs and other motile animals may be found out of water during low tides. Intertidal fishes that come out of water have been characterized by their emergence behaviors into three groups: the skippers, the tidepool emergers, and the remainers (Martin, 1995). All three types fully emerge from the water and breathe air; however, when in water, none of these types breathes air. Intertidal fishes that are skippers may actively emerge out of pools, shuttling back and forth between water and air (Gordon *et al.,* 1969; Zander, 1972; Graham, 1973; Graham *et al.,* 1985; Sayer and Davenport, 1991). Tidepool emerging fishes may leave the water if aquatic conditions become inhospitable (Wright and Raymond, 1978; Martin, 1991; Yoshiyama *et al.,* 1995). A third strategy exists; a remainer fish may become out of water passively, simply by remaining in a site that is usually covered with water, but during a low tide is exposed to air (Horn and Riegle, 1981; Martin, 1995). In this case the site may be protected from desiccation by cover, such as a boulder or seaweed, and there may be an extremely shallow layer of water on the substrate. Species of fishes found out of water under these conditions during low tides may also occur in tidepools at low tides, but they are much more numerous in the emerged habitat. Table 1 lists the species of intertidal fishes that have shown evidence of air-breathing ability, by behaviors and by respirometry.

Skippers, including groups called the mudskippers (Gobiidae) and the rockskippers (Blenniidae), are active emergers, seen out of water routinely and very active terrestrially. Members of this group engage in feeding, courtship, and dominance display behaviors out of water (Brillet, 1984). The blenny *Alticus saliens* performs its courtship rituals and nests above the water line in the Red Sea (Magnus, 1966). The dramatic and extrovert behaviors of skippers, and their obvious ease in navigating terrestrially (Zander, 1972; Graham *et al.,* 1985), have inspired numerous studies of air-breathing in these fishes [see Clayton (1993) and Graham (1997) for reviews]. Skippers show morphological and locomotor adaptations for survival and movement out of water and are found in the supralittoral zone on rocky shores (rockskippers, usually blennies) and in mud flats and estuaries (mudskippers, usually gobies). Closely related species lower in the intertidal zone may show some similar characteristics, but to a lesser degree; for example, *Boleophthalmus* emerges less frequently and is less active terrestrially than *Periophthalmus* (Tamura *et al.,* 1976). Although these fishes are obviously very active and visible out of water, they consistently return to pools or the ocean briefly every few minutes to prevent desiccation (Graham *et al.,* 1985; Brown *et al.,* 1992). Graham (1997) suggests that among members of the Oxudercine mudskipper subfamily, all 12 species of *Periophthalmus,* all 3 species of *Periophthalmodon,* all 5 species of *Boleophthalmus,* all 4 species of *Scartelaos, Pseudapocryptes lanceolatus,* and *Apocryptes bata* show amphibious behavior and are presumed to breathe air.

Remainers are present in numerous families of intertidal fishes, including the Stichaeidae (Daxboeck and Heming, 1982; Horn and Riegle, 1981; Riegle, 1976; Edwards and Cech, 1990), Blenniidae (Pelster *et al.,* 1988), Cottidae (Cross, 1981; Martin, 1996), Gobiesocidae (Ebeling *et al.,* 1970; Eger, 1971; Martin, 1993), Pholididae, and Batrachoididae (Martin, 1993). Typically these fishes can be found emerged under boulders or in crevices during low tides, occasionally in shallow pools in water that does not cover their bodies. The entire body of the fish is emerged, including the head and gills, for the duration of the low tide. Desiccation may occur (Horn and Riegle, 1981; Bridges, 1993a). These fishes are not active terrestrially on a routine basis, although they are able to move rapidly about for short periods of time to avoid predators (Horn and Gibson, 1988). Some intertidal fishes remain in the intertidal zone guarding nests of eggs that are also emerged during low tides, for example the stichaeid *Xiphister* (Coleman, 1992), some sculpins (Marliave and DeMartini, 1977), and the toadfish *Porichthys notatus* (Crane, 1981). For additional information on intertidal spawners, see DeMartini (Chapter 8, this volume).

Some of the fishes normally found in tidepools are able to emerge actively into air. Tidepool emergers are fish that are typically aquatic but may leap or crawl out of water when aquatic hypoxia occurs. Cottids provide the best studied examples of this type of emergence behavior (Wright and Raymond, 1978; Davenport and Woolmington, 1981;

**Table 1.** Intertidal Fishes from Studies of Air Emergence and/or Aerial Respirometry, Listed by Family and Species

| Family<br>*Species* | Classi-fica-tion | Respi-rometry in air | Emergence in wild and/or in lab | Source |
|---|---|---|---|---|
| Acanthoclinidae ($N = 1$) | | | | |
| *Acanthoclinus fuscus* | T | Y | W, LF | Mayr and Berger, 1992; Hill *et al.*, 1996 |
| Batrachoididae ($N = 1$) | | | | |
| *Porichthys notatus* | R | Y | W, LF | Crane, 1981; Martin, 1993 |
| Blenniidae ($N = 9$) | | | | |
| *Alticus kirki* | S | Y | W, LV | Magnus, 1966; Zander, 1972; Martin and Lighton, 1989; Brown *et al.*, 1992 |
| *A. saliens* | S | N | W | Magnus, 1966; Abel, 1973 |
| *Blennius pholis* (*Liphophrys*) | R | Y | W, LF | Laming *et al.*, 1982; Pelster *et al.*, 1988; Davenport and Woolmington, 1981 |
| *B. trigloides* | R | N | W | Zander, 1983; Louisy, 1987 |
| *Coryphoblennius galerita* | R | N | W | Zander, 1983; Louisy, 1987 |
| *Entomacrodus nigricans* | S | N | W | Graham *et al.*, 1985 |
| *E. vermiculatus* | S | N | W | Abel, 1973 |
| *Hypsoblennius gilberti* | Other | Y | LF | Luck and Martin, 1998 |
| *Istiblennius edentulus* | R | N | W | Zander, 1972 |
| Cottidae ($N = 11$) | | | | |
| *Artedius lateralis* | R | N | W | Lamb and Edgell, 1986 |
| *Ascelichthys rhodorus* | R | Y | W, LV, LF | Cross, 1981; Martin, 1996; Yoshiyama and Cech, 1994; Yoshiyama *et al.*, 1995 |
| *Clinocottus acuticeps* | T | N | W | Yoshiyama and Cech, 1994 |
| *C. analis* | T | Y | W, LV, LF | Martin, 1991 |
| *C. globiceps* | T | Y | W, LV, LF | Martin, 1993; Yoshiyama and Cech, 1994; Lamb and Edgell, 1986; Yoshiyama *et al.*, 1995 |
| *C. recalvus* | T | Y | W, LV, LF | Wright and Raymond, 1978; Martin, 1993 |
| *Leptocottus armatus* | T | Y | LF | Martin, 1993 |
| *Oligocottus maculosus* | T | Y | W, LV, LF | Martin, 1993; Yoshiyama and Cech, 1994; Yoshiyama *et al.*, 1995 |
| *O. rimensis* | T | N | W | Yoshiyama and Cech, 1994 |
| *O. snyderi* | T | Y | W, LV, LF | Martin, 1993; Yoshiyama and Cech, 1994; Yoshiyama *et al.*, 1995 |
| *Taurulus bubalis* (*Acanthocottus*) | T | Y | LF | Davenport and Woolmington, 1981 |
| Fundulidae (Cyprinodontidae) ($N = 1$) | | | | |
| *Fundulus heteroclitus* | T | Y | W, LF | Halpin and Martin, unpublished |
| Gobiesocidae ($N = 6$) | | | | |
| *Gobiesox maeandricus* | R | Y | W, LF | Cross, 1981; Martin, 1993 |
| *G. pinniger* | R | N | W, LF | Eger, 1971 |
| *Pherallodiscus funebris* | R | N | W, LF | Eger, 1971 |
| *Sicyases sanguineus* | R | Y | W, LF | Gordon *et al.*, 1970; Ebeling *et al.*, 1970 |
| *Tomicodon boelkei* | R | N | W, LF | Eger, 1971 |
| *T. humeralis* | R | N | W, LF | Eger, 1971 |
| Gobiidae ($N = 16$) | | | | |
| *Boleophthalmus pectinirostris* (formerly *chinensis*) | S | Y | W, LV, LF | Tamura *et al.*, 1976 |
| *B. boddarti* | S | Y | W, LV, LF | Teal and Carey, 1967 |

**Table 1.** *Continued*

| Family<br>*Species* | Classi-fica-tion | Respi-rometry in air | Emergence in wild and/or in lab | Source |
|---|---|---|---|---|
| *Gillichthys mirabilis* | Other | Y | Not emerging | Barlow, 1961; Todd and Ebeling, 1966 |
| *G. seta* | Other | Y | Not emerging | Barlow, 1961; Todd and Ebeling, 1966 |
| *Gobionellus sagitulla* | S | N | W | Todd, 1976 |
| *Kelloggella cardinalis* | S | N | W | Larson, 1983 |
| *Periophthalmodon freycineti* (*australis*) | S | N | W | Bandurski *et al.*, 1968; Milward, 1974 |
| *Pn. schlosseri* | S | N | W | Schöttle, 1931 |
| *Periophthalmus gracilis* | S | N | W | Nursall, 1981 |
| *P. minutus* | S | N | W | Gregory, 1977 |
| *P. modestus* (*cantonensis*) | S | Y | W, LV, LF | Stebbins and Kalk, 1961; Tamura *et al.*, 1976; Gordon *et al.*, 1978 |
| *P. argentilinatus* (*barbarus, dipus, kalolo, koelreuteri, sobrinus*, or *vulgaris*) | S | Y | W, LV, LF | Schöttle, 1931; Gordon *et al.*, 1969; Milward, 1974; Teal and Carey, 1967; Hillman and Withers, 1987 |
| *P. novaeguineaensis* | S | Y | W, LV, LF | Milward, 1974 |
| *P. novemradiatus* (*pearsei*) | S | N | W, LF | Singh and Datta Munshi, 1969 |
| *P. chrysospilos* | S | Y | W, LV, LF | Natarajan and Rajulu, 1983; Lee *et al.*, 1987 |
| *Scartelaos histophorus* (as *Boleophthalmus*) | S | Y | W, LV, LF | Tamura *et al.*, 1976; Milward, 1974; Gregory, 1977 |
| Kyphosidae (Girellidae) (*N* = 1) | | | | |
| *Girella nigricans* | T | Y | LF | Martin, 1993; R. Orton, pers. comm. |
| Labrisomidae (*N* = 2) | | | | |
| *Dialommus fuscus* | S | N | W | Graham, 1970 |
| *Mnierpes macrocephalus* | S | Y | W, LF | Graham, 1970, 1973 |
| Pholididae (*N* = 5) | | | | |
| *Apodichthys flavidus* | R | N | W | Cross, 1981; Lamb and Edgell, 1986 |
| *Pholis gunnellus* | R | Y | W, LF | Bridges, 1988; Laming, 1983 |
| *P. laeta* | R | N | W | Lamb and Edgell, 1986 |
| *Xererpes fucorum* | R | Y | W, LF | Horn and Riegle, 1981; Lamb and Edgell, 1986; Martin, 1993 |
| *Zoarces viviparous* | R | Y | W, LF | Hartvig and Weber, 1984 |
| Stichaeidae (*N* = 4) | | | | |
| *Anoplarchus purpurescens* | R | Y | W, LF | Cross, 1981; Horn and Riegle, 1981; Martin, 1996; Yoshiyama and Cech, 1994; Lamb and Edgell, 1986 |
| *Cebidichthys violaceus* | R | Y | W, LF | Reigle, 1976; Horn and Riegle, 1981; Edwards and Cech, 1990; Martin, 1993 |
| *Xiphister atropurpureus* | R | Y | W, LF | Daxboeck and Heming, 1982; Lamb and Edgell, 1986; Martin, 1993 |
| *X. mucosus* | R | Y | W, LF | Horn and Riegle, 1981; Lamb and Edgell, 1986; Martin, 1993 |
| Tripterygiidae (*N* = 2) | | | | |
| *Forsterygion* sp. | T | Y | W, LF | Berger and Mayr, 1992; Hill *et al.*, 1996 |
| *Helcogramma medium* | R | Y | W, LF | Innes and Wells, 1985; Hill *et al.*, 1996 |

*Note.* Fish are classified as remainers (R, *N* = 24), skippers (S, *N* = 19), or tidepool emergers (T, *N* = 12). "Other" (*N* = 4) in this column indicates nonemerging air-breathers (see Section IV). Respirometry in air is shown as Y (yes) or N (not studied) and indicates measurement of oxygen consumption by fishes out of water. Emergence behavior is indicated by W (seen in fishes in the wild), LF (survives forced emergence in the laboratory), or LV (voluntarily emerges from water in the laboratory). A total of 60 species in 12 families are listed.

Martin, 1991, 1996; Yoshiyama *et al.,* 1995), but other fishes also do this, including the Acanthoclinid *Acanthoclinus fuscus* and the Tripterygiid *Forsterygion* sp. (Hill *et al.,* 1996). Tidepool emergers generally are relatively inactive when out of water, but can locomote on land to some extent, perhaps to escape predators or to move between tidepools. They may emerge by jumping out with a tail flip, or by crawling out balanced on the pectoral fins and the posterior region of the body. Tidepool emergers have been observed out of water during low tides in the field, in caves (Wright and Raymond, 1978), and nocturnally (Martin, 1993). Other physical factors that may trigger emergence could be temperature, salinity, pH, and increased carbon dioxide levels, but these have so far not been shown to influence emergence (Davenport and Woolmington, 1981).

Active emergence from water is a behavior shown by tidepool emergers but emergence behavior is not shown by some other fishes that occur in tidepools (Davenport and Woolmington, 1981), nor by closely related fishes that occur lower in the subtidal zone (Congleton, 1980; Martin, 1996). The fish species that do emerge from pools can be routinely observed in tide pools, but are not regularly emerged under boulders during low tides, as are the remainers.

A fourth category of air-breathing intertidal fish may exist. Some tidepool fishes can breathe air if removed from water and can tolerate well a prolonged emergence of several hours. However, these fishes are not observed emerged in nature and do not emerge from hypoxic water when tested in the laboratory. The facultative air-breathing ability they possess may be useful under circumstances of truly accidental stranding, perhaps if a fish is excluded from a tidepool as the tide ebbs, or if a crevice chosen as a refuge is above the water line rather than below it when the tide is out. Examples of fish of this type include juvenile *Girella nigricans* (Martin, 1993) and adult *Hypsoblennius gilberti* (Luck and Martin, 1998). Intertidal fishes of this type can survive prolonged emergence during a low tide and have the facultative ability to breathe air, but they do not appear to exercise any behavioral component or "choice" in the matter.

By contrast, there are other species of fishes found in tidepools and subtidally that do not emerge by themselves and cannot survive a forced emergence of the duration of a low tide. These fishes are unable to meet their metabolic needs through air-breathing. Tidepool fishes that lack the ability to emerge or breathe air include clinids (Congleton, 1980), the cottid *Ciliata mustela* (Davenport and Woolmington, 1981), and possibly the juveniles of open water fishes. Typically these fishes are found in tidepools diurnally, but not during nocturnal low tides, when hypoxia is most likely to occur (Congleton, 1980). Marine cottid fishes with a lower vertical distribution, in the lowest intertidal or subtidally and below, do not breathe air and do not emerge from hypoxic water (Martin, 1996).

## IV. Respiratory Organs of Intertidal Fishes

Gills are the primary respiratory organs of fishes. They use a flowthrough pattern for the respiratory medium (Piiper and Scheid, 1975) and countercurrent blood flow for efficient extraction of oxygen from water. Gills are thin and filamentous, ensuring the high total surface area and short diffusion distances needed to promote exchange of respiratory gases between the water and the blood. When fish are out of water, gills tend to collapse and this causes a reduction in surface area available for respiration. Air-breathing marine fishes cope with this effect of air emergence in two ways. One way is to strengthen and support the gills through adaptation for use in air. The other is to employ alternative respiratory surfaces such as the skin, which are not subject to collapse.

Modifications of the gills are seen in some skipper-type marine air-breathing fishes, including a thickening of the gill epithelium, a reduction in surface area of the secondary lamellae, and the presence of cartilagenous rods to stiffen and support the primary lamellae (Low *et al.,* 1988, 1990; Brown *et al.,* 1992). These morphological adaptations minimize the effects of gill collapse upon emergence of the fish into air. On the other hand a reduction in number of secondary lamellae along with an increase in epithelial thickness would tend to have a negative effect on aquatic respiratory ability. Thus an increase in air-breathing ability may force a trade-off in the form of a reduction to some extent of aquatic-breathing ability. By improving its gills for air-breathing, the fish worsens its ability to remain in hypoxic water, thus increasing the likelihood that it will be obliged to emerge during a low tide.

Modification and reduction of the gills are seen in some freshwater air-breathing fishes. Frequently these modifications reflect a reduction in the use of the gills even while the fish remains aquatic (Graham, 1997), because of the reliance on a separate air-breathing organ for oxygen uptake. Loss of gill structure to the point of loss of function in water is a strategy that amphibious marine fishes cannot afford, since they must rely on the gills as their primary respiratory surfaces in both water and air.

Almost no work has been done to describe the structure of gills and other respiratory structures in remainer and tidepool emerger types of marine air-breathing fishes. Since their physiology and behavior differ from nonemerging fishes (Martin, 1996), it is likely that their morphology differs to some extent as well, and this could be a fruitful area for further research. The remainer *Xiphister atropurpureus,* a stichaeid of the Pacific coast of North America, has cartilaginous rods within its primary gill lamellae (Sanders and Martin, unpublished).

When gills are insufficient, the skin can be called into service as a respiratory organ for some fishes, even in water (Nonnotte and Kirsch, 1978). Cutaneous surfaces are ectodermal in origin and juxtaposed against the respiratory medium, creating the potential for an "infinite pool" gradient for respiratory diffusion (Piiper and Scheid, 1975). In air, cutaneous surfaces are in direct contact with the respiratory medium and need only be kept moist and well vascularized for use as a respiratory organ (Feder and Burggren, 1985). In addition, the skin itself must be well vascularized and relatively free of scales or other obstructions, at least in the areas that are used for gas exchange. Also, to promote diffusion of oxygen into the blood, it must be kept moist. Intertidal fishes roll over onto the lateral surfaces and return to pools frequently, apparently to maintain moisture on the skin surfaces (Graham, 1973; Graham *et al.,* 1985; Brown *et al.,* 1992).

Other organs may be used for respiratory gas exchange by intertidal air-breathing fishes, but this is speculative at this time. One potential site is the vascularized linings of the opercular and buccal cavities, and/or the esophagus (see Bridges, 1993b). Laming *et al.* (1982) suggested that *Blennius pholis* may use the esophagus as a site for gas exchange, on the basis of evidence from the presence of gas bubbles that had entered the esophagus by swallowing. Pelster *et al.* (1988) found no evidence for increased cellular carbonic anhydrase activity in the region of the esophagus, but Jermann (1987) conclusively showed that the mucosal area of the esophagus was increased by folding in the caudal part. Local circulation is enhanced by the presence of many capillaries. A sphincter allows the esophagus to be closed off from the rest of the digestive tract. Many freshwater fish species have accessory air-breathing organs for air ventilation, but structures dedicated to air-breathing are most unusual in marine air-breathing fishes. In practice, it is difficult to show whether the mouth area is actually used for gas exchange in these small fishes. Air-breathing organs exist in a wide variety among freshwater air-breathing fishes (Graham, 1997), derived from modifications of the skin, opercular cavity, intestines, swim bladder, and stomach. Such specializations have rarely been described in amphibious marine fishes and are more typically adaptations for air-breathing in fishes that remain in water rather than emerging (Graham, 1976). Very few marine fishes remain aquatic while air-breathing. The few that do are estuarine or brackish water residents and they use air-breathing organs modified for the task. The tarpon *Megalops atlanticus* uses its swim bladder (Wade, 1962) and the longjaw mudsucker *Gillichthys mirabilis* uses its highly vascularized buccopharyngeal epithelium (Todd and Ebeling, 1966) as accessory air-breathing organs while swimming in water. *Gillichthys seta* and another Gulf of California gobiid, *Quietula saggitula,* also appear to gulp air at the surface of hypoxic water (Todd and Ebeling, 1966).

The partitioning of gas exchange between the gills and the skin has been examined for several species of intertidal fishes in air (Teal and Carey, 1967; Tamura *et al.,* 1976; Natarajan and Rajulu, 1983; Martin, 1991) and for several species of aquatic fishes in water (Nonnotte and Kirsch, 1978). However, only Tamura *et al.* (1976) have compared this division for any one species when respiring in water to the division that occurs when this same species respires in air. The skin is an important respiratory organ in air for amphibious fishes (Schottle, 1931; Zander, 1972), but the proportion of gas exchange by the skin varies with taxa. Some intertidal fishes show increased reddening of certain cutaneous surfaces on emergence (Zander, 1972; Freitag, 1995). Long, eel-like fishes may tend to rely more heavily on cutaneous respiration than shorter fishes in water, but there are insufficient data to test this relationship among air-breathing intertidal fish (Figure 1). In addition, the am-

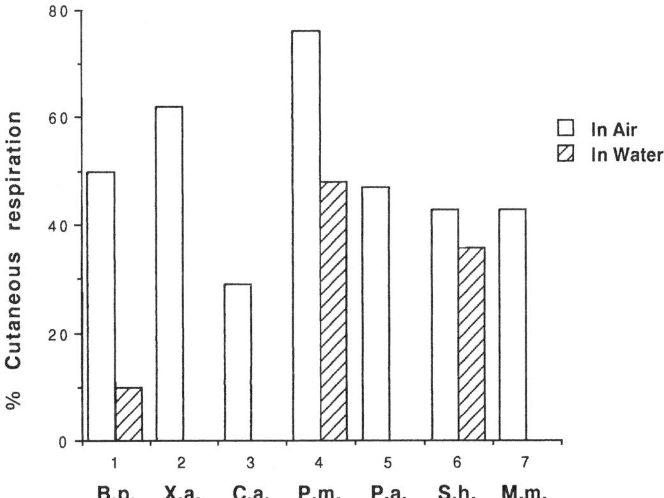


**Figure 1.** Partitioning of cutaneous and gill respiration in fishes in and out of water. Data from the following species are shown: B.p., *Blennius pholis* [Pelster *et al.* (1988) in air; Nonnotte and Kirsch (1978) in water]; X.a., *Xiphister atropurpureus* (Sanders and Martin, unpublished); C.a., *Clinocottus analis* (Martin, 1991); P.m., *Periophthalmus modestus* (as *P. cantonensis*) (Tamura *et al.*, 1976); P.a., *Periophthalmus argentilineatus* (as *P. sobrinus*) (Teal and Carey, 1967); S.h., *Scartelaos histophorus* (as *Boleophthalmus chinensis*) (Tamura *et al.*, 1976); and M.m., *Mnierpes macrocephalus* (Graham, 1973).

phibious fishes *Periophthalmus modestus* and *Scartelaos histophorus* decrease total oxygen consumption in air, but continue to obtain approximately the same amount of oxygen via the skin, thereby increasing the percentage component of cutaneous respiration in air (Tamura *et al.,* 1976). Thus the skin may be more important as a respiratory organ during air-breathing than it is under aquatic conditions (Bridges, 1993b).

## V. Respiratory Gas Exchange in Water and Air

For many facultatively air-breathing marine fish, oxygen consumption rates are similar in air and in water (Graham, 1976; Bridges, 1988). In many cases the ratio of aerial to aquatic rates is near unity. This is in contrast to the case with freshwater air-breathing fishes, which often drastically reduce oxygen consumption when out of water (Johansen, 1970; Graham, 1997). The same organs, gills and skin, are used in both respiratory media and as long as these surfaces are moist and not reduced in area by collapse (see above), it is reasonable that they would function well in both media. In general, the ratio of aerial to aquatic metabolic rates is between 0.6 and 1.5 (Bridges, 1988).

There are some blennies that have a decreased metabolic rate in air, including the very amphibious *Alticus kirki* (Brown *et al.,* 1992) and the tidepool fish *Hypsoblennius gilberti* (Luck and Martin, 1998). It is not clear how this decrease is accomplished. Evidence for a prolonged "diving response" in reverse for most intertidal fish emerging into air does not exist, although a transient drop in heart rate may be observed in *Blennius* (= *Lipophrys*) *pholis* on emergence (Pelster *et al.,* 1988) and *Periophthalmodon freycineti* (as *Periophthamus australis*) when placed into water from air (Garey, 1962). The clingfish *Sicyases sanguineus* reduces the heart rate by one-half when out of water and is extremely inactive terrestrially (Gordon *et al.,* 1970), although metabolic rates of smaller fish are higher out of water. The respiratory pump in water may take 10 to 22% of a resting fish's metabolism (Shelton, 1970); this demand is likely to be much reduced when breathing air (Hill *et al.,* 1996).

Anaerobiosis would be one mechanism for fish to meet metabolic requirements if oxygen consumption in air were insufficient. Of the intertidal and amphibious marine fishes that have been tested, however, none undergoes anaerobiosis while in air under resting or routine circumstances (Graham, 1976; Martin, 1991, 1995). Aquatic oxygen consumption rates are equal before and after long periods of emergence (Martin, 1991; Steeger and Bridges, 1995), indicating that no oxygen debt has been incurred. On the other hand, subtidal sculpins lacking amphibious capabilities do increase production of lactate when artificially emerged in air (Martin, 1996). Anaerobiosis is apparently not necessary for the survival of inactive marine amphibious fishes in air, although some accumulation of lactate does occur over time of emergence for the clingfish *Sicyases sanguineus* (Gordon *et al.,* 1970) and *Blennius pholis* (Pelster *et al.,* 1988). It is clear that anaerobiosis is employed when mudskippers undergo hypoxia, either experimentally or perhaps in a burrow (Garey, 1962; Ip and Low, 1990; Ip *et al.,* 1990; Low *et al.,* 1993).

Very few studies have been done on marine fishes during activity out of water. This is in part because many of these fishes, particularly the remainers, are not very active when emerged (Horn and Gibson, 1988; Martin, 1993). Amphibious skipper fishes are highly active terrestrially and can increase oxygen consumption during activity to compensate for this increased metabolism (Hillman and Withers, 1987; Martin and Lighton, 1989). However, some anaerobiosis may be involved during some kinds of activity in air (Martin and

Lighton, 1989). The mudskippers *Periophthalmus argentilineatus* (as *P. sobrinus*), *P. chrysospilos, Periophthalmodon schlosseri,* and *Boleophthamlus boddarti* show increased lactate concentrations under asphyxic conditions (Teal and Carey, 1967; Ip and Low, 1990; Ip *et al.,* 1990). Remainers can move rapidly to escape predators and may incur an oxygen debt during this time. The remainer *Xiphister atropurpureus* can consume oxygen in air at double its resting aquatic rate under confinement (Daxboeck and Heming, 1982; Martin, 1993), indicating that resting aerial gas exchange is not maximal.

Release of carbon dioxide into air is straightforward for marine air-breathing fishes and does not diminish from aquatic levels (Graham, 1976; Bridges, 1988, 1993a; Martin, 1993; Steeger and Bridges, 1995). One can calculate the respiratory exchange ratio by dividing the rate of carbon dioxide released by the rate of oxygen consumed. Under normal conditions the ratio at any aerobically respiring cell is approximately 0.7 to 1.0 (Withers, 1992). The whole-animal ratio for air-breathing intertidal fishes is also within this range (Martin, 1993; Bridges, 1993a; Steeger and Bridges, 1995), indicating that carbon dioxide is released as it is formed and does not accumulate in the body, in contrast to the condition that occurs for emerged freshwater air-breathing fishes (Graham, 1976, 1997). It has been suggested that it is more difficult to release carbon dioxide into air than into water (Randall *et al.,* 1981), but this does not seem to be the case for marine air-breathing fishes. Instead, the structure of the respiratory surfaces is an advantage in this regard. Many freshwater air-breathing fishes have enclosed air-breathing organs, such as lungs, that are ventilated tidally and infrequently (Randall *et al.,* 1981; Graham, 1997). Thus between breaths carbon dioxide can accumulate at the respiratory epithelium, with accompanying diffusion gradients and effects on acid–base equilibrium (Heisler, 1982). In contrast, for marine amphibious fishes, the cutaneous and gill surfaces are in direct contact with the outside respiratory medium, so carbon dioxide can be released continuously by diffusion across these surfaces and no accumulation need occur. Even in fishes that may seal the operculae and hold their "breath" while emerged into air, cutaneous surfaces would still be available for continuous release of carbon dioxide.

## VI. Ventilation

Fluctuations of temperature, oxygen, and carbon dioxide tensions in the water of intertidal rock pools, brought about by both tidal and diurnal influences, make a number of demands on the gas exchange system of intertidal fishes. Large changes in aquatic ventilation may occur as a response to temperature and water oxygenation state, and even more extreme transformations may occur during air-breathing or emersion. Many of the aquatic ventilatory adjustments made by intertidal fishes are common to other fish species that have been studied in more detail (for review, see Randall and Daxboeck, 1984). One of the major problems in making physiological measurements on intertidal fish is their small size (Bridges, 1988). In general, therefore, ventilatory rate has been measured in the laboratory (Bridges, 1988) and the field (Harder, 1995), although ventilatory flow would give more detail.

In normoxic water, most intertidal fishes show continuous ventilation. Rates of 63.3 ± 6.4 beats min$^{-1}$ in *Blennius pholis* at 12.5°C (Pelster, 1985), 17.3 ± 15.1 at 12.5°C in *Gobius cobitis,* and 80–82 beats min$^{-1}$ in *Pholis gunnellus* at 13°C (Laming, 1983) have been measured in these rocky shore intertidal fishes. In the mudskipper *Periophthalmus argentilineatus* (as *P. barbarus*) some differences are seen (Freitag, 1995). Figure 2 illus-

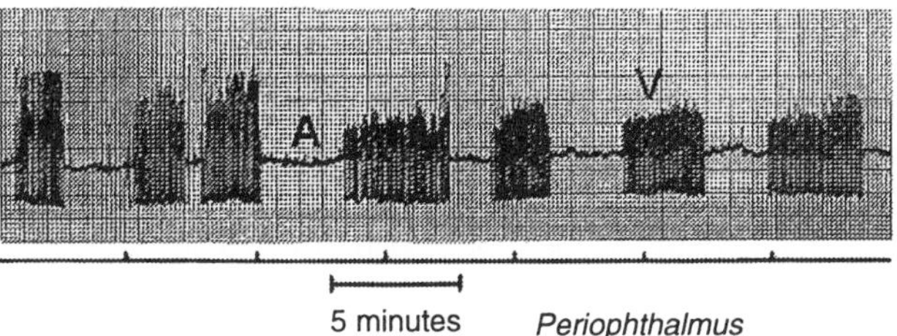


**Figure 2.** A typical impedance recording of ventilation in *Periophthalmus argentilineatus* in normoxic water at 25°C. Note that there are ventilatory periods (V) and apnoeic periods (A).

trates its ventilatory pattern in normoxic water, however normally these fish emerge into air. When kept under water at 25°C, only 48.6 ± 14.1% of the time is spent ventilating (Freitag, 1995) with long apnoeic periods in between. Mudskippers carry out a large proportion of their gas exchange via the skin (Clayton, 1993), so continous gill ventilation is not necessary. During gill ventilatory bouts, ventilatory frequencies reach approximately 20.1 ± 5.5 beats $min^{-1}$ (Freitag, 1995).

Ventilatory rate increases with temperature, as does metabolic rate in fishes (Heath and Hughes, 1973). An increase in ventilation rate of 2.8 beats $°C^{-1}$ occurs with temperature in *Pholis gunnellus* (Laming, 1983). The *Q*10 value for ventilatory rate in the rainbow trout was 1.4 (Heath and Hughes, 1973) compared with 2.4 in the intertidal fish *Gobius cobitis* (Berschick *et al.,* 1987) as ventilatory frequency changed from 17 to 50 beats $min^{-1}$ with a temperature change of 12.5°C (Figure 3). This parallels changes in metabolism as the *Q*10

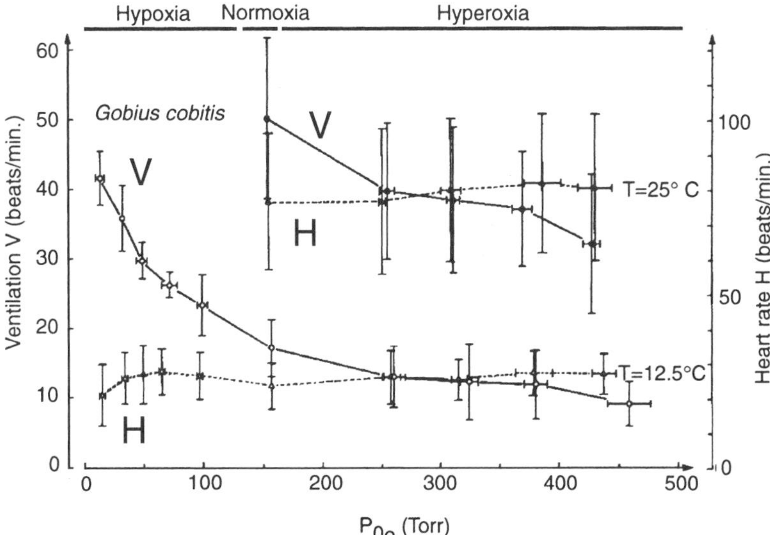


**Figure 3.** Effects of hyperoxia at 12.5 and 25°C and hypoxia at 12.5°C on the ventilatory frequency (V) and heart rate (H) of *Gobius cobitus*. Values are means ±1 SD. Modified with permission from Berschick *et al.* (1987). The influence of hyperoxia, hypoxia, and temperature on the respiratory physiology of the intertidal rockpool fish *Gobius cobitis* Pallas. *J. Exp. Biol.* **130,** 369–387, Company of Biologists Ltd.

value for oxygen consumption averages around 2.3 in intertidal fishes (Bridges, 1988, 1993a), with values between 1.7 and 3.6 (Bridges, 1997).

Hyperoxia may occur in rock pools in daylight from photosynthesis, with oxygen partial pressure ($PO_2$) above 400 Torr, and possibly increased temperature (Bridges, 1988). The return from hyperoxia to normoxia occurs rapidly when the incoming tide washes out the rock pools (Bridges, 1993a, 1997). Metabolic rate itself is not influenced by hyperoxia in the intertidal fishes that have been measured to date (Berschick *et al.,* 1987). Oxygen consumption in *Gobius cobitis* remained constant over a $PO_2$ range of 150 to 450 Torr at both 12.5 and 25°C. Similar results were seen in freshwater fish (Dejours, 1973), with no change in oxygen consumption when *Catostomus commersoni* was exposed to $PO_2$ values over 500 Torr (Wilkes *et al.,* 1981).

Aquatic hyperoxia causes hypoventilation, reducing both ventilation rate and ventilation volume in fishes. A two- to fivefold decrease in the ventilatory requirement occurs in a number of marine fishes under hyperoxia (Dejours *et al.,* 1977) while oxygen extraction efficiency remains the same. In *G. cobitis* ventilatory frequency decreases to 9.2 from 17.3 beats $min^{-1}$ at 460 Torr and 12.5°C (Figure 3). A change of similar magnitude occurs at 25°C with ventilation frequency decreasing by approximately 50% (Berschick *et al.,* 1987). In the laboratory, ventilation decreased in *B. pholis* by 19% during hyperoxia (Bridges *et al.,* 1984), and this response has been confirmed in field studies (Harder, 1995). When *P. argentilineatus* (as *P. barbarus*) is exposed to hyperoxic water with a $PO_2$ greater than 300 Torr, after a few minutes normal ventilation ceases completely. Long apnoeic periods continue until the $PO_2$ is decreased to normoxic values (Freitag, 1995), indicating that in this species ventilatory drive is influenced by cutaneous gas exchange.

As described earlier in this chapter, hypoxia occurs most severely during nighttime low tides in tidepools. It occurs together with hypercapnia but is usually not accompanied by temperature extremes. The return to normoxic conditions may be rapid with the incoming tide refreshing the rock pool. Oxygen consumption in intertidal fishes remains constant as $PO_2$ values decrease down to the critical oxygen tension, $P_c$. At this point, oxygen consumption decreases and becomes dependent on oxygen tension. $P_c$ values for intertidal fishes are low in comparison with those for other teleosts (Hughes *et al.,* 1983) and represent an adaptation to the hypoxic conditions found within the intertidal zone. For rocky shore species, Congleton (1980) reported $P_c$ values in the range of 16 to 24 Torr for *Clinocottus analis,* 20 to 26 Torr for *Paraclinus intergrippinis,* and 26 to 30 Torr for *Gibbonsia elegans.* Innes and Wells (1985) report a $P_c$ of 30 to 40 Torr in the triple-finned blenny *Helcogramma medium* and Pelster (1985) found a $P_c$ of 20 to 30 Torr for *B. pholis.* In *G. cobitis* the $P_c$ ranged between 30 and 40 Torr (Berschick *et al.,* 1987). Congleton (1974) observed a $P_c$ value of 9 to 16 Torr in *Typhlogobius californiensis* and a value of 16 to 25 Torr for *Gillichthys mirabilis.* Both of these species colonize the burrows of ghost shrimps where hypoxic conditions may prevail. After the $P_c$ is reached, anaerobic metabolism appears to play a greater role. Pelster (1985) found that the energy charge remained constant down to a $PO_2$ of 40 Torr in *B. pholis* and that lactic acid accumulated only at a $PO_2$ below the $P_c$, increasing in both blood and muscle compartments by about 2.6 to 3.0 m*M*.

Ventilatory responses to aquatic hypoxia are normally larger than those seen during hyperoxia. In *G. cobitis,* hyperventilation occurred as the $PO_2$ decreased below normoxia, as ventilatory frequency rose (Figure 3) nearly 150% (Berschick *et al.,* 1987). Ventilatory frequency increased 14 to 19% in *B. pholis* on exposure to hypoxia in the laboratory (Bridges *et al.,* 1984) and 100% in the field (Laming *et al.,* 1982; Pelster, 1985). A fourfold change occurred in the ventilatory rate of *H. medium* on exposure to hypoxia (Innes and

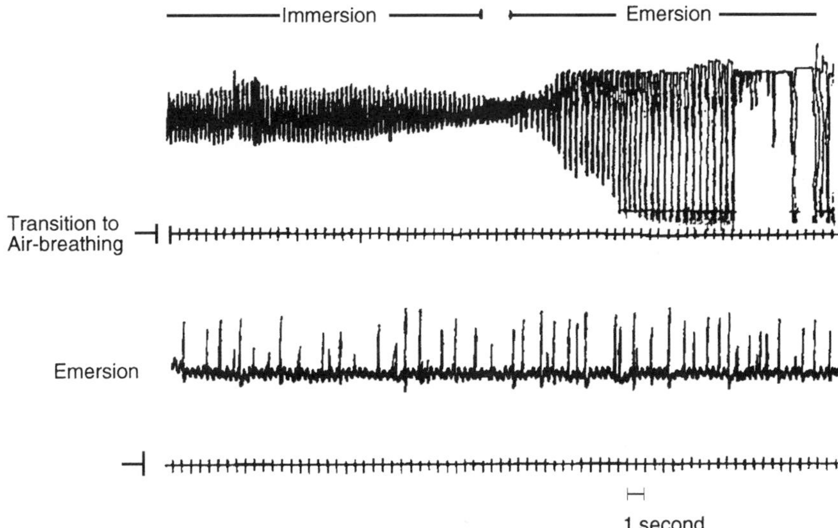


**Figure 4.** Impedance recording of opercular moments of *B. pholis* on the transition from immersion to and during aerial exposure. Modified from *Respiration Physiology* **71,** Pelster *et al.* Physiological adaptations of the intertidal rockpool teleost *Blennius pholis* L., to aerial exposure. Pp. 355–374, © 1988 with permission from Elsevier Science.

Wells, 1985). In *Periophthalmus argentilineatus* (as *P. barbarus*), ventilation increased by 31 to 50% during hypoxia (Freitag, 1995; Harder, 1995) and the percentage of time spent ventilating increased from 48 to 83% (Freitag, 1995). Fishes that normally emerge into air to avoid aquatic hypoxia may be forced to reduce surface area and branching of gills, resulting in a reduced ability to obtain oxygen during aquatic hypoxia.

Ventilatory activity on emersion has been reviewed by Graham (1976) and Bridges (1988, 1993a). The most typical response during emersion is that of "gulping," reported in numerous species (Laming *et al.,* 1982; Graham *et al.,* 1985). *Periophthalmus argentilineatus* (as *P. barbarus*) exchanges the air in the branchial chamber every 1 to 5 min in air by extreme opercular movements and gulping (Freitag, 1995). At all times the gap between the opercula and the body wall is kept closed by the branchiostegal membrane. Sealing the operculae has also been observed in *Entomacrodus nigricans* (Graham *et al.,* 1985) and *B. pholis* (Pelster *et al.,* 1988). Figure 4 is an impedance recording of ventilation during the transition from water to air-breathing in *B. pholis,* which shows a short period of increased opercular movement, mainly an increase in stroke volume, followed after 1 or 2 min by a reduction in the ventilatory frequency. The mode of ventilation also changes from flow-through to tidal. During the tidal ventilatory mode in air, the mouth remains slightly open and the operculum is closed. The buccal pumps are used to oscillate the air within the buccopharyngeal cavity (Pelster *et al.,* 1988).

## VII. Circulation

General changes in metabolism due to activity may affect cardiac peformance (Northcott *et al.,* 1990), and circulatory changes can influence heart rate, stroke volume, and tissue or organ perfusion. *Periophthalmus argentilineatus* (as *P. barbarus*) shows rhythmic ventilation under normoxic aquatic conditions, with a coupling of ventilation and circulation rates (Freitag, 1995). Ventilatory apnea (Figure 2) is terminated by an extreme opercular movement followed by the commencement of ventilation. At the same time heart rate increases from around 62.6 ± 13.6 to 73.7 ± 14.8 beats $min^{-1}$ and blood flow amplitude in the bulbus increases by more than 40%. On resumption of apnea, heart rate and amplitude return to their previous levels.

Metabolism in ectotherms generally increases with temperature, and heart rate increases as well (Withers, 1992). A *Q*10 of 2.6 was calculated for the heart rate of *G. cobitis* (Figure 3; Berschick *et al.,* 1987). This value corresponds to the general range for heart rate *Q*10 values found in acclimated fish hearts (Farrell, 1993). In general it is difficult to distinguish between intrinsic and extrinsic factors affecting heart rate. How the pacemaker system responds to temperature is not known (Farrell and Jones, 1992).

Cardiac responses to hyperoxia are little studied in intertidal fishes. There is no significant change in heart rate in *G. cobitis* under hyperoxia (Figure 3) at 12.5 or 25°C (Berschick *et al.,* 1987). In *B. pholis* the heart rate, relative blood flow, and blood pressure in the ventral aorta remained unchanged when the $PO_2$ was greater than 500 Torr (Eickelmann, 1991; Figure 5). In *Periophthalmus argentilineatus* (as *P. barbarus*) exposed to $PO_2$ greater than 300 Torr, heart rate and relative amplitude of blood flow remained unchanged (Figure 5). The lack of response by intertidal fishes to hyperoxia appears to be different from that of freshwater fishes; for example, *C. commersoni* decreases aortic pressure and heart frequency when exposed to hyperoxia (Wilkes *et al.,* 1981).

In hypoxia, cardiac responses typically involve a bradycardia. Heart rate decreases by 50% in *B. pholis* (Figure 5) with corresponding decreases in blood flow of 25% and blood pressure of 10% (Eickelmann, 1991). In *G. cobitis* no change in heart rate may occur with hypoxia because heart rate is already low (Figure 3). This may be adaptive in that *G. cobitis,* living in high shore rock pools (Gibson, 1972), is often exposed to hypoxia and maintaining a heart rate unchanged over normoxic levels when $PO_2$ was greater may be energetically more efficient than making constant adjustments. In *Periophthalmus argentilineatus* (as *P. barbarus*), the heart rate was significantly lower during hypoxia, but blood pressure and blood flow remained unchanged compared to normoxic levels (Figure 5).

Cardiac responses to aerial exposure are well documented (Graham, 1976; Bridges, 1988, 1993a). Clingfish show a "diving syndrome" with a bradycardia on air exposure that lasts several hours and rapid recovery of heart rate on return to water (Gordon *et al.,* 1970). *B. pholis* shows a transient bradycardia on exposure to air (Daniel, 1971; Laming *et al.,* 1982; Pelster *et al.,* 1988) with a rapid recovery to aquatic levels (Figure 6). Blood pressure does not change on aerial exposure in *B. pholis* (Pelster *et al.,* 1988; Eickelmann, 1991), and heart frequency and relative blood flow in the ventral aorta are not significantly altered from aquatic levels (Figure 5a). On the other hand, the mudskipper *Periophthalmodon freycineti* (as *P. australis*) shows an increased heart rate on emergence into air and a decrease in water (Garey, 1962). Heart frequency and the relative blood flow amplitude in the dorsal aorta of the mudskipper *Periophthalmus argentilineatus* (as *P. barbarus*) show no change after the transition from water to air-breathing (Freitag, 1995). Heart rate in the mudskipper *P. sobrinus* was unaffected by emersion (Gordon *et al.,* 1969) but Garey (1962) reported a bradycardia in *P. freycineti* exposed to water and much lower heart rates when the animal was in a burrow. *Cebidichthys violaceus* does not show a bradycardia on exposure to air but an initial tachycardia is observed with a gradual decrease over the emersion period (Riegle, 1976).

Other circulatory adjustments are thought to occur during emersion to help cutaneous

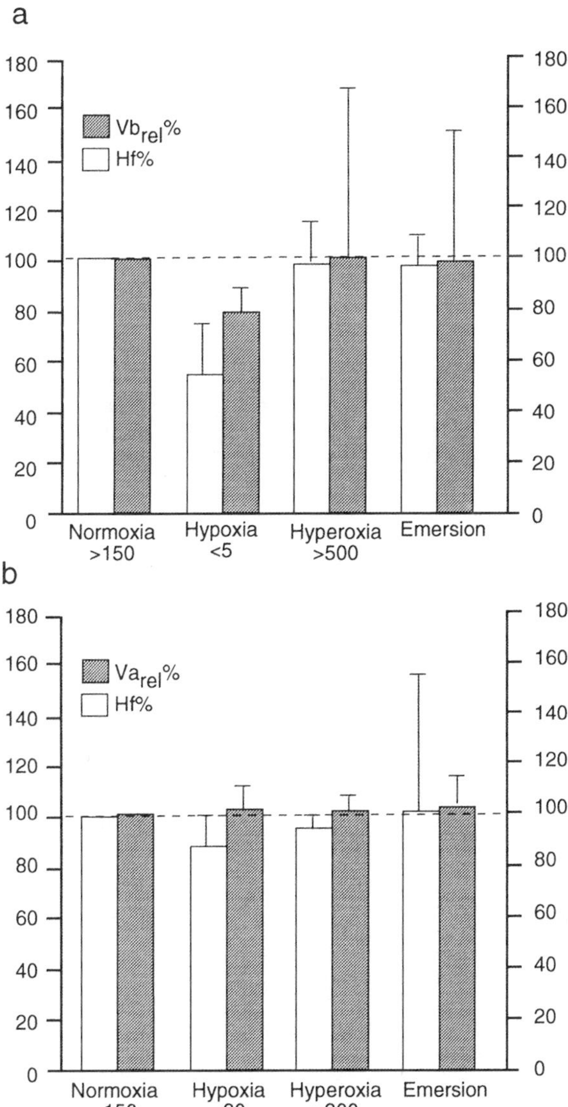


**Figure 5.** Influence of hypoxia, hyperoxia, and emersion on the standardized heart rate Hf (normoxia = 100%), relative blood flow ($Vb_{rel}$%), and stroke amplitude ($Va_{rel}$%), for (a) *B. pholis* at 12.5°C and (b) *Periophthalmus argentilineatus* at 25°C. Values are means ±1 SD and * denotes a significant difference from normoxic levels ($P < 0.05$). Oxygen tensions ($PO_2$) are shown in Torr for aquatic treatments.

respiration (see above). Microsphere perfusion of the gills of *B. pholis* indicated no difference in gill perfusion during aerial exposure but increased perfusion of the skin (Bridges, 1993b).

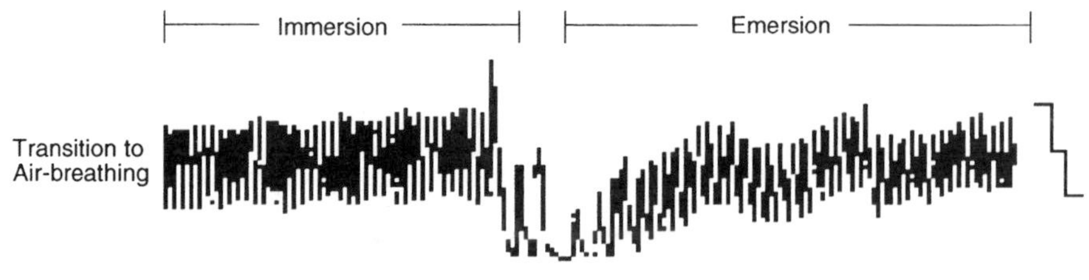


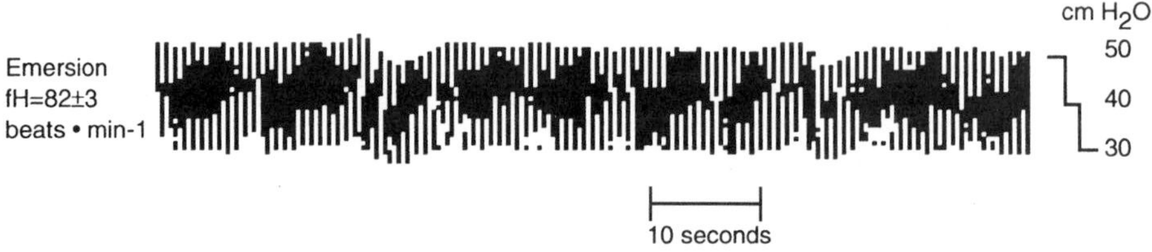


**Figure 6.** The influence of the transition from immersion to emersion on the cardiac activity of *B. pholis*. Measurements are pressure recordings (cm $H_2O$) made in the bulbus arteriosus using indwelling catheters. Modified from *Respiration Physiology* **71,** Pelster *et al.* Physiological adaptations of the intertidal rockpool teleost *Blennius pholis* L., to aerial exposure. Pp. 355–374, © 1988 with permission from Elsevier Science.

## VIII. Acid–Base Balance

Three major acid–base problems arise in intertidal rock pools during emersion: (a) hypoxia and hypercapnia at night, (b) hyperoxia and hypocapnia during daylight, and (c) the concomitant fall and rise in temperature (Bridges, 1993a,b). Measurements of acid–base parameters are scarce in intertidal fish due to their small size (Bridges *et al.,* 1984; Pelster *et al.,* 1988). Figure 7a depicts the changes measured in venous blood pH, $PCO_2$, and $HCO_3$ during normoxia, hypoxia, and hyperoxia using indwelling catheters in *B. pholis.* During hypoxia, ventilation increases (see above), and a respiratory alkalosis is seen as more carbon dioxide is released. The nonbicarbonate buffer line is followed by changes in pH. During hyperoxia at 12.5°C and 25°C the reverse is true; ventilation decreases as a result of the abundance of oxygen in the medium. The change in pH is exacerbated at 25°C and again almost all of the change is respiratory, following roughly the nonbicarbonate buffer lines. The results shown in Figure 7a represent equilibrations in the absence of environmental carbon dioxide; the changes *in situ* will be influenced by aquatic hypercapnia during hypoxia, and hypocapnia during hyperoxia, reducing and increasing the pH gradient, respectively (Bridges, 1988).

Figure 7b summarizes the changes on transition from aquatic to aerial respiration. After 3 h emersion the venous blood $PCO_2$ rose by approximately 1 Torr and was accompanied at the same time by a mild metabolic acidosis (1.5 mmol $L^{-1}$ lactate). Inhibition of carbonic anhydrase activity with methazolamide resulted in $CO_2$ retention and a three-fold increase in $PCO_2$ over control levels. Obviously a new steady state is established in air to maintain $CO_2$ release (see above) and is dependent on the carbonic anhydrase activity of the red blood cells.

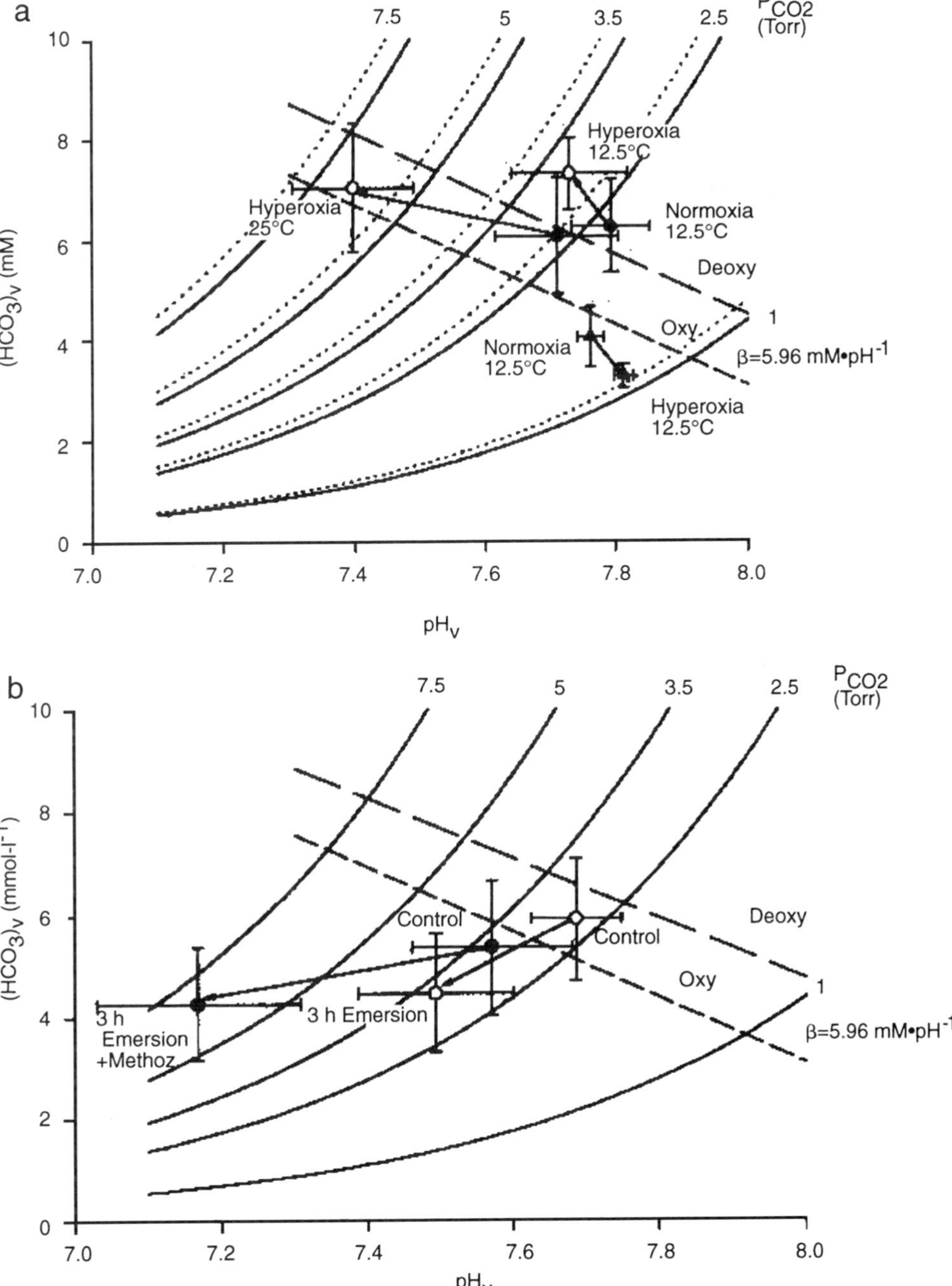


**Figure 7.** (a) Acid–base changes in the venous blood of *B. pholis* sampled via indwelling catheters during exposure from normoxia to hypoxia (30 Torr) at 12.5°C; normoxia to hyperoxia (410 Torr) at 12.5°C; and normoxia at 12.5°C to hyperoxia (420 Torr) at 25°C. $PCO_2$ isobars are calculated from values for $p$K = 6.124 and $\delta CO_2$ = 0.0584 mmol L$^{-1}$ Torr$^{-1}$ at 12.5°C and $p$K = 5 and $\delta CO_2$ = 0.040 mmol L$^{-1}$ Torr$^{-1}$ at 25°C. The nonbicarbonate buffer values for oxy- and deoxygenated blood are taken from Pelster *et al.* (1988). (b) Changes in acid–base status of venous blood on the transition from aquatic to aerial respiration (3 h emersion) at 12.5°C. A second series of experiments was performed with fish that had been injected with methazolamide (final concentration $10^{-4}$ $M$ kg$^{-1}$) and compared with sham-injected fish. Other parameters are as above or modified from Pelster *et al.* (1988).

**Table 2.** Blood Respiratory Properties of Some Air-Emerging Intertidal Fishes

| Family / *Species* | $Hb_4$ Conc. (mmol $L^{-1}$) | Hct (%) | $ATP/Hb_4$ | $GTP/Hb_4$ | Temp. (°C) | pH | $P_{50}$ (Torr) | $n_{50}$ | Bohr Coeff. ($\delta \log P_{50}/\delta pH$) | $\delta H^a$ (kJ $mol^{-1}$) | Source |
|---|---|---|---|---|---|---|---|---|---|---|---|
| Blenniidae | | | | | | | | | | | |
| *Blennius pholis* | 0.59 | 18.3 | $2.49^b$ | $0.39^b$ | 12.5 | 7.9 | $16.2^b$ | 2.0 | $0.85^b$ | | Bridges *et al.*, 1984 |
| *B. pholis* | | | | | 25.0 | 7.9 | $20.6^b$ | | $0.86^b$ | $-14.4^b$ | Bridges *et al.*, 1984 |
| Cottidae | | | | | | | | | | | |
| *Taurulus bubalis* | 0.45 | 14.5 | 1.33 | 0.66 | 12.5 | 7.9 | 8.5 | 1.4 | −1.25 | | Kirchhof, 1983 |
| *T. bubalis* | | | | | 25.0 | 7.9 | 25.0 | 1.5 | −0.85 | −61 | Kirchhof, 1983 |
| Fundulidae (Cyprinodontidae) | | | | | | | | | | | |
| *Fundulus heteroclitus* | — | 23 | 1.3 | — | 20.5 | 7.4 | 5.0 | — | — | −66 | Powers, 1980 |
| Pholididae | | | | | | | | | | | |
| *Zoarces viviparous* | — | — | <1.2 | <0.2 | 10 | 7.5 | 23.0 | 1.9 | $-0.85^b$ | | Hartvig and Weber, 1984 |
| Tripterygiidae | | | | | | | | | | | |
| *Helcogramma medium* | .47 | 11.6 | 2.38 | — | 15 | 7.7 | 19.0 | 1.2 | −1.06 | | Innes and Wells, 1985 |

*Note.* Additional data on hemoglobin and hematocrit are available (see Graham, 1997) but the analyses are not as complete as those listed.

[a] Calculated for a constant pH.

[b] Calculated from authors' data.

## IX. Blood Oxygen Affinity

Bridges (1988) has reviewed the respiratory properties of the blood of intertidal fish and the modified data are shown in Table 2. Recently, these data have been compared with those from other bimodal air breathers (Morris and Bridges, 1994). Two strategies appear to have evolved for bimodal breathing in fish: (a) a decrease in hemoglobin oxygen affinity, or (b) changes in hematological properties. The first trend can be seen in Table 2. *B. pholis, H. medium,* and *Zoarces viviparous* are all bimodal breathers and show a low oxygen affinity in comparison to *Fundulus heteroclitus* and *Taurulus bubalis,* which are predominantly aquatic. In general the Bohr coefficient ($\delta \log PO_2/\delta$pH), which describes the pH sensitivity of oxygen affinity, is large in all intertidal species, whereas the $\delta H$, which indicates the temperature sensitivity of oxygen affinity, is low only in *B. pholis* and higher in the more aquatic forms, similar to other fish species (Powers, 1980). High amounts of the organic phosphates ATP and GTP, which control oxygen affinity, pH, and temperature sensitivity (Bridges *et al.*, 1984; Bridges, 1988), are present in intertidal species. Bridges *et al.* (1984) and Bridges (1988) have modeled the effects of hyperoxia and hypoxia during daylight and nighttime on the oxygen delivery system. A right shift of the oxygen dissociation curve brought about by the decrease in pH (Figure 7a) and the large Bohr coefficient allows the venous reserve to be used during hyperoxia. During hypoxia a left shift of the curve due to an increase in pH and perhaps due to changes in organic phosphates (see Bridges *et al.*, 1984; Lykkeboe and Weber, 1978) will increase oxygen affinity and help protect oxygen loading at the gills as arterial and venous oxygen tensions are reduced. The hemoglobins of intertidal fish are highly flexible in their responses to environmental stresses.

## X. Future Work

The study of air-breathing and air emergence in intertidal fishes has not received the attention that has been given to crabs or to air-breathing fishes that live in freshwater systems and remain aquatic. Much of the work that has been done has focused on relatively few common or readily obtained species. It is clear that air-breathing ability and behaviors associated with air emergence among fishes of the rocky intertidal zone are complex and variable features that occur in many different taxa and are correlated to some extent to the vertical distribution or tidal height of these species on shore. A phylogenetic comparison of air-breathing intertidal fishes with non-air-breathing relatives could show the importance of evolutionary history and ecology in the distribution of this ability.

Partitioning of respiration among the gills, skin, and other organs needs further study to ascertain the roles of each organ, and their relative contributions to respiratory gas exchange when the fish is emerged and when it is aquatic. Enzymatic assays for carbonic anydrase, microsphere perfusion, and histological examinations will aid the elucidation of these structures and their functions. Biochemistry of the hemoglobins that are present in fish blood could illuminate the affinities to oxygen, the sensitivity to acid–base disturbances, and possibly ontogenetic changes in expression of different forms of hemoglobin, particularly for species in which the juvenile is in tidepools and the adult is in deeper water. Comparisons between the different behavioral types of emergers for differences in physiology could be instructive in this regard as well, particularly in terms of tolerance for long-term air emergence and terrestrial activity.

Acid–base balance merits further study, both from respiratory and metabolic sources.

Blood pH and buffering capabilities should be examined during emergence over the time of a tidal cycle or longer, and following activity in water and in air. It would be intriguing to examine the possibility that metabolic acidosis caused by anaerobic terrestrial activity could be compensated for by respiratory release of $CO_2$ at the gills, as in mammals.

More detailed examination of the natural history and ecology of these intertidal air-breathing fishes is also needed, particularly field observations at night. It would be fruitful to document patterns of distribution of local populations of air-breathing species during low tides, whether in pools or emerged, and the potential intraspecific size differences in those distributions.

## Acknowledgments

The work of K.L.M. was supported in part by U.S. National Science Foundation Grant DBI-96-05062 and by the Pepperdine University Research Council. The work of C.R.B. was supported by the Deutsche Forschungsgemeinschaft and the State of North-Rhine-Westphalia. Thanks are also due to the Marine Biological Stations in the Isle of Man, U.K., and Roscoff, Brittany, France, for experimental facilities and animals. We thank Michael Chotkowski, Patricia Halpin, and William Davison for helpful comments on the manuscript.

## References

Abel, E. F. (1973). Zur Öko-Ethologie des amphibisch lebenden Fisches *Alticus saliens* (Förster) und von *Entomacrodus vermiculatus* (Val.) (Blennioidea, Salariidae), unter besonderer Berucksichtigung des Fortpflanzungsverhaltens. *Sitzungsber. Österr. Akad. Wiss. Math. (Nat. Kl., Abt. I.)* **181,** 137–153.

Bandurski, R. S., Bradstreet, E., and Scholander, P. F. (1968). Metabolic changes in the mud-skipper during asphyxia or exercise. *Comp. Biochem. Physiol.* **24,** 271–274.

Barlow, G. W. (1961). Intra- and interspecific differences in rate of oxygen consumption in gobiid fishes of the genus *Gillichthys. Biol. Bull.* **121,** 209–229.

Berger, M., and Mayr, A. (1992). Ecological studies on two intertidal New Zealand fishes, *Acanthoclinus fuscus* and *Forsterygion nigripenne robustum. N. Z. J. Mar. Freshw. Res.* **26,** 359–370.

Berschick, P., Bridges, C. R., and Grieshaber, M. K. (1987). The influence of hyperoxia, hypoxia and temperature on the respiratory physiology of the intertidal rockpool fish *Gobius cobitis* Pallas. *J. Exp. Biol.* **130**, 369–387.

Bridges, C. R. (1988). Respiratory adaptations in intertidal fish. *Am. Zool.* **28,** 79–96.

Bridges, C. R. (1993a). Adaptation of vertebrates to the intertidal environment. *In* "The Vertebrate Gas Transport Cascade—Adaptations to the Environment and Mode of Life" (J. Eduardo and P. W. Bicudo, Eds.), pp. 12–22. CRC Press, Boca Raton, FL.

Bridges, C. R. (1993b). Ecophysiology of intertidal fish. *In* "Fish Ecophysiology" (J. C. Rankin and F. B. Jensen, Eds.), pp. 375–400. Chapman and Hall, London.

Bridges, C. R. (1997). Ecophysiological adaptations in intertidal rockpool fishes. *In* "Water/Air Transitions in Biology" (B. Eddy, A. K. Mittal, and J. S. Datta Munshi, Eds.). Academic Press.

Bridges, C. R., Taylor, A. C., Morris, S. J., and Grieshaber, M. K. (1984). Ecophysiological adaptations in *Blennius pholis* (L.) blood to intertidal rockpool environments. *J. Exp. Mar. Biol. Ecol.* **77,** 151–167.

Brillet, C. (1984). The habitat and behavior of the amphibious fish *Periophthalmus koelreuteria africanus* at Tulear, Madagascar: Comparison with the sympatric *Periophthalmus sobrinus. Rev. Ecol. Terre Vie* **39**(3), 337–346.

Brown, C. R., Gordon, M. S., and Martin, K. L. M. (1992). Aerial and aquatic oxygen uptake in the amphibious Red Sea rockskipper fish, *Alticus kirki* (Family Blenniidae). *Copeia* **1992,** 1007–1013.

Clayton, D. A. (1993). Mudskippers. *Oceanogr. Mar. Biol. Annu. Rev.* **31,** 507–577.

Coleman, R. (1992). Reproductive biology and female parental care in the cockscomb prickleback, *Anoplarchus purpurescens* (Pisces: Stichaeidae). *Environ. Biol. Fish.* **41,** 177–186.

Congleton, J. L. (1974). The respiratory response to asphyxia of *Typhlogobius californiensis* (Teleostei: Gobiidae) and some related gobies. *Biol. Bull.* **146,** 186–205.

Congleton, J. L. (1980). Observations on the responses of some southern California tidepool fishes to nocturnal hypoxic stress. *Comp. Bioch. Physiol. A* **66,** 719–722.

Crane, J. (1981). Feeding and growth by the sessile larvae of the teleost *Porichthys notatus. Copeia* **1981**(4), 895–897.

Cross, J. N. (1981). "Structure of a Rocky Intertidal Fish Assemblage." Ph.D. dissertation, University of Washington, Seattle.

Daniel, M. J. (1971). Aspects of the physiology of the intertidal teleost, *Blennius pholis* (L.). Ph.D. thesis, University of London.

Davenport, J., and Woolmington, A. D. (1981). Behavioural responses of some rocky shore fish exposed to adverse environmental conditions. *Mar. Behav. Physiol.* **8,** 1–12.

Daxboeck, C., and Heming, T. A. (1982). Bimodal respiration in the intertidal fish, *Xiphister atropurpureus* (Kittlitz). *Mar. Behav. Physiol.* **9,** 23–33.

Dejours, P. (1973). Problems of control of breathing in fishes. *In* "Comparative Physiology, Locomotion, Respiration, Transport and Blood" (L. Bolis, W. Schmidt-Nielsen, and S. H. P. Maddrell, Eds.), pp. 117–133. North Holland/Elsevier, Amsterdam/New York.

Dejours, P. (1994). Environmental factors as determinants in bimodal breathing: An introductory overview. *Am. Zool.* **34,** 178–183.

Dejours, P., Toulmond, A., and Truchot, J. P. (1977). The effect of hyperoxia on the breathing of marine fishes. *Comp. Biochem. Physiol. A* **58,** 409–411.

Ebeling, A. W., Bernal, P., and Zuleta, A. (1970). Emersion of the amphibious Chilean clingfish, *Sicyases sanguineus. Biol. Bull.* **139,** 115–137.

Edwards, D. G., and Cech, J. J., Jr. (1990). Aquatic and aerial metabolism of the juvenile monkeyface prickleback, *Cebidichthys violaceus,* an intertidal fish of California. *Comp. Biochem. Physiol. A* **96,** 61–65.

Eger, W. H. (1971). "Ecological and Physiological Adaptations of Intertidal Clingfishes (Teleostei: Gobeisocidae) in the Northern Gulf of California." Ph.D. dissertation, University of Arizona.

Eickelmann, P. (1991). "Zur Kreislaufphysiologie des Felstumpelfisches *Blennius pholis* unter extremen Bedingungen." Diplomarbeit, University of Düsseldorf.

Farrell A. P. (1993). Cardiovascular system. *In* "The Physiology of Fishes" (D. H. Evans, Ed.), Chap. 8. CRC Press, Boca Raton, FL.

Farrell A. P., and Jones, D. R. (1992). The heart. *In* "Fish Physiology" Vol XIIA, "The Cardio-Vascular System." Academic Press, San Diego.

Feder, M. E., and Burggren, W. W. (1985). Cutaneous gas exchange in vertebrates: Design, patterns, control and implications. *Biol. Rev.* **60,** 1–45.

Freitag, J. (1995). "Untersuchungen zur Kreislaufphysiologie und Ventilation bei *Blennius pholis* und *Periophthalmus barbarus:* Ein vergleich zweier amphibisch lebender Teleosteer." Diplomarbeit, University of Düsseldorf.

Garey, W. F. (1962). Cardiac responses of fishes in asphyxic environments. *Biol. Bull.* **122,** 362–368.

Gibson, R. N. (1972). The vertical distribution and feeding relationships of intertidal fish on the Atlantic coast of France. *J. Anim. Ecol.* **41,** 189–207.

Gordon, M. S., Boetius, I, Evans, D. H., McCarthy, R., and Oglesby, L. C. (1969). Aspects of the physiology of terrestrial life in amphibious fishes. I. The mudskipper *Periophthalmus sobrinus. J. Exp. Biol.* **50,** 141–149.

Gordon, M. S., Fischer, S., and Tarifeño, E. (1970). Aspects of the physiology of terrestrial life in amphibious fishes. II. The Chilean clingfish, *Sicyases sanguineus. J. Exp. Biol.* **53,** 559–572.

Gordon, M., Ng, W. W., and Yip, A. Y. (1978). Aspects of the physiology of the terrestrial life in amphibious fishes. III. The Chinese mudskipper *Periophthalmus cantonensis. J. Exp. Biol.* **72,** 57–75.

Graham, J. B. (1970). Preliminary studies on the biology of the amphibious clinid *Mnierpes macrocephalus. Mar. Biol.* **5,** 136–140.

Graham, J. B. (1973). Terrestrial life of the amphibious fish *Mnierpes macrocephalus. Mar. Biol.* **23,** 83–91.

Graham, J. B. (1976). Respiratory adaptations of marine air-breathing fishes. *In* "Respiration in Amphibious Vertebrates" (G. M. Hughes, ed.), pp. 165–187. Academic Press, London.

Graham, J. B. (1997). "Air-Breathing Fishes: Evolution, Diversity and Adaptation." Academic Press, San Diego.

Graham, J. B., Jones, C. B., and Rubinoff, I. (1985). Behavioural, physiological, and ecological aspects of the amphibious life of the pearl blenny *Entomacrodus nigricans* Gill. *J. Exp. Mar. Biol. Ecol.* **89,** 255–268.

Graham, J.B., Rosenblatt, R. H., and Gans, C. (1978). Vertebrate air breathing arose in fresh waters and not in the oceans. *Evolution* **32,** 459–463.

Gregory, R. B. (1977). Synthesis and total excretion of waste nitrogen by fish of the *Periophthalmus* (mudskipper) and *Scartelaos* families. *Comp. Biochem. Physiol. A* **57,** 33–36.

Steeger, H.-U., and Bridges, C. R. (1995). A method for long-term measurement of respiration in intertidal fishes during simulated intertidal conditions. *J. Fish Biol.* **47,** 308–320.

Tamura S. O., Morii, H., and Yuzuriha, M. (1976). Respiration of the amphibious fishes *Periophthalmus cantonensis* and *Boleophthalmus chinensis* in water and on land. *J. Exp. Biol.* **65,** 97–107.

Teal, J. M., and Carey, F. G. (1967). Skin respiration and oxygen debt in the mudskipper *Periophthalmus sobrinus. Copeia* **1967,** 677–679.

Todd, E. S. (1976). Terrestrial grazing by the eastern tropical Pacific goby *Gobionellus sagittula. Copeia* **1976,** 374–377.

Todd, E. S., and Ebeling, A. W. (1966). Aerial respiration in the longjaw mudsucker *Gillichthys mirabilis* (Teleostei: Gobiidae). *Biol. Bull.* **130,** 265–288.

Truchot, J. P., and Duhamel-Jouve, A. (1980). Oxygen and carbon dioxide in the marine intertidal environment: Diurnal and tidal changes in rockpools. *Respir. Physiol.* **39,** 241–254.

Wade, R. A. (1962). The biology of the tarpon *Megalops atlanticus,* with emphasis on larval development. *Bull. Mar. Sci.* **12,** 545–622.

Wilkes, P. R. H., Walker, R. L., McDonald, D. G., and Wood, C. M. (1981). Respiratory, ventilatory, acid-base and ionoregulatory physiology of the white sucker *Catostomus commersoni:* The influence of hyperoxia. *J. Exp. Biol.* **91,** 239–254.

Withers, P. C. (1992). "Comparative Animal Physiology." Saunders, New York.

Wright, W. G., and Raymond, J. A. (1978). Air-breathing in a California sculpin. *J. Exp. Zool.* **203,** 171–176.

Yoshiyama, R. M., and Cech, J. J., Jr. (1994). Aerial respiration by rocky intertidal fishes of California and Oregon. *Copeia* **1994,** 153–158.

Yoshiyama, R. M., Valey, C. J., Schalk, L. L., Oswald, N. M., Vaness, K. K., Lauritzen, D., and Limm, M. (1995). Differential propensities for aerial emergence in intertidal sculpins (Teleostei; Cottidae). *J. Exp. Mar. Biol. Ecol.* **191,** 195–207.

Zander, C. D. (1972). Beziehungen zwischen Körperbau und Lebensweise bei Blenniidae (Pisces) aus dem Roten Meer. I. Äußere Morphologie. *Mar. Biol.* **13,** 238–246.

Zander, C. D. (1983). Terrestrial sojourns of two Mediterranean Blennioid fish (Pisces, Blennioidei, Blenniidae). *Senckenberg. Marit.* **15,** 19–26.

# 5

# Osmoregulation, Acid–Base Regulation, and Nitrogen Excretion

David H. Evans

*Mt. Desert Island Biological Laboratory, Salsbury Cove, Maine*
*Department of Zoology, University of Florida, Gainesville, Florida*

J. B. Claiborne

*Mt. Desert Island Biological Laboratory, Salsbury Cove, Maine*
*Department of Biology, Georgia Southern University, Statesboro, Georgia*

Gregg A. Kormanik

*Mt. Desert Island Biological Laboratory, Salsbury Cove, Maine*
*Department of Biology, University of North Carolina, Asheville, North Carolina*

## I. Introduction

Periodic emersion of the intertidal zone produces an environment in which physical and chemical characteristics are semiterrestrial (see Gibson, Chapter 2, this volume). In addition to potential thermal and respiratory challenges, intertidal fishes may face dehydration, salinity changes (evaporation vs. rainfall), and changes in pH in the relatively restricted volumes of tidepools or burrows. Thus, fishes inhabiting this zone must maintain consistent internal osmolarity, specific ion levels, and pH, and eliminate excess nitrogen via mechanisms somewhat different from those utilized by fishes whose skin and gills are bathed in a relatively homogeneous aqueous medium.

## II. Osmoregulation

Osmoregulation is the maintenance of consistent blood and intracellular volume in the face of changing environmental osmolarity. Since osmoregulation always involves the transport of ions across epithelial membranes, ion regulation is intimately associated with osmoregulation. Ion regulation most commonly involves transport of $Na^+$ and $Cl^-$, because these are the dominant cation and anion in the extracellular fluids and the aqueous environments

*Intertidal Fishes: Life in Two Worlds*

of fishes. Osmoregulatory strategies of fishes have been reviewed a number of times (see Evans, 1979, 1984b, 1993, 1995, for general reviews) and it is clear that the branchial epithelium of the gills plays a major role, both as the site of diffusional ion and osmotic water movements and as the site of the active transport steps that oppose these passive movements. In addition, the skin, intestinal epithelium, renal tubules, and urinary bladder provide sites for water and salt transport to maintain internal osmotic consistency.

Despite facing a net diffusional influx of salt from seawater across ion-permeable gills, marine teleost fishes maintain their plasma $Na^+$ and $Cl^-$ below that of seawater by extruding salt across the branchial epithelium via a Na–K–2Cl cotransport system in specialized cells of the branchial epithelium, the specifics of which are described in the reviews cited above. Gill extrusion of unwanted salt is necessitated by the kidney's inability to produce urine that is hypertonic to the blood. The osmotic loss of water across the gills (and possibly the skin) is balanced by oral ingestion of seawater, adding to the diffusional salt load. Renal loss of water is minimized by extremely low urine production, associated with (in some species) the absence of renal glomeruli, the site of ultrafiltration of the blood to form urine.

Brackish-water or freshwater teleosts (which are relevant here because intertidal fishes may experience periods of brackish water in tidepools during rainstorms) face the opposite osmotic and ionic problems—osmotic gain of water and diffusional loss of ions—because their plasma contains more salt per liter than the surrounding medium. Needed salt is extracted from the medium by the branchial epithelium and the osmotic influx of water is balanced by a substantial urine flow. The salt concentration of the urine is less than that of the plasma because of renal tubular and urinary–bladder extraction of salt before the urine is voided.

A priori, one might expect that some intertidal fishes must be able to tolerate relatively long periods of emersion during low tides. There are some published data on survival times of intertidal fishes maintained in air (Table 1), and it is quite clear that these species can be out of water for prolonged periods of time, as long as they are kept in high relative humidity (RH). In addition, although relative humidity is a parameter of major importance for a particular species (e.g., *Periophthalmus sobrinus* and *Entomacrodus nigricans*), it cannot account for the apparent difference between the survival times of *Periophthalmus cantonensis* and *P. sobrinus* and other species maintained at the same RH. Moreover, it is not known if the limiting factor of emersion is gas exchange, dehydration, pH regulation, or nitrogen balance. (Thermal death was avoided in these experiments by maintaining the fishes either in the shade or in the laboratory under constant temperature.) The size specificity of the tolerances (Table 1) and the fact that evaporative water loss is inversely related to body length in the intertidal fishes *Cebidichthys violaceus, Xiphister mucosus, Xiphister atropurpureus, Anoplarchus purpurescens,* and *Xererpes fucorum* (Horn and Riegle, 1981) suggest that either dehydration or gas exchange may be limiting because, in theory, smaller individuals would have a larger surface-to-volume ratio, although we do not know the relationship between gill surface area and body size in these species. However, Horn and Riegle (1981) demonstrated that the water loss in the five species of their study was directly related to body surface area. One might argue that body surface area and branchial surface area are positively correlated, but their data may also suggest the importance of evaporative water loss from the skin of intertidal fishes. Such a proposition is supported by the finding that various intertidal species utilize the skin for gas exchange (Nonnotte and Kirsch, 1978; see also Martin and Bridges, Chapter 4, this volume)

The few data on the rates of evaporative water loss by intertidal fishes (Table 2) suggest relative consistency across taxa. However, it is clear that the experimental conditions in the three studies in Table 2 are so disparate that comparisons between species are difficult. The

**Table 1.** Survival Times of Intertidal Fishes

| Species | Conditions | Survival time | Reference |
|---|---|---|---|
| *Periophthalmus sobrinus* (<15 g) | <95% RH[a] | 37 h | Gordon *et al.*, 1969 |
| | 70–80% RH | 24 h | Gordon *et al.*, 1969 |
| | Sun | 50 min | Gordon *et al.*, 1969 |
| *P. catonensis* (<5 g) | <95% RH[a] | 22–60 h | Gordon *et al.*, 1978 |
| *Sicyases sanguineus* (5–110 g) | 60–75% RH[a] | 17–40 h[b] | Gordon *et al.*, 1970 |
| *Cebidichthys violaceus* (5–20 g) | <95% RH[a] | 15–35 h[b] | Horn and Riegle, 1981 |
| *Xiphister mucosus* (10–20 g) | <95% RH[a] | 17–23 h[b] | Horn and Riegle, 1981 |
| *X. atropurpureus* (7–20 g) | <95% RH[a] | 10–23 h[b] | Horn and Riegle, 1981 |
| *Anoplarchus purpurescens* (6–10 g) | <95% RH[a] | 15–25 h[b] | Horn and Riegle, 1981 |
| *Xererpes fucorum* (5–13 g) | <95% RH[a] | 5–20 h[b] | Horn and Riegle, 1981 |
| *Entomacrodus nigricans* (>10 g) | 95% RH[a] | 2.25 h | Graham *et al.*, 1985 |
| | 89–90% RH[a] | 0.75 h | Graham *et al.*, 1985 |
| *Mnierpes macrocephalus* (0.5–8 g) | 60–70% RH | 1.5–4 h | Graham, 1973 |

[a] In shaded containers with a moist substrate or in environmental chambers.
[b] Directly related to size.

**Table 2.** Rates of Evaporative Water Loss in Intertidal Fishes

| Species | Conditions | Rate[a] | Reference |
|---|---|---|---|
| *Periophthalmus sobrinus* | 70–80% RH[b] | 1.6 | Gordon *et al.*, 1969 |
| | sun[c] | 8 | |
| *P. catonensis* | 40–55% RH[b] | 5 | Gordon *et al.*, 1978 |
| | 51–57% RH[d] | 14 | |
| | 55% RH, sun[e] | 23.3 | |
| *Sicyases sanguineus* | 65–76% RH[b] | 0.25 | Gordon *et al.*, 1970 |
| | 65–76% RH[f] | 1.2 | |
| | sun[g] | 4 | |
| *Cebidichthys violaceus* | <95% RH[h] | 2.5[i] | Horn and Riegle, 1981 |
| *Xiphister mucosus* | <95% RH[h] | 3.0[i] | Horn and Riegle, 1981 |
| *X. atropurpureus* | <95% RH[h] | 3.0[i] | Horn and Riegle, 1981 |
| *Anoplarchus purpurescens* | <95% RH[h] | 2.0[i] | Horn and Riegle, 1981 |
| *Xerepes fucorum* | <95% RH[h] | 2.7[i] | Horn and Riegle, 1981 |
| *Mnierpes macrocephalus* | 60–70% RH[j] | 5–15 | Graham, 1973 |

[a] % Change in body weight per hour; calculated from published figures.
[b] Maintained in still air at 23–30°C (*P.s.*), 23–24°C (*P.c.*), 13–16°C (*S.s.*).
[c] Ambient temperature was 43–44°C at time of death.
[d] Maintained in moving air (12–15 km $h^{-1}$) at 27–28°C.
[e] Maintained in moving air (5–12 km $h^{-1}$) at 30–31°C.
[f] Maintained in moving air (13–15 km $h^{-1}$) at 13–16°C.
[g] Maintained in moving air (0–8 km $h^{-1}$) at 15–16°C.
[h] Maintained in still air at 15°C.
[i] Data calculated from regression data at 9 cm standard length.
[j] Maintained at 30–32°C.

rate of evaporative water loss by the freshwater frog, *Rana tigrina,* is 7.8% (of the body water per hour) in still air and shade (24–26°C, 52–56% RH; Gordon *et al.,* 1978), so it appears that all these fish species maintain an evaporative water permeability equivalent to that of amphibians under similar conditions. This is presumably crucial for a semiterrestrial existence, even for short periods of time.

There is some indication that *P. sobrinus* (Stebbins and Kalk, 1961) and *P. cantonensis* (Gordon *et al.,* 1978) may reduce the effects of evaporative water loss by carrying water in the buccal cavity during emersion, although another study of a different population of *P. sobrinus* found no evidence for buccal water reserves (Gordon *et al.,* 1968). A priori, buccal water retention would seem to be a beneficial adaptation, unless it interferes with feeding or gas exchange. It would be useful to know if intertidal fishes also utilize water stored in urinary bladders, as terrestrial toads do (e.g., Alvarado, 1979), although urine flows are severely reduced in marine teleosts (e.g., Evans, 1993), so bladder volume would be relatively small.

It is interesting to note that dehydration produced by emersion of intertidal fishes is merely an extension of the osmotic loss of water chronically faced by any marine teleost. Thus, compensation upon reimmersion would involve physiological processes (drinking the medium, extrusion of salt) already expressed in the fishes when in seawater. The mechanism of salt extrusion by the branchial epithelium has not been specifically studied in intertidal fishes, but there is no a priori reason to suspect that any unique mechanisms are present, and retention of water in the buccal cavity (if it occurs) would provide a reservoir into which salts could be excreted. However, it would also provide a source for net salt uptake into the fish down diffusional gradients. One might wonder then if retention of water in the buccal cavity is especially adaptive for intertidal species.

Recent data suggest that auxiliary sites of salt secretion may be present in some species. Yokoya and Tamura (1992) described mitochondrion-rich cells in the skin of *P. cantonensis,* and similar, potential transport sites also have been described in *Gillichthys mirabilis* (Marshall, 1977, 1995) and *P. modestus* (Yokota *et al.,* 1997). In these species, a high density of mitochonrion-rich cells has been identified in the skin of the jaw and pectoral fin, respectively, and electrophysiological experiments with *in vitro* skin preparations have shown that these epithelia possess the classic Na–K–2Cl cotransport system, which has been shown to extrude salt across the branchial epithelium in other marine teleosts (e.g., Evans, 1993, Marshall, 1995).

Intertidal fishes do ingest the seawater medium to offset the osmotic loss of water when submerged (Evans, 1967b, 1969; Dall and Milward, 1969). At least two species (*Xiphister atropurpureus* and *Pholis gunnellus*) have extraordinarily low drinking rates when compared to other teleosts (Dall and Milward, 1969), suggesting a low gill (skin?) permeability to water, which might be adaptive when emersed (see above). However, the survival time and rate of evaporative water loss of emersed *Xiphister* does not seem to be different from that of *P. sobrinus* (Tables 1 and 2), whose congener, *P. vulgaris,* has a drinking rate 10 times higher than that of *Xiphister* (Dall and Milward, 1969). Thus, a low drinking rate during immersion (correlated with osmotic permeability) may not necessarily mean a low evaporative water loss during emersion.

Because the salinity of intertidal pools or burrows may be diluted by rainfall or even concentrated by evaporation, one might expect resident fishes to be tolerant of a wide salinity range, i.e., be euryhaline. This seems to be the case (Figure 1), but, as always, the database is quite limited. There are no published investigations of the mechanism of salt uptake by intertidal fishes in hypoosmotic salinities, but it appears that such mechanisms

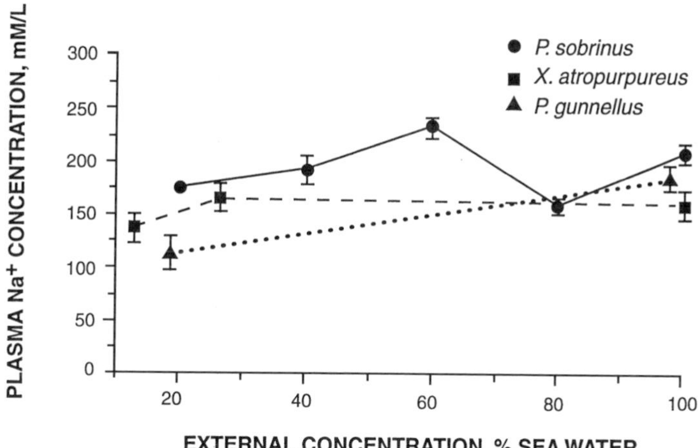


**Figure 1.** Plasma $Na^+$ concentration in three species of intertidal fishes acclimated to various concentrations of seawater. Data are taken from Gordon *et al.* (1965; *Periophthalmus sobrinus*), Evans (1967a; *Xiphister atropurpureus*), and Evans (1969; *Pholis gunnellus*). Each data point is the mean of 5–12 fish ± SE.

may be expressed in the branchial epithelium of even marine fish species and involved in acid–base regulation (see below). Thus, a priori, one may suggest that intertidal species may be capable of ion regulation in lower salinities, produced by dilution of their environment by rainfall.

## III. Acid–Base Regulation

Control of internal pH is also of critical importance to biochemical pathways and cellular function in fishes and other vertebrates. Fluctuations in pH are initially reduced by internal buffers (bicarbonates, phosphates, hemoglobin) and then excess acid or base is thought to be regulated by transfers of $H^+$ and/or $HCO_3^-$ between the fish and the water (see reviews by Heisler, 1986; Claiborne, 1998). The skin and renal contribution to the net transfer of acid–base relevant molecules in marine fishes is negligible and most ion movements are via the gills. Several gill mechanisms for the transport of $H^+$ and $HCO_3^-$ have been postulated. Current models for acid excretion usually include the electroneutral exchange of $Na^+$ in the ambient water for internal $H^+$. The exchange may be driven by an apical $Na^+/H^+$ antiporter of the NHE family (Tse *et al.,* 1993), or by a $H^+$–ATPase [especially in freshwater fishes; reviewed by Lin and Randall (1995)]. $HCO_3^-$ excretion may be functionally linked to the uptake of external $Cl^-$ via a "band-3" (or other electroneutral) $Cl^-/HCO_3^-$ exchange, or a $Cl^-$–$HCO_3^-$ ATPase (Bornancin *et al.,* 1980; Schuster, 1991; Sullivan *et al.,* 1996). Morphological changes in the gills also may play a role in modifying the rate of $HCO_3^-$ excretion by covering or uncovering the specific populations of gill cells containing the exchangers (Goss *et al.,* 1995, 1998).

Interestingly, while these mechanisms generate the loss of $H^+$ or $HCO_3^-$, they also exacerbate the ion load already faced by marine fishes. Nevertheless, a number of seawater species are thought to carry out these transfers (Evans, 1984a; Claiborne and Evans, 1988; Tang *et al.,* 1989; Milligan *et al.,* 1991); this is an indication of the importance of acid–

base regulation. Evans (1984b) has calculated that the additional NaCl uptake may increase the total ion load by about 10% and it is likely that this extra salt is excreted, by the mechanisms described above, after the acid–base disturbance has passed. The linkage between salt and acid–base transfers also implies that alterations of external salinity (e.g., dilution due to freshwater runoff or dehydration and concentration of tidepools) may have an influence on acid–base transfers. For example, recovery from acidosis is normally much faster in marine species than in freshwater-adapted fishes, both in conspecifics (Tang *et al.,* 1989) and among species [see Evans (1982) versus Claiborne and Heisler (1986)]. This is due in part to the availability of ambient NaCl for the ion-exchange mechanisms (see also Heisler, 1986). Exposure of the marine long-horned sculpin (*Myoxocephalus octodecimspinosus*) to dilutions of the water to less than 10% seawater caused an inhibition of normal $H^+$ excretion or even a reversal of $H^+$ excretion to a net base loss (Claiborne *et al.,* 1994). *M. octodecimspinosus* preloaded with acid display a net loss of $H^+$ even if the ambient water is diluted to 20% seawater, but this excretion also stops when the fish are in 4% seawater (Claiborne *et al.,* 1997). Thus, an intertidal fish subjected to extreme tidepool dilution may face not only the obvious osmoregulatory stress, but also inhibition of acid excretion until it is able to return to normal seawater.

It appears that of the several species of freshwater bimodal breathers studied, all exhibit a significant extracellular acidosis driven by retention of plasma $PCO_2$ when air-breathing (Daxboeck *et al.,* 1981; Heisler, 1982; Ishimatsu and Itazawa, 1983). $PCO_2$ in *Synbranchus marmoratus* increases by ~4-fold over several days, and plasma pH drops by more than 0.6 of a unit (Heisler, 1982). Intracellular pH appears to be "protected" at the expense of extracellular pH by transfer of $HCO_3^-$ from the ECF (extracellular fluid) to the ICF (intracellular fluid), so that the ICF pH is nearly completely restored. Although fishes breathing water would normally excrete excess $H^+$ to the medium across the branchial epithelium (and thus increase ECF $[HCO_3^-]$), these air breathers have a much reduced (100-fold) contact time of water with the gill. Thus, the role played by gill ion-exchange mechanisms is very limited, and the ECF pH cannot be compensated by $HCO_3^-$ accumulation (see review by Heisler, 1993).

In contrast to freshwater air breathers, marine amphibious fishes and low-tide-emerged intertidal fishes appear to be capable of maintaining $CO_2$ excretion even when in air for several hours or more (Martin, 1993; see also Martin and Bridges, Chapter 4, this volume). While these fishes may be able to avoid the dramatic respiratory acidosis observed in freshwater bimodal breathers, some respiratory and metabolic acidosis is likely while active and/or out of water. In one of the few studies to measure *in vivo* blood acid–base status in an intertidal species, Pelster *et al.* (1988) found that when *Blennius pholis* was exposed to a 3-h emersion period, a pH drop of ~0.2 unit (to pH 7.49) was induced. Blood $PCO_2$ increased by ~1.1 mm Hg (43%) while plasma $[HCO_3^-]$ decreased by 1.5 mmol $L^{-1}$ (25%). The authors calculated that 88% of the total proton elevation was metabolically derived and the majority of this metabolic load could be accounted for by lactate released to the blood. Several other species also exhibit elevations in whole body or muscle lactate concentrations following bouts of terrestrial activity (see review by Martin, 1995), but this is not always the case (Martin, 1991, 1996; Van Winkle and Martin, 1995).

Currently, there are no data describing the net rate of $H^+$ transfer to the environment in intertidal fishes following air exposure. It is likely that any moderate metabolic and respiratory acidosis developed in an intertidal fish during air emersion could be compensated for upon return to the water via the gill exchange mechanisms described above, in combination with the normally high rate of diffusive $CO_2$ excretion to the water.

Intertidal species have been subjected to selective pressures of their environment more extreme than most marine fishes: small tidepool volumes with fluctuating temperatures and salinities, the respiratory requirement for air emergence, and the cyclical nature of these challenges. It is intriguing to speculate that active intertidal species may have an enhanced ability to quickly regulate acid–base balance that goes beyond the scope of "normal" aquatic fishes. They must be capable of buffering ECF and ICF acid–base changes during the hours seawater is unavailable and branchial $Na^+/H^+$ and $Cl^-/HCO_3^-$ systems cannot be utilized. Upon return to the water, they must then rapidly transfer $H^+$ across the gills to regain normal acid–base homeostasis before the next air emersion occurs. Whether the mechanisms driving this transfer are similar to those of other fishes or have been enhanced to increase the efficiency of pH recovery is unknown, but this is certainly an area that warrants further investigation.

## IV. Nitrogen Excretion

In addition to salt, water, and pH regulation during emersion, intertidal teleost fishes must maintain consistent processing of waste nitrogen. During immersion, they may use the amplified surface area of the gills to eliminate most of their excess nitrogen, as do most aquatic animals. Gills, however, require a large flow of water and are usually ill-suited for use in a terrestrial environment. The mechanisms by which intertidal teleost fishes process and eliminate waste nitrogen during bouts of emersion are not well described.

The literature regarding modes of nitrogen metabolism, nitrogenous waste excretion, and their implications for aquatic lower vetebrates, especially teleosts, has been reviewed both recently and extensively (see Mommsen and Walsh, 1991, 1992; Walsh and Henry, 1991; Wood, 1993; Korsgaard *et al.,* 1995; Walsh, 1997). These reviews discuss ammonia and urea toxicity, metabolism, excretion, and the transition from ammonotelism to ureotelism. Physiological responses of amphibious fishes can be compared to those of the anuran amphibians (e.g., Gordon *et al.,* 1969). Goldstein and Forster (1965) have stated that ammonotelism and ureotelism, as originally defined by Baldwin (1964), can be somewhat misleading since these terms emphasize only the chemical form of excretory nitrogen that predominates and say nothing regarding metabolic origins. Teleosts typically eliminate both ammonia and urea; the relative amounts, however, vary with species and physiological status. This discussion is limited to aspects of nitrogen excretion related to the manner by which intertidal fishes shift between the aquatic and the terrestrial environment. Although numerous studies have examined nitrogen excretion in intertidal fishes, approaches using more current physiological and biochemical techniques are lacking.

Several questions arise regarding the handling of nitrogenous waste during the aquatic to terrestrial transition by intertidal teleosts. First, what are the molecular forms and proportions of nitrogenous waste production? While ammonia and urea predominate, usually they are the only excretory nitrogen forms measured by investigators; other compounds (see below) may be involved. Second, does overall nitrogenous waste production (i.e., metabolism) by semiterrestrial fishes continue at the same rate as that of aquatic fishes, or is it modified by emersion? Third, does nitrogen elimination via ammonia shift to urea elimination (or vice versa) because of a shift in excretory effectors (e.g., gill to skin, kidney/cloaca, etc.) or because of a shift in the end product of metabolism? If the latter, does the shift occur via the ornithine–urea cycle, or are other metabolic pathways involved? Fourth, does nitrogenous waste excretion continue during emersion, or are most if not all of these products stored until reimmersion? If these products are stored, in what form are they stored, in what compartments are they sequestered, and how are potential toxic effects ameliorated? Last, what routes of excretion are used (e.g., gill/head region, gut, skin, kidney), and how do their contributions vary during the transition from the aquatic to the terrestrial environment? This transition likely involves several if not all of these possibilities. Few of these aspects have been thoroughly investigated.

Behavior, habitat, and diet can be expected to affect the ways intertidal fishes produce and handle nitrogenous wastes. Some fishes are forced by the tides to move from an aquatic to a terrestrial environment and back in the course of a day. Some fishes exploit the intertidal habitat (e.g., mudflats) for feeding and predator avoidance. Some mudskippers, like *Boleophthalmus boddaerti,* burrow on the lower regions of the intertidal zone, while *Periophthalmodon schlosseri* occupies higher ground. The former stays in its burrow during high tide, while the latter will swim along shore's edge. *Periophthalmus chrysospilos* rests on land during high and low tides and can be found frequently emersed, clinging to mangrove roots. Branchial structure is correlated with degree of terrestriality (Low *et al.,* 1988). Feeding behavior is frequently different, with some species acting as herbivores (*B. boddaerti*), some as carnivores (*P. schlosseri*), and some as omnivores (*P. chrysospilos;* Low *et al.,* 1988). Herbivorous fishes feeding on algal mats might be expected to face a correspondingly different nitrogen load, compared to carnivores. In addition, arginine, an essential amino acid for teleosts (Halver and Shanks, 1960), can be degraded to urea by the ubiquitous arginase (Gregory, 1977). Availability of dietary arginine may vary for herbivorous and carnivorous fishes. Tissue levels of arginine (and therefore production of urea via arginolysis) would be affected by starvation and diet. Thus, considerations of nitrogen metabolism of intertidal fishes must include habitat, diet, and behavior.

The salinity of the environment also requires consideration. Gordon *et al.* (1965) suggest that nitrogen excretion shifts from ammonotelism in lower salinities to ureotelism in higher salinities. In contrast, Sayer and Davenport (1987a) showed that urea excretion by the blenny, *Blennius pholis,* accounts for a larger percentage of nitrogen excretion in freshwater (52.2%) versus seawater (22.5%) acclimated fishes, while ammonia excretion was not significantly altered. Iwata *et al.* (1981) showed that muscle ammonia and free amino acids (FAA) increase with increasing salinities. Higher salinities limit osmotic water availability, so seawater teleosts in hyperosmotic environments typically produce less urine than freshwater teleosts (see above).

In most teleost fishes, ammonia is the major end product of nitrogen metabolism, arising mainly from catabolism of proteins and nucleic acids (see Wood, 1993; Mommsen and Walsh, 1992; Anderson, 1995). Excess amino acids may be deaminated or transaminated, and amide groups of asparagine or glutamine may be hydrolyzed, producing ammonia at relatively little metabolic cost. The purine nucleotide cycle may be involved as a deamination route (see Wood, 1993). However, Chew and Ip (1987) found that enzymes involved in transdeamination were sufficient to account for ammonia production in *B. boddaerti* and *P. schlosseri,* with aspartate as the major substrate for ammonia production. The purine nucleotide cycle in these fishes was reported to be of no importance (see also Mommsen and Walsh, 1992).

In most aquatic teleosts, excretion of ammonia occurs both in the uncharged form, $NH_3$, which behaves as a respiratory gas, and the charged form, $NH_4^+$, which behaves as an ion (see Mommsen and Walsh, 1992; Wood, 1993). Ammonia, as used here, refers to the sum of $NH_3$ and $NH_4^+$; $NH_3$ is generally considered to be the more toxic form, but $NH_4^+$ exhibits toxicity as well since it is capable of substituting for other ions (e.g., $K^+$ in excitable tissues; Hille, 1973). Both forms are highly soluble in water. Whatever the form, excretion takes advantage of copious water flow over the highly amplified surface area of the gills. Water flow may be unavailable (or greatly reduced) to a semiterrestrial intertidal fish. Determination of the relative amounts, routes, diffusional gradients, and transport or exchange mechanisms for $NH_3$ and $NH_4^+$ used by teleost fishes has stimulated a considerable literature (for review, see Wood, 1993). Suffice it to say that few experiments have specifically addressed intertidal fishes; they likely use the same suite of mechanisms for ammonia excretion available to aquatic teleosts.

Several pathways can serve to generate urea in teleosts, including uricolysis, arginolysis, and less commonly, the ornithine–urea cycle (OUC) (see Mommsen and Walsh, 1991; Anderson and Walsh, 1995). The OUC has been well documented in elasmobranchs and is found in some species of teleosts as well, notably those exhibiting a terrestrial lifestyle (Saha and Ratha, 1990) and some tilapia living in highly alkaline lakes (Randall *et al.*, 1989; Wood *et al.*, 1989, 1994). Evidence also exists for the presence of OUC enzymes in larval rainbow trout, suggesting that the genes are retained in teleosts but expressed in species exposed to environmental extremes, e.g., terrestriality, highly alkaline waters, or during larval development in conditions that could compromise nitrogen excretion as ammonia (Wright *et al.*, 1995). While some teleost fishes, e.g., those living in environmental extremes, possess a complete OUC that can synthesize urea from ammonia (see Mommsen and Walsh, 1992; Anderson, 1995, for reviews), most do not, or they express OUC enzymes at very low titers. The use of appropriate assay conditions is important for the assessment of the OUC and origins of urea. Recent literature (Anderson, 1991) indicates the need to examine tissues for the presence of carbamoylphosphate synthetase III (CPSIII), which requires *N*-acetylglutamate, uses the amide group of glutamine as a nitrogen source, and in fishes leads to urea production via OUC. Gregory (1977) was unable to find a complete set of OUC enzymes in *P. expeditionium, P. gracilis,* or *S. histophorus.* Gregory did detect arginase, a ubiquitous enzyme in fishes (Cvancara, 1969) that is capable of converting arginine to urea, and also ornithine carbamoyltransferase. Thus, there is no evidence that these species possess a complete OUC and are therefore capable of synthesis of urea. Gregory (1977), however, used ammonia rather than glutamine (he did use *N*-acetylglutamate, a positive allosteric effector) in the CPS assay and therefore would not have detected at least low levels of CPSIII, the enzyme form found in teleosts (Anderson, 1995). He calculated that the liver arginase as well as urate oxidase activities were sufficient to account for urea excretion by *P. gracilis* and *P. expeditionium.* The presence of a complete OUC cycle, under appropriate physiological conditions, should be demonstrated (Wood, 1993; Anderson, 1995).

Urea production can occur through purine catabolism via urate oxidase, allantoinase, and allantoicase (Goldstein and Forster, 1965; Gregory, 1977), or arginine via arginase and arginolysis (Gregory, 1977), although degradation of non-excess arginine, an essential amino acid for teleosts (Halver and Shanks, 1960), does not seem logical. Because several metabolic pathways can generate urea, the assumption expressed by some authors that intertidal fishes switch from ammonotelism to ureotelism by synthesis of urea from ammonia during terrestrial sojourns (Gordon *et al.*, 1969; see below) should be accompanied by supporting biochemical evidence concerning the origin of urea.

Urea synthesis (at least via the OUC) may also be involved in acid–base balance as a mechanism for $HCO_3^-$ removal (Atkinson, 1992; Atkinson and Camien, 1982). However, its potential role in both mammals and fishes is controversial (see Meijer, 1995; Walser, 1986). Acid–base balance involves different constraints in aquatic and terrestrial teleost fishes. The former likely achieve little if any benefit from urea synthesis via the OUC (Walsh and Henry, 1991). Acid–base balance appears to be independent of urea synthesis in some ureogenic teleosts (Barber and Walsh, 1993; Wood *et al.*, 1994) while to other species, buffering and $HCO_3^-$ processing may be of some use (Atkinson, 1992; Wood, 1993).

**Table 3.** Ammonia and Urea Excretion by Some Intertidal Teleosts in Seawater (SW) and after Exposure to Air, with Subsequent Return to Seawater (rates are expressed in millimole N $kg^{-1}$ $h^{-1}$)

| | Ammonia | Urea | Reference |
|---|---|---|---|
| *Sicyases sanguineus* | | | |
| In SW | 0.27 | 0.76 | Gordon *et al.*, 1970 |
| 20 h air, 1–4 h SW | 0.21 | 2.62 | Gordon *et al.*, 1970 |
| 36 h air, 1–4 h SW | 0.16 | 8.58 | Gordon *et al.*, 1970 |
| *Periophthalmus sobrinus* | | | |
| 40% SW | 0.77 | 0.12 | Gordon *et al.*, 1965 |
| 100% SW | 0.49 | 0.72 | Gordon *et al.*, 1965 |
| 12 h air, rtn to SW | 1.03 | 2.04 | Gordon *et al.*, 1969 |
| *Periophthalmus cantonensis* | | | |
| 25% SW | 0.32 | 0.033 | Morii *et al.*, 1978 |
| SW, starved 9.5 days | 6.3 | 7.4 | Gordon *et al.*, 1978 |
| 16.5 h air, rtn to SW | 12 | 22 | Gordon *et al.*, 1978 |
| *Boleophthalmus pectinirostris* | | | |
| 25% SW | 0.101 | 0.012 | Morii *et al.*, 1978 |
| *Pholis gunnellus* | | | |
| SW | 0.19 | 0.077 | Kormanik and Evans, 1987 |
| 24 h air, rtn to SW | 0.36 | 0.19 | Kormanik and Evans, 1987 |

Urea (after ammonia) likely accounts for most of the remainder of nitrogen excretion in teleost fishes (Wood, 1993). In some intertidal fishes, urea excretion predominates (Table 3). Urea is highly soluble and less toxic than ammonia, and therefore is more readily stored. Urea, while low in concentration, could also play a role as an osmolyte in teleosts (Griffith, 1991) but the benefits are not yet fully assessed.

Little consideration has been given to other forms of nitrogen excreta, although a study by Morii *et al.* (1978) of the mudskippers *Periophthalmus cantonensis* and *Boleophthalmus pectinirostris* shows that their contribution (e.g., uric acid, creatine, creatinine, trimethylamine, trimethylamine *n*-oxide, and amino acids) is minimal, totaling 3–4% of nitrogen excretion. Gregory (1977) detected no uric acid excretion by either of the mudskippers *Scartelaos histophorus* or *Periophthalmus expeditionium.* Ammonia and trimethylamine are the only forms of nitrogen exhibiting volatility that might enhance gaseous excretion into air, presumably via the skin or branchial cavity.

The gills typically predominate for ammonia and urea excretion in aquatic teleosts (Wood, 1993). Nevertheless, the normal mode of ammonia and urea elimination into the copious flow of water over ventilating gills is minimized when fishes leave the water. Some fishes may sequester water in the branchial chambers (e.g., *Sicyases sanguineus,* the chilean clingfish; Ebeling *et al.,* 1970). Some mudskippers may move water over gills and branchial chambers during skimming/feeding behavior (Low *et al.,* 1988). Thick mucus also covers the gills, and, with proton trapping (Wright *et al.,* 1989) may provide a means by which a local gradient for ammonia excretion as $NH_3$ may be temporarily maintained.

The role of the kidney of intertidal teleosts in nitrogen excretion is seldom directly assessed, but likely is of some importance, at least during immersion (Wood, 1993). Sayer and Davenport (1987b) found that the kidney was as least as important as the gill for ammonia excretion of *B. pholis* in seawater. Terrestriality, with the potential for dehydration, likely reduces an already minimal urine flow in seawater teleosts. Determining true urinary output in fishes is often difficult technically, since cannulation to collect urinary output

**Table 4.** Blood Levels of Ammonia and Urea in Several Species of Intertidal Teleosts in Seawater (SW), and after Emersion (data are expressed in millimole N liter$^{-1}$)

| | Ammonia | Urea | Reference |
|---|---|---|---|
| *Periophthalmus cantonensis* | | | |
| 25% SW | 4.2–4.6 | 7–4 | Morii *et al.*, 1979 |
| 24–48 h emersion | 4.6–5.2 | 4–3 | Morii *et al.*, 1979 |
| *Boleophthalmus pectinirostris* | | | |
| 25% SW | 2.3–2.6 | 1.5 | Morii *et al.*, 1979 |
| 24–48 h emersion | 3.1–4.6 | 1.5–2.3 | Morii *et al.*, 1979 |
| *Sicyases sanguineus* | | | |
| SW | 0.2 | 6.4 | Gordon *et al.*, 1970 |
| *Pholis gunnellus* | | | |
| SW | 0.51 | 16.4 | Kormanik and Evans, 1987 |
| 24 h emersion | 0.70 | 31.4 | Kormanik and Evans, 1987 |

typically eliminates the bladder's contribution to urine modification; a ureteral/bladder urine mix is typically collected with patent cannulation. We have found that bladder urine, collected from anesthetized *Pholis gunnelis,* contains both ammonia and urea (1.1 and 22 m*M* nitrogen, respectively; Kormanik, personal observation) at concentrations somewhat higher than those for blood (Table 4). Intermittent release of urine (micturation) may lead to irregular nitrogen excretion rates for whole animals observed by some investigators but cannot account for variable rates observed in cannulated or excretion-partitioned animals (Davenport and Sayer, 1986; Sayer, 1988). Bursts of urea and ammonia excretion observed in *B. pholis* (Sayer, 1988) may indicate that the fishes can regulate release of these nitrogenous waste products (Korsgaard *et al.,* 1995).

Little is known about the gut and its role in nitrogenous waste excretion. Davenport and Sayer (1986) found that 15% of the urea and 2% of the ammonia excreted by emersed *B. pholis* occurred in the feces, about equal to that eliminated via the urine.

The skin of teleost fishes is usually considered to be relatively impermeable and of little use in general salt transport or nitrogenous waste excretion, yet there are few data to support any firm conclusion. Some experiments designed to partition nitrogenous waste excretion have been performed, but the results are by no means definitive, since most attempts do not effectively separate skin output from cloacal/kidney. Morii *et al.* (1978) attempted to separate the anterior region (including skin/gills) from posterior skin and cloaca of *P. cantonensis* and *B. pectinirostris.* Skin and cloaca were responsible for both ammonia and urea excretion. Nevertheless, emersed fishes eliminate only about 10% of the ammonia excreted by fishes in water. Davenport and Sayer (1986) found that the major route of nitrogen elimination by *B. pholis* during emersion was the body surface (including gills/buccal cavity, head, flanks, and tail). Nearly 90% of the ammonia and 65% of the urea excreted was found in the mucus. Mucus secretions from both skin and gills is enhanced during emersion (Laming *et al.,* 1982). The two sites were not separated in the experiments of Davenport and Sayer (1986).

Several investigators have concluded that emersion causes accumulation of nitrogenous waste products, as measured by increases in blood and tissue ammonia and urea (Table 4). Some fishes simply store ammonia and urea produced during emersion (Figure 2). Kormanik *et al.* (unpublished) found that total body pools of ammonia and urea doubled after 24 h of emersion of *P. gunnellus,* yet while blood urea doubled, blood

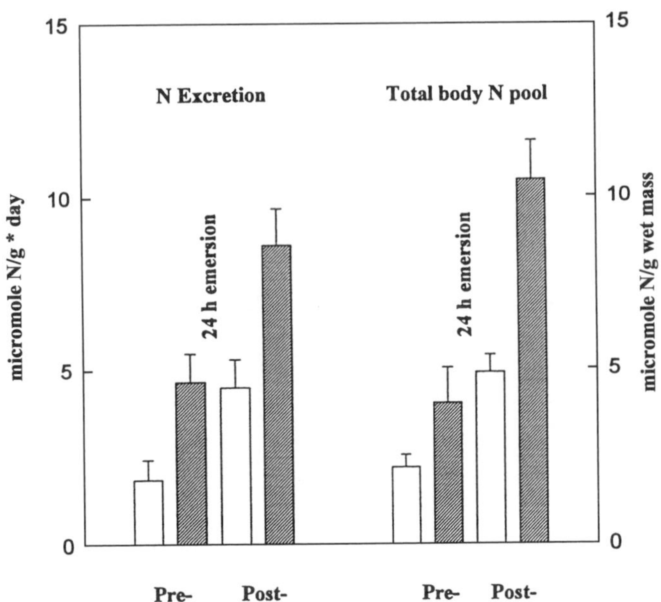


**Figure 2.** Ammonia (filled bars) and urea (open bars) excretion and total body pools in *Pholis gunnellus* in pre- and post-24-h emersion fish, showing that total body ammonia and urea double after 24 h of emersion and that the excretion rate over the 24-h period following reimmersion doubles, which can account for all of the ammonia and urea accumulated during emersion (Kormanik *et al.*, unpublished).

ammonia did not increase significantly (Table 4) and is therefore not a good indicator of total body ammonia load. The usual caveats to measuring this labile compound apply (see Wood, 1993). Blood ammonia levels found by Morii *et al.* (1979) for *P. cantonensis* and *B. pectinirostris* were high in fishes both in and out of water (2–7 m*M*). In contrast, Gordon *et al.* (1970) suggest that *S. sanguineus* does not accumulate ammonia in body fluids while emersed, since urea but not ammonia excretion is enhanced following reimmersion. In *B. pectinirostris,* blood ammonia and urea were higher for emersed fishes than for fishes in water (see Table 4); concentrations increased with increased emersion (Morii *et al.*, 1979). In emersed *P. cantonensis,* however, the opposite was observed with blood ammonia similar and urea slightly lower compared to that of fishes in water (Morii *et al.*, 1979). Urea decreased with longer emersion times. It is difficult to derive clear trends from these data. Most of the accumulated ammonia was eliminated within the first 6 h of reimmersion into water (Morii *et al.,* 1979). While not all of the ammonia accumulated appeared to be eliminated, the authors have no evidence that any of the ammonia was converted to urea during emersion. Iwata (1988) found that urea levels did not increase with ammonia loading in *P. cantonensis* and suggests that this mudskipper does not modify its metabolism toward ureotelism and that urea production is not used to detoxify ammonia during emersion. Iwata *et al.* (1981) and Iwata (1988) also found that nonessential FAA increase with emersion and therefore could contribute to ammonia detoxification as well as play a role in osmoregulation, since FAA concentrations also increased with increasing salinities.

Iwata *et al.* (1981) and Iwata (1988) found that muscle levels of ammonia increased substantially with emersion of *P. cantonensis,* approaching 20 m*M*/kg tissue water after 72 h, while urea increased to a lesser extent, if at all. Wilkie and Wood (1995) have demonstrated (by manipulation of pH and ammonia gradients) that white muscle can serve as an ammonia reservoir during periods of reduced excretion in *O. mykiss,* which is followed by a "rapid washout" upon return to normal conditions. Thus small changes in extracellular fluid ammonia concentrations may accompany large changes in intracellular fluid ammonia stores (Wilkie and Wood, 1995).

If the high values reported for blood ammonia in some species are accurate (Table 4), then several intertidal teleost fishes tolerate concentrations of ammonia that are lethal to other marine fishes (see Korsgaard *et al.,* 1995). These high blood levels should be confirmed, since most aquatic teleosts typically have lower values (ca. 200–600 $\mu M$; see Wood, 1993) and true circulating values for ammonia are difficult to measure (Wood, 1993). Nevertheless, blood and tissue levels may be substantial and indicate a far greater ability of fishes to tolerate ammonia compared to mammals (Korsgaard *et al.,* 1995). Few investigators have directly measured nitrogen excretion during emersion, or were able to separate effectively the contribution of the branchial/gill region from that of the skin, gut/cloaca, and kidney. Morii *et al.* (1978) reported that both *P. cantonensis* and *B. pectinirostris* were able to excrete nitrogen-containing compounds while emersed, albeit at reduced rates compared to fishes in water.

Nitrogen accumulation during emersion should result in enhanced excretion rates after reimmersion into seawater following air exposure. Morii *et al.* (1978) found that ammonia excretion into water by *P. cantonensis* and *B. pectinirostris* increased following an exposure to air (emersion) but urea excretion did not increase. They concluded that emersion causes an accumulation of nitrogen. In *P. gunnellus,* an equivalent amount of the ammonia and urea normally produced during immersion appears to be stored during a bout of emersion, to be excreted upon subsequent reimmersion, at rates substantially elevated over those of preemersion fishes (Figure 2; Kormanik *et al.,* unpublished; Kormanik and Evans, 1987). Gordon *et al.* (1969) found that rates for ammonia and urea excretion into water by *P. sobrinus* more than doubled, compared to controls, for fishes replaced into water after a 12-h exposure to air (see Table 3).

Based on the considerably elevated rate of urea excretion observed when fish were returned to water, Gordon *et al.* (1970) concluded that most if not all of nitrogenous waste produced by *S. sanguineus* during emersion is retained, and waste nitrogen production during emersion was shifted toward urea. Of note are the extremely high total body excretion rates for urea observed in *S. sanguineus* after emersion (Gordon *et al.,* 1970), and for both ammonia and urea in *P. cantonensis,* both before and after emersion (Gordon *et al.,* 1978), which exceed values measured by other investigators, in the same species, by one to two orders of magnitude (Table 3). If these high rates are accurate, they indicate a prodigious ability of some of these fishes to produce, store, and excrete nitrogenous waste.

It is possible that emersion may result in a lowered metabolic rate (i.e., lowered measured oxygen consumption during emersion; Tamura *et al.,* 1976), with the potential for reduced nitrogen production. However, Gordon *et al.* (1969) found no significant change in metabolic rate of *P. sobrinus* measured as oxygen consumption in fish immersed and emersed for periods up to 8.5 h. Gordon *et al.* (1969) calculated that the rate of ammonia and urea production increased 2- and 3.5-fold, respectively, during air exposure and that the ammonia : urea ratio shifted toward urea production during the period out of water.

## V. Summary and Future Directions

It is clear that intertidal fishes are presented with unique problems in osmoregulation, acid–base regulation, and nitrogen excretion, but homeostatic strategies have evolved that utilize physiological control systems expressed in their aquatic predecessors. Much is yet to be learned about these strategies. For instance, it is intuitively obvious that intertidal species must maintain relatively low evaporative water permeabilities, and extant data support this hypothesis, but few comparative data exist, especially of more aquatic species. Is there a major modification from their aquatic ancestors, or merely a slight decline in water permeability? Buccal water retention appears to be adaptive, but does it exacerbate osmoregulatory problems because the solution is hypertonic to the plasma? Does the urinary bladder provide a reservoir for fluids in intertidal fishes as it does in amphibians? Is euryhalinity a requisite for intertidal life? Since recovery from acid–base disturbances requires branchial extrusion mechanisms that may be coupled with ions in the medium, are intertidal fishes limited in their responses to such disturbances in dilute salinities? During "dry" periods, is there a respiratory acidosis in intertidal fishes? Do most intertidal species face high blood lactate levels? Does the intertidal lifestyle demand major intracellular and plasma buffering, and relatively rapid onset of acid extrusion mechanisms when the animal is immersed? Does emersion and reimmersion dictate substantial changes in the metabolism of nitrogen or pathway of its excretion? Do intertidal species exhibit larger tolerances to either ammonia or urea than their more aquatic ancestors? Is there a relation between herbivory or salinity and nitrogen metabolism? What role does mucus production play? Do intertidal species have a reduced metabolic rate to decrease nitrogen production? It is hoped that such questions will prompt the interest of readers who may help to increase our knowledge of the physiology of these interesting organisms.

## Acknowledgments

The writing of this chapter was supported by the following grants from the National Science Foundation: IBN-9306997, IBN-9604824 (D.H.E.), IBN-9419849 (J.B.C.), and IBN-9507456 (G.A.K.).

## References

Alvarado, R. H. (1979). Amphibians. *In* "Comparative Physiology of Osmoregulation in Animals" (G. M. O. Maloiy, Ed.), Vol. 1, pp. 261–303. Academic Press, London.

Anderson, P. M. (1991). Glutamine-dependent urea synthesis in elasmobranch fishes. *Biochem. Cell Biol.* **69,** 317–319.

Anderson, P. M. (1995). Urea cycle in fish: Molecular and mitochondrial studies. *In* "Cellular and Molecular Approaches to Fish Ionic Regulation" (C. M. Wood and T. J. Shuttleworth, Eds.), Vol. 14, pp. 57–84. Academic Press, San Diego.

Anderson, P. M., and Walsh, P. J. (1995). Subcellular localization and biochemical properties of the enzymes of carbamoyl phosphate and urea synthesis in the batrachoidid fishes *Opsanus beta, Opsanus tau* and *Porichthys notatus. J. Exp. Biol.* **198,** 755–766.

Atkinson, D. D., and Camien, M. N. (1982). The role of urea synthesis in the removal of metabolic bicarbonate and the regulation of blood pH. *Curr. Top. Cell. Regul.* **21.**

Atkinson, D. E. (1992). Functional roles of urea synthesis in vertebrates. *Physiol. Zool.* **65,** 243–267.

Baldwin, E. (1964). "An Introduction to Comparative Biochemistry." Cambridge University Press, London.

Barber, M. L., and Walsh, P. J. (1993). Interactions of acid-base status and nitrogen excretion and metabolism in the ureogenic teleost *Opsanus beta. J. Exp. Biol.* **185,** 87–105.

Bornancin, M., De Renzis, G., and Naon, R. (1980). $Cl^-$-$HCO_3^-$-ATPase in gills of the rainbow trout: Evidence for its microsomal localization. *Am. J. Physiol.* **238,** R251–R259.

Chew, S. F., and Ip, Y. K. (1987). Ammoniogenesis in mudskippers *Boleophthalmus boddaerti* and *Periophthalmus schlosseri. Comp. Biochem. Physiol. B* **87,** 941–948.

Claiborne, J. B. (1997). Acid-base regulation. *In* "The Physiology of Fishes" (D. H. Evans, Ed.), 2nd ed., pp. 177–198. CRC Press, Boca Raton, FL.

Claiborne, J. B., and Evans, D. H. (1988). Ammonia and acid-base balance during high ammonia exposure in a marine teleost (*Myoxocephalus octodecimspinosus*). *J. Exp. Biol.* **140,** 89–105.

Claiborne, J. B., and Heisler, N. (1986). Acid-base regulation and ion transfers in the carp (*Cyprinus carpio*): pH compensation during graded long- and short-term environmental hypercapnia, and the effect of bicarbonate influsion. *J. Exp. Biol.* **126,** 41–62.

Claiborne, J. B., Perry, E., Bellows, S., and Campbell, J. (1997). The mechanisms of acid excretion across the gills of a marine fish. *J. Exp. Zool.* **279,** 509–520.

Claiborne, J. B., Walton, J. S., and Compton-McCullough, D. (1994). Acid-base regulation, branchial transfers and renal output in a marine teleost fish (the long-horned sculpin *Myoxocephalus octodecimspinosus*) during exposure to low salinities. *J.Exp. Biol.* **193,** 79–95.

Cvancara, V. A. (1969). Studies on tissue arginase and ureogenesis in freshwater teleosts. *Comp. Biochem. Physiol.* **30,** 489–496.

Dall, W., and Milward, N. E. (1969). Water intake, gut absorption and sodium fluxes in amphibious and aquatic fishes. *Comp. Biochem. Physiol.* **30,** 247–260.

Davenport, J., and Sayer, M. D. J. (1986). Ammonia and urea excretion in the amphibious teleost, *Blennius pholis* (L) in seawater and in air. *Comp. Biochem. Physiol.* **84,** 189–194.

Daxboeck, C., Barnard, D. K., and Randall, D. J. (1981). Functional morphology of the gills of the bowfin, *Amia calva* L. with special reference to their significance during air exposure. *Respir. Physiol.* **43,** 349–364.

Ebeling, A. W., Bernal, P., and Zuleta, A. (1970). Emersion of the amphibious Chilean clingfish, *Sicyases sanguineus. Biol. Bull.* **139,** 115–137.

Evans, D. H. (1967a). Sodium, chloride and water balance in the intertidal teleosts, *Xiphister atropurpureus.* I. Regulation of plasma concentration and body water content. *J. Exp. Biol.* **47,** 513–517.

Evans, D. H. (1967b). Sodium, chloride and water balance in the intertidal teleosts, *Xiphister atropurpureus.* II. The role of the kidney and the gut. *J. Exp. Biol.* **47,** 519–24.

Evans, D. H. (1969). Sodium, chloride and water balance of the intertidal teleost, *Pholis gunnelis. J. Exp. Biol.* **50,** 179–190.

Evans, D. H. (1979). Fish. *In* "Comparative Physiology of Osmoregulation in Animals" (G. M. O. Maloiy, Ed.), Vol. 1, pp. 305–390. Academic Press, Orlando, FL.

Evans, D. H. (1982). Mechanisms of acid extrusion by two marine fishes: The teleost, *Opsanus beta,* and the elasmobranch, *Squalus acanthias. J. Exp. Biol.* **97,** 289–299.

Evans, D. H. (1984a). Gill Na/H and $Cl/HCO_3$ exchange systems evolved before the vertebrates entered fresh water. *J. Exp. Biol.* **113,** 464–470.

Evans, D. H. (1984b). The roles of gill permeability and transport mechanisms in euryhalinity. *In* "Fish Physiology" (W. S. Hoar and D. J. Randall, Eds.), Vol. X, Part B, pp. 239–283. Academic Press, Orlando, FL.

Evans, D. H. (1993). Osmotic and ionic regulation. *In* "The Physiology of Fishes" (D. H. Evans, Ed.), pp. 315–341. CRC Press, Boca Raton, FL.

Evans, D. H. (1995). The roles of natriuretic peptide hormones (NPs) in fish osmoregulation and hemodynamics. *In* "Advances in Environmental and Comparative Physiology—Mechanisms of Systemic Regulation in Lower Vertebrates. II. Acid-Base Regulation, Ion Transfer and Metabolism" (N. Heisler, Ed.), pp. 119–152. Springer-Verlag, Heidelberg.

Goldstein, L., and Forster, R. P. (1965). The role of uricolysis in the production of urea by fishes and other aquatic vertebrates. *Comp. Biochem. Physiol.* **14,** 567–576.

Gordon, M. S., Boetius, J., Boetius, I., Evans, D. H., McCarthy, R., and Oglesby, L. C. (1965). Salinity adaptation in the mudskipper fish *Periophthalmus sobrinus. Hvalradets Skrifter* **48,** 85–93.

Gordon, M. S., Boetius, I., Evans, D. H., McCarthy, R., and Oglesby, L. C. (1969). Aspects of the physiology of terrestrial life in amphibious fishes. I. The mudskipper, *Periophthalmus sobrinus. J. Exp. Biol.* **50,** 141–149.

Gordon, M. A., Boetius, J., Evans, D. H., and Oglesby, L. C. (1968). Additional observations on the natural history of the mudskipper, *Periophthalmus sobrinus. Copeia* **1968,** 853–857.

Gordon, M. S., Fischer, S., and Tarifeno, E. (1970). Aspects of the physiology of terrestrial life in amphibious fishes. II. The Chilean clingfish, *Sicyases sanguineus. J. Exp. Biol.* **53,** 559–572.

Gordon, M. S., Wilson, W. N., and Alice, Y. Y. (1978). Aspects of the physiology of terrestrial life in amphibious fishes. III. The Chinese mudskipper, *Periophthalmus cantonensis. J. Exp. Biol.* **72,** 57.

Goss, G., Perry, S., Fryer, J., and Laurent, P. (1998). Gill morphology and acid-base regulation in freshwater fishes. *Comp. Biochem. Physiol.* **119A,** 107–115.

Goss, G., Perry, S., and Laurent, P. (1995). Ultrastructural and morphometric studies on ion and acid-base transport processes in freshwater fish. *In* "Cellular and Molecular Approaches to Fish Ionic Regulation" (C. M. Wood and T. J. Shuttleworth, Eds.), Vol. 14, pp. 257–284. Academic Press, San Diego.

Graham, J. B. (1973). Terrestrial life of the amphibious fish *Mnierpes macrocephalus. Mar. Biol.* **23,** 83–91.

Graham, J. B., Jones, C. B., and Rubinoff, I. (1985). Behavioural, physiological, and ecological aspects of the amphibious life of the pearl blenny *Entomacrodus nigricans* Gill. *J. Exp. Mar. Biol. Ecol.* **89,** 255–268.

Gregory, R. B. (1977). Synthesis and total excretions of waste nitrogen by fish of the Periophthalmus (mudskipper) and Scartelaos families. *Comp. Biochem. Physiol. A* **57,** 33–36.

Griffith, D. (1991). Guppies, toadfish, lungfish, coelocanths and frogs: A scenario for the evolution of urea retention in fishes. *Envron. Biol. Fish.* **32,** 199–218.

Halver, J. E., and Shanks, W. E. (1960). Nutrition of salmonoid fishes. VII. Indispensible amino acids for sockeye salmon. *J. Nutr.* **72,** 340.

Heisler, N. (1982). Intracellular and extracellular acid-base regulation in the tropical fresh-water teleost fish *Synbranchus marmoratus* in response to the transition from water breathing to air breathing. *J. Exp. Biol.* **99,** 9–28.

Heisler, N. (1986). Acid-base regulation in fishes. *In* "Acid-Base Regulation in Animals" (N. Heisler, Ed.), pp. 309–356. Elsevier, Amsterdam.

Heisler, N. (1993). Acid-base regulation. *In* "The Physiology of Fishes" (D. H. Evans, Ed.), pp. 343–378. CRC Press, Boca Raton, FL.

Hille, B. (1973). Potassium channels in myelinated nerve. *J. Gen. Physiol.* **61,** 669–682.

Horn, M. H., and Riegle, K. C. (1981). Evaporative water loss and intertidal vertical distribution in relation to body size and morphology of stichaeoid fishes from California. *J. Exp. Mar. Biol. Ecol.* **50,** 273–288.

Ishimatsu, A., and Itazawa, Y. (1983). Blood oxygen levels and acid-base status flollowing air exposure in an air-breathing fish, *Channa argus:* The role of air ventilation. *Comp. Biochem. Physiol. A* **74,** 787–793.

Iwata, K. (1988). Nitrogen metabolism in the mudskipper, *Periophthalmus cantonensis:* Changes in free amino acids and related compounds in various tissues under conditions of ammonia loading, with special reference to its high ammonia tolerance. *Comp. Biochem. Physiol. A* **91,** 499–508.

Iwata, K., Kakuta, I., Ikeda, M., Kimoto, S., and Wada, N. (1981). Nitrogen metabolism in the mudskipper, *Periophthalmus cantonensis:* A role of free amino acids in detoxification of ammonia produced during its terrestrial life. *Comp. Biochem. Physiol. A* **68,** 589.

Kormanik, G. A., and Evans, D. H. (1987). Nitrogenous waste excretion in the intertidal rock gunnel (*Pholis gunnellus* L.): Effects of emersion. *Bull. Mt. Desert Isl. Biol. Lab.* **27,** 33–35.

Korsgaard, B., Mommsen, T. P., and Wright, P. A. (1995). Nitrogen excretion in teleostean fish: Adaptive relationships to environment, ontogenesis, and viviparity. *In* "Nitrogen Metabolism and Excretion" (P. J. Walsh and P. Wright, Eds.), pp. 259–287. CRC Press, Boca Raton, FL.

Laming, P. R., Funston, C. W., Roberts, D., and Armstrong, M. J. (1982). Behavioural, physiological and morphological adaptations of the shanny (*Blennius pholis*) to the intertidal habitat. *J. Mar. Biol. Assoc. U.K.* **62,** 329–338.

Lin, H., and Randall, D. J. (1995). Proton pumps in fish gills. *In* "Cellular and Molecular Approaches to Fish Ionic Regulation" (C. M. Wood and T. J. Shuttleworth, Eds.), Vol. 14, pp. 229–255. Academic Press, San Diego.

Low, W. P., Lane, D. J. W., and Ip, Y. K. (1988). A comparative study of terrestrial adaptations of the gills in three mudskippers—*Periophthalmus chrysospilos, Boleophthalmus boddaerti,* and *Periophthalmus schlosseri. Biol. Bull.* **175,** 434–438.

Marshall, W. S. (1977). Transepithelial potential and short-circuit current across the isolated skin of *Gillichthys mirabilis* (Teleostei: Gobiidae) acclimated to 5% and 100% seawater. *J. Comp. Physiol.* **114,** 157–165.

Marshall, W. S. (1995). Transport processes in isolated teleost epithelia: Opercular epithelium and urinary bladder. *In* "Cellular and Molecular Approaches to Fish Ionic Regulation" (C. M. Wood and T. J. Shuttleworth, Eds.), Vol. 14, pp. 1–23. Academic Press, San Diego.

Martin, K. (1991). Facultative aerial respiration in an intertidal sculpin, *Clinocottus analis* (Scorpaeniformes: Cottidae). *Physiol. Zool.* **64,** 1341–1355.

Martin, K. L. M. (1993). Aerial release of $CO_2$ and respiratory ratio in intertidal fishes out of water. *Environ. Biol. Fish.* **37,** 189–196.

Martin, K. L. M. (1995). Time and tide wait for no fish—Intertidal fishes out of water. *Environ. Biol. Fishes* **44,** 165–181.

Martin, K. L. M. (1996). An ecological gradient in air-breathing ability among marine cottid fishes. *Physiol. Zool.* **69,** 1096–1113.

Meijer, A. J. (1995). Urea synthesis in mammals. *In* "Nitrogen Metabolism and Excretion" (P. J. Walsh and P. Wright, Eds.), pp. 193–204. CRC Press, Boca Raton, FL.

Milligan, C. L., McDonald, D. G., and Prior, T. (1991). Branchial acid and ammonia fluxes in response to alkalosis and acidosis in two marine teleosts: Coho salmon (*Oncorhynchus kisutch*) and starry flounder (*Platichthys stellatus*). *Physiol. Zool.* **64,** 169–192.

Mommsen, T. P., and Walsh, P. J. (1991). Urea synthesis in fishes: Evolutionary and biochemical perspectives. *In* "Biochemical and Molecular Biology of Fishes," Vol. 1, "Phylogenetic and Biochemical Perspectives" (P. W. Hochachka and T. P. Mommsen, Eds.), pp. 137–163. Elsevier, New York.

Mommsen, T. P., and Walsh, P. J. (1992). Biochemical and environmental perspectives on nitrogen metabolism in fishes. *Experientia* **48,** 583–593.

Morii, H., Nishikata, K., and Tamura, O. (1978). Nitrogen excretion of mudskipper fish *Periophthalmus cantonensis* and *Beleophthalmus pectinirostris* in water and on land. *Comp. Biochem. Physiol. A* **60,** 189–193.

Morii, H., Nishikata, K., and Tamura, O. (1979). Ammonia and urea excretion from mudskipper fishes *Periophthalmus cantonensis* and *Beleophthalmus pectinirostris* transferred from land to water. *Comp. Biochem. Physiol. A* **63,** 23–28.

Nonnotte, G., and Kirsch, R. (1978). Cutaneous respiration in seven sea-water teleosts. *Respir. Physiol.* **35,** 111–118.

Pelster, B., Bridges, C. R., and Grieshaber, M. K. (1988). Physiological adaptations of the intertidal rockpool teleost *Blennius pholis* L., to aerial exposure. *Respir. Physiol.* **71,** 355–374.

Randall, D. J., Wood, C. M., Perry, S. F., Bergman, H., Maloiy, G. M. O., Mommsen, T. P., and Wright, P. A. (1989). Ureotelism in a completely aquatic teleost fish: A strategy for survival in an extremely alkaline environment. *Nature* **337,** 165–166.

Saha, N., and Ratha, B. K. (1990). Alterations in excretion pattern of ammonia and urea in a freshwater air-breathing teleost, *Heteropneustes fossilis* (Bloch) during hyper-ammonia stress. *Indian J. Exp. Biol.* **28,** 597–599.

Sayer, M. D. J. (1988). An investigation into the pattern of nitrogen excretion in *Blennius pholis. Comp. Biochem. Physiol. A* **89,** 359.

Sayer, M. D. J., and Davenport, J. (1987a). Ammonia and urea excretion in the amphibious teleost *Blennius pholis* exposed to fluctuating salinity and pH. *Comp. Biochem. Physiol. A* **87,** 851.

Sayer, M. D. J., and Davenport, J. (1987b). The relative importance of the gills to ammonia and urea excretion in five seawater and one freshwater teleost species. *J. Fish Biol.* **31,** 561.

Schuster, V. L. (1991). Cortical collecting duct bicarbonate secretion. *Kidney Int.* **33,** S47–S50.

Stebbins, R. C., and Kalk, M. (1961). Observations on the natural history of the mudskipper, *Periophthalmus sobrinus. Copeia* **1961,** 18–27.

Sullivan, G. V., Fryer, J. N., and Perry, S. F. (1996). Localization of mRNA for proton pump ($H^+$-ATPase) and $Cl^-/HCO_3^-$ exchanger in rainbow trout gill. *Can. J. Zool.* **74,** 2095–2103.

Tamura, S. O., Morii, H., and Yuzuriha, M. (1976). Respiration of the amphibious fishes *Periophthalmus cantonensis* and *Boleophthalmus chinensis* in water and on land. *J. Exp. Biol.* **65,** 97–107.

Tang, Y., McDonald, D. G., and Boutilier, R. G. (1989). Acid-base regulation following exhaustive exercise: A comparison between freshwater- and seawater-adapted rainbow trout (*Salmo gairdneri*). *J. Exp. Biol.* **141,** 407–418.

Tse, M., Levine, S., Yun, C., Brant, S., Counillon, L. T., Pouyssegur, J., and Donowitz, M. (1993). Structure/function studies of the epithelial isoforms of the mammalian $Na^+/H^+$ exchanger gene family. *J. Membr. Biol.* **135,** 93–108.

Van Winkle, R. C., and Martin, K. L. M. (1995). A field study of anaerobiosis during terrestrial spawning in the grunion. *Am. Zool.* **35,** 98A.

Walser, M. (1986). Roles of urea production, ammonium excretion and amino acid oxidation in acid-base balance. *Am. J. Physiol.* **250,** F181–F188.

Walsh, P. J. (1997). Nitrogen excretion and metabolism. *In* "The Physiology of Fishes" (D. H. Evans, Ed.), 2nd ed., pp. 199–214. CRC Press, Boca Raton, FL.

Walsh, P. J., and Henry, R. P. (1991). Carbon dioxide and ammonia metabolism and exchange. *In* "Biochemistry and Molecular Biology of Fishes" (P. W. Hochachka and T. P. Mommsen, Eds.), Vol. 1, p. 181. Elsevier, New York.

Wilkie, M. P., and Wood, C. M. (1995). Recovery from high pH exposure in the rainbow trout: White muscle ammonia storage, ammonia washout, and the restoration of blood chemistry. *Physiol. Zool.* **68,** 379–401.

Wood, C. M. (1993). Ammonia and urea metabolism and excretion. *In* "The Physiology of Fishes" (D. H. Evans, Ed.), pp. 379–425. CRC Press, Boca Raton, FL.

Wood, C. M., Bergman, H. L., Laurent, P., Maina, J. N., Narahara, A., and Walsh, P. J. (1994). Urea production, acid-base regulation and their interactions in the Lake Magadi tilapia, a unique teleost adapted to a highly alkaline environment. *J. Exp. Biol.* **189,** 13–36.

Wood, C. M., Perry, S. F., Wright, P. A., Bergman, H. L., and Randall, D. J. (1989). Ammonia and urea dynamics in the Lake Magadi tilapia, a ureotelic teleost fish adapted to an extremely alkaline environment. *Respir. Physiol.* **77,** 1–20.

Wright, P. A., Felskie, A., and Anderson, P. M. (1995). Induction of ornithine-urea cycle enzymes and nitrogen metabolism and excretion in rainbow trout (*Oncorhynchus mykiss*) during early life stages. *J. Exp. Biol.* **198,** 127–135.

Wright, P. A., Randall, D. J., and Perry, S. F. (1989). Fish gill water boundary layer: A site of linkage between carbon dioxide and ammonia excretion. *J. Comp. Physiol. B* **158,** 627–635.

Yokota, S., Iwata, K., Fujii, Y., and Ando, M. (1997). Ion transport across the skin of the mudskipper *Periophthalmus modestus. Comp. Biochem. Physiol.* **118A,** 903–910.

Yokoya, S., and Tamura, O. S. (1992). Fine structure of the skin of the amphibious fishes, *Boleophthalmus pectinirostris* and *Periophthalmus cantonensis,* with special reference to the location of blood vessels. *J. Morphol.* **214,** 287–297.

# 6

# Movement and Homing in Intertidal Fishes

R. N. Gibson

*Centre for Coastal and Marine Science, Dunstaffnage Marine Laboratory, Oban, Argyll, Scotland*

## I. Introduction

Marine animals living in the intertidal zone are subject to environmental conditions that change at a variety of time scales. Motile species, such as fishes, can respond to these fluctuating conditions in two ways. Some may seek shelter from adverse conditions during emersion and remain within the intertidal zone. Others leave the zone completely until more favorable conditions return or, conversely, only enter it when it is submerged. Whichever strategy is employed, the changing conditions impose a periodicity on an individual's movements such that it regularly changes its location. The frequency, timing, and extent of locational change depend on the nature of the cycle to which it is responding and its current physiological needs. At one end of the scale are relatively gradual but ultimately extensive shifts in position related to seasonal changes in physical factors such as temperature, salinity, or turbulence. At the other are local, short-term, tidally related foraging excursions. This chapter describes the various movement patterns that have been observed for fishes in different types of intertidal environments. It concentrates on tidally related migrations and the evidence for the existence of homing but also considers the functions of these movements and the mechanisms involved in their control. Earlier reviews of the subject are given by Gibson (1969, 1982, 1988, 1992).

## II. Spatial and Temporal Scales of Intertidal Movement

The spatial scales over which fishes move in the course of their intertidal movements are dependent on several factors. The first is their residential status. Intertidal fishes can be broadly categorized as either residents or visitors, depending on whether they spend most of their life in the intertidal zone or only visit it on a regular basis. Further discussion of residential status is given elsewhere (Gibson, 1982, 1988; Gibson and Yoshiyama, Chapter 13, this volume). Resident intertidal species, by definition, remain within the intertidal zone and tend to be restricted in their movements. Consequently they do not move

*Intertidal Fishes: Life in Two Worlds*

great distances. In contrast, visitors regularly enter and leave the intertidal zone and in so doing may cover considerable distances, sometimes several hundred meters in each tide (Kuipers, 1973; Wolff *et al.,* 1981; Wright *et al.,* 1990). A second factor determining scale of movement is body size. Resident intertidal species are generally small, benthic forms with limited powers of locomotion whereas visitors are frequently, but not always, larger pelagic or demersal species with good locomotory abilities and hence capable of covering greater distances more rapidly. The third factor, behavior, is related to the first two. Resident species, in addition to being small, are often cryptic, usually thigmotactic and sometimes territorial in their behavior. All these behavioral traits tend to restrict movement to small areas. The visitors, on the other hand, possess the opposite behavioral tendencies and forage widely over the bottom or in open water.

The time scales over which intertidal fishes move can also be considered with respect to their residential status. Resident species are generally active at a particular state of the tide, although these tidally phased movements can be modulated by the diel cycle so that some species may only be active, for example, on high tides that occur during the night. Visitors present a more complicated picture because, although their movements are also basically of tidal frequency, they are modulated by a wider range of cycles of lower frequency. Individuals may migrate intertidally on each tide, on every other tide depending on whether they are diurnal or nocturnal, or only on day or night spring tides. They may enter the zone as juveniles in spring or summer, stay there for several months, during which time they are tidally active, and then leave when conditions become unsuitable in winter or as they grow and mature. Examples of all these variations in patterns of movement are described in the next section.

## III. Intertidal Movements in Different Habitats

### A. Rocky Shores

The complex nature of rocky shores means that a wide range of refuges are available in the form of tidepools, rock crevices, and spaces beneath boulders and algae where suitably adapted fishes can shelter over the low-tide period. Consequently, intertidal fish communities on rocky shores contain a high proportion of resident species. The accessibility of many of these species at low tide has enabled their movement patterns to be studied in considerable detail both directly and indirectly, although direct observation of small, cryptic fishes in the often turbulent waters of the intertidal zone can be difficult, especially at high tide. Nevertheless, it has been attempted on several occasions. For example, the almost complete upshore shift of Californian populations of the opaleye (*Girella nigricans*) from the tidepools in which they are found at low tide has been observed by Williams (1957) and Norris (1963). Williams (1957) was also able to watch and describe similar movements made by the sculpin *Clinocottus analis.* With the advance of the tide the smallest fish move out first and are then followed by larger individuals, many moving "in excess of ten meters." The related *Oligocottus maculosus* has also been seen to move out of tidepools on the rising tide and, here too, the smallest individuals tend to travel furthest (Green, 1971a). The extent of movement, always less than 10 m, and the number of fish leaving pools is strongly influenced by the degree of turbulence. In sheltered areas all fish leave the pools at high tide whereas this is not the case in exposed areas. When the water is very turbulent, activity may even be restricted to the low-tide period (Green, 1971b). Turbulence also alters the behavior of small *C. analis.* In calm weather they swim from the top of one boulder to another but in rougher conditions limit their activity to crevices and the bases of the boulders (Williams, 1957). Such observations can mostly be made for relatively short periods of time only during the day, although *O. maculosus* has been observed emerging from cover at night (Green, 1971a) and Fishelson (1963) describes the nocturnal upshore movement of the blenny *Blennius pavo* on shores in the Mediterranean. To record *in situ* movements on a longer time scale other methods are necessary. Unfortunately, most resident intertidal fishes are too small to allow the attachment of electronic tagging devices that would enable their movements to be studied remotely. There is one species, however, where this technique has been possible. The monkeyface prickleback (*Cebidichthys violaceus*) reaches a maximum length of ~70 cm and individuals >20 cm are common intertidally. Attachment of ultrasonic transmitters to six individuals demonstrated that they were mainly active on the rising tide and moved within very restricted areas of generally less than 1 $m^2$ (Ralston and Horn, 1986; Figure 1).

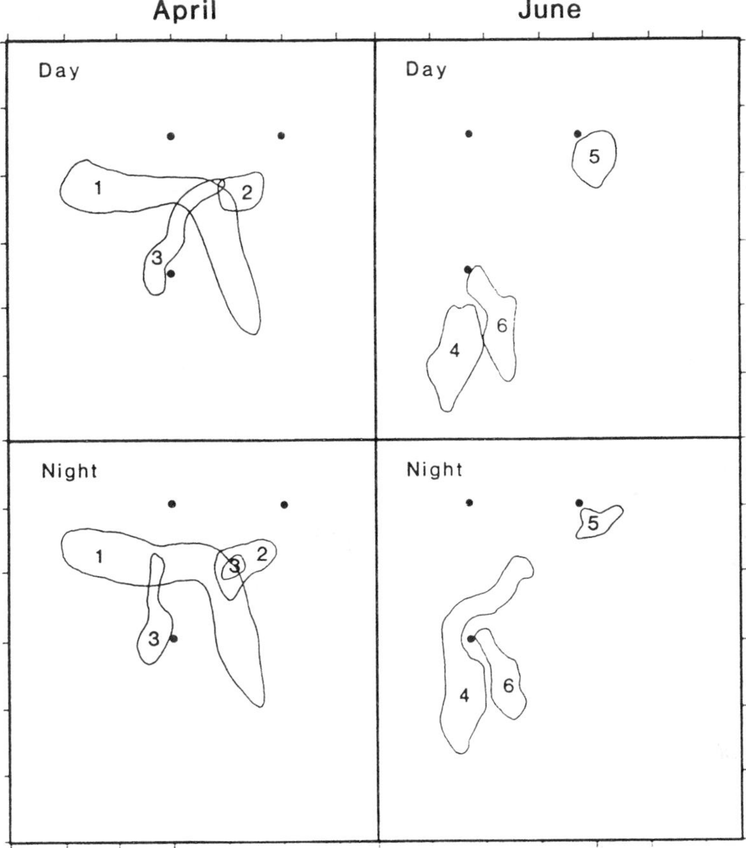


**Figure 1.** Restricted movement of three individuals of *Cebidichthys violaceus* (1, 2, 3, left-hand panels) during the day and night in April and three others (4, 5, 6, right-hand panels) in June. The three hydrophones used to receive signals from the ultrasonic tag attached to each fish are shown by solid circles. The edges of the panels are marked in 5-m divisions. Reprinted from *J. Exp. Mar. Biol. Ecol.* **98,** Ralston and Horn, High tide movements of the temperate-zone herbivorous fish *Cebidichthys violaceus* (Girard) as determined by ultrasonic telemetry, pp. 35–50 (1986), with kind permission of Elsevier Science-NL, Sara Burgerhartstraat 25, 1055 KV Amsterdam, The Netherlands.

**Table 1.** Comparison of Studies on Restricted Movements of Rocky Shore Intertidal Fishes

| Species | Habitat | Total number marked | % Never recaptured | % Recaptured in original location | % Recaptured elsewhere | Longest residence time | Sampling interval | Reference |
|---|---|---|---|---|---|---|---|---|
| *Acanthocottus bubalis* | Tidepools | 37 | 57* | 43 | ? | 12 weeks | 2 weeks | Gibson, 1967 |
| *Blennius pholis* | Tidepools | 101 | 48* | 52 | ? | 14 weeks | 2 weeks | Gibson, 1967 |
| *Clinocottus analis* | Tidepools | 367 | ? | 5–30 | ? | >20 weeks | ~2 weeks | Richkus, 1978 |
| *Clinocottus globiceps* | Tidepools | 22 | 50 | 27 | 33 | ? | Irregular over 2 months | Yoshiyama *et al.*, 1992 |
| *Clinocottus globiceps* | Tidepools | 39 | 16* | 94 | ? | >6 months | Irregular for >1 month | Green, 1973 |
| *Clinus superciliosus* and *C. cottoides* | Tidepools | 43 | 37 | ~53 | ~13 | ? | Irregular over 3 months | Marsh *et al.*, 1978 |
| *Oligocottus maculosus* | Tidepools | 35 | 72* | 28 | ? | >10 days | 10 days | Gersbacher and Denison, 1930 |
| *Oligocottus maculosus* | Tidepools | 48 | 48 | 31 | 21 | ? | Irregular over 2 months | Yoshiyama *et al.*, 1992 |
| *Oligocottus snyderi* | Tidepools | 49 | >49 | 33 | >4 | ? | Irregular over 2 months | Yoshiyama *et al.*, 1992 |
| *Forsterygion n. robustum* | Under stones | 89 | 85 | 8 | 7 | 4 months | 1 month | Berger and Mayr, 1992 |
| *Pholis gunnellus* | Under stones | 156 | 79 | 13 | 8 | ? | Daily over 6 days | Koop and Gibson, 1991 |
| *Acanthoclinus fuscus* | Under stones | 391 | 66 | 24 | 10 | 6 months | 1 month | Berger and Mayr, 1992 |
| *Xererpes fucorum* | Under stones | 32 | 44 | 56 | ? | ? | 1 tide | Burgess, 1978 |

*Note.* ? = No data available in original reference. * = Studies in which only original locations were searched for marked fish; values for % never recaptured are therefore higher than in those studies where fish were found outside the original location.

Most indirect evidence for the restricted movements of resident species comes from field studies in which individuals are marked and returned to their original location. Some of these studies also represent the earliest published examples of investigations on intertidal fishes. Hubbs (1921) observed that individuals of the viviparous perch *Amphigonopterus* (=*Micrometrus*) *aurora* occupied the same tidepools on successive low tides and Gersbacher and Denison (1930), in the first study of *Oligocottus maculosus,* showed that at least 28% of tagged fish were in the same pool 10 days later. Subsequent more detailed studies have confirmed this early picture (Table 1), and the results are remarkably consistent among species and areas. In most studies of tidepool species about half of marked fish are never recaptured at their original location, suggesting that they have moved elsewhere. Where searches for marked individuals have been made away from the original tidepool, recaptured fish are rarely more than a few meters away from the place where they were first marked, and most movement takes place between adjacent pools (Richkus, 1978; Figure 2). Fish that are absent on one or more sampling occasion can often be found there subsequently (e.g., Gibson, 1967). A further common finding is that repeated sampling of the same pool almost always reveals the presence of unmarked fish that have arrived since the last sampling occasion (Gibson, 1967; Burgess, 1978; Marsh *et al.,* 1978; Richkus, 1978; Berger and Mayr, 1992; Pfister, 1995). Nevertheless, the period of residence of individuals in a particular tidepool can be considerable, up to several months (Table 1) or even more than a year in the case of *O. maculosus* (Green, 1971a) and *Hypsoblennius gilberti* (Stephens *et al.,* 1970). Most individuals, however, stay in the same pool for much shorter periods, often less than 2 weeks and, at least in *Blennius* (=*Lipophrys*) *pholis,* the original occupants of a pool, that is, those encountered on the first sampling occasion, stay longer than subsequent arrivals (Gibson, 1967). Younger fishes are probably more mobile than the older age groups and are often the first to colonize defaunated pools. Other investigations that have concentrated on different aspects have produced similar results. The overall picture that emerges from such studies is one of movement over a limited area, a

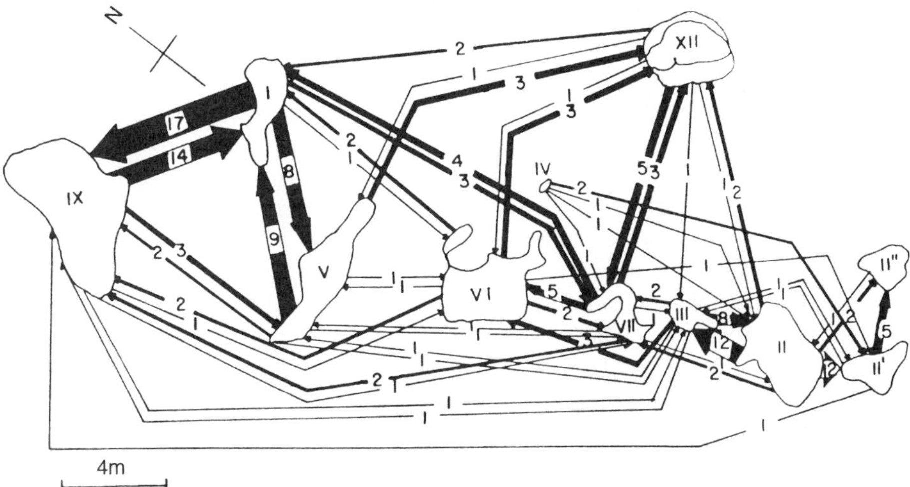


**Figure 2.** Diagram illustrating movements between tidepools of *Clinocottus analis*. Tidepools are numbered with roman numerals. The numbers within arrows and the width of the arrows indicate the number of fish moving between pools. Reproduced from *Mar. Biol.* **49,** A quantitative study of intertidepool movement of the wooly sculpin *Clinocottus analis*, Richkus, 277–284, Fig. 5 (1978), © Springer-Verlag, with permission.

home range, at high tide and the return to one, or one of a series, of adjacent pools at each low tide.

There are, however consistent differences among species in the same habitat that reflect individual behavioral traits (compare for example, *Acanthoclinus fuscus* and *Forsterygion n. robustum* in Table 1) and local conditions. Turbulence and the stability of pool topography seem to be important regulators of residence time (Gibson, 1967; Green, 1971a; Richkus, 1978). Habitat instability, or perhaps greater habitat uniformity, may also be the cause of the greater mobility of the elongate species living under stones on boulder shores without pools (Table 1). Barton (1973) and Koop and Gibson (1991) also found that other eel-like species tend to move at random, and Williams (1957) noted that several eel-shaped species recolonized defaunated pools more rapidly than typical tidepool species.

In contrast to the restricted movement of the resident species, tidal visitors to rocky shores enter and leave the intertidal zone with each tide and may move considerable distances, depending on the extent of the intertidal area available. Few detailed studies of this type of movement have been made, perhaps because of the difficulties of sampling in this environment. Observations while diving provide some evidence for this movement (e.g., Carlisle, 1961; Gibson, 1972; Whoriskey, 1983; Black and Miller, 1991) but netting studies give more quantitative information. Using such techniques, Black and Miller (1991) found that 18 species moved into the intertidal zone on rising tides in Nova Scotia. One of the common intertidal visitors there was the pollock (*Pollachius virens*), which was subsequently studied in more detail by Rangeley and Kramer (1995a,b) in New Brunswick. In this location schools of juveniles move distances of about 150 m into open intertidal areas and then disperse into weed patches. This influx is followed by a rapid retreat out of the upper intertidal areas on the falling tide and the fish reform schools in the lower intertidal and subtidal areas. A similar extensive migration takes place in the surgeonfish *Acanthurus fuscus* in the Gulf of Aqaba. In some areas where the intertidal feeding grounds are separated from refuges by distances of 500–600 m, the fish migrate daily between the two sites. In other areas, where food and shelter are close together, the fish move only a few meters each day (Fishelson *et al.,* 1987). In most cases it is not known whether individual fish that move into the intertidal zone with each tide return to the same place on each occasion but there are several instances where this has been recorded. In American Samoa, another surgeonfish, *Acanthurus lineatus,* moves into the intertidal zone and establishes feeding territories over the high-tide period. In this species fidelity to individual territories is almost complete (Craig, 1996). On some English shores, bass (*Morone labrax*) and grey mullet (*Mugil labrosus*) have also been seen moving onto an intertidal reef with the rising tide where they defend intraspecific territories and leave about 1 h after high tide (Carlisle, 1961). In the same area, a proportion of the population of the wrasse *Crenilabrus melops* builds nests in intertidal rock crevices and presumably returns to them on each tide during the breeding season (Potts, 1985). Another wrasse, the western Atlantic *Tautoglabrus adspersus,* congregates in intertidal refuges about 2 h before high water and leaves again about 3 h later. Fish make feeding forays from these refuges over the high-tide period but although the same refuges are occupied on each high tide, it is not known whether individuals always return to the same one on each tide (Whoriskey, 1983).

*Crenilabrus melops* provides one example of a fish using the intertidal zone as a breeding ground and, although such behavior is the rule in resident species, it is relatively rare in transient visiting species. Migrations into the rocky intertidal zone for spawning are also known in the Pacific herring (*Clupea harengus pallasi*), whose eggs are deposited on vegetation on sheltered areas of the shore in vast numbers (Haegele and Schweigert, 1985). Similar behavior, but on a smaller scale, occurs in the Japanese pufferfish *Takifugu ni-*

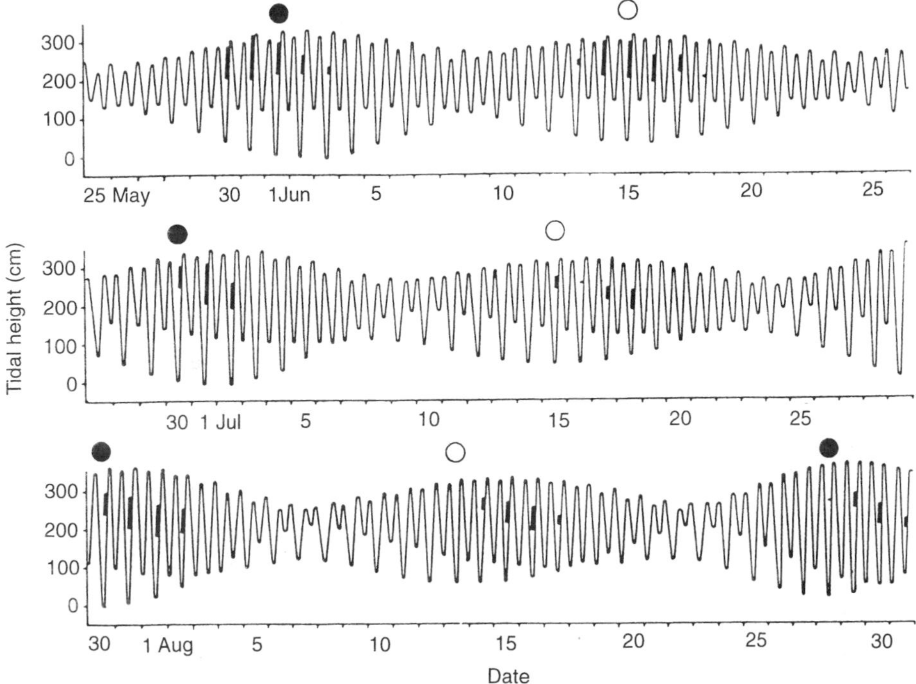


**Figure 3.** Relationship between the time of spawning of *Takifugu niphobles* and tidal phase. The continuous line shows variation in tidal height and the black bars the occurrence of spawning. Filled and open circles represent the times of new and full moons, respectively. Reproduced from *Environ. Biol. Fish.* **40** (1994), 255–261, Combined effects of tidal and diurnal cycles on spawning of the puffer, *Takifugu niphobles* (Tetraodontidae), Yamahira, Fig. 2, © Kluwer Academic Publishers, with kind permission .

*phobles*. During the breeding season this species lays its eggs intertidally on shores consisting of pebble and gravel on rising tides at the time of full and new moons. The time of each daily spawning run depends mainly on the time of the high tide but is modified by the diel cycle so that spawning ends earlier when high tides are close to sunset (Yamahira, 1994; Figure 3). A more northerly species, the capelin (*Mallotus villosus*) has a similar semilunar rhythm of intertidal egg deposition on gravel shores (Therriault *et al.*, 1996).

On a longer time scale, many species undergo seasonal changes in their distribution. These movements are generally related to seasonal variation in environmental conditions and hence are most pronounced at higher latitudes. In Maine, for example, fish are virtually absent from the intertidal zone in the winter but return in the spring (Moring, 1990). The most pronounced movements are usually made by the juveniles of species that use rocky shores as nursery grounds. In these cases, the young stages recruit to the shore in the summer and move into deeper water in the winter. Such movements generally result in lower abundances and community diversity in winter (Gibson and Yoshiyama, Chapter 13, this volume).

## B. Sandy Beaches

The mobile and relatively homogeneous nature of the substratum on sandy shores means that few refuges are available for fishes to remain intertidally over the low-tide period. Consequently those individuals that do so either occupy shallow pools left by the receding tide or belong to species that are capable of remaining buried in the sediment (e.g., members of the Ammodytidae). Occupants of tidepools are usually small individuals that live in shallow water and represent only a small fraction of the total fish community present at high water. Whereas it is probable that some individuals may be involuntarily stranded in such pools, there are species that use such pools as permanent habitats for part of their life history or possess mechanisms that ensure their escape from pools on the ebb tide. An example of the first category is the plaice (*Pleuronectes platessa*), whose larvae in some areas settle in intertidal pools, metamorphose, and remain there as juveniles until temperatures and oxygen concentrations become unsuitable. The juveniles then develop a pattern of migration in and out of the intertidal zone with the tides and gradually move their low-water refuge into deeper and deeper water as they grow (van der Veer and Bergman, 1986; Berghahn, 1987a,b). Presumably they must also develop a mechanism for escaping from pools at the same time. Such escape behavior has only been described for one species, *Fundulus majalis,* in pools on the North Carolina coast. This species continually examines the depth of water at the pool outlet and leaves when the water reaches a critical depth. If the outlet is blocked, the fish escape overland by leaving the water and reaching the sea by means of a series of directed leaps (Mast, 1915).

The great majority of species that live on sandy beaches, however, enter and leave the intertidal zone with each tide. In comparison with rocky shores, sheltered sandy beaches are relatively easy to sample and it is therefore surprising that only a few detailed studies of fish movements in this habitat have been made. The two species that have received most attention in this respect are both flatfishes. In the Bay of Fundy, American plaice (*Pseudopleuronectes* (=*Pleuronectes*) *americanus*) move into the intertidal zone just after low

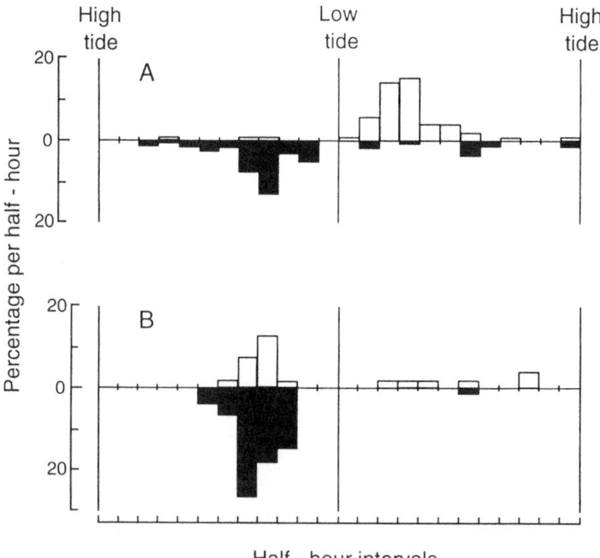


**Figure 4.** Intertidal movements of *Pseudopleuronectes (Pleuronectes) americanus* in Passamaquoddy Bay, New Brunswick. Numbers of fish are expressed as a percentage of the mean number moving past a fixed point over a 12-h tidal cycle. (A) Fish >10 cm, (B) Fish <10 cm. Upshore movements are shown by white columns above the horizontal axis and downshore movements by black columns below the axis. Modified from Tyler (1971) and reproduced with permission.

water and stay there to feed for about 7 h. The movements of fish longer than 10 cm are more directional than the smaller fish but all sizes move in and out together (Tyler, 1971; Figure 4). Several other species show similar tidally phased movements in the same area but there are also others whose activity does not seem to be related to the tidal cycle. This behavior pattern is thus very similar to that of the European plaice (*Pleuronectes platessa*), which has been studied in considerable detail. The juveniles of this species have a well-developed pattern of tidal migration in all areas where tidal ranges are significant (Kuipers, 1973; Gibson, 1973; Burrows *et al.,* 1994). In areas where subtidal regions are poor and intertidal areas are rich in food, as on the extensive sand flats off the Dutch coast, virtually the whole population migrates with each tide (Kuipers, 1973). In other areas where most food is concentrated in the lower intertidal and subtidal areas only a proportion of the population moves inshore, and the smaller fish tend to move further than the larger and older ones (Ansell and Gibson, 1990). A similar difference in the extent of movement of age groups has been described for the intertidal feeding migrations of *Takifugu niphobles,* in which the smallest fish are found in much shallower water at high tide than the larger individuals (Yamahira *et al.,* 1996). In many species there is a strong diel modulation of the tidal pattern whereby fish move into shallow water at night independently of the tide. In the plaice this movement is particularly marked so that the overall pattern is the result of a complex interplay between the relative times of the tide and the light/dark cycle (Burrows *et al.,* 1994; Figure 5). In other species, entry into the intertidal zone seems to be inhibited

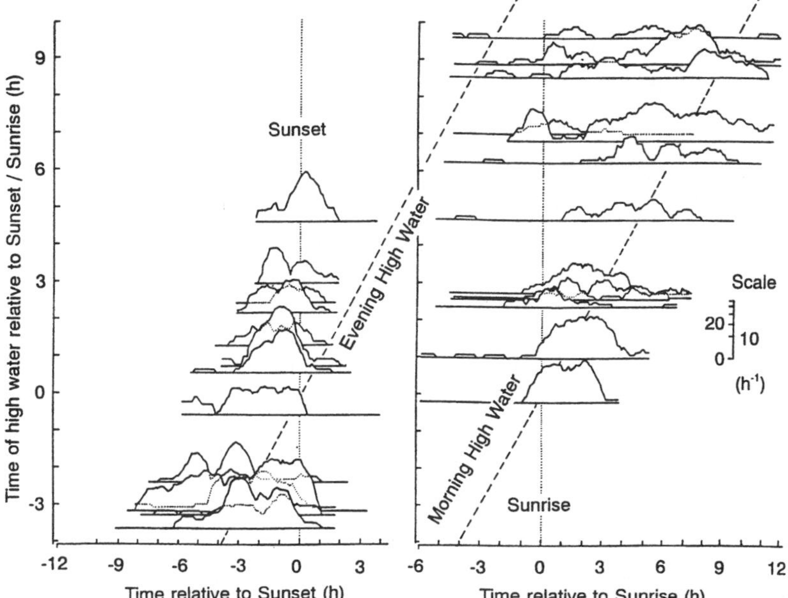


**Figure 5.** Intertidal movements of juvenile *Pleuronectes platessa* on a sandy beach in Scotland illustrating the complex nature of the timing of the movements. Each individual graph plots the running average number of fish moving beneath an underwater TV camera on separate dates relative to the times of sunset (left) and sunrise (right). The diagonal dashed lines show how the times of high tide change relative to sunset and sunrise. Reprinted from *J. Exp. Mar. Biol. Ecol.* **177,** Burrows *et al.*, Temporal patterns of movement in juvenile flatfishes and their predators: underwater television observations, 251–268 (1994), with kind permission of Elsevier Science-NL, Sara Burgerhartstraat 25, 1055 KV Amsterdam, The Netherlands.

during the day and they are only present at night. Juvenile gadoids are good examples of such nocturnal inshore movements (Burrows *et al.*, 1994), and comparison of the number of other species present at high tide during the day and night on a Scottish sandy beach showed that more than twice as many species were present at night (Gibson *et al.*, 1996).

In comparison with these onshore–offshore movements, longshore movements on sandy beaches seem to be much more restricted although studies of such movements are few. Marked young European plaice, for example, in a sandy bay in Wales, moved less than 500 m laterally over several months (Riley, 1973). A similar study of juvenile American plaice in Massachusetts showed that 90% of marked fish had moved less than 100 m from the release site in 3 weeks (Saucerman and Deegan, 1991).

A few species use sandy beaches for spawning purposes. During the reproductive season fish move into the intertidal zone during the high waters of spring tides and deposit their eggs in the sand, where they develop and hatch when covered by the next series of spring tides. These tidally synchronized movements of normally pelagic sublittoral species often involve large numbers of fish and are consequently very conspicuous events. The best known examples are the Californian grunion (*Leuresthes tenuis*), which spawns during nighttime high waters (Walker, 1952), and the related species *Leuresthes sardina,* the Gulf grunion, which spawns mainly during the day (Thomson and Muench, 1976).

In common with fish movements on rocky shores, movements of fishes on sandy beaches can be also detected on a longer, seasonal time scale. In temperate regions these movements usually take the form of an offshore migration in late summer and autumn. Numerous examples of such movements have been described for individual species, including the flatfish *Parophrys vetulus* in California (Toole, 1980) and for communities in, for example, Scotland (Gibson *et al.,* 1993) and the Azores (Santos and Nash, 1995). These movements result in a decline in both numbers and diversity in the community.

## C. Estuaries and Mudflats

Estuaries are dynamic environments where tidal fluctuations are accentuated by the restriction of water movements between their banks. Estuarine fish communities are also diverse and strongly seasonal and individuals may be very abundant so that fish movement patterns within estuaries are variable. Movements of estuarine fishes are often extensive and always closely coupled to the tidal cycle. They can be roughly grouped as movements either into and out of the estuary or the intertidal regions flooded by the incoming tide, although an interesting case of colonization of Australian supralittoral estuarine pools is described by Russell and Garrett (1983). It is the intertidal movements that are relevant here and rather than outlining similar patterns for several species, one species is used as an example of the nature and complexity of movement patterns that can arise.

The flounder (*Platichthys flesus*) is a common inhabitant of European estuaries in all its life stages and its movements have been studied in considerable detail (Wirjoatmodjo and Pitcher, 1980; Raffaelli *et al.*, 1990). Ultrasonic tracking of individuals for up to 12 h in the estuary of the River Bann in Northern Ireland showed that fish over this relatively short period remained within 400 m of their release point. They follow the rising tide upstream and move into intertidal areas where they meander over the inundated tidal flats at high tide. As the tide starts to fall, their movements increase in frequency and become directed downstream. At low tide the fish make rarer but longer downstream movements and eventually bury in the mud. As the tide starts to rise again the cycle is repeated although movement is more likely at dusk. This cycle of upstream and downstream movement keeps

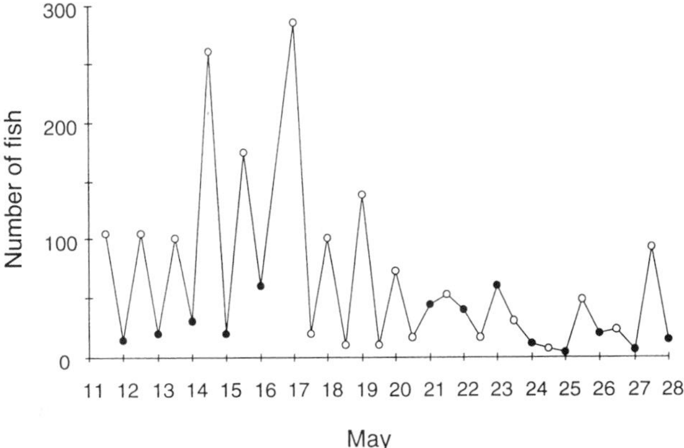


**Figure 6.** Intertidal movements of *Platichthys flesus* in a Scottish estuary showing marked differences in catches between successive tides. Filled circles night tides; open circles, day tides. Modified from Raffaelli *et al.* (1990) and reproduced with permission.

individuals in a relatively restricted area of the estuary (Wirjoatmodjo and Pitcher, 1980). A complex pattern of movement has been described for a population of flounders in another estuary, that of the River Ythan in Scotland. Here flounders also move onto the intertidal mudflats to feed but only distances of a few tens of meters. The smaller individuals in the population do so more at night than the older and larger fish, a difference that may be related to the greater risk of predation for the small fish by diurnally feeding birds. A puzzling feature of this movement is that at certain times of the semilunar cycle catches in intertidal fyke nets, presumably reflecting the number of fish moving off the flats, can differ by an order of magnitude on alternate tides (Figure 6). The fish in the higher catches are smaller on average than those in the low catches and there is a rough correspondence between catch size and the duration of the flood tide, which can vary by almost an hour (Raffaelli *et al.,* 1990).

Mudflats outside estuaries are also used as habitats by fishes but these environments have been relatively little studied perhaps because of the difficulties of access and sampling in the soft sediment. They are, however, utilized by numerous species and populations can be large and diverse. Studies of catches by traditional intertidal fish traps in Kuwait, for example, demonstrated that at least 76 species in 36 families entered the intertidal zone in this region. The catches were dominated numerically by grey mullet (Mugilidae) but about half of the Carangidae known in the area were found in the intertidal zone (Abou-Seedo, 1992). More detailed studies in the same area using small trawls and seines caught a total of only 37 species but revealed a complex picture of fish movements. Most species were concentrated near the low-water mark and did not change their position greatly over the tidal cycle. When the water was clear in spring, shoals of very young grey mullet (*Liza carinata*) followed the edge of the rising tide both day and night but 2 months later only formed such shoals on daytime rising tides. In turbid conditions any tidal effects on catches were obscured (Abou-Seedo *et al.,* 1990). One of the other common species in this area, *Leiognathus decoratus,* also shows a clear movement into the intertidal zone of all size groups but not all individuals in the population make this migration, which can cover a distance of up to 2 km (Figure 7). In contrast, the mudskippers that are active over the

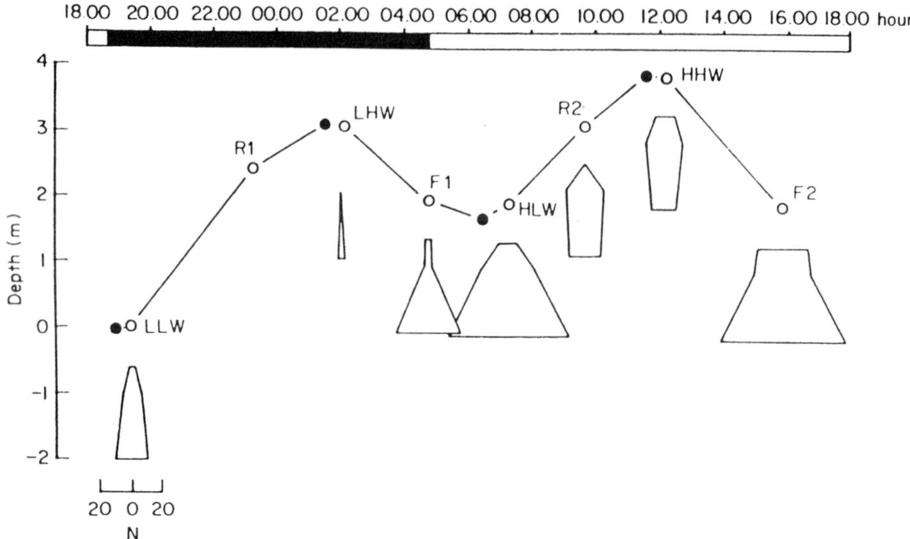


**Figure 7.** Intertidal movements of *Leiognathus decorus* in Kuwait. The kite diagrams show the catch at depths of 0.6, 1, and 2 m at eight times in the tidal cycle. The height of the tide is shown by the continuous line and the times of day and night are at the top of the figure. Reproduced with permission from Wright *et al.* (1990).

mudflat at low tide were rarely caught in the catches, suggesting that they stay in the burrows in the mud over the high-tide period (Wright *et al.,* 1990). In Japan, the mudskipper *Periophthalmus modestus* also feeds on mudflats at low tide but migrates upshore with the incoming tide (Ikebe and Oishi, 1996).

## D. Salt Marshes and Mangroves

In contrast to most rocky shores and sandy beaches, salt marshes are very sheltered environments. Although the marsh surfaces may flood and drain on each tidal cycle, there are often sufficient pools and small drainage channels in which fish can remain over the low-tide period and therefore be classed as resident species. Most knowledge of salt marsh and creek fishes originates from studies in the extensive marshes of the Atlantic and Gulf coasts of North America. Few studies have been made elsewhere. In North American marshes, the commonest resident fishes belong to the genus *Fundulus* whose movement patterns seem to vary between areas. On Long Island, as elsewhere, *Fundulus heteroclitus* move into smaller channels from the main stream at high tide and spread out over the flooded marsh, returning to the main stream as the tide ebbs. In studies in which fish were marked, the recapture rate was generally <10% and decreased daily, suggesting that individuals move gradually over the marsh at random (Butner and Brattstrom, 1960). Studies of the same species in Delaware, however, revealed a much more restricted pattern of movement. Here, marked fish rarely moved outside a home range of about 36 m along the banks of the creeks and if they were experimentally displaced from one side of the creek to the other, a distance of 8 m at low tide, they returned to their original side (Lotrich, 1975). In Georgia, the largest individuals travel furthest into the marsh and on each tide overtake the smaller fish on each flood and ebb tide (Kneib and Wagner, 1994). In New Jersey *F. heteroclitus* overwinter in marsh pools where, although only 7.6% of marked fish were recaptured over a 7-month period, 85% of these were in the pool where the fish were first marked (Smith and Able, 1994). In the one study of a European marsh in the Netherlands, few fish species were found to enter the marsh creeks and most stayed in the main channels. The larvae and juveniles of two species (flounder, *Platichthys flesus,* and the common goby, *Pomatoschistus microps*) concentrate in pools at low tide and the latter can be considered the ecological equivalent of *Fundulus* (Cattrijsse *et al.,* 1994).

Numerous species from a variety of families can be regarded as tidal visitors to marshes (Moyle and Cech, 1982) and show many different patterns of movement. The Atlantic silverside, *Menidia menidia,* moves into creeks at night in New Jersey (Rountree and Able, 1993), as does the silver perch *Bairdiella chrysura* in South Carolina (Kleypas and Dean, 1983) whereas in the same area, the bay anchovy *Anchoa mitchelli* is most abundant in creeks during the day (Reis and Dean, 1981). In Louisiana, salt marsh fishes can be grouped into four types depending on the extent of their use of the marsh surface: (1) interior marsh residents that remain in flooded depressions on the marsh surface at low tide; (2) interior marsh users that move onto the marsh at high tide but return to the edge of the creeks at low tide; (3) edge marsh users, resident and transient species that occupy only the margins of marshes at the edge of creeks; and (4) a group of subtidal species that do not move onto the flooded marsh surface at all (Peterson and Turner, 1994). Gross movements such as these are usually inferred from differences in net catches at different stages of the tide because direct observation of fishes is difficult in the turbid waters of creeks and channels. Direct location of fishes and the ability to follow them for long distances can be achieved by attaching ultrasonic transmitters to individuals large enough to accommodate them. Using such techniques, young American eels (*Anguilla rostrata*) have been followed moving up Georgia tidal creeks and onto the marsh at night at high tide and moving off again shortly after the tide changes. During the day they usually remain in the deepest part of the creek system (Helfman *et al.,* 1983). Comparable tidal movements up creeks have been detected in larger individuals of juvenile summer flounder (*Paralichthys dentatus*) in New Jersey, also using ultrasonic tags. In this species, movement patterns vary among individuals but many move into the creeks with the tide to feed (Szedlmayer and Able, 1993; Figure 8). The absence of the youngest stages in net catches early in the year suggests that they do not migrate with the tides and that this behavior develops later in the season. If this proves to be the case, there is a clear parallel between the behavior of this species in marsh creeks and that of *Pleuronectes platessa* in pools on sandy beaches (van der Veer and Bergman, 1986).

The vegetation in salt marshes acts as a substratum for spawning for some species, and fish migrate into the high intertidal zone to lay their eggs among the plants. Two species have been studied in detail. *Menidia menidia* times its spawning runs into the intertidal zone to coincide precisely with the time of daytime high tides and most spawning runs are associated with the higher tides at new and full moons (Middaugh, 1981). The other abundant salt marsh species, *Fundulus heteroclitus,* has similar habits (Taylor *et al.,* 1979).

In common with most other intertidal ecosystems, seasonal movements can also be detected such that the high abundance and diversity of the summer months are reduced in the winter. These seasonal changes have been described, for example, in South Carolina (Cain and Dean, 1976), New Jersey (Rountree and Able, 1992), and the Netherlands (Cattrijsse *et al.,* 1994).

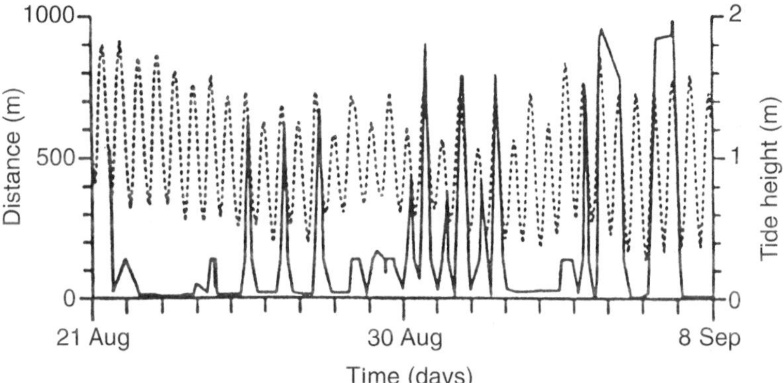


**Figure 8.** Pattern of movement of an individual *Paralichthys dentatus* in a New Jersey creek. The distance moved by the fish away from the mouth of the creek is shown by the continuous line and the tidal height by the dotted line. Reproduced with permission from Szedlmayer and Able (1993).

Mangroves form a habitat equivalent to salt marshes in the tropics and are important nursery grounds for many fish and crustacean species in these areas. The complex nature of the habitat makes these ecosystems difficult to sample and even more difficult to observe fish directly. Consequently, studies of fish movements in the intertidal areas of mangroves are relatively few and mostly consist of trapping at high tide (Sheridan, 1992, in Florida) or netting fish as they leave on the ebbing tide (Morton, 1990; Robertson and Duke, 1990; Laegdsgaard and Johnson, 1995; and Vance *et al.,* 1996, in Australia). The exceptions are studies of mudskippers that are possible because the fishes are active out of water on the mud surface when it is exposed. The best studied species is *Periophthalmus sobrinus* whose activity patterns are generally similar from place to place but vary in some of the details. In Kenya, the fish is active both day and night on the surfaces of the banks of mangrove-lined creeks. On spring tides all fish are active and move down the banks closely following the water line on the early ebb. As the tide continues to recede the fish confine their activity to the central portion of the intertidal zone and retreat in front of the waterline as it advances on the flood tide (Figure 9). At night this pattern remains the same except that the fish move further down the bank than they do during the day. On neap tides only about half the population is active, the other half remain resting on their nests. The movements of those that are active are much less extensive and they remain close to the water line (Colombini *et al.,* 1995). In Madagascar, the fish are active in the same tidally synchronized pattern but only during the day (Brillet, 1975).

## E. Seagrass Beds

Relatively little seems to be known about movements of fishes into intertidal seagrass beds. In such habitats most movement appears to be related to the diel cycle with some species being more common in beds at night or during the day. Where tidal fluctuations are sufficient to make stranding or exposure to avian predators a possibility, then fishes present at high tide tend to leave on the ebbing tide. In Florida, for example, gill net catches of several species were high as the tide was falling, reflecting their movement off the seagrass bank at this time (Sogard *et al.,* 1989) and in Australia the flathead (*Platycephalus laevigatus*),

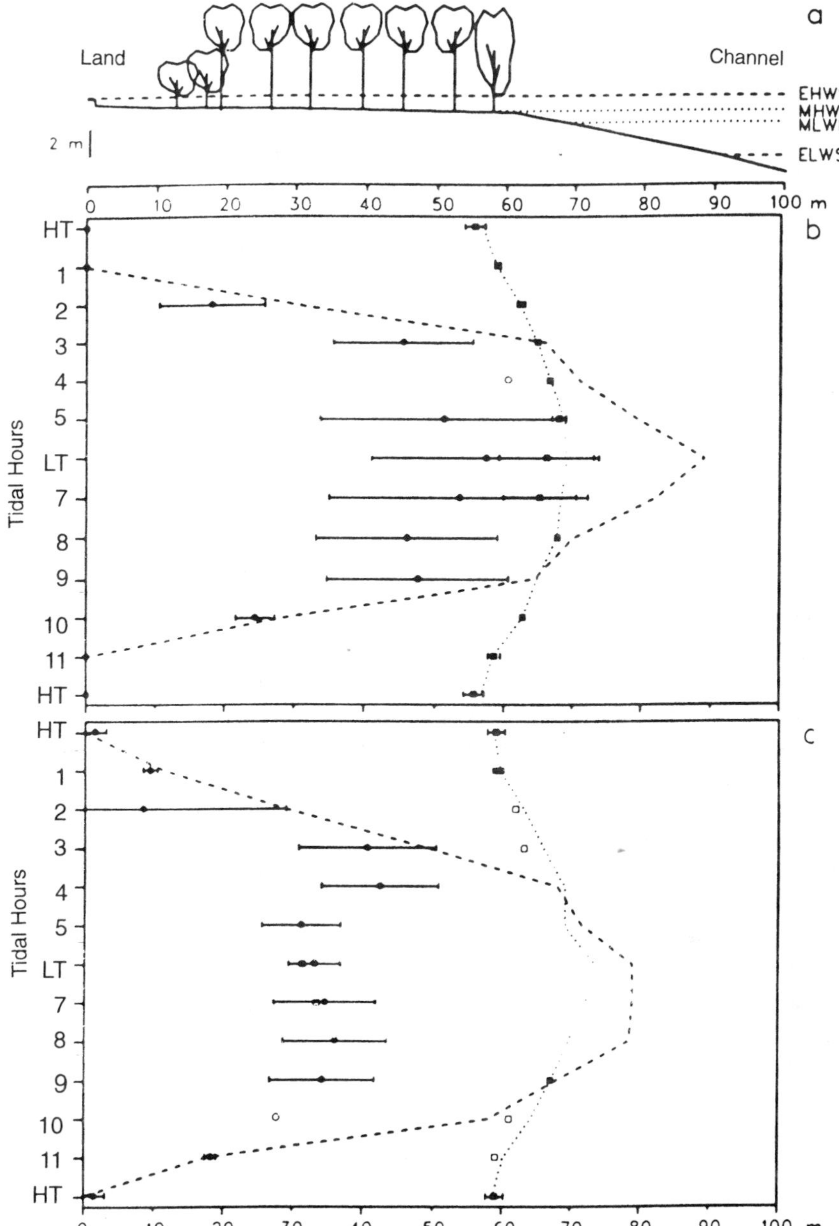


**Figure 9.** Tidal movements of the mudskipper *Periophthalmus sobrinus* in a Kenyan mangrove. The average positions (±95% confidence limits) of the fish at different tidal states are indicated by the circles (spring tides) and squares (neap tides). The water level on spring and neap tides is shown by the dashed and dotted lines, respectively. The profile of the mangrove creek is shown diagrammatically at the top of the figure. Reprinted from *J. Exp. Mar. Biol. Ecol.* **190,** Colombini *et al.*, Environmental factors influencing the zonation and activity patterns of a population of *Periophthalmus sobrinus* Eggert in a Kenyan mangrove, 135–149 (1995) with kind permission of Elsevier Science-NL, Sara Burgerhaststraat 25, 1055 KV Amsterdam, The Netherlands.

a major predator of crabs and small fishes, moves into and out of seagrass beds on evening and nighttime high tides (Klumpp and Nichols, 1983).

## IV. Homing

The observations that many tidepool fishes move out of their pool at high tide and yet can be found there on several subsequent occasions (Table 1) indicates they are able to return to their original low-tide location after an absence. Such behavior could be accounted for if the fish only move over a small home range containing a few adjacent pools but it also suggests that the fish are familiar with the topography of their home range and use this knowledge to return to specific low-tide locations. The latter behavior can be considered as a form of homing, defined by Gerking (1959) as "the return to a place formerly occupied instead of going to other equally probable places." Convincing evidence for homing proper, that is, a directed homeward movement rather than the regular occupation of selected parts of the home range, is provided by the return of fish after displacement. Such experiments have been carried out with a wide variety of species and demonstrate that many can home over considerable distances after displacement outside their current home range (*Blennius* (=*Parablennius*) *sanguinolentus, Oligocottus maculosus, Oligocottus snyderi, Clinocottus globiceps,* and *Tripterygion tripteronotus;* Table 2). Furthermore, the frequency of return of displaced fish to their home pool is always much greater than the movement to the home pool of (a) fish from other shores released in the same pool as the displaced fish (Gibson, 1967) and (b) the residents of the pool in which the displaced fish were released (Santos, 1986). Homing success, the percentage of displaced fish returning at least once to their point of capture, varies widely within and between species although detailed comparison is hampered by differences in the methods and criteria of success employed. Nevertheless, two general findings are that there are no sex differences in homing success and that success is generally greater in larger individuals. The latter observation has been interpreted as evidence that, at least in *O. maculosus,* the younger fish, which are generally more extensive in their movements, acquire a spatial knowledge of their habitat at this time and this knowledge is used when homing to a restricted area later in life (Craik, 1981). It seems that such knowledge can be remembered for weeks in *Bathygobius soporator* (Beebe, 1931) and months in *O. maculosus* (Green, 1971a). Differences in the homing success between populations of *O. maculosus* and *C. globiceps* (Green, 1971a, 1973; Yoshiyama *et al.,* 1992) suggest that physical factors such as shore topography and exposure to wave action and turbulence can also have an effect on an individual's ability to home. Furthermore, because the environment varies over time (e.g., seasonally, or episodically due to storms) it might also be expected that the ability of individuals to home will also vary temporally, apart from the influence of age.

## V. Functions of Intertidal Movements and Homing

The patterns of movement that have been described serve a variety of functions some of which may differ according to the phase of the tidal cycle. Movement into the intertidal zone by visitors or away from their low-tide refuge by residents, for example, may be for feeding, whereas the return journey represents the need to avoid stranding in unfavorable locations by the ebbing tide. For convenience, the possible functions of intertidal movements can be considered under four headings (feeding, avoidance, the search for a physiologically optimum environment, and spawning) while recognizing that these functions are not necessarily mutually exclusive.

## A. Feeding

The intertidal zone is often a rich source of food for fishes but as a feeding ground it has the disadvantage that it is only periodically accessible. Many of the movements of fishes into (visitors) or around (residents) the zone represent foraging excursions to areas that are only available to them at high tide (Figure 10). Again, the exceptions to this statement are the mudskippers that tend to forage intertidally when the tide is out (Brillet, 1975; Columbini *et al.,* 1996). Evidence for intertidal feeding can be obtained from direct observation (e.g., Gibson, 1980) or, more usually, by comparing the stomach fullness of individuals entering and leaving the intertidal zone. In such studies, which have been carried out in a variety of habitats including rocky shores (Black and Miller, 1991; Rangeley and Kramer, 1995a), sandy beaches (Wells *et al.,* 1973; Healey, 1971; Ansell and Gibson, 1990), salt marshes (Weisberg *et al.,* 1981; Kleypas and Dean, 1983; Rountree and Able, 1992; Cattrijsse *et al.,* 1994), and estuaries, fish leaving the zone on the ebb tide frequently, but not always, have more food in their stomachs than when they entered on the flood. Presumably the effort expended in these foraging movements is more than compensated for by the energy obtained from feeding (Miller and Dunn, 1980; Wolff *et al.,* 1981). A consequence of fish feeding intertidally is the export of considerable quantities of energy as biomass from the intertidal zone into sublittoral areas, not only by fishes that visit on each tide (Kleypas and Dean, 1983) but also by those species whose juveniles use the intertidal zone as a nursery ground and subsequently move into deeper water.

## B. Avoiding Predators and Adverse Conditions

The observation that in some cases feeding in the intertidal zone is no more intensive than elsewhere (Black and Miller, 1991; Rangeley and Kramer, 1995a) or that inshore movement actually takes fish away from an abundant food supply (Ansell and Gibson, 1990) suggests that a second important function of the movements of many intertidal fishes is the avoidance of predators. For young fishes that undertake tidal feeding migrations, the movement up the shore close to the advancing water line keeps them in shallow water and reduces the risk of predation from larger predators that also move in with the tide. Juvenile *Pollachius virens,* for example, enter the intertidal zone in schools on the rising tide, disperse into patches of algae at high tide, and then, as the tide starts to fall, reform schools and rapidly move into deeper water (Rangeley and Kramer, 1995b). Where cover is available as a refuge, many species have reduced this risk further by remaining resident either for their whole lives or just during the vulnerable juvenile stages. In this strategy, risk of predation by aquatic predators is limited to the high-tide period. For fully resident species, their restricted movement and intimate knowledge of their surroundings can both be seen as aids in avoiding predators and there is good evidence for small fishes that the presence of, and access to, suitable shelter is important in reducing predation (Markel, 1993). In some extreme cases, species such as the amphibious rocky shore blennies *Alticus, Mnierpes,* and *Entomacrodus* restrict their movements to the shallow turbulent region at the water's edge. This region is difficult to exploit by aerial and aquatic predators alike (Brown *et al.,* 1991; Graham, 1970; Graham *et al.,* 1985).

**Table 2.** Examples of Homing in Resident Intertidal Fishes

| Family<br>*Species* | No. released | % Homing | Sampling interval | Displacement distance (m) | Habitat | Remarks | Reference |
|---|---|---|---|---|---|---|---|
| Plesiopidae | | | | | | | |
| *Acanthoclinus fuscus* | 19 | 31<br>58 | 1 day<br>1 week | 5 | Rocky shore | Species lives under stones | Berger and Mayr, 1992 |
| Blenniidae | | | | | | | |
| *Blennius pholis* | 83 | 50 | At intervals over 4 months | 2–17 | Rocky shore | | Gibson, 1967 |
| *Blennius* (*Parablennius*) *sanguinolentus* | 53 | 83 | ? | 10–120 | Rocky shore | | Santos, 1986 |
| *Hypsoblennius gilberti* | 100 | 53 | ? | 2–46 | Rocky shore | % homing declines at distances >12 m | Stephens *et al.*, 1970 |
| Cottidae | | | | | | | |
| *Acanthocottus bubalis* | 12 | 58 | At intervals over 4 months | 2–17 | Rocky shore | | Gibson, 1967 |
| *Clinocottus analis* | 37 | 8 | ? | 2 | Rocky shore | | Williams, 1957 |
| *Clinocottus globiceps* | 43 | 30 | Up to 30 days | 30–44 | Rocky shore | | Yoshiyama *et al.*, 1992 |
| *Clinocottus globiceps* | 47 | 94 | Up to 1 month | Average of 10 m | Rocky shore | | Green, 1973 |
| *Oligocottus maculosus* | 26 | 19 | Up to 30 days | 27–76 | Rocky shore | % homing increases with fish length | Yoshiyama *et al.*, 1992 |
| *Oligocottus maculosus* | 469<br>445<br>446 | 36<br>20<br>4 | 2 weeks | 30–122 m | Rocky shore | Untreated fish<br>Blinded fish<br>Anosmic fish | Khoo, 1974 |
| *Oligocottus maculosus* | 142 | 68% of tagged, 88% of recaptured | Days to months | Up to 102, mostly 30–60 | Rocky shore | Some return within 1–2 tides | Green, 1971a |
| *Oligocottus maculosus* | 426 | <10 to >70 | 2 weeks | ~60 | Rocky shore | % homing to home range is length dependent | Craik, 1981 |

| | | | | | | | |
|---|---|---|---|---|---|---|---|
| *Oligocottus snyderi* | 107 | Average of 46 | Up to 30 days | 27–76 | Rocky shore | % homing varies with site and distance from 14 to 75% | Yoshiyama *et al.*, 1992 |
| Fundulidae | | | | | | | |
| *Fundulus heteroclitus* | 60 | 37% of total tagged, 92% of recaptures | Days to weeks | 8–15 | Salt marsh creek | Fish displaced to opposite bank of creek | Lotrich, 1975 |
| Gadidae | | | | | | | |
| *Ciliata mustela* | 24 | 46 | At intervals over 4 months | 2–17 | Rocky shore | | Gibson, 1967 |
| Girellidae | | | | | | | |
| *Girella nigicans* | 23 | 56 | ? | 2 | Rocky shore | | Williams, 1957 |
| Gobiidae | | | | | | | |
| *Bathygobius soporator* | 6 | 83 | 1 day | "Many yards" | Rocky shore | | Beebe, 1931 |
| *Periophthalmus sobrinus* | 403 | 69 | Immediate | 7–10 | Mangrove creek | Fish released in mid channel and return to home bank observed visually | Berti *et al.*, 1994 |
| Pleuronectidae | | | | | | | |
| *Pleuronectes platessa* | 4480<br>5480 | 3<br>3 | Days to weeks | 3500 offshore to 7 m depth | Sandy bay | | Riley, 1973 |
| Tripterygiidae | | | | | | | |
| *Tripterygion tripteronotus* | 11 | 64 | ? | 100–200 m | Rocky sub-littoral, 0.5–1 m deep | Homed in 27–78 h | Heymer, 1977 |

*Note.* All experiments involved displacement of marked fish various distances from their point of capture. Except where indicated, homing in rocky shore species refers to returns to the home pool where the fish were first captured.

**Figure 10.** Evidence for intertidal feeding by the flounder *Platichthys flesus*. The photograph shows the trail of circular depressions (bite marks) made by a feeding fish as it moved over a mudflat at high tide. Reproduced with permission from Summers (1980).

Movement on the rising tide away from the location occupied at low tide poses the problem to both visitors and residents of how to avoid being stranded in an unfavorable position on the shore when the tide ebbs. For visitors, avoidance simply takes the form of downshore migration with the falling tide. For residents, the need to avoid adverse low-tide conditions requires that they return to their low-tide refuge before they are stranded by the outgoing tide. The restricted extent of the movements of such residents, their ability to learn the topography of their environment, and the use of this knowledge to return (home) to suitable low-tide refuges can all be inferred as adaptations to avoid being stranded in unsuitable locations at low tide. How such refuges are initially chosen remains to be discovered but it may be a matter of chance and if the refuge remains suitable then the fish continue to return to it. In this context, it has been observed on the ebbing tide that pools that do not drain at low tide contain many fishes whereas those that dry out have none (Williams, 1957). Homing can therefore be seen partly as a behavioral mechanism that aids the return to suitable low-tide locations. Nevertheless, all observations on homing and fidelity to specific sites have shown that homing is not 100% efficient (Table 2) and fidelity is never complete. Homing and restriction of movement are therefore sufficiently flexible to allow for changes in a fish's currently occupied low-tide refuge. Consequently, rather than remain in, or continue to home to, a site that has been rendered unsuitable, by storms for example, an individual can relocate its center of activity to another site within its usual range of movement. Such flexibility of behavior seems essential for species living in an environment where changes in structure and topography may be frequent. Furthermore, incomplete homing after accidental displacement or straying outside the normal home range may result in the occupation of other equally suitable low-tide refuges whose location the fish will then learn and to which they will subsequently home. In this sense, homing may also serve to distribute the population over all the suitable habitat and prevent concentration in a few particularly favorable areas (Green, 1971a; Yoshiyama *et al.,* 1992).

## C. Searching for a Physiologically Optimum Environment

In a general sense, movement from one place to another is only necessary if the environment or an individual's physiological needs change. If both remain constant no benefit is gained from a change in location. Constancy of conditions is rarely met, particularly in intertidal environments, so that some patterns of movement can be considered as attempts to remain in conditions that are as close to their physiological optimum as possible, a process termed "behavioral enviroregulation" by Neill (1984). Such behavior is normally thought of in terms of physical factors such as temperature or salinity but it could also be extended to biological factors such as food abundance. Indeed, physical and biological optima may be related such that maintaining position in one results in the attainment of another (Miller and Dunn, 1980). Changes in position of intertidal fishes have not often been viewed from this perspective but the observation that young summer flounders in marsh creeks tend to stay within a narrow range of temperatures, salinities, and oxygen concentrations led to the suggestion that their movements partly result from behavioral thermoregulation (Rountree and Able, 1992). Movements interpreted as behavioral thermoregulation have also been recorded in mudskippers. *Periophthalmus koelreuteri* and *Boleophthalmus boddarti* in Kuwait avoid low winter temperatures by remaining in their burrows and reduce their body temperatures in summer by selecting areas of the mudflat where evaporative cooling is high (Tytler and Vaughan, 1983). In Japan, more *Periophthalmus modestus* remain out of water at high tide in summer, possibly to avoid the low oxygen concentrations present in the warmer water at that time (Ikebe and Oishi, 1996).

On a longer time scale, the seasonal movements of many species can be seen as a similar process that allows them to avoid adverse conditions in winter. The probable function of such movements is to reduce unnecessary expenditure of energy in physiologically suboptimal habitats, on osmoregulation for example, and to remain as far as possible in conditions that promote maximal growth.

### D. Spawning

The spawning migrations of several essentially subtidal species into a range of intertidal habitats appears to represent an interesting example of convergent evolution in response to the problem of egg predation. All such species described in the previous sections make their spawning runs on spring tides and deposit their eggs on vegetation or bury them in the substratum high in the intertidal zone. The eggs then develop and the larvae hatch when inundated by the next set of spring tides. It is assumed that such a reproductive strategy places the eggs in areas where predation is likely to be low and/or the eggs are deposited in such numbers as to swamp the predators' ability to inflict significant mortality.

## VI. Control Mechanisms

Underlying all the movement patterns shown by intertidal fishes, whatever their temporal or spatial scale, is the question of the mechanisms controlling such movements. Two basic questions need to be addressed: (1) What cues (Harden Jones, 1984) are used to time movement? and (2) What clues do fish use to determine their direction of movement? Unfortunately in most cases, the answers to these questions must remain speculative because relatively little is known of the appropriate stimuli.

### A. Temporal Cues

Numerous environmental factors are capable of providing information that could be used to signal the appropriate time to begin or end activity and it is likely that no one factor acts in isolation. Visitors are likely to be in continual movement but the rate of movement may vary with the state of the tide (Gibson, 1980) particularly if they are migrating long distances. The stimuli governing the timing of the movements of visitors into and out of the intertidal zone are unknown but for benthic species the rate and direction of change of numerous cues associated with the changing depth of water above them, such as temperature, light intensity, or pressure, could be used for timing purposes. Fishes on or close to the bottom and with reference to a fixed point may be able to use some form of rheotactic response to judge the changing tidal phase. For resident species, the signals for starting and finishing activity may be much more distinct as, for example, when the tide floods a pool or submerges a previously emersed fish. The relative importance of the many stimuli present on these occasions is mostly unknown although in *Oligocottus maculosus* changes in light, temperature, and turbulence are all important, whereas salinity is not (Green, 1971b).

External stimuli no doubt play an important role in determining the timing of activity; in addition, many species from all the main intertidal families possess endogenous rhythms of tidally phased activity (Gibson, 1982; Northcott *et al.*, 1990; Northcott, 1991), providing them with the ability to judge tidal phase without reference to external conditions. It is likely, therefore, that timing and level of activity at any one instant, or indeed whether a fish should become active at all, is determined by reference to both internal and external stimuli.

### B. Directional Clues

Species that enter and leave the intertidal zone on each tide are likely to travel in a net direction that is approximately at right angles to the shore. The simplest hypothesis to account for such movements is passive onshore/offshore transport, but the use of such a mechanism seems unlikely because direct observations of several species have indicated that their tidally synchronized migrations are caused by active and directed swimming (Tyler, 1971; Wirjoatmodjo and Pitcher, 1980; Gibson, 1980; Burrows *et al.,* 1994; Rangeley and Kramer, 1995a,b). It may be that fish swimming off the bottom do make partial use of tidal currents to aid in this movement and that benthic species simply follow the tide by maintaining a constant depth (Gibson, 1973). Observations on demersal (*Tautogolabrus,* Whoriskey, 1983) and pelagic (*Pollachius;* Rangeley and Kramer, 1995a,b) species have shown, however, that they did not select a constant depth and their inshore and offshore movements were in advance of the ebb tide, or lagged behind the flood. The directionality of movement of resident species may not be as great as that of visitors because their activity is limited to the intertidal zone, although field observations on *Clinocottus analis* (Williams, 1957) and *Girella nigricans* (Williams, 1957; Norris, 1963) show that their movements on the rising tide are directed strongly upshore and vice versa. In contrast, the pattern of activity of the sluggish herbivore *Cebidichthys violaceus* obtained by ultrasonic position fixing showed no evidence of up- or downshore direction in its limited movements (Ralston and Horn, 1986).

Some indication of relevant stimuli for at least one species can be gained by the results of homing experiments with *Oligocottus maculosus.* Comparison of the homing success of displaced blinded and anosmic fish with that of untreated controls (Table 2) showed that removal of the visual and olfactory senses impaired the ability to home (Khoo, 1974). The effect of blinding fish was less than that of cauterizing the olfactory organs and in some cases there was no difference between blinded and untreated fish. These results suggest that *O. maculosus* uses both visual and olfactory cues to return to its home pool but that the latter are the most important. Other evidence for the use of visual cues is provided by experiments on the tidepool goby *Bathygobius soporator,* which appears to learn the topography of its immediate environment by swimming over it at high tide. The fish then uses this knowledge to make directed jumps from one pool to another at low tide when disturbed (Aronson, 1951, 1971). Visual reference to the sun's position may also be responsible for the ability of *Periophthalmus sobrinus* to return to its home bank when released in the middle of channels in mangrove creeks whose banks are topographically similar, particularly as this ability is inversely related to the sun's altitude (Berti *et al.,* 1994).

Although it is possible to speculate on the many possible mechanisms such as route following, predator avoidance, foraging into areas of greater food abundance, or the use of photo-, baro- or thermotaxic behaviors, it has to be concluded that very little is known of the directional sign stimuli used by intertidal fishes in their movements or whether the behavioral mechanisms employed are similar to those identified in other more wide-ranging species (McCleave *et al.,* 1984; Quinn and Dittman, 1992).

## VII. Modeling Intertidal Movements

Modeling techniques have been used to explore the implications of the various putative mechanisms involved in large-scale migrations of fish (DeAngelis and Yeh, 1984) with valuable results (Arnold and Cook, 1984). These techniques have rarely been applied to the smaller scale movements discussed in this chapter, even though the regularity of such movements means they can also be considered as forms of migration, using Neill's (1984) definition of migration as "a regular, cyclical pattern of temporal change in spatial distribution." One exception is the model developed for the intertidal movements of young plaice on sandy beaches (Burrows, 1994). In this model, the state of individual fish is characterized by their vertical position on the shore, the fullness of their gut, and the quantity of their energy reserves. Vertical position on the shore in relation to prey and predator abundance determines the probability of acquiring further energy (food) and the risk of predation. Assuming that the fish are visual predators and can only feed during the day, the model computes the optimum position for a fish to occupy at particular tidal states and times of day in relation to different combinations of predator and prey distributions. Although it takes no account of variation in physical factors such as gradients of light and temperature, the model still successfully predicts behavior patterns observed under a range of natural conditions.

## VIII. Conclusions

It will be clear from the numerous examples outlined in the foregoing sections that fishes occupy virtually all intertidal habitats and use them for a variety of purposes, the most common of which are feeding, refuging, and spawning. The extent to which fishes occupy any one habitat depends mainly on the amount of cover available and, for this reason, rocky shores have the most resident species. In other areas where cover is limited, resident species are rare and the intertidal zone is only invaded by fishes at high tide. On sediment shores where the substratum is suitable one group, the mudskippers, have overcome the problem of lack of cover by constructing their own in the form of burrows in the mud.

The dynamic nature of the extent and timing of this occupation must always be considered in any study of intertidal fish ecology. In particular, the timing of sampling will have a considerable effect on the size and species composition of the catches. Fishing at high tide during the day, for example, may give very different results from fishing at low tide or at night. Such differences must obviously be taken into account when attempting to obtain accurate estimates of abundance or describe predator–prey interactions within intertidal fish communities.

Although many more examples of movement patterns of intertidal fishes are likely to be encountered in the future, those already described are sufficient to reveal our ignorance of their controlling mechanisms. Although generally small, such fishes are reasonably accessible and can provide the subjects for numerous experiments in the field of orientation and homing given the increased miniaturization of sophisticated tagging devices and underwater television. Such techniques could allow the design and execution of experiments that are not practical with larger, wider ranging species and make valuable contributions to the field of migration as a whole.

Finally, it will be clear that the world's intertidal zones are extensively used for many purposes by a great range of fishes. The accessibility of these zones and the remarkable adaptations of their inhabitants make them attractive areas for study. Unfortunately, their accessibility also renders them vulnerable to pressure from industrial and recreational use such that their value as a habitat can become seriously degraded. Apart from the intrinsic scientific interest of the resident species, which are usually of little commercial value, all intertidal zones and their adjacent subtidal areas represent nursery grounds for numerous other species that eventually recruit to coastal fisheries. Destruction or degradation of such areas therefore has serious consequences for both the economy and quality of life of many nations. It thus behooves all interested in intertidal fishes, for whatever reason, to make such information widely available in the hope that intertidal zones can be recognized for the invaluable assets they are, rather than simply as sites for effluent discharge or uncontrolled development.

## References


Abou-Seedo, F. S. (1992). Abundance of fish caught by stake traps (hadra) in the intertidal zone in Doha, Kuwait Bay. *J. Univ. Kuwait (Science)* **19,** 91–99.

Abou-Seedo, F., Clayton, D. A., and Wright, J. M. (1990). Tidal and turbidity effects on the shallow-water fish assemblage of Kuwait Bay. *Mar. Ecol. Prog. Ser.* **65,** 213–223.

Ansell, A. D., and Gibson, R. N. (1990). Patterns of feeding and movement of juvenile flatfishes on an open sandy beach. *In* "Trophic Relationships in the Marine Environment, Proceedings of the 24th European Marine Biology Symposium" (M. Barnes and R. N. Gibson, Eds.), pp. 191–207. Aberdeen University Press, Aberdeen.

Arnold, G. P., and Cook, P. H. (1984). Fish migration by selective tidal stream transport: First results with a computer simulation model for the European continental shelf. *In* "Mechanisms of Migration in Fishes" (J. D. McCleave *et al.,* Eds.), pp. 227–261. Plenum Press, New York.

Aronson, L. R. (1951). Orientation and jumping behaviour in the gobiid fish *Bathygobius soporator. Am. Mus. Novit.* **1486,** 1–22.

Aronson, L. R. (1971). Further studies on orientation and jumping behaviour in the gobiid fish, *Bathygobius soporator. Ann. N. Y. Acad. Sci.* **188,** 378–392.

Barton, M. G. (1973). "Studies on the Intertidal Vertical Distribution, Food Habits, and Movements of Five Species of Eel Blennies (Pisces: Stichaeidae and Pholidae) at San Simeon, California." PhD thesis, California State University.

Beebe, W. (1931). Notes on the gill-finned goby. *Zoologica (N. Y.)* **12,** 55–66.

Berger, A., and Mayr, M. (1992). Ecological studies on two intertidal New Zealand fishes, *Acanthoclinus fuscus* and *Forsterygion nigripenne robustum. N. Z. J. Mar. Freshw. Res.* **26,** 359–370.

Berghahn, R. (1987a). The Wadden Sea as a nursery for fish and crustacean species. *In* "Proceedings of the 5th International Wadden Sea Symposium" (S. Tougaard and S. Asbirk, Eds.), pp. 69–85. National Forest and Nature Agency, Esbjerg.

Berghahn, R. (1987b). Effects of tidal migration on growth of 0-group plaice (*Pleuronectes platessa* L.) in the north Frisian Wadden Sea. *Meeresforschung* **31,** 209–226.

Berti, R., Colombini, I., Chelazzi, L., and Ercolini, A. (1994). Directional orientation in Kenyan populations of *Periophthalmus sobrinus* Eggert: Experimental analysis of the operating mechanisms. *J. Exp. Mar. Biol. Ecol.* **181,** 135–141.

Black, R., and Miller, R. J. (1991). Use of the intertidal zone by fish in Nova Scotia. *Environ. Biol. Fish.* **31,** 109–121.

Brillet, C. (1975). Relations entre territoire et comportement agressif chez *Periophthalmus sobrinus* Eggert (Pisces, Periophthalmidae) au laboratoire et en milieu naturel. *Z. Tierpsychol.* **39,** 283–331.

Brown, C. R., Gordon, M. S., and Chin, H. G. (1991). Field and laboratory observations on microhabitat selection in the amphibious Red Sea rockskipper fish *Alticus kirki* (Family Blenniidae). *Mar. Behav. Physiol.* **19,** 1–13.

Burgess, T. J. (1978). The comparative ecology of two sympatric polychromatic populations of *Xererpes fucorum* Jordan and Gilbert (Pisces: Pholididae) from the rocky intertidal zone of central California. *J. Exp. Mar. Biol. Ecol.* **35,** 43–58.

Burrows, M. T. (1994). An optimal foraging and migration model for juvenile plaice. *Evol. Ecol.* **8,** 125–149.

Burrows, M. T., Gibson, R. N., Robb, L., and Comely C. (1994). Temporal patterns of movement in juvenile flatfishes and their predators: Underwater television observations. *J. Exp. Mar. Biol. Ecol.* **177,** 251–268.

Butner, A., and Brattstrom, B. H. (1960). Local movement in *Menidia* and *Fundulus. Copeia* **1960,** 139–140.

Cain, R. L., and Dean, J. M. (1976). Annual occurrence, abundance and diversity of fish in a South Carolina intertidal creek. *Mar. Biol.* **36,** 369–379.

Carlisle, D. B. (1961). Intertidal territory in fish. *Anim. Behav.* **9,** 106–107.

Cattrijsse, A., Makwaia, E. S., Dankwa, H. R., Hamerlynck, O., and Hemminga, M. A. (1994). Nekton communities of an intertidal creek of a European estuarine brackish marsh. *Mar. Ecol. Prog. Ser.* **109,** 195–208.

Colombini, I., Berti, R., Ercolini, A., Nocita, A., and Chelazzi, L. (1995). Environmental factors influencing the zonation and activity patterns of a population of *Periophthalmus sobrinus* Eggert in a Kenyan mangrove. *J. Exp. Mar. Biol. Ecol.* **190,** 135–149.

Colombini, I., Berti, R., Ercolini, A., Nocita, A., and Chelazzi, L. (1996). Foraging strategy of the mudskipper *Periophthalmus sobrinus* Eggert in a Kenyan mangrove. *J. Exp. Mar. Biol. Ecol.* **197,** 219–236.

Craig, P. (1996). Intertidal territoriality and time-budget of the surgeonfish, *Acanthurus lineatus,* in American Samoa. *Env. Biol. Fish.* **46,** 27–36.

Craik, G. J. S. (1981). The effects of age and length on homing performance in the intertidal cottid, *Oligocottus maculosus* Girard. *Can. J. Zool.* **59,** 598–604.

DeAngelis, D. L., and Yeh, G. T. (1984). An introduction to modeling migratory behaviour of fishes. *In* "Mechanisms of Migration in Fishes" (J. D. McCleave *et al.,* Eds.), pp. 445–469. Plenum Press, New York.

Fishelson, L. (1963). Observations on littoral fishes of Israel. I. Behaviour of *Blennius pavo* Risso (Teleostei, Blenniidae). *Isr. J. Zool.* **12,** 67–80.

Fishelson, L., Montgomery, L. W., and Myrberg, A. H. (1987). Biology of surgeonfish *Acanthurus nigrofuscus* with emphasis on changeover in diet and annual gonadal cycles. *Mar. Ecol. Prog. Ser.* **39,** 37–47.

Gerking, S. D. (1959). The restricted movement of fish populations. *Biol. Rev.* **34,** 221–242.

Gersbacher, W. M., and Denison, M. (1930). Experiments with animals in tide pools. *Pub. Puget Sound Biol. Stn* **7,** 209–215.

Gibson, R. N. (1967). Studies of the movements of littoral fish. *J. Anim. Ecol.* **36,** 215–234.

Gibson, R. N. (1969). The biology and behaviour of littoral fish. *Oceanogr. Mar. Biol. Annu. Rev.* **7,** 367–410.

Gibson, R. N. (1972). The vertical distribution and feeding relationships of intertidal fish on the Atlantic coast of France. *J. Anim. Ecol.* **41,** 189–207.

Gibson, R. N. (1973). The intertidal movements and distribution of young fish on a sandy beach with special reference to the plaice (*Pleuronectes platessa* L.). *J. Exp. Mar. Biol. Ecol.* **12,** 79–102.

Gibson, R. N. (1980). A quantitative description of the behaviour of wild juvenile plaice (*Pleuronectes platessa*). *Anim. Behav.* **28,** 1202–1216.

Gibson, R. N. (1982). Recent studies on the biology of intertidal fishes. *Oceanogr. Mar. Biol. Annu. Rev.* **20,** 363–414.

Gibson, R. N. (1988). Patterns of movement in intertidal fishes. *In* "Behavioural Adaptations to Intertidal Life" (G. Chelazzi and M. Vannini, Eds), pp. 55–63. Plenum Press, New York.

Gibson, R. N. (1992). Tidally-synchronised behaviour in marine fishes. *In* "Rhythms in Fishes" (M. A. Ali, Ed.), pp. 55–63. Plenum Press, New York.

Gibson, R. N., Ansell, A. D., and Robb, L. (1993). Seasonal and annual variations in abundance and species composition of fish and macrocrustacean communities on a Scottish sandy beach. *Mar. Ecol. Prog. Ser.* **98,** 89–105.

Gibson, R. N., Robb, L., Burrows, M. T., and Ansell, A. D. (1996). Tidal, diel and longer term changes in the distribution of fishes on a Scottish sandy beach. *Mar. Ecol. Prog. Ser.* **130,** 1–17.

Graham, J. B. (1970). Preliminary studies on the biology of the amphibious clinid *Mnierpes macrocephalus. Mar. Biol.* **6,** 136–140.

Graham, J. B., Jones, C. B., and Rubinoff, I. (1985). Behavioural, physiological and ecological aspects of the amphibious life of the pearl blenny *Entomacrodus nigricans* Gill. *J. Exp. Mar. Biol. Ecol.* **89,** 255–268.

Green, J. M. (1971a). High tide movements and homing behaviour of the tidepool sculpin *Oligocottus maculosus. J. Fish. Res. Bd Can.* **28,** 383–389.

Green, J. M. (1971b). Field and laboratory activity patterns of the tidepool cottid *Oligocottus maculosus* Girard. *Can. J. Zool.* **49,** 255–264.

Green, J. M. (1973). Evidence for homing in the mosshead sculpin (*Clinocottus globiceps*). *J. Fish. Res. Bd Can.* **30,** 129–130.

Haegele, C. W., and Schweigert, J. F. (1985). Distribution and characteristics of herring spawning grounds and description of spawning behaviour. *Can. J. Fish. Aquat. Sci.* **42,** 39–55.

Harden Jones, F. R. (1984). A view from the ocean. *In* "Mechanisms of Migration in Fishes" (J. D. McCleave *et al.,* Eds.), pp. 1–26. Plenum Press, New York.

Healey, M. C. (1971). The distribution and abundance of sand gobies, *Gobius minutus,* in the Ythan estuary. *J. Zool.* **163,** 177–229.

Helfman, G. S., Stoneburner, D. L., Bozeman, E. L., Christian, P. A., and Whalen, R. (1983). Ultrasonic telemetry of American eel movements in a tidal creek. *Trans. Am. Fish. Soc.* **112,** 105–110.

Heymer, A. (1977). Experiences subaquatiques sur les performances d'orientation et de retour au gite chez *Tripterygion tripteronotus* et *Tripterygion xanthosoma* (Blennioidei, Trypterygiidae). *Vie Milieu* **27,** Ser A, 425–435.

Hubbs, C. L. (1921). The ecology and life-history of *Amphigonopterus aurora* and other viviparous perches of California. *Biol. Bull.* **40,** 181–209.

Ikebe, Y., and Oishi, T. (1996). Correlation between environmental parameters and behaviour during high tides in *Periophthalmus modestus. J. Fish Biol.* **49,** 139–147.

Khoo, H. W. (1974). Sensory basis of homing in the intertidal fish *Oligocottus maculosus* Girard. *Can. J. Zool.* **52,** 1023–1029.

Kleypas, J., and Dean, J. M. (1983). Migration and feeding of the predatory fish, *Bairdiella chrysura* Lacépède, in an intertidal creek. *J. Exp. Mar. Biol. Ecol.* **72,** 199–209.

Klumpp, D. W., and Nichols, P. D. (1983). A study of food chains in seagrass communities. II. Food of the rock flathead, *Platycephalus laevigatus* Cuvier, a major predator in a *Posidonia australis* seagrass bed. *Aust. J. Mar. Freshw. Res.* **34,** 745–754.

Kneib, R. T., and Wagner, S. L. (1994). Nekton use of vegetated marsh habitats at different stages of tidal inundation. *Mar. Ecol. Prog. Ser.* **106,** 227–238.

Koop, J. H., and Gibson, R. N. (1991). Distribution and movements of intertidal butterfish *Pholis gunnellus. J. Mar. Biol. Assoc. U.K.* **71,** 127–136.

Kuipers, B. (1973). On the tidal migration of young plaice (*Pleuronectes platessa*) in the Wadden Sea. *Neth. J. Sea Res.* **6,** 376–388.

Laegdsgaard, P., and Johnson, C. R. (1995). Mangrove habitats as nurseries: Unique assemblages of juvenile fish in subtropical mangroves in eastern Australia. *Mar. Ecol. Prog. Ser.* **126,** 67–81.

Lotrich, V. A. (1975). Summer home range and movements of *Fundulus heteroclitus* (Pisces: Cyprinodontidae) in a tidal creek. *Ecology* **56,** 191–198.

Markel, R. W. (1993). An adaptive value of spatial learning and memory in the blackeye goby, *Coryphopterus nicholsi. Anim. Behav.* **47,** 1462–1464.

Marsh, B., Crowe, T. M., and Siegfried, W. R. (1978). Species richness and abundance of clinid fish (Teleostei: Clinidae) in intertidal rock pools. *Zool. Afr.* **13,** 283–291.

Mast, S. O. (1915). The behaviour of *Fundulus,* with especial reference to overland escape from tide-pools and locomotion on land. *J. Anim. Behav. (Boston)* **5,** 341–350.

McCleave, J. D., Arnold, G. P., Dodson, J. J., and Neill, W. H. (1984). "Mechanisms of Migration in Fishes." Plenum Press, New York.

Middaugh, D. P. (1981). Reproductive ecology and spawning periodicity of the Atlantic silverside, *Menidia menidia* (Pisces: Atherinidae). *Copeia* **1981,** 766–776.

Miller, J. M., and Dunn, M. L. (1980). Feeding strategies and patterns of movement in juvenile estuarine fish. *In* "Estuarine Perspectives" (V. S. Kennedy, Ed.), pp. 437–448. Academic Press, New York.

Moring, J. R. (1990). Seasonal absence of fishes in tidepools of a boreal environment (Maine, USA). *Hydrobiologia* **194,** 163–168.

Morton, R. N. (1990). Community structure, density and standing crop of fishes in a subtropical Australian mangrove. *Mar. Biol.* **105,** 385–394.

Moyle, P. B., and Cech, J. J. (1982). "Fishes: An Introduction to Ichthyology." Prentice Hall, Englewood Cliffs, NJ.

Neill, W. H. (1984). Behavioural enviroregulation's role in fish migration. *In* "Mechanisms of Migration in Fishes" (J. D. McCleave *et al.,* Eds), pp. 61–66. Plenum Press, New York.

Norris, K. S. (1963). The functions of temperature in the ecology of the percoid fish *Girella nigricans* (Ayres). *Ecol. Monogr.* **33,** 23–62.

Northcott, S. J. (1991). A comparison of circatidal rhythmicity and entrainment by hydrostatic pressure cycles in the rock goby, *Gobius paganellus* L. and the shanny, *Lipophrys pholis* (L.). *J. Fish Biol.* **39,** 25–33.

Northcott, S. J., Gibson, R. N., and Morgan, E. (1990). The persistence and modulation of endogenous circatidal rhythmicity in *Lipohrys pholis* (Teleostei). *J.Mar. Biol. Assoc. U.K.* **70,** 815–827.

Peterson, G. W., and Turner, R. E. (1994). The value of salt marsh edge vs interior as a habitat for fish and decapod crustaceans in a Louisiana tidal marsh. *Estuaries* **17,** 235–262.

Pfister, C. A. (1995). Estimating competition coefficients from census data: A test with field manipulations of tidepool fishes. *Am. Nat.* **146,** 271–291.

Potts, G. W. (1985). The nest structure of the corkwing wrasse, *Crenilabrus melops* (Labridae: Teleostei). *J. Mar. Biol. Assoc. U.K.* **65,** 531–546.

Quinn, T. P., and Dittman, A. H. (1992). Fishes. *In* "Animal Homing" (F. Papi, Ed.), pp. 145–211. Chapman and Hall, London.

Raffaelli, D., Richner, H., Summers, R., and Northcott, S. (1990). Tidal migrations in the flounder (*Platichthys flesus*). *Mar. Behav. Physiol.* **16,** 249–260.

Ralston, S. L., and Horn, M. H. (1986). High tide movements of the temperate-zone herbivorous fish *Cebidichthys violaceus* (Girard) as determined by ultrasonic telemetry. *J. Exp. Mar. Biol. Ecol.* **98,** 35–50.

Rangeley, R. W., and Kramer, D. L. (1995a). Use of rocky intertidal habitats by juvenile pollock *Pollachius virens. Mar. Ecol. Prog. Ser.* **126,** 9–17.

Rangeley, R. W., and Kramer, D. L. (1995b). Tidal effects on habitat selection and aggregation by juvenile pollock *Pollachius virens* in the rocky intertidal zone. *Mar. Ecol. Prog. Ser.* **126,** 19–29.

Reis, R. R., and Dean, J. M. (1981). Temporal variation in the utilization of an intertidal creek by the bay anchovy (*Anchoa mitchilli*). *Estuaries* **4,** 16–23.

Richkus, W. A. (1978). A quantitative study of intertidepool movement of the wooly sculpin *Clinocottus analis. Mar. Biol.* **49,** 277–284.

Riley, J. D. (1973). Movements of 0-group plaice *Pleuronectes platessa* L. as shown by latex tagging. *J. Fish Biol.* **5,** 323–343.

Robertson, A. I., and Duke, N. C. (1990). Mangrove fish communities in tropical Queensland, Australia: Spatial and temporal patterns in densities, biomass and community structure. *Mar. Biol.* **104,** 369–379.

Rountree, R. A., and Able, K. W. (1992). Foraging habits, growth and temporal patterns of salt-marsh creek habitat use by young-of-the-year summer flounder in New Jersey. *Trans. Am. Fish. Soc.* **121,** 765–776.

Rountree, R. A., and Able, K. W. (1993). Diel variations in decapod crustacean and fish assemblages in New Jersey polyhaline marsh creeks. *Estuar. Coast. Shelf Sci.* **37,** 181–201.

Russell, D. J., and Garrett, R. N. (1983). Use by juvenile barramundi, *Lates calcarifer* (Bloch), and other fishes of temporary supralittoral habitats in a tropical estuary in northern Australia. *Aust. J. Mar. Freshw. Res.* **34,** 805–811.

Santos, R. S. (1986). Capacidade de retorno à área vital, padrão de dispersão e organização social em *Blennius sanguinolentus* Pallas (Pisces: Blenniidae) durante a época de reprodução. *Psicologia* **5,** 121–131.

Santos, R. S., Almada, V. C., and Santos, A. J. F. (1987). Field experiments and observations on homing and territoriality in intertidal blennies. *In* "Ethoexperimental Approaches to the Study of Behaviour" (R. J. Blanchard *et al.,* Eds.), pp. 623–632. Kluwer Academic, Dordrecht.

Santos, R. S., and Nash, R. D. M. (1995). Seasonal changes in a sandy beach fish assemblage at Porto Pim, Faial, Azores. *Estuar. Coast. Shelf Sci.* **41,** 579–591.

Saucerman, S. E., and Deegan, L. A. (1991). Lateral and cross-channel movement of young-of-the-year winter flounder (*Pseudopleuronectes americanus*) in Waquoit Bay, Massachusetts. *Estuaries* **14,** 440–446.

Sheridan, P. F. (1992). Comparative habitat utilization by estuarine macrofauna within the mangrove ecosystem of Rookery Bay, Florida. *Bull. Mar. Sci.* **50,** 21–39.

Smith, K. J., and Able, K. W. (1994). Salt-marsh tide pools as winter refuges for the mummichog, *Fundulus heteroclitus,* in New Jersey. *Estuaries* **17,** 226–234.

Sogard, S. M., Powell, G. V. N., and Holmquist, J. G. (1989). Utilization by fishes of shallow, seagrass-covered banks in Florida Bay. 2. Diel and tidal patterns. *Environ. Biol. Fish.* **24,** 81–92.

Stephens, J. S., Johnson, R. K., Jr., Key, G. S., and McCosker, J. E. (1970). The comparative ecology of three sympatric species of California blennies of the genus *Hypsoblennius* Gill (Teleostomi, Blenniidae). *Ecol. Monogr.* **40,** 213–233.

Summers, R. W. (1980). The diet and feeding behaviour of the flounder *Platichthys flesus* (L.) in the Ythan estuary, Aberdeenshire, Scotland. *Estuar. Coast. Mar. Sci.* **11,** 217–232.

Szedlmayer, S. T., and Able, K. W. (1993). Ultrasonic telemetry of age-0 summer flounder, *Paralichthys dentatus,* movements in a southern New Jersey estuary. *Copeia* **1993,** 728–736.

Taylor, M. H., Leach, G. J., DiMichele, L., Levitan, W. M., and Jacob, W. F. (1979). Lunar spawning cycle in the mummichog, *Fundulus heteroclitus* (Pisces: Cyprinodontidae). *Copeia* **1979,** 291–297.

Therriault, T. W., Schneider, D. C., and Methven, D. A. (1996). The timing of spawning in capelin (*Mallotus villosus* Muller) at a coastal location in eastern Newfoundland. *Polar Biol.* **16,** 201–207.

Thomson, D. A., and Muench, K. A. (1976). Influence of tides and waves on the spawning behaviour of the Gulf of California grunion, *Leuresthes sardinia* (Jenkins and Evermann). *Bull. S. Calif. Acad. Sci.* **75,** 198–203.

Toole, C. L. (1980). Intertidal recruitment and feeding in relation to optimal utilization of nursery areas by juvenile English sole (*Parophrys vetulus:* Pleuronectidae). *Environ. Biol. Fish.* **5,** 383–390.

Tyler, A. V. (1971). Surges of winter flounder, *Pseudopleuronectes americanus,* into the intertidal zone. *J. Fish. Res. Bd Can.* **28,** 1727–1732.

Tytler, P., and Vaughan, T. (1983). Thermal ecology of the mudskippers, *Periophthalmus koelreuteri* (Pallas) and *Boleophthalmus boddarti* (Pallas) of Kuwait Bay. *J. Fish Biol.* **23,** 327–337

Vance, D. J., Haywood, M. D. E., Healy, D. S., Kenyon, R. A., Loneragan, N. R., and Pendrey, R. C. (1996). How far do prawns and fish move into mangroves? Distribution of juvenile banana prawns *Penaeus merguiensis* and fish in a tropical mangrove forest in northern Australia. *Mar. Ecol. Prog. Ser.* **131,** 115–124.

van der Veer, H. W., and Bergman, J. N. (1986). Development of tidally related behaviour of a newly settled 0-group plaice (*Pleuronectes platessa*) population in the western Wadden Sea. *Mar. Ecol. Prog. Ser.* **31,** 121–129.

Walker, B. W. (1952). A guide to the grunion. *Calif. Fish Game* **38,** 409–420.

Weisberg, S. B., Whalen, R., and Lotrich, V. A. (1981). Tidal and diurnal influence on food consumption of a saltmarsh killifish *Fundulus heteroclitus. Mar. Biol.* **61,** 243–246.

Wells, B., Steele, D. H., and Tyler, A. V. (1973). Intertidal feeding of winter flounders (*Pseudopleuronectes americanus*) in the Bay of Fundy. *J. Fish. Res. Bd. Can.* **30,** 1374–1378.

Whoriskey, F. C., Jr. (1983). Intertidal feeding and refuging by cunners, *Tautogolabrus adspersus* (Labridae). *Fish. Bull.* **81,** 426–428.

Williams, G. C. (1957). Homing behavior of California rocky shore fishes. *Univ. Calif. Pub. Zool.* **59,** 249–284.

Wirjoatmodjo, S., and Pitcher, T. J. (1980). Flounders follow the tides to feed: Evidence from ultrasonic tracking in an estuary. *Estuar. Coast. Shelf Sci.* **19,** 231–241.

Wolff, W. J., Manders, M. A., and Sandee, A. J. J. (1981). Tidal migration of plaice and flounders as a feeding strategy. *In* "Feeding and Survival Strategies of Estuarine Organisms" (N. V. Jones and W. J. Wolff, Eds), pp. 159–171. Plenum Press, London.

Wright, J. M., Clayton, D. A., and Bishop, J. M. (1990). Tidal movements of shallow water fishes in Kuwait Bay. *J. Fish Biol.* **37,** 959–974.

Yamahira, K. (1994). Combined effects of tidal and diurnal cycles on spawning of the puffer, *Takifugu niphobles* (Tetraodontidae). *Environ. Biol. Fish.* **40,** 255–261.

Yamahira, K., Kikuchi, T., and Nojima, S. (1996). Age specific food utilization and spatial distribution of the puffer, *Takifugu niphobles,* over an intertidal sandflat. *Environ. Biol. Fish.* **45,** 311–318.

Yoshiyama, R. M., Gaylord, K. B., Philippart, M. T., Moore, T. R., Jordan, J. R., Coon, C. C., Schalk, L. L., Valpey, C. J., and Tosques, I. (1992). Homing behaviour and site fidelity in intertidal sculpins (Pisces: Cottidae). *J. Exp. Mar. Biol. Ecol.* **160,** 115–130.

# 7

# Sensory Systems

Kurt Kotrschal

*Zoologisches Institut der Universität Wien and Konrad-Lorenz-Forschungsstelle, Grünau, Austria*

## I. What Sensory Systems Are "Typical" in Intertidal Fish?

Unfortunately, it is no simple task to produce a reasonable contribution on the sensory systems of intertidal fishes. As is true for any single trait in any living organism, sensory and nervous systems of fish are a product of two ingredients, evolutionary traditions and more recent adaptations. This explains why we do not observe a single common solution for the perceptual apparatus in all intertidal fishes. All over the world, intertidal habitats are utilized by a range of families, each with group-specific traditional traits. Moreover, the intertidal zone provides a range of habitats, which are not used the same way by all. Some residents endure the changing conditions within their zones, others move with the tides; some dwell on rocks, pounded by heavy wave action, while others prefer sheltered areas. Some semi-intertidal fishes leave during low tide, others retreat to tidepools. There, the water is clear and calm, and predators may be sparse. When threatened, some species of combtooth blennies (Abel, 1962), mudskippers (Harms, 1929), and other fish from the water–air interface try to escape downward, some horizontally, and still others may even leap out of the water. Thus, we cannot expect to find ecomorphologically homogeneous trends for all the intertidal fishes, even though specific constraints may have led to some sensory convergence.

In addition to complexities arising from evolutionary history and from habitat diversity, our knowledge of the sensory world of intertidal fishes is patchy, at best. These fishes live near the air–water interface, in the presence of turbulence and obstructed vision, salinity and temperature fluctuations, and other intertidal features that not only require physiological adaptations, but certainly constrain sensory capabilities. Furthermore, intertidal fishes may, depending on their habitat, face different physicoenvironmental challenges (in particular, see Chapters 3–6, this volume). Information is scattered widely over the literature and it is difficult to read trends from these materials. Therefore, instead of constructing a nonexistent "typical" resident intertidal fish, I sketch some aspects of diversity and constraints on sensory equipment in intertidal fishes.



*Intertidal Fishes: Life in Two Worlds*

## II. Approaches

In the search for general principles of relationships between habitat and sensory equipment, two comparative approaches are used: either the investigation of *divergence* within closely related species along a habitat cline or the investigation of *convergence* of species from distantly related taxa living in the same habitats (Goldschmid and Kotrschal, 1989; Motta and Kotrschal, 1992). The latter is done by examining shared and analogous features of species living in a particular habitat, for example, in the surge zone. In the search for divergence, speciose taxa are needed, out of which different representatives can be found in different intertidal and subtidal habitats, thereby creating the opportunity to show correlated differentiation of homologous characters. These requirements are met by only a few taxa, among them the suborder Blennioidei (6 families, 732 species), especially the families Blenniidae (345 species) and Chaenopsidae (56 species), the family Gobiidae (>212 genera, 1875 species), which, in the mudskippers, also has a number of supratidal fishes), the family Cottidae (>300 species), and the family Gobiesocidae (120 species; Nelson, 1994).

Sensory ecomorphology has not been particularly well investigated in any of these candidate groups. The most information is available for the blennioids and, in the case of Neuromasts, for the gobiids. But even there, coverage of the field is more a matter of well-educated guesswork than knowledge, and substantial speculation is necessary. For this reason, even the deep sea (!) has sneaked into this review on intertidal fishes. Still, a benevolent editor and patient readers provided, it is worth trying.

## III. Toward the Surface

The water–air interface of the intertidal zone is one of the most abrupt physical and ecological transitions on earth (Ricketts and Calvin, 1968). In sheltered areas, this may be a stable line, shifting with the tides. However, wave exposure makes the surge zone a high-energy, high-noise, turbulent environment. From a comparison of closely related species (Figure 1), it appears that heads become more bulky (Kotrschal, 1988a, 1989), bodies shorten, and head appendages as well as fins are reduced in size from sheltered toward exposed areas or toward the surface in wave-exposed habitats. These changes probably serve drag reduction, but also leave less body surface for external chemosensory and mechanosensory organs. Furthermore, some of the sensory barbels, cirri, etc., would be of little use anyway (see below). This principle is exemplified by two species of *Hypsoblennius,* the mainly subtidal bay blenny, *H. gentilis,* which avoids extreme surge, and the intertidal barnaclebill, *H. brevipinnis,* which utilizes empty barnacle shells in zones of heavy surge (Thomson *et al.,* 1979). The former shows an elongated body and long supraorbital tentacles (cirri, which are skin protrusions), whereas the body of the latter is stout and tentacles are short. The same trend can be found in *Acanthemblemaria* blennies (Figure 1). Stoutness is not a general adaptation related to emergence, because other fishes with a semiamphibious lifestyle, such as the algae-scraping salariin blennies (Magnus, 1963), and the gobiid mudskippers, retain the elongated body shapes typical of their groups (Figure 2).

Phylogenetic inertia certainly contributes to body shapes. In the Mediterranean, two sympatric combtooth blennies occur in the rocky, wave-exposed intertidal zone: *Coryphoblennius galerita,* with an elongated and slender body, and *Parablennius trigloides,* with a

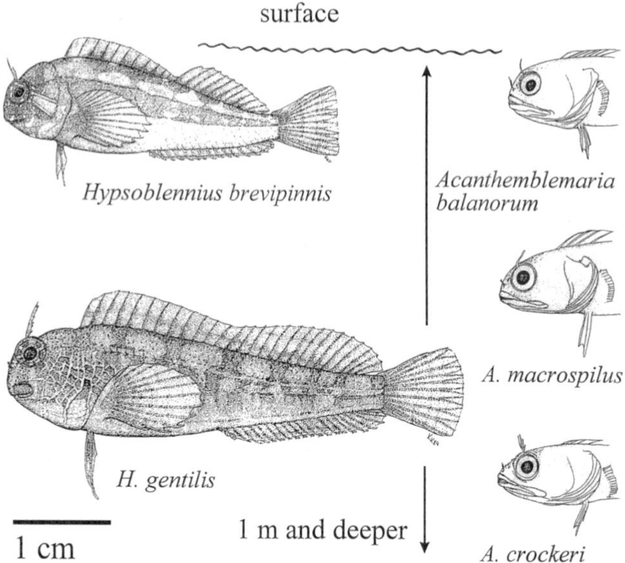


**Figure 1.** Representatives of the fully wave-exposed, high intertidal zone are bulkier and have shorter tentacles and fins than species from the lower intertidal zone or from sheltered areas. This is exemplified by two species of *Hypsoblennius* combtooth blennies (left) and three *Acanthemblemaria* tube blennies from the Gulf of California (Thomson *et al.*, 1979; compare Kotrschal, 1988a, 1989). Approximate scale valid for all species. All drawings by the author.

large head and stout body. *C. galerita* may avoid the snorkeler by leaping out of the water. *P. trigloides,* however, takes refuge in submerged crevices (Abel, 1962). In both species, head tentacles are small (*C. galerita*) or absent (*P. trigloides*), which is quite a contrast to subtidal relatives of the latter, such as *P. gattorugine* or *P. tentacularis,* which show formidable supraorbital tentacles (Figure 3).

## A. Chemosenses of the Skin

In general, external taste buds are only moderately developed in the perciform fishes. This seems particularly true for intertidal fishes exposing themselves to surge or for species that have adopted an amphibious lifestyle. No external taste buds have been found in mudskippers, even though these gobies are plentifully equipped with taste buds within the oral cavity. In contrast, hillocked taste buds are present on nasal and supraorbital tentacles of the semiemergent *Salarias* (Blenniidae; Harms, 1929).

The reduction of head appendages may be caused by the need to reduce drag in the high-energy environment and by the fact that these flaps of skin collapse when emerged. Supraorbital tentacles are sexually dimorphic and may play a primary role in species recognition, but they also carry chemosensory cells (Von Bartheld and Meyer, 1985). Therefore, such tentacles may, secondarily, also function as chemosensory antennae.

All the intertidal combtooth blennies and tube blennies utilize empty and often tightly fitting holes, cavities created by invertebrates (Hastings, 1986; Kotrschal, 1988b). Thus, one could argue that a reduction in size of head tentacles may enable these species to fit

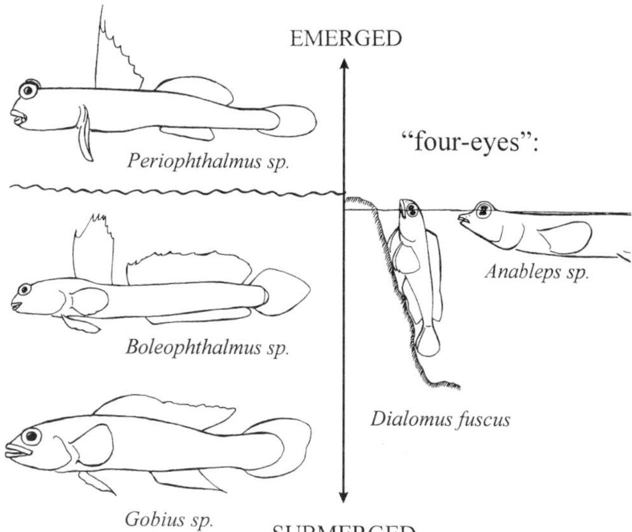


**Figure 2.** The position of eyes and the development of the anterior dorsal fin as a visual signal in eastern Asiatic Gobiidae varies with lifestyle: submerged in *Gobius* sp., at the interface in *Boleophthalmus* sp., and amphibic in *Periophthalmus* sp. (Harms, 1929). Interface specialists, such as the cyprinodontiform *Anableps,* have developed a bipartite eye, for simultaneous aerial and aquatic vision. The Galapagos clinid *Dialomus* is not adapted for interface vision (Munk, 1969), but for aerial vision (Stevens and Parsons, 1980). Fish approximately one-third of natural size, not drawn to scale. All drawings by the author.

into these shelters. However, subtidal or calm-water blennioids dwelling in holes also show formidable tentacles (Fig. 3). Therefore, the reduction in tentacle size may rather be attributed to turbulence than to a specific use of shelters.

## B. Olfaction

Although many intertidal fishes, particularly perciforms, seem microsmathic (Zeiske *et al.,* 1992), this only means that olfaction as a chemosense used to detect distant stimulus sources is of limited use in a turbulent environment (Atema, 1988). However, olfaction still seems important in pheromonal communication between mates (Laumen *et al.,* 1974; Liley, 1982; Losey, 1969).

In combtooth blennies as well as in tube blennies (Hastings, 1986; Kotrschal, 1988b), for example, egg-guarding males develop fin glands (Kotrschal *et al.,* 1984a), which may either provide an anti-fouling agent for the protection of the eggs or produce pheromones to attract females or motivate them to enter the hole and spawn. In the emerged *Periophthalmus,* the nasal organ is still well developed and fish in air responded to airborne olfactory stimuli over a distance of some 10 cm (Harms, 1929). The semiemerged combtooth blenny *Salarias* sp. has a spacious nasal sac with well-vascularized linings, even though the olfactory mucosum is small. Therefore, the nasal cavity probably functions as an accessory olfactory organ in these blennies (Harms, 1929). Potentially high variability in the

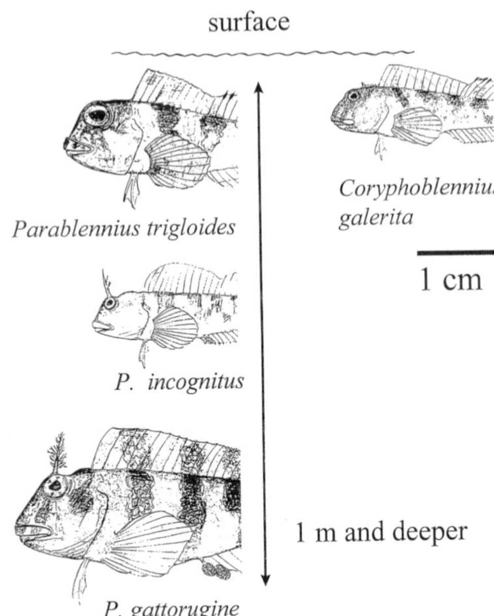


**Figure 3.** Four species of Mediterranean combtooth blennies (Blenniidae). Both *P. trigloides* and *C. galerita* occupy the high intertidal and show no supraorbital tentacles. Among the three species of *Parablennius*, *P. trigloides* from the high intertidal has the bulkiest head and no tentacles, whereas species from the lower intertidal (*P. incognitus*) or from the subtidal (*P. gattorugine*) show large supraorbital tentacles. Approximate scale bar valid for all species. All drawings by the author.

environmental salinity of intertidal habitats does not necessarily affect the functioning of olfactory receptor cells (Shoji *et al.*, 1994).

## C. Hair Cell Senses

In general, the system of free neuromasts, which is, among other functions, used in locating prey (Bleckmann *et al.*, 1989; Montgomery, 1989), seems only moderately developed in intertidal fishes from high-energy habitats, but may be highly developed in intertidal fishes from low-energy environments (Vischer, 1990). Free neuromasts located within mucus-filled epidermal pits still seem to be present in the emerged and semiemerged *Boleophthalmus* and *Periophthalmus* (Harms, 1929). In general, the skin of these fishes is characterized by adaptations for aerial respiration rather than by sensory differentiation (Bridges, 1993; Rauther, 1910). The functional significance of (free) neuromasts in an amphibious fish remains unclear.

The reduction of the posterior lateral line canal to the anterior half of the body (Figure 4) may be attributed to dwelling in tight holes. From the posterior end of the trunk canal, a series of free neuromasts extends toward the base of the tail. Only the anterior third of the body sticks out of a hole and is therefore useful as a lateral line (or chemosensory) antenna.

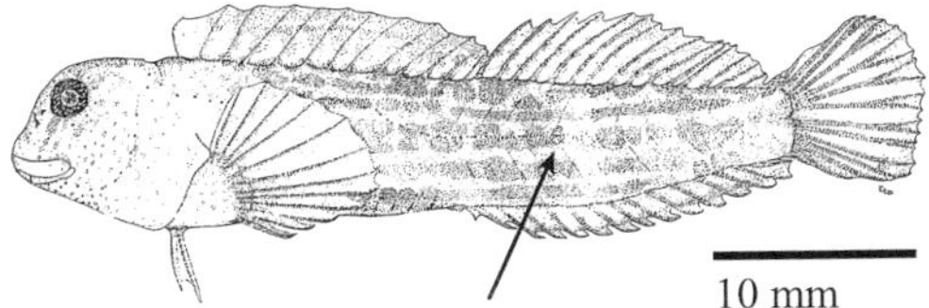


**Figure 4.** A Mediterranean combtooth blenny (Blenniidae, female) showing the pores of head and trunk lateral line canals. As typical for this family, there is a simple head canal system with narrow channels and pores and the trunk canal covers only the anterior half of the body. The posterior end of the trunk canal is marked with an arrow. The trunk canal is supported by bone rings, which are scale rudiments, whereas the skin is otherwise devoid of scales. Scale bar approximate. Drawing by the author.

Consequently, all species of blennies known feature the standard version of tubular lateral line head canals (Bath, 1965, 1976). In most intertidal species, these canals are not widened or diverticulated, as is the case in a number of pelagic and calm-water species (Coombs *et al.,* 1988; Coombs and Janssen, 1989; Webb, 1989). Exceptions are the stichaeids, where the lateral line canals are highly branched (Bleckmann and Munz, 1990).

Hardly anything is known about hearing capacities in intertidal fishes. Acute hearing was, for example, reported in mudskippers (Harms, 1929). It can be assumed (Evans, 1940; Popper and Coombs, 1980) that in the wave-exposed intertidal zone acoustic signals (Ladich, 1991) are masked by noise and, therefore, hearing is hampered. This is certainly not true for fishes from sheltered areas. Hearing and lateral line seem particularly important in low-noise environments, hence their predominance in the deep sea (Fine *et al.,* 1987; see Section VI). A peculiar solution to avoid the masking of a signal by turbulence-generated noise is to use the substrate for communication, as, for example, the riffle-spawning freshwater mottled sculpin does (Whang and Janssen, 1990). Despite the lack of data, it may be speculated that the transmission of vibrations via the substrate could also be an option for sound-producing fish in the intertidal zone, such as toadfish (Bass, 1990).

## D. Vision

Although potentially useful for benthic fishes, the directional components of lateral line and taste stimuli (Moore and Atema, 1991) may be lacking in highly turbulent environments, forcing these fishes to rely mainly on vision. In accordance with the general "Grundfischkonvergenzen" (convergent morphologies in benthic fishes; Abel, 1961), eyes may assume an extremely dorsal position. This is generally true in intertidal fishes, most of which are benthic. This is particularly true for some emergent fish, such as the mudskippers (Figure 2; Harms, 1929; Walls, 1967). This high-up positioning of the eye increases the visual field and may improve binocular vision. In fact, when comparing three genera of gobies, the submerged *Gobius,* the episodically emergent *Boleophthalmus,* and the nearly permanently emerged *Periophthalmus* (Figure 2), eyes increasingly assume a position on top of the head. Some species that are particularly adapted for the water surface have

resolving power in comparison with three larger species (Zaunreiter, 1990). Such visual constraints due to small size may particularly hamper activity and feeding at dusk and dawn. The visual systems of some extremely small blennies, such as the Mediterranean *Lipophrys,* the Caribbean tube blennies in general, or the Gulf of California *Coralliozetus* may likewise be affected by such size constraints.

## V. The Chemosensory Anterior Dorsal Fin in Rocklings: A Case Study

Rocklings are epibenthic, cigar-shaped, gadid fishes of 20–40 cm standard length. Two genera, *Ciliata* and *Gaidropsarus* can be found in the rocky intertidal zone of the North Atlantic coast of Europe, which is rich in tidepools; hence their common name. Rocklings are crepuscular or nocturnal. In comparison with other Gadidae, we would expect vision in rocklings to be of minor importance (Evans, 1940). However, they are plentifully equipped with chemosenses, olfaction, and external taste. Taste buds are spread all over the body surface and aggregate at barbels and at some barbel-like, free rays of the ventral fins. There is a well-developed lateral line system. And there is also another sensory organ, which makes rocklings unique among fishes.

All rocklings have two dorsal fins: the posterior is a conventional, unpaired teleost fin, whereas the anterior dorsal fin (ADF) has evolved into a peculiar chemosensory organ (Kotrschal *et al.,* 1984b; Figure 5). In contrast to the body surface and to the barbel-like ventral fins, where taste buds abound, the epidermis of 60–80 minute, vibratile rays of the ADF is packed with solitary chemosensory cells (SCCs), aggregated at densities of approximately 100,000 per square millimeter (Kotrschal and Whitear, 1988). These rays are only a few millimeters long, 1/5 to 1/10 mm in diameter, and are connected by a fin membrane only at their base (Figure 5). The ADF does not contain taste buds, except for the enlarged, anteriormost ray.

SCCs resemble chemosensory cells in taste buds (Whitear, 1965, 1971), and occur within the epidermis of virtually all primary aquatic vertebrates: agnathans, all groups of jawed fishes, and some amphibians (Kotrschal, 1991, 1996; Whitear, 1992). SCC densities in freshwater teleosts may vary between a few hundreds and a few thousands per square millimeter (Kotrschal, 1992; Kotrschal *et al.,* 1997). In comparison to taste buds, the small size of sensory cells and their scattered distribution have hampered investigation of this, the second epidermal chemosensory system. Therefore, the rockling ADF has gained some importance as a model system for SCC research. In the ADF, SCCs are innervated by fine-caliber fibers of the recurrent facial nerve (Kotrschal and Whitear, 1988; Kotrschal, 1991, 1996; Kotrschal *et al.,* 1990b, 1993a; Whitear and Kotrschal, 1988). A few hundred SCCs converge onto a single nerve fiber, indicating a low threshold of this system. Only nerve fibers from SCCs of the ADF terminate in a peculiar, dorsal subdivision of the brain stem facial lobe (Kotrschal and Whitear, 1988). Secondary brain connections, however, were hardly qualitatively different from the taste bud system (Kotrschal and Finger, 1996), supporting the idea that SCCs may be considered a taste subsystem.

The vibratory, lateral 10–20 Hz movements of single ADF rays are coordinated into a regular, sine-wave undulation, running caudally (Kotrschal *et al.,* 1993b; Peters *et al.,* 1987). At rest, the fin is regularly turned on and off every few seconds. When the fish is aroused (by any stimulus) or moves, the fin is permanently undulating. This activity does not contribute to locomotion and can be interpreted as a sampling movement, because

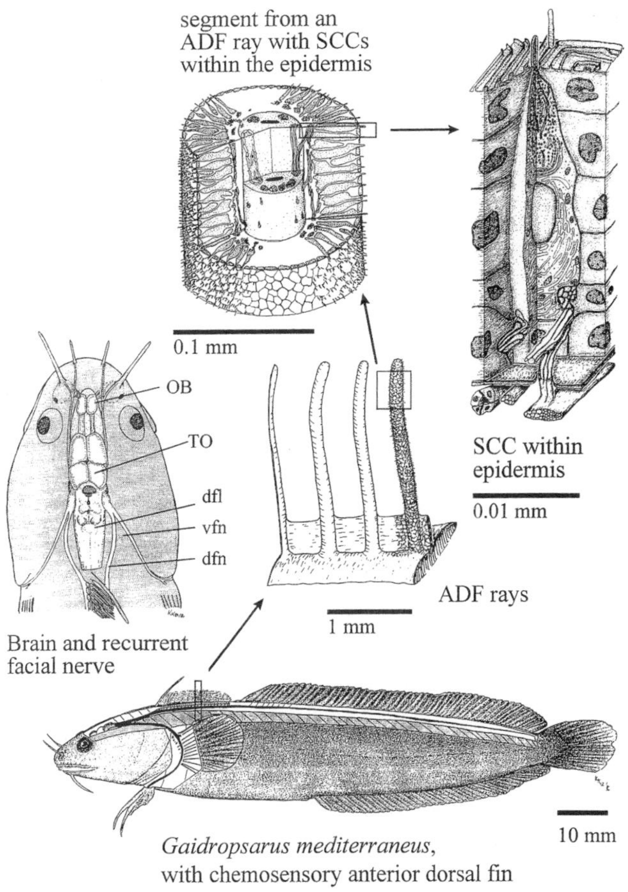


**Figure 5.** The chemosensory anterior dorsal fin (ADF) in intertidal rocklings. Bottom: Habitus of the three-bearded rockling, *Gaidropsarus mediterraneus* showing the ventral branch of the recurrent facial nerve, which innervates pectoral and pelvic fins, and its dorsal branch, which innervates the ADF and the skin of the trunk. Middle left: Dorsal view of the brain of the five-bearded rockling, *Ciliata mustela,* showing large olfactory bulbs (OB) and a moderately sized tectum opticum (TO). The bean-shaped swelling of the dorsal brain stem (cerebellum removed), the dorsal facial lobe (dfl), is the exclusive primary representation area of the nerve fibers innervating the ADF (Kotrschal and Whitear, 1988; habitus and *in situ* brain redrawn from Kotrschal *et al.,* 1984b); vfn, ventral recurrent facial nerve; dfn, dorsal recurrent facial nerve. Semi-diagram shows a few ADF rays (center). Note the connection of fin rays with only a rudimentary fin web. A segment of a ray (block diagram top left) illustrates the tight packing of solitary chemosensory cells (SCCs) along the rostral and caudal faces of the ray epidermis. The block diagram top right shows fine structural details of a single SCC, its position within the epidermis, and connections with afferent fibers of the facial nerve. The villus on top of the cell contacts the ambient water; facial nerve fibers make 1–3 synapses with the SCC base. Scale bars approximative. All drawings by the author (see Kotrschal, 1996; Kotrschal and Finger, 1996; Kotrschal *et al.,* 1990b, for more extensive coverage of the topic).

responses to chemical stimulation were only recorded via a silver electrode, implanted into the recurrent facial nerve, when the fin was undulating (Peters *et al.,* 1987, 1991).

In contrast to taste buds, the system responds only to a small range of natural stimuli, mainly body mucus dilutions of other fish and dilutions of fish bile, but not to amino acids (Kotrschal *et al.,* 1996). Only rocklings with an intact anterior dorsal fin respond to fish mucus dilutions with an immediate decrease in breathing frequency (as judged from opercular movements, a general arousal response in fish) and with an increase of time the ADF was sampling (Kotrschal *et al.,* 1993b). These responses can also be elicited by mechanical disturbance (tapping at the wall of the tank), but not by stimulation with shrimp extract or other natural mixtures of amino acids.

These results support the conclusion that the aggregated SCC system in the rockling ADF in particular, and possibly also the distributed SCC systems of other fish, may function to direct the attention of individuals toward the upstream presence of other fish (Kotrschal, 1996). This information probably has to be refined (conspecific, heterospecific, friend or foe) by integration with the olfactory input (Kotrschal, 1995). If olfaction is blocked by plugging the nostrils, fish cease to respond behaviorally to the specific SCC stimuli (Essler and Kotrschal, 1995; Kotrschal *et al.,* 1993b; Kotrschal and Essler, 1995), even when the SCC system remains intact. We therefore dispute the earlier belief that obliteration of the olfactory system is a useful method to discriminate between taste and olfactory functions, because this procedure may affect not only the sensory periphery, but also central mechanisms.

Our limited knowledge of the ADF organs in rocklings limits us to speculating that this may be a key innovation (*sensu* Liem and Wake, 1985) for the successful colonization of certain intertidal habitats, but its usefulness is not necessarily limited to the intertidal zone. Rocky tidepools are generally highly structured by boulders and crevices, and water currents may periodically change directions. It can therefore be speculated that it is beneficial for a fish to stay within its shelter and still collect information about the upstream presence of fish (competitors, predators, mates) before leaving. However, most rocklings are subtidal and the ADF may have enabled the rockling genera *Ciliata* and *Gaidropsarus* to be successful there. A chemosensory antenna such as the ADF is certainly useful in all habitats where structure limits the transmission of visual and mechanical stimuli.

## VI. Clues from the Deep Sea?

Another inference toward sensory systems of intertidal fish may be made by examining environmental extremes opposite to intertidal habitats. Probably the best example is the deep sea. Whereas the intertidal zone is characterized by the highest possible variability and stochasticity in physical parameters of all marine habitats known, the deep sea is characterized by the highest possible stability and predictability. The intertidal zone is a very "noisy" environment, where currents, turbulence, obstacles such as rocks, and salinity and temperature differences severely interfere with the transmission and directional information of stimuli. The deep sea is the opposite.

We may, therefore, consider an equation with a single unknown variable, that is, the specifics of the sensory equipment of intertidal fishes. Known are the habitat characteristics in the deep pelagial, and the sensory adaptations found there (Figure 6). Known also are the major habitat characteristics of the intertidal zone. What we do not know, but may be able to reasonably infer from this equation, are possible convergence rules for sensory equipment in the intertidal zone.

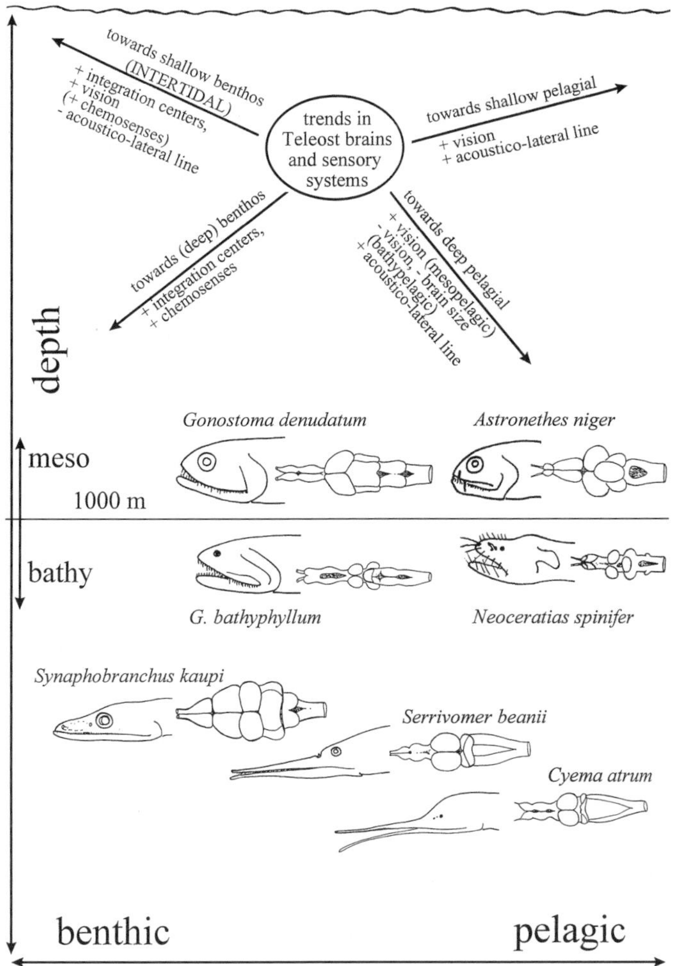


**Figure 6.** Brains reflect sensory orientation in fish (Evans, 1940; Gadidae; Kotrschal and Palzenberger, 1992, Cyprinidae; Huber *et al.,* 1997, cichlidae; Kotrschal and Huber, 1998, review; Van Staaden *et al.,* 1994/1995, Cichlidae). From general trends in the relationships between habitats and brains, it can be inferred that integration and visual centers should prevail in the wave-exposed intertidal. In sheltered areas, chemosenses and/or lateral line may be developed in addition. A comparison with deep-sea fishes may be revealing with respect to the brain and sensory systems of intertidal fishes. In physical parameters, the intertidal is the opposite of the deep sea. Thus, opposite trends can also be expected for brains and sensory systems. Mesopelagic fishes (*G. denudatum, A. niger*) show highly developed acoustico-lateralis centers and a small telencephalon (integration center). The brains of bathypelagic species (*G. bathyphyllum, N. spinifer*) are among the relatively smallest of all vertebrates. The brains are reduced in virtually all their subdivisions except for highly developed acoustico-lateralis centers. A comparison of bathybenthic to bathypelagic deep-sea eels (bottom) shows that the brain itself, particularly the olfactory bulbs and the telencephalon, is well developed in a benthic species (*S. kaupi*), but reductions parallel to other bathypelagic species can be found in the pelagic representatives (*C. atrum*). Fishes and brains redrawn and rearranged from Marshall (1967b, 1979).

Whereas mesopelagic fishes may show reasonably sized brains, with well-developed optic tecta, brains of bathypelagic fishes (below 1000 m of depth) are generally reduced in size (Marshall, 1957, 1967a; Figure 6). This is in line with the general reduction in body tissues as an adaptive response to low food density. Also, due to the stability of their habitat, reflexes and simple stereotyped behaviors requiring only a small brain may suffice for the major tasks of life, such as feeding, predator avoidance, and mating. Relative to their body size, bathypelagic fishes in fact have the smallest brains of all known vertebrates (Fine *et al.,* 1987). Particularly reduced are the integration centers, telencephalon, tectum opticum, and diencephalon. The olfactory complex (olfactory mucosa, bulbus olfactorius, and olfactory telencephalon) of some deep-sea teleosts are distinctly larger in males than in females, probably in species featuring male competition for mates (Marshall, 1967b). Particularly well developed are the rhombencephalic centers for hearing and the lateral line (Fine, 1987). The brain stem taste lobes are barely discernible in the bathypelagic fishes, but in parallel to olfaction, may be of some importance in bathybenthic representatives, for example, taste in the macrourids, olfaction in the deep-sea eels (Marshall, 1967b; Figure 6).

In contrast to the deep-sea fishes, which, aside from a few exceptions (Munk, 1982) have only (in cases highly specialized) rod retinae, the retinae of intertidal fishes are dominated by cones. Therefore, most diurnal intertidal fishes may have elaborate color vision, at least trichromatic, but most likely tetrachromatic, with another receptor in the ultraviolet range, as is the case in many shallow-water fishes (Lythgoe, 1988). This, together with a regular retinal mosaic, would also allow these fishes to see patterns of polarized light (Hawryshyn, 1992) and provide a mechanism for the excellent homing abilities shown in some intertidal fishes (Abel, 1962; Gibson, 1967, 1982).

In a comparative investigation of the brains of European cyprinids, Kotrschal and Palzenberger (1992) found that toward the benthos, species tended to have well-developed chemosenses, particularly taste, but less-developed vision and acoustico-lateralis lobes (Schellart, 1992). Inverse trends were found toward the open water. In essence, the same relationships between habitats and sensory orientation were found by Huber *et al.* (1997) and Van Staaden *et al.* (1994/1995) in African cichlids. However, in contrast to the cyprinids, the most variable part of the brain of cichlids was found to be the integrative (not covarying with olfactory bulb size) telencephalon, which was large in fishes from complex, benthic habitats, but smaller in fishes from soft bottoms.

If general trends, opposite to the deep-sea fishes, but parallel to the shallow-water cyprinids and cichlids, also apply to the intertidal fishes, representatives from wave-swept coasts should have relatively large brains with large centers for vision and integration. Lateral line and hearing as well as olfaction and taste should be of minor importance. Even though potentially useful for benthic fishes, these senses are constrained in wave-exposed areas by the destruction of directional information in the stimuli by turbulence. This does not apply to sheltered areas, particularly at soft bottoms where, besides vision, chemosenses and lateral line may also be of some importance.

## VII. Conclusions: Does the Intertidal Produce Convergent Sensory Systems?

Yes and no. There are a few trends in turbulent, high-energy habitats and at the water–air interface. Body appendages that may serve sensory functions, such as tentacles, barbels, and fins, are reduced in size for three major reasons: (1) Appendages contribute to frictional drag. Probably for the same reason, fish in wave-exposed habitats show a convergence toward short, stout bodies. Also, fins on the ventral sides (pectoral, ventral, and anal rays), even when elongated, are needed for clinging to the substrate and are thus covered with a tough "cuticle" (Whitear, 1986), rendering these fins unlikely places for carrying sensory equipment, such as taste buds or solitary chemosensory cells. (2) The turbulence in the surf zone creates an excessive background noise and destroys directional information in the domains of chemical stimuli and pressure waves. Thus far-range and close-range chemosenses, such as olfaction and external taste, as well as the mechanosensory lateral line system may be of limited use. Thus, these systems are generally reduced in intertidal fishes from surge habitats. (3) The taste buds and neuromasts at the body surface of fishes function only when immersed in water. Therefore, fish that are episodically emergent or that have adopted an amphibious lifestyle, such as mudskippers and some salariin blennies, generally have reduced their typically aquatic sensory systems in favor of vision.

Thus, physical constraints at the exposed intertidal zone have probably selected for visual predominance. This may also explain why only species out of visually oriented groups, such as gobies and blennies, are able to utilize the wave-swept shore, but not fishes, such as catfish or eels, lacking "preadaptation" in the visual domain.

As the chemosensory specialization of the Atlantic rocklings and the free neuromasts in gobies show, the considerations pertinent to the wave-swept shore may not apply to sheltered areas or to intertidal fishes that use tidepools at low tide. Tidepools are clear-water, low-noise habitats and may be utilized by fishes with any sensory equipment and diurnal or nocturnal activities. Thus, tidepools are utilized by a range of fishes, intertidal or subtidal, and are often nurseries for subtidal species such as sparids, mullids, and others. Therefore, common trends in the sensory systems of fishes may be found for certain habitats, but not for the intertidal zone in general.

## Acknowledgments

The crucial help of A. Goldschmid in search of some cryptic literature and the permanent support of the Verein der Förderer der Konrad Lorenz Forschungsstelle are gratefully acknowledged. The contributions of two reviewers substantially improved this chapter.

## References

Abel, E. (1961). Über die Beziehung mariner Fische zu Hartbodenstrukturen. *Sitzungsber. Österr. Akad. Wiss., Abt. I* **170,** 223–263.

Abel, E. (1962). Freiwasserbeobachtungen an Fischen im Golf von Neapel als Beitrag zur Kenntnis ihrer Ökologie und ihres Verhaltens. *Int. Rev. Ges. Hydrobiol.* **47,** 219–290.

Atema, J. (1988). Distribution of chemical stimuli. *In* "Sensory Biology of Aquatic Animals" (J. Atema, R. R. Fay, A. N. Popper, and W. N. Tavolga, Eds.), pp. 29–56. Springer Verlag, New York.

Bass, A. H. (1990). Sounds from the intertidal zone: Vocalizing fish. *Am. Sci.* **40,** 249–258.

Bath, H. (1965). Vergleichende biologisch-anatomische Untersuchungen über die Leistungsfähigkeit der Sinnesorgane für den Nahrungserwerb, ihre gegenseitige Abhängigkeit und ihre Beziehung zum Bau des Gehirns bei *Blennius gattorugine* Brünn., *Blennius galerita* L. und *Gobius cruentatus* L., GM. *Z. Wiss. Zool.* **172,** 347–375.

Bath, H. (1976). Revision der Blenniini (Pisces: Blenniidae). *Seuckenberg Biol.* **57,** 167–234.

Bleckmann, H., and Munz, H. (1990). Physiology of lateral line mechanoreceptors in a teleost with highly branched lateral line. *Brain Behav. Evol.* **35,** 240–250.

Bleckmann, H., Tittel, G., and Blübaum-Gronau, E. (1989). Lateral line system of surface-feeding fish: Anatomy, physiology and behavior. *In* "The Mechanosensory Lateral Line. Neurobiology and Evolution" (S. Coombs, P. Görner, and H. Münz, Eds.), pp. 501–526. Springer Verlag, New York.

Bone, Q., and Marshall, N. B. (1982). "The Biology of Fishes." Blackie & Son, Glasgow.

Bridges, C. R. (1993). Ecophysiology of intertidal fish. *In* "Fish Ecophysiology" (J. C. Rankin and F. B. Jensen, Eds.), pp. 375–400. Chapman & Hall, London.

Collin, S. P., and Pettigrew, J. D. (1988). Retinal topography in reef teleosts. I. Some species with well-developed areae, but poorly-developed streaks. *Brain Behav. Evol.* **31,** 269–282.

Coombs, S., and Janssen, J. (1989). *In* "The Mechanosensory Lateral Line. Neurobiology and Evolution" (S. Coombs, P. Görner, and H. Münz, Eds.), pp. 299–322. Springer Verlag, New York.

Coombs, S., Janssen, J., and Webb, J. C. (1988). *In* "Sensory Biology of Aquatic Animals" (J. Atema, R. R. Fay, A. N. Popper, and W. N. Tavolga, Eds.), pp. 553–594. Springer Verlag, New York.

Engström, K. (1963). Cone types and cone arrangements in teleost retinae. *Acta Zool.* **44,** 23–65.

Essler, H., and K. Kotrschal, K. (1995). Fische verändern ihr Schwimmverhalten bei Wahrnehmung von Nahrungs- und Feindgeruch: Versuche an Elritzen (*Phoxinus phoxinus* L.). *Ö. Fisch.* **48,** 83–89.

Evans, H. M. (1940). "Brain and Body of Fish. A Study of Brain Patterns in Relation to Hunting and Feeding in Fish." The Technical Press, London.

Fernald, R. (1984). Vision and behavior in African Cichlid fish. *Am. Sci.* **72,** 58–65.

Fine, M. L., Horn, M. H., and Cox, B. (1987). *Acanthonus armatus,* a deep-sea teleost with a minute brain and large ears. *Proc. R. Soc. London B* **230,** 257–265.

Gibson, R. N. (1967). Studies of the movements of littoral fish. *J. Anim. Ecol.* **36,** 215–234.

Gibson, R. N. (1982). Recent studies on the biology of intertidal fishes. *Oceanogr. Mar. Biol. Annu. Rev.* **20,** 363–414.

Goldschmid, A., and Kotrschal, K. (1989). Ecomorphology: Developments and concepts. *Fortschr. Zool., Suppl.* **35,** 501–512.

Guthrie, D. M. (1986). Role of vision in fish behaviour. *In* "The Behaviour of Teleost Fishes" (T. J. Pitcher, Ed.), pp. 75–113. Croom & Helm, London.

Harms, J. W. (1929). Die Realisation von Genen und die consecutive Adaptation. I. Phasen in der Differenzierung der Anlagenkomplexe und die Frage der Landtierwerdung. *Z. Wiss. Zool.* **133,** 211–397.

Hastings, P. A. (1986). Habitat selection, sex ratio and sexual selection in *Coralliozetus angelica* (Blennioidea: Chaenopsidae). *In* "Indo-Pacific Fish Biology" (T. Uyeno, R. Arai, T. Taniuchi, and K. Matsura, Eds.), pp. 785–793. Proc. 2nd Int. Conf. Indo-Pacific Fishes.

Hawryshyn, C. W. (1992). Polarization vision in fish. *Am. Sci.* **80,** 164–175.

Huber, R., van Staaden, M., Kaufman, L. S., and Liem, K. F. (1997). Microhabitat use, trophic patterns and the evolution of brain structure in African cichlids. *Brain Behav. Evol.,* **50,** 167–182.

Illich, I., and Kotrschal, K. (1990). Depth distribution and abundance of northern Adriatic littoral rocky reef blennioid fishes. *P.S.Z.N.I.: Mar. Ecol.* **11,** 277–289.

Karsten, H. (1923). Das Auge von *Periophthalmus koelreuteri. Jenaische Zeitschr. Naturwiss.* **59,** 60–77.

Kishida, R. (1979). Comparative study on the teleostean optic tectum. Lamination and cytoarchitecture. *J. Hirnforsch.* **20,** 57–67.

Kotrschal, K. (1981). "Die Anatomie des Gehirns von *Blennius incognitus* (Bath 1968) (Blenniidae, Perciformes, Teleostei) unter besonderer Berücksichtigung des aminergen Systems." PhD thesis, University of Salzburg.

Kotrschal, K. (1988a). A catalogue of skulls and jaws of eastern tropical Pacific blennioid fishes (Blennioidei: Teleostei): A proposed evolutionary sequence of morphological change. *Z. Zool. Syst. Evol.-forsch.* **26,** 442–466.

Kotrschal, K. (1988b). Blennies and endolithic bivalves: Differential utilization of shelter in Adriatic Blenniidae (Pisces: Teleostei). *P.S.Z.N.I.: Mar. Ecol.* **9,** 253–269.

Kotrschal, K. (1989). Trophic ecomorphology in eastern Pacific blennioid fishes: Character transformation of oral jaws and associated change of their biological roles. *Environ. Biol. Fish.* **24,** 199–218.

Kotrschal, K. (1991). Solitary chemosensory cells—Taste, common chemical sense or what? *Rev. Fish Biol. Fish.* **1,** 3–22.

Kotrschal, K. (1992). Quantitative scanning electron microscopy of solitary chemoreceptor cells in cyprinids and other teleosts. *Environ. Biol. Fish.* **35,** 273–282.

Kotrschal, K. (1995). Ecomorphology of solitary chemosensory cell systems in fish: A review. *Environ. Biol. Fish.* **44,** 143–155.

Kotrschal, K. (1996). Solitary chemosensory cells: Why do primary aquatic vertebrates need another taste system? *TREE* **11,** 110–114.

Kotrschal, K., Weisl, H., and Goldschmid, A. (1984a). Mehrzellige Drüsen in der Epidermis der unpaaren Flossen bei den Blenniiden (Blenniidae, Teleostei). Polycellular glands in the median fins of blennies (Blenniidae, Teleostei). *Z. Mikrosk.-anat. Forsch., Leipzig* **98,** 184–192.

Kotrschal, K., Adam, H., and Whitear, M. (1984b). Morphology and functional features of the first dorsal fin in *Gaidropsarus mediterraneus* (Gadidae, Teleostei). *Zoomorphology* **104,** 365–372.

Kotrschal, K., and Whitear, M. (1988). Chemosensory anterior dorsal fin in rocklings (*Gaidropsarus* and *Ciliata,* Teleostei, Gadidae): Somatotopic representation of the Ramus recurrens facialis as revealed by transganglionic transport of HRP. *J. Comp. Neurol.* **268,** 109–120.

Kotrschal, K., and Junger, H. (1988). Patterns of brain morphology in mid-European cyprinids (Cyprinidae, Teleostei): A quantitative histological study. *J. Hirnforsch.* **29,** 341–353.

Kotrschal, K., Adam, H., Brandstätter, R., Junger, H., Zaunreiter, M., and Goldschmid, A. (1990a). Larval size constraints determine directional ontogenetic shifts in the visual system of teleosts. A mini-review. *Z. Zool. Syst. Evol.-forsch.* **28,** 166–182.

Kotrschal, K., Kinnamon, J. C., and Royer, S. M. (1990b). High-voltage electron microscopy and 3-D reconstruction of solitary chemosensory cells and Di-I labeling of primary afferent nerves. *Proc. XII. Int. Congr. Electron Microsc.,* 412–413.

Kotrschal, K., and Palzenberger, M. (1992). Neuroecology of cyprinids (Cyprinidae, Teleostei): Comparative, quantitative histology reveals diverse brain patterns. *Environ. Biol. Fish.* **33,** 135–152.

Kotrschal, K., Whitear, M., and Finger, T. (1993a). Spinal and facial innervation of the skin in the gadid fish *Ciliata mustela* (Teleostei). *J. Comp. Neurol.* **330,** 1–11.

Kotrschal, K., Peters, R., and Atema, J. (1993b). Sampling and behavioral evidence for mucus detection in a unique chemosensory organ: The anterior dorsal fin in rocklings (*Ciliata mustela,* Gadidae, Teleostei). *Zool. Jb. Physiol.* **97,** 47–67.

Kotrschal, K., and Essler, H. (1995). Goals and approaches in the analysis of locomotion in fish, with a focus on laboratory studies. *Rev. Fisher. Sci.* **3,** 171–200.

Kotrschal, K., and Finger, T. (1996). Secondary connections of the dorsal and ventral facial lobes in a teleost fish, the rockling (*Ciliata mustela*). *J. Comp. Neurol.* **370,** 415–426.

Kotrschal, K., Peters, R. C., and Doving, K. B. (1996). Chemosensory and tactile nerve responses from the anterior dorsal fin of a rockling, *Gaidropsarus vulgaris* (Gadidae, Teleostei). *Primary Sensory Neuron* **1,** 297–309.

Kotrschal, K., Krautgartner, W.-D., and Hansen, A. (1997). Ontogeny of the solitary chemosensory cells in zebrafish, *Danio rerio. Chem. Senses* **22,** 111–118.

Kotrschal, K., and Huber, R. (1998). Fish brains: Evolution and environmental relationships. *J. Fish Biol. Fish,* in press.

Ladich, F. (1991). Fische schweigen nicht. Lautbildung, Hören und akustische Kommunikation bei Fischen. *Naturwiss. Rdsch.* **44,** 379–384.

Laumen, J., Pern, U., and Blüm, V. (1974). Investigations on the function and hormonal regualtion of the anal appendices in *Blennius pavo* (Risso). *J. Exp. Zool.* **190,** 47–56.

Liem, K. F., and Wake, D. B. (1985). Morphology: Current approaches and concepts. *In* "Functional Vertebrate Morphology" (M. Hildebrand, D. M. Bramble, K. F. Liem, and D. B. Wake, Eds.), pp. 366–377. Belknap-Harvard University Press, Cambridge, MA.

Liley, N. R. (1982). Chemical communication in fish. *Can. J. Fish. Aquat. Sci.* **39,** 22–35.

Losey, G. S., Jr. (1969). Sexual pheromone in some fishes of the genus *Hypsoblennius* Gill. *Science* **163,** 181–183.

Lythgoe, J. N. (1988). Light and vision in the aquatic environment. *In* "Sensory Biology of Aquatic Animals" (J. Atema, R. R. Fay, A. N. Popper, and W. N. Tavolga, Eds.), pp. 57–82. Springer Verlag, New York.

Magnus, D. B. E. (1963). *Alticus saliens,* ein amphibisch lebender Fisch. *Nat. Mus.* **93,** 128–132.

Marshall, N. B. (1957). "Tiefseebiologie." Gustav Fischer Verlag, Jena.

Marshall, N. B. (1967a). The organization of deep-sea fishes. *Stud. Trop. Oceanogr. Miami* **5,** 473–479.

Marshall, N. B. (1967b). The olfactory organs of bathypelagic fishes. *Symp. Zool. Soc. London* **19,** 57–70.

Marshall, N. B. (1979). "Developments in Deep-Sea Biology." Blandford Press, Poole.

Montgomery, J. C. (1989). Lateral line detection of planktonic prey. *In* "The Mechanosensory Lateral Line. Neurobiology and Evolution" (S. Coombs, P. Görner, and H. Münz, Eds.), pp. 561–574. Springer Verlag, New York.

Moore, P. A., and Atema, J. (1991). Spatial information in the three-dimensional fine structure of an aquatic odor plume. *Biol. Bull.* **181,** 408–418.

Motta, P. J., and Kotrschal, K. (1992). Correlative, experimental, and comparative evolutionary approaches in ecomorphology. *Neth. J. Zool.* **42,** 400–415.

Munk, O. (1969). The eye of the four-eyed fish *Dialomus fuscus* (Pisces, Blennioidea, Clinidae). *Vidensk. Meddr. Dansk Naturh.Foren.* **132,** 7–24.

Munk, O. (1982). Cones in the eye of the deep-sea teleost *Diretmus argenteus. Vision Res.* **22,** 179–181.

Nelson, J. S. (1994). "Fishes of the World," 3rd ed. Wiley Interscience, New York.

Peters, R. C., Kotrschal, K., and Steenderen, G. W. (1987). A chemoreceptive function for the anterior dorsal fin in rocklings (*Gaidropsarus* and *Ciliata:* Teleostei, Gadidae): Electrophysiological evidence. *J. Mar. Biol. Assoc. U.K.* **67,** 819–823.

Peters, R., Kotrschal, K., and Krautgartner, W. D. (1991). Solitary chemoreceptor cells of *Ciliata mustela* (Gadidae, Teleostei) are tuned to mucoid stimuli. *Chem. Senses* **16,** 31–42.

Popper, A. N., and Coombs, S. (1980). Auditory mechanisms in teleost fishes. *Am. Sci.* **68,** 429–440.

Rauther, M. (1910). Die accessorischen Atmungsorgane der Knochenfische. *Periophthalmus koelreuteri. Erg. Fortschr. Zool.* **2,** 531–537.

Ricketts, E. F., and Calvin, J. (1939/1968). "Between Pacific Tides," 4th ed. Stanford University Press, Stanford, CA.

Ridet, J.-M., and Bauchot, R. (1990). Analyse quantitive de l'encèphale des Tèlèostèens: caractères evolutifs et adaptatifs de l'encèphalisation. II. Le grandes subdivisions encèphaliques. *J. Hirnforsch.* **31,** 433–458.

Schellart, N. A. M. (1992). Interrelations between the auditory, the visual and the lateral line systems of teleosts; a mini-review of modelling sensory capabilities. *Neth. J. Zool.* **42,** 459–477.

Shoji, T., Fujita, K., Ban, M., Hiroi, O., Ueda, H., and Kurihara, K. (1994). Olfactory responses of chum salmon to amino acids are independent of large differences in salt concentrations between fresh and sea water. *Chem. Senses* **19,** 609–615.

Sivak, J. G. (1988). Optics of amphibious eyes in vertebrates. *In* "Sensory Biology of Aquatic Animals" (J. Atema, R. R. Fay, A. N. Popper, and W. N. Tavolga, Eds.), pp. 467–486. Springer Verlag, New York.

Stevens, L. K., and Parsons, K. E. (1980). A fish with a double vision. *Nat. Hist.* **89,** 62–67.

Thomson, D. A., Findley, L. T., and Kerstitch, A. N. (1979). "Reef Fishes of the Sea of Cortez. The Rocky Shore Fishes of the Gulf of California." Wiley, New York.

Van Staaden, M., Huber, R., Kaufman, L., and Liem, K. (1994/1995). Brain evolution in cichlids of the African Great Lakes: Brain and body size, general patterns and evolutionary trends. *Zoology* **98,** 165–178.

Vischer, H. A. (1990). The morphology of the lateral line in three species of Pacific cottoid fishes. *Experientia* **46,** 244–250.

Von Bartheld, C. S., and Meyer, D. L. (1985). Trigeminal and facial innervation of cirri in three teleost species. *Cell Tiss. Res.* **241,** 615–622.

Walls, G. L. (1967). "The Vertebrate Eye and Its Adaptive Radiation." Hafner Publ., New York.

Webb, P. W. (1984). Form and function in fish swimming. *Sci. Am.* **251,** 72–82.

Webb, J. F. (1989). Gross morphology and evolution of the mechanoreceptive lateral-line system in teleost fishes. *Brain Behav. Evol.* **33,** 34–53.

Whang, A., and Janssen, J. (1990). Sound production through the substrate during reproductive behavior in the mottled sculpin, *Cottus bairdi* (Cottidae). *Environ. Biol. Fish.* **40,** 141–148.

Whitear, M. (1965). Presumed sensory cells in fish epidermis. *Nature (London)* **208,** 703–704.

Whitear, M. (1971). Cell specialization and sensory function in fish epidermis. *J. Zool. London* **163,** 237–264.

Whitear, M. (1986). Epidermis. *In* "The Biology of the Integument," Vol. 2, "Vertebrates" (J. Bereiter-Hahn, A. G. Matoltsy, and K. S. Richards, Eds.), pp. 8–38. Springer-Verlag, Berlin.

Whitear, M. (1992). Solitary chemoreceptor cells. *In* "Chemoreception in Fishes" (T. J. Hara, Ed.), pp. 103–125. Chapman and Hall, London.

Whitear, M., and Kotrschal, K. (1988). Chemosensory anterior dorsal fin in rocklings (*Gaidropsarus* and *Ciliata,* Teleostei, Gadidae): Activity, fine structure and innervation. *J. Zool. (London)* **216,** 339–366.

Winkelmann, E., and Winkelmann, L. (1968). Vergleichend histologische Untersuchungen zur funktionellen Morphologie des Tectum opticum verschiedener Teleostier. *J. Hirnforsch.* **10,** 1–16.

Zander, C. D. (1972). Beiträge zur Ökologie und Biologie von Blenniiden (Pisces) des Mittelmeeres. *Helgol. Wiss. Meeresunters.* **23,** 193–231.

Zander, C. D. (1983). Terrestrial sojourns of two Mediterranean blennioid fish (Pisces, Blennioidei, Blenniidae). *Senckenberg. Marit.* **15,** 19–26.

Zaunreiter, M. (1990). "Zwischenartliche Differenzierung des optischen Systems bei Mediterranen Blenniiden (Blenniidae, Perciformes, Teleostei)." Ph.D. thesis, University of Salzburg.

Zaunreiter, M., Kotrschal, K., Goldschmid, A., and Adam, H. (1985). Ecomorphology of the optic system in 5 species of blennies (Blenniidae, Perciformes, Teleostei). *Fortschr. Zool.* **30,** 731–734.

Zaunreiter, M., Junger, H., and Kotrschal, K. (1991). Ecomorphology of the cyprinid retina: A quantitative histological study on ontogenetic shifts and interspecific variation. *Vis. Res.* **31,** 383–394.

Zeiske, E., Theisen, B., and Breucker, H. (1992). Structure, development and evolutionary aspects of the peripheral olfactory system. *In* "Fish Chemoreception" (T. J. Hara, Ed.), pp. 11–39. Chapman & Hall, London.

# 8

# Intertidal Spawning

Edward E. DeMartini

*National Marine Fisheries Service, Southwest Fisheries Science Center, Honolulu, Hawaii*

## I. Introduction

Fishes that spawn intertidally can be divided into two basic functional groups: (1) intertidal residents that live between the tide marks continuously during all or part of their lives and (2) migrants. True residents live exclusively in the intertidal after benthic settlement from planktonic larvae, whereas partial residents inhabit the intertidal continuously during only a part of their postsettlement lives (Gibson, 1969). Most true resident intertidal fishes are small bodied (<10 cm long) and site attached (Gibson, 1982); these two factors are likely related because small body size in teleosts is in part an adaptation to the size distribution of available shelter (Miller, 1979). Intertidal fishes also include tidal migrants and seasonal visitors that emigrate subtidally to avoid stranding at lower tide levels (Gibson, 1969). Many of these more vagile fishes are larger bodied and relatively wide ranging (Gibson, 1982). Most resident species, including partial residents that live intertidally as adults during the breeding season, spawn within the intertidal zone. Other species migrate into the intertidal to spawn during higher water levels, including some otherwise strictly subtidal species (Gibson, 1982). The distinction between residence and tidal/seasonal migration is key to understanding the life histories, including reproduction and spawning, of the resident intertidal and other species of fishes that spawn intertidally.

I review herein the intertidal spawning patterns of the fishes of rocky shores and other intertidal habitats. Nearly all of the latter are tidally migrant subtidal species, with the exception of some cyprinodontids like *Fundulus heteroclitus* (Kneib, 1987) that shelter in intertidal pools during low tides and gobiid mudskippers (Clayton, 1993) that remain active atop tropical mudflats when the habitat is emergent. Several famous tidal migrants are known, including species that spawn in sandy surf zones: California and Gulf grunions, *Leuresthes* spp. (Moffatt and Thomson, 1978); on cobble beaches: capelin, *Mallotus villosus* (Templeman, 1948); and on salt marsh vegetation: mummichogs and killifishes, *Fundulus* spp. (Taylor, 1984). Several tidal migrant spawners (e.g., grunion, capelin) are economically valuable. Migrant spawners in the rocky intertidal also include some species (herring, *Clupea pallasi*) of great economic importance (Blaxter and Hunter, 1982).

In this review, I consider the spawning patterns of resident and tidal/seasonal migrant species; I draw on examples of the latter to better illustrate general patterns. My specific

*Intertidal Fishes: Life in Two Worlds*

objectives in this chapter are, first, to review the taxonomic distribution of intertidal spawning by teleost fishes. I next summarize relevant reproductive biology and identify major environmental factors influencing the occurrence of intertidal spawning by resident and migrant species. I also evaluate available evidence for the relative importance of these factors. In conclusion, I discuss evidence for the likely ultimate causes of intertidal spawning and suggest needed topics and avenues of research.

## II. Taxonomic Survey

Resident intertidal fishes primarily comprise specialized Perciformes whose basic characteristics include small body size, compressed or depressed body shape, marked reduction or loss of swim bladders, thick skin, copious mucus, and fins specialized for clinging or attaching to substrates (Gibson, 1969; Horn and Gibson, 1988). A taxonomic listing of resident species of intertidal spawners is largely equivalent to a list of intertidal residents, because virtually all intertidal residents spawn intertidally (Gibson, 1982; Table 1).

Tidal migrant spawners comprise many of the same higher order taxa to which resident species belong, but in addition include larger bodied and more vagile taxa of Perciformes

**Table 1.** Taxonomic Distribution of Intertidal Spawning by Resident Intertidal and Tidal/Seasonal Migrant Subtidal Fishes

| Order | Family | Resident | Migrant |
|---|---|---|---|
| Osmeriformes | Galaxiidae | | x |
| | Osmeridae | | x |
| Clupeiformes | Clupeidae | | x |
| Cyprinidontiformes | Cyprinodontidae | x | x |
| Atheriniformes | Atherinidae | | x |
| Gasterosteiformes | Gasterosteidae | x | x |
| Scorpaeniformes | Hexagrammidae | | x |
| | Cottidae | x | x |
| | Liparidae | x | |
| | Cyclopteridae | | x |
| Perciformes | Embiotocidae | x | x |
| | Labridae | | x |
| | Trichodontidae | | x |
| | Blenniidae | x | |
| | Chaenopsidae | x | |
| | Clinidae | x | |
| | Labrisomidae | x | |
| | Pholidae | x | |
| | Stichaeidae | x | |
| | Tripterygiidae | x | x |
| | Gobiidae | x | x |
| | Gobiesocidae | x | |
| | Zoarcidae | | x |
| Tetraodontiformes | Tetraodontidae | | x |
| Batrachoidiformes | Batrachoididae | | x |

*Note.* Assignment of families within orders is based on Nelson (1994).

(e.g., Labridae) and other teleosts (e.g., puffers, Tetraodontiformes; Table 1) that emigrate subtidally at lower water levels.

The taxonomic representations of resident and tidal migrant fishes thus greatly overlap, but small-bodied, highly specialized Perciformes (Blennioidea in particular) are disproportionately represented in the residents (Table 1).

# III. Reproductive Biology

## A. Modes of Reproduction

Intertidal fishes, like teleosts in general, exhibit a diversity of reproductive modes ranging from oviparity with no parental care ("egg scatterers") or care of demersal eggs ("egg guarders") to viviparity ("internal brooders": Balon, 1975). Viviparity has independently evolved from oviparity in numerous teleost groups (Wourms and Lombardi, 1992). Most of the fishes that reproduce intertidally, however, are oviparous (Table 2). Among oviparous intertidal spawners, most taxa provide some type of parental care, ranging from the fanning and mouthing of eggs for sanitation or ventilation to the physical defense of eggs against cannibals or extraspecific predators (reviewed by Coleman, Chapter 9, this volume).

**Table 2.** Distribution of Reproductive Modes among Families of Intertidal Spawning Fishes

| Family | Oviparous | Viviparous | Source |
|---|---|---|---|
| Galaxiidae | x | | McDowall, 1968 |
| Osmeridae | x | | Loosanoff, 1937 |
| Clupeidae | x | | Jones, 1972 |
| Cyprinodontidae | x | | Taylor and DiMichele, 1983 |
| Atherinidae | x | | Middaugh, 1981 |
| Gasterosteidae | x | | MacDonald *et al.*, 1995a |
| Hexagrammidae | x | | DeMartini, 1986 |
| Cottidae | x | | Koya *et al.*, 1994; Munehara *et al.*, 1989 |
| Liparidae | x | | Marliave and Peden, 1989 |
| Cyclopteridae | x | | Goulet and Green, 1988 |
| Embiotocidae | | x | Hubbs, 1921 |
| Labridae | x | | Potts, 1985 |
| Trichodontidae | x | | Marliave, 1980 |
| Plesiopidae | x | | Jillett, 1968 |
| Blenniidae | x | | Patzner, 1983 |
| Chaenopsidae | x | | Hastings, 1986 |
| Clinidae | x | x | Coyer, 1982; Veith, 1979 |
| Labrisomidae | x | | Petersen, 1988 |
| Pholidae | x | | Hughes, 1986 |
| Stichaeidae | x | | Coleman, 1992 |
| Tripterygiidae | x | | Berger and Mayr, 1992 |
| Gobiidae | x | | Miller, 1984 |
| Gobiesocidae | x | | Allen, 1979 |
| Zoarcidae | | x | Yonge, 1949 |
| Tetraodontidae | x | | Yamahira, 1994 |
| Batrachoididae | x | | Arora, 1948 |

Several taxa of resident and migrant intertidal spawners are viviparous—the extreme development of parental care. The best known examples are live-bearing blennioids of the genus *Clinus* in South Africa (Prochazka, 1994) and several embiotocids (*Micrometrus* spp.) of the west coast of North America that are resident in intertidal pools (Hubbs, 1921; Warner and Harlan, 1982) or that migrate intertidally to copulate or parturate (*Cymatogaster aggregata:* Gordon, 1965; Wiebe, 1968; DeMartini, 1988a). The eelpout *Zoarces viviparus* also migrates seasonally into the intertidal to give birth to live young (Yonge, 1949). Clinid viviparity is typified by superfoetation (multiple concurrent broods in varying stages of embryonic development: Veith, 1979; Gunn and Thresher, 1991; Prochazka, 1994). All embiotocids, including the intertidal breeders, produce a single brood per reproductive season (Baltz, 1984).

Apparently only one family of Scorpaeniformes (Cottidae) has evolved copulation with oviparity, perhaps as a bridge between oviparity and viviparity (Hubbs, 1966). Males in many genera of Pacific cottids have intromittent organs (Bolin, 1947) that are used in copulation (e.g., see Morris, 1952, 1956; Krejsa, 1964). For at least two species, eggs are not fertilized until they are laid even though sperm are stored in the female's genital tract (Munehara *et al.,* 1989, 1991). For other copulating species of cottids, fertilization may be internal (Hubbs, 1966; Ragland and Fischer, 1987).

Viviparous and other copulating species are pair spawners by necessity. Most small-bodied, site-attached intertidal fishes that are oviparous also are pair spawners (but see below). Many tidal migrant species, however, are group spawners, perhaps in part because many of these more vagile fishes (e.g., atherinids) typically school in midwaters.

The males in several specialized groups (Gobiidae, demersal egg-laying blennioids) have evolved highly specialized testicular glands, sperm ducts, or other accessory reproductive organs whose functions are incompletely understood. Hypothesized functions of these structures include sperm storage, the production of sexual pheromones and secretions for sperm nourishment, facilitation of sperm–egg contact during fertilization, adherence of fertilized eggs, and the production of biocides to protect developing spawn (Miller, 1984; Fishelson, 1991; Rasotto, 1995). An association with paternal care is suggested in several species having dimorphic males, in which the accessory glands are hypertrophied during the spawning season in nest-tending males. In these species, accessory glands are undeveloped in other male types (see below), which have relatively huge testes (de Jonge *et al.,* 1989; Ruchon *et al.,* 1995).

Many taxa of marine and freshwater fishes are recognized as having multiple male spawning types that include satellites, sneakers, or both (Taborsky, 1994). Whether such males are genetically distinct from territorial, pair-spawning males is inadequately understood for most species. It has been hypothesized that the reproductive success of these nonterritorial male types varies in a frequency-dependent manner that is inverse to the success of territorial males (Taborsky, 1994; Gross, 1996). Intertidal spawners appear similar in this regard, and examples include both resident intertidal species such as *Blennius sanguinolentus* (Santos and Almada, 1988; Ruchon *et al.,* 1995) and tidal/seasonal migrant spawners like the batrachoidid *Porichthys notatus* (Brantley and Bass, 1994). Multiple male types are likely to have evolved in many lineages of resident intertidal fishes, because small-bodied, site-attached fishes with polygynous mating systems and pair spawning predominate in the group (Coleman, Chapter 9, this volume).

The mapping of mating types and spawning modes on the phylogenies (Brooks and McLennan, 1991) of intertidally spawning fishes is beyond the scope of this review. Such potentially informative studies are needed and encouraged.

## B. Demersal Egg and Spawn Types

### 1. Egg Types

All known species of resident and intertidal migrant intertidal fishes, if oviparous, spawn demersal eggs. The majority of subtidal fishes spawn planktonic eggs (Wootton, 1994), although many small-bodied species of shallow, structured subtidal habitats spawn demersal eggs (e.g., see Hart, 1973; Thresher, 1984). The conspicuous lack of planktonic spawners among intertidal fishes has been noted by Taylor (1984, 1990), who speculated that water-column spawning has not evolved in intertidal fishes because planktonic eggs cannot endure the turbulence and other physical rigors of the intertidal zone. Denny and Shibata (1989) argue, based on theoretical models, that constraints on fertilization success limit water-column spawning in intertidal invertebrates. To date no one has experimentally tested whether turbulence precludes planktonic spawning by fishes in the intertidal zone.

Among demersal spawners, the eggs of both intertidal and subtidal spawners range greatly in quantity (size, number) and in the relation between the quantity of eggs and size of parent (Duarte and Alcaraz, 1989; Elgar, 1990). Subtidal spawners are represented by two extremes: small-bodied fishes that produce few relatively large eggs, and large-bodied species that produce many relatively small eggs (Barlow, 1981; Duarte and Alcaraz, 1989; Wootton, 1992). Intertidal spawners in general produce relatively large, hence relatively few, eggs (Elgar, 1990), compared to subtidal spawners. Among small-bodied, structure-associated fishes, batch fecundity ranges from <100 to $10^5$ eggs with a median of $10^3$, and egg size varies from <0.5 mm in greatest dimension in several small-bodied gobies (Miller, 1979, 1984) to >0.5 cm diameter in toadfishes like *Porichthys notatus* (DeMartini, 1990, 1991). Eggs are commonly >1 mm diameter in intertidal species that spawn demersal eggs.

The shapes of demersal eggs are their most obvious morphological adaptations. The eggs of most intertidal and subtidal spawners are spherical; in intertidal spawners, spherical eggs are perhaps a preadaptation that minimizes surface area-to-volume ratios and thereby reduces desiccation while spawns are emergent. Included in both intertidal and subtidal spawners, though, are gobiids and gobiesocids that produce ellipsoidal eggs (Breder, 1943), perhaps as adaptations to shed sediment during development or to promote water flow during fanning by male tenders (Miller, 1984). Some taxa, most notably Gasterosteiformes and Atheriniformes, certain blenniids and clinids, and some Gobiidae, have demersal eggs with entangling threads or stalks used to attach extraembryonic membranes to vegetation or other habitat structure.

Besides their relatively large size, the most notable attribute of the demersal eggs of many species (e.g., the Cottidae) are bright and conspicuous yolks whose colors range from pale to brilliant yellow, orange, red, and purple (Orton, 1955). Intense yolk coloration in demersal eggs is usually due to the presence of carotenoids (Mikulin and Soin, 1975; Balon, 1975). The function of carotenoids in teleost eggs has been reviewed (Mikulin and Soin, 1975) and suggested as facilitating oxygen transport during respiration (Fishelson, 1976; Balon, 1977). Embryos within thick egg masses typically experience low, inefficient gas exchange (Strathmann and Chaffee, 1984; Cohen and Strathmann, 1996), and depressed respiration can impede development as well as survivorship (Jones, 1972; Giorgi and Congleton, 1984). Carotenoids may be especially prevalent as a respiratory adaptation in the demersal eggs of fishes that spawn in crevices or other microhabitats with reduced water flows and gas exchange (Fishelson, 1976), such as might occur beneath intertidal stones.

Bright colors also may be aposematic to warm-blooded vertebrate (Pillsbury, 1957) and fish predators. The eggs of at least four species of teleosts—the blennioid *Stichaeus grigorjewi* (Asano, 1964; Hatano and Hashimoto, 1974), the cottid *Scorpaenichthys marmoratus* (Hubbs and Wick, 1951; Fuhrman *et al.,* 1969, 1970; Hashimoto *et al.,* 1976), and the puffers *Fugu niphobles* (Suyama and Uno, 1957) and *Canthigaster valentini* (Gladstone, 1987)—are known to be toxic. The eggs of intertidally spawning *S. marmoratus* are notably conspicuous (Orton, 1955; Pillsbury, 1957).

Attributes such as the manner and type of attachment and the shapes of individual eggs and spawns represent obvious morphological solutions to physiological challenges. Additional, more specific adaptations (e.g., for desiccation resistance, greater temperature tolerance, or aerial respiration) have not been described for intertidal spawns, as they have been for adult intertidal fishes (reviewed by Bridges, 1988, 1993). Although a few papers qualitatively describe desiccation resistance and the ability of embryos to arrest development (e.g., Taylor *et al.,* 1977), quantitative studies of the biochemical and biophysical adaptations of intertidal fish embryos are conspicuously lacking. In particular, the respiratory and perhaps other functions of carotenoids in brightly colored, demersal fish eggs deserve further study.

## 2. Spawn Types

Demersal spawners in the intertidal, like fishes that spawn demersal eggs subtidally, produce eggs in masses that are attached to one another or to physical or biological substrates. Both intertidal and subtidal spawners include taxa with adhesive and nonadhesive eggs. Adhesive eggs adhere to one another, to other substrates, or to both. Adhesive spawns vary in shape from single eggs, each attached in a monolayer to an inclined or vertical substrate surface (e.g., blenniids: Patzner *et al.,* 1986) or to the horizontal internal roofs of habitat structure (gobiids, batrachoidids, some gobiesocids: Breitburg, 1987; DeMartini, 1988b; Shiogaki and Dotsu, 1971); to clumps of eggs that are several to many layers thick and that adhere to one another and conform to substrate irregularities (e.g., many cottids; DeMartini and Patten, 1979; Koya *et al.,* 1994). Pholid and stichaeid blennioids lay spherical masses of eggs that are adherent to one another but are not attached to the substrate (Kimura *et al.,* 1989; Coleman, 1992). Many different physical and biological substrates are used for egg attachment (reviewed by Potts, 1984; examples in Figure 1).

In most taxa with demersal spawns, the eggs hatch as planktonic larvae (Pfister, Chapter 10, this volume) that likely disperse widely from hatching sites (but see Marliave, 1986).

**Figure 1.** Representative habitat features of intertidal spawning fishes. (A) Male *Dictyosoma burgeri* (Stichaeidae), a resident intertidal species, tending (wrapped around) ball of eggs in an aquarium. Reproduced, with permission, from Shiogaki and Dotsu (1972a). (B) Eggs of seasonal migrant *Porichthys notatus* (Batrachoididae) attached to undersurface of overturned stone, with attendant male (indicated by arrow) partly buried in gravel beneath stone on beach at Hood Canal, Washington. Photograph by E. DeMartini. (C) Tidal migrant *Fundulus heteroclitus* (Cyprinodontidae) eggs (arrow) spawned within dead mussel (*Geukensia*) valves in the high intertidal within a Delaware salt marsh. Reprinted by permission of the Estuarine Research Federation, ©Estuarine Research Federation, from Able and Castagna (1975). (D) Tidal (seasonal) migrant *Leuresthes tenuis* (Atherinidae) female (arrow) laying eggs buried in beach sand, surrounded by males fertilizing eggs as they are laid. Reproduced, with permission, from Walker (1952).

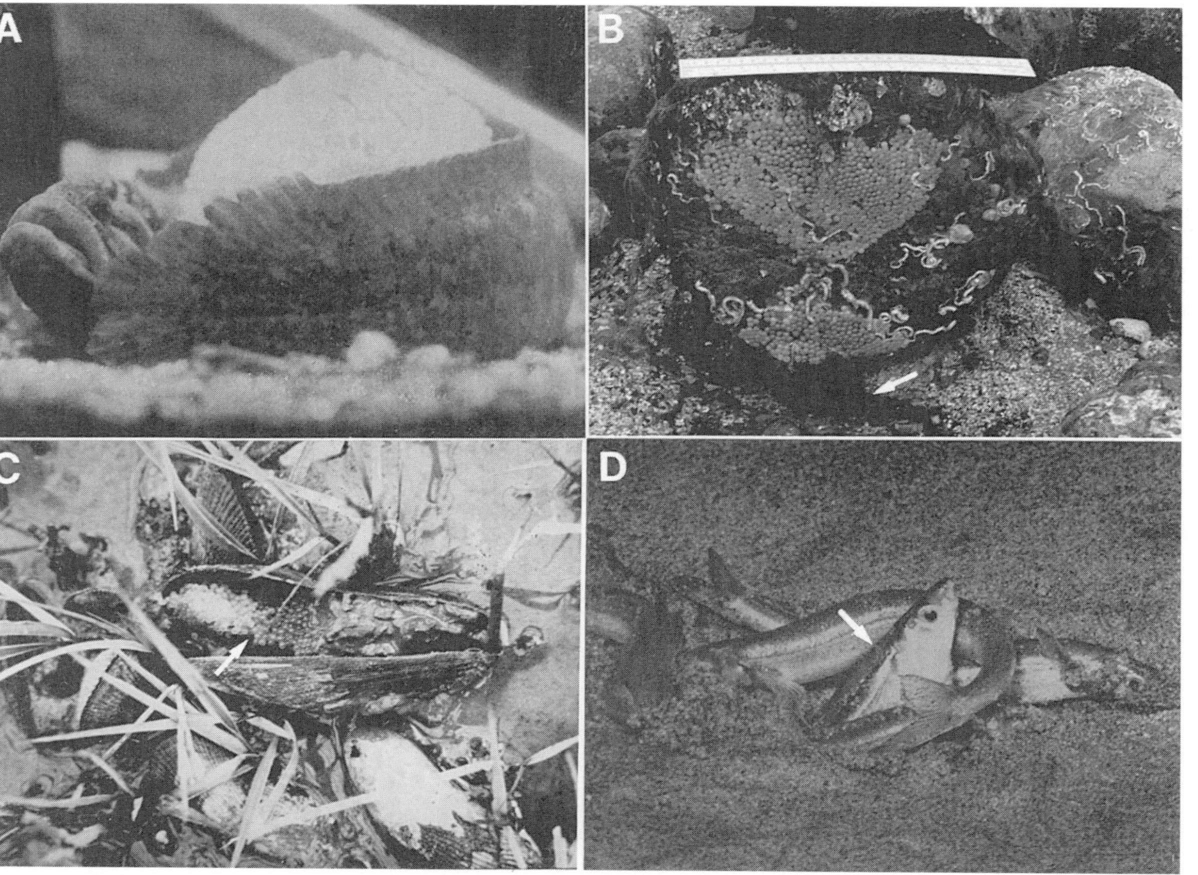
A
B
C
D

Toadfishes (Batrachoididae) may be unique among intertidal demersal spawners in producing juveniles directly from extremely large eggs (Arora, 1948; Crane, 1981; DeMartini, 1988b, 1991).

## 3. Accessory "Nest" Structures

Some species of oviparous fishes, including intertidal species, sequester eggs in accessory structures or "nests" of modified habitat or constructed materials (Potts, 1984). These range from simple scrapes in soft substrate and excavations made beneath bivalve shells on soft sediments (e.g., the gobiid *Pomatoschistus microps*) to burrows in soft sediments that are shared with, or that have been appropriated from, invertebrates [the gobiids *Typhlogobius californiensis* (MacGinitie, 1939) and *Clevelandia ios* (MacDonald, 1975)] or that are excavated by the fish [the goby *Ilypnus gilberti* (Brothers, 1975)]. Other nests are simply constructed of eggs and algal fragments "glued" together by male gland secretions (e.g., the stickleback *Gasterosteus aculeatus*) or more complex, like the bird-like algal nests of the labrid genus *Crenilabrus* (*Symphurus*) (Table 3).

**Table 3.** Distribution of "Nest" Types among Families of Intertidal Spawning Fishes

| Family | None | Sheltered | Constructed | Source |
|---|---|---|---|---|
| Galaxiidae | x | | | McDowall, 1968 |
| Osmeridae | x | | | Loosanoff, 1937 |
| Clupeidae | x | | | Jones, 1972 |
| Cyprinodontidae | x | x | | Able, 1984; Taylor, 1984, 1990 |
| Atherinidae | x | x | | Middaugh, 1981; Moffatt and Thomson, 1978 |
| Gasterosteidae | x | | x | MacDonald *et al.,* 1995a |
| Hexagrammidae | | x | | DeMartini, 1986 |
| Cottidae | | x | | Koya *et al.,* 1994 |
| Liparidae | | x | | Marliave and Peden, 1989 |
| Cyclopteridae | | x | | Goulet and Green, 1988 |
| Embiotocidae | x | | | Baltz, 1984 |
| Labridae | | | x | Potts, 1985 |
| Plesiopidae | | x | | Jillett, 1968 |
| Blenniidae | | x | | Patzner *et al.,* 1986 |
| Trichodontidae | | x | | Marliave, 1980 |
| Chaenopsidae | | x | | Hastings, 1986 |
| Clinidae | | x | x | Lindquist, 1981; Coyer, 1982 |
| Labrisomidae | | x | | Petersen, 1988 |
| Pholidae | | x | | Hughes, 1986 |
| Stichaeidae | | x | | Marliave and DeMartini, 1977 |
| Tripterygiidae | | x | | Berger and Mayr, 1992 |
| Gobiidae | | x | x | Hesthagen, 1977; MacDonald, 1975 |
| Gobiesocidae | | x | | Shiogaki and Dotsu, 1971 |
| Tetraodontidae | x | | | Yamahira, 1994 |
| Batrachoididae | | x | | Crane, 1981 |

## IV. Environmental Influences

### A. Spatial Factors

#### 1. Horizontal Features

Habitat discontinuities represent one obvious spatial feature that importantly influences spawning in the intertidal zone. Examples include rocky outcrops on sandy beaches and patches of algal or salt marsh vegetation on unconsolidated substrates. Although resident intertidal species usually occupy distinct habitats, many tidal migrant spawners utilize structure on otherwise unconsolidated substrates, including several species that use a variety of attachment substrates that may differ markedly within and among different spawning populations (e.g., *Menidia menidia:* Middaugh, 1981; Conover and Kynard, 1984; the mummichog *Fundulus heteroclitus* and the killifishes *F. grandis* and *F. similis:* Able, 1984; Taylor, 1984, 1990). Tidal migrants also include egg scatterers that spawn eggs over (the tetraodontid *Takifugu niphobles:* Yamahira, 1994) or on (the osmerid *Hypomesus pretiosus:* Loosanoff, 1937) unconsolidated substrates. Other egg scatterers bury their eggs (grunion: Thomson and Muench, 1976; capelin: Templeman, 1948).

Microhabitat represents another type of equally important spatial feature. Resident intertidal species perhaps best exemplify the extreme degree to which uses of very specific spawning microhabitats have evolved. Examples include cottids that spawn egg masses on exposed, vertical rock faces (DeMartini, 1978; Lauth, 1989; Koya *et al.,* 1994), the undersurfaces of rocks (Matarese and Marliave, 1982), or extremely specific habitats like rock faces draped with algae (Marliave, 1981), empty barnacle tests (Ragland and Fischer, 1987), or sponge cavities (Munehara, 1991). Interestingly, no one has yet evaluated whether eggs laid in or on sponges benefit either from enhanced water flow and gas exchange or proximity to predation-deterring toxins. Species that spawn on the undersides of rocks and shells on soft substrates (e.g., many gobiids) excavate a depression in the substrate to provide activity space for the egg-tending male (see below). Some blennioids (stichaeids and pholids) sequester unattached balls of eggs under and between rocks. Other blennioids (blenniids, chaenopsids, clinids, labrisomids, tripterygiids) use physical crevices or biological features such as oyster shells, algae, and the tubes of polychaetes and vermetid and pholadid gastropods (Patzner *et al.,* 1986; Kotrschal, 1988; Peters, 1981; Lindquist, 1985; Shiogaki and Dotsu, 1972b; Petersen, 1988; Wirtz, 1978).

#### 2. Vertical Zonation

Tidal height has an obviously important influence on the spawning distributions of both resident intertidal and tidal migrant species. Many resident fishes inhabit narrow intertidal subzones that are representative of the local population. These subzones, however, are not necessarily species-wide if the exposure gradient or latitudinal range of the species is great (Berger and Mayr, 1992). The subzones inhabited by intertidal populations also often change ontogenetically (Horn and Riegle, 1981) and therefore can vary seasonally with growth. Examples also include cases in which the vertical distributions of resident adults shift seasonally within the intertidal and spawning occurs within a segment of the adult distribution (e.g., the blennioid *Pholis laeta:* Hughes, 1986). Tidal migrant and subtidal species usually spawn at characteristic tidal heights. Cyprinodontids (*Fundulus* spp.) and atherinids (*Menidia menidia*) typically spawn at higher high-tide levels (Taylor, 1984, 1990; Conover and Kynard, 1984; Middaugh *et al.,* 1984). Intertidal spawning in many shallow subtidal, reef-associated fishes occurs in the extremely low intertidal zone or "infralittoral fringe" (Ilich and Kotrschal, 1990; MacPherson, 1994).

## B. Temporal Patterns

### 1. Seasonal

Many or most species of intertidal (and subtidal) spawners are seasonally reproductive, particularly those that spawn at temperate and higher latitudes. The timing of reproduction typically tracks the seasonal availability of food resources for offspring in most organisms including fishes (Bye, 1984), and intertidal spawners are no exception (Gibson, 1969).

Many seasonal migrations by fishes into the intertidal are not directly reproductive. Numerous tidal visitors migrate intertidally at higher water levels (some seasonally) to feed (Gibson, 1982), and juveniles of these and other species shelter intertidally either year-round or seasonally (Christensen, 1978; Gibson, 1988). Many bothids and pleuronectids, for example, are common tidal migrants as young-of-the-year only in certain seasons (Gibson, 1969, 1986).

One interesting, additional temporal pattern in intertidal fish spawning is response to seasonal variation in physical rigor of the spawning habitat. In one of the few such evaluations that exist, Marliave (1975) showed that the paternal-caring (Marliave and DeMartini, 1977) stichaeid *Xiphister atropurpureus* spawns during late winter (when conditions are stormy) at semiprotected locations but later in the breeding season (when conditions are predictably calmer) at more exposed locations.

### 2. Diel

Data on the natural diel activity patterns of resident intertidal fishes are incomplete, undoubtedly because of the difficulty in directly observing or monitoring (Ralston and Horn, 1986) activity patterns in the intertidal other than during calm periods of daytime low tides. The available data suggest that most resident intertidal fishes are diurnal and are active and capable of spawning only when immersed at higher tide levels (Gibson, 1982, 1992). In fact, most species of structure-associated, demersal-spawning fishes are diurnal and spawn during daylight hours (e.g., coral reef fishes: Hobson, 1972; Thresher, 1984). Intertidal and subtidal fishes likely differ little in diurnality except in situations where tidal height and time of day are linked. In such cases, diurnal activity may be precluded by emersion at times of daytime low tides (Gibson, 1992). Certain mudskippers are conspicuous exceptions to the general pattern in being less active when immersed, regardless of time of day (Clayton, 1993; Colombini *et al.,* 1995).

### 3. Lunar/Tidal

Monthly (circalunar) and bimonthly (semilunar) spawning at times of new or full moons has been observed for numerous fishes, especially tropical coral reef species (Robertson *et al.,* 1990). Intertidal spawners provide some of the best known examples of lunar periodicity, undoubtedly because of the controlling influence of the moon and sun on tides, water levels, and activity patterns. Among the best known examples are mummichogs and other cyprinodontids (*Fundulus* spp.: Taylor *et al.,* 1979; Taylor and DiMichele, 1983; Taylor, 1984, 1990) and the atherinids, *Menidia menidia* (Middaugh, 1981; Conover and Kynard, 1984; Middaugh *et al.,* 1984) and *Leuresthes* spp. (Thomson and Muench, 1976; Moffatt and Thomson, 1978). Fortnightly (semilunar) spawning in *M. menidia* may be limited to mid-morning times of higher high tides when preferred spawning substrates are available (immersed), although an interaction with celestial (lunar) and endogenous factors may be involved (Conover and Kynard, 1984). Semilunar spawning of the mummichog, *Fundulus heteroclitus,* also likely represents an interaction between celestial and tidal and other environmental factors (Taylor *et al.,* 1979; Taylor, 1986). Semilunar intertidal spawning by the subtidal puffer, *Takifugu niphobles,* is restricted to higher (spring) high tides during new and full moons, and tidal and diel cycles interact so that spawning occurs only at higher high tides near dusk (Yamahira, 1994). Apparent semilunar spawning has been recently described for the rocky tidepool blenny *Istiblennius enosimae* (Sunobe *et al.,* 1995). Many of the well-known examples are tidal migrants; however, some populations of the mummichog, *F. heteroclitus,* are residents of intertidal salt marshes (Taylor, 1990).

## 4. Endogenous Rhythms

Many examples of lunar, semilunar, and tidal reproductive rhythms are recognized, but the extent to which rhythms are entrained (endogenous) is generally unknown because the causes and effects of intertidal spawning behaviors are obscure and mechanisms are poorly understood. Most resident intertidal species (and apparently all rocky intertidal fishes) must spawn when immersed at higher tide levels. Mudskippers are a conspicuous exception to the general pattern because their activity levels peak when emergent at lower tide levels (Clayton, 1993; Colombini *et al.,* 1995; but see Ikebe and Oishi, 1996).

## 5. Tidal Stranding

As a general rule, egg-guarding male fishes are strongly site attached to eggs at their spawning sites (Coleman, Chapter 9, this volume). Egg guarders of most intertidal resident fishes, however, remain attendant at lower tide levels only if sufficient water covers the parent and spawning site at these times. Blennioids tending unattached egg masses beneath stones tend to strand with their eggs at low tide (Schultz and DeLacy, 1932; Marliave and DeMartini, 1977). Most other taxa temporarily desert their eggs during lower water levels (see DeMartini, 1978), unless a damp crevice (some blenniids: Patzner *et al.,* 1986) is present to buffer temperature and desiccation stress on the emergent parent. The adults of some intertidal residents like *Blennius pholis* (Laming *et al.,* 1982) and the gobiesocid *Sicyases sanguineus* (Gordon *et al.,* 1970) have specialized physiological adaptations that slow metabolic rates (bradycardia) in addition to cutaneous vascularization for aerial respiration (stichaeids and pholids: Yoshiyama and Cech, 1994; cottids: Yoshiyama *et al.,* 1995). Other species like the weaverfish *Trachinus vipera* (Trachinidae; Lewis, 1976) and the sandlance *Ammodytes hexapterus* (Liem and Scott, 1966) have the ability to remain buried in damp emergent sand at low tide. Interestingly, the demersal eggs of the trichodontid sandfish *Trichodon trichodon,* which buries in shallow sandy bottoms as an adult, have been found in the exposed rocky intertidal, about 8-km distant from the nearest sandy intertidal habitat (Marliave, 1980), suggesting that adults either migrate alongshore or onshore from the sandy subtidal. Amphibious mudskippers are generally immune to stranding at low tide; even the male egg guarders of most mudskipper species are emergent and active atop tidal flats at low tide while embryos remain sequestered within burrows (Kobayashi *et al.,* 1971; Clayton, 1993).

## V. Ultimate Factors and Intertidal Spawning

### A. Processes Affecting Patterns

Factors affecting the survivorship of offspring and the subsequent survival and future expected reproductive success of adults should have paramount importance in the evolution of spawning patterns for any organism. Two general classes of factors appear particularly relevant for intertidally spawning fishes. These include the biological processes of predation and the coevolved antipredatory responses that affect survival of parents and offspring. In addition, intertidal spawners (unlike subtidal spawners) need to confront the greater physical demands on developing young that are related to the cyclic shifts between terrestrial and aquatic environments. Table 4 lists the likely or possible benefits and costs affecting fishes that spawn intertidally. These include physical and biological factors that either compete with or reinforce one another (e.g., relative mortality due to aquatic and terrestrial predators, better oxygenation and faster embryonic development while spawns are emergent in air).

Few studies have examined the importance of the major physical (temperature stress, differential gas exchange in water and air, desiccation) and biological (predation) factors that influence intertidal spawning. To date no study has rigorously evaluated the relative significance of physical and biological factors in terms of both adult and offspring survivorship.

### B. Case Studies

There are some studies, however, that have successfully evaluated one or the other (usually physical) suite of factors. Among the most notable of these are studies of cyprinodontids

**Table 4.** Known or Likely Costs and Benefits of Intertidal Spawning by Intertidal Resident and Tidal/Seasonal Migrant Subtidal Fishes

| Costs | Benefits |
|---|---|
| Intertidal residents | |
| Time and energy expended in site defense | Mate sequestering at site |
| Foraging of adult limited by tending spawn | Temporal and spatial predation refuge for adult spawners from aquatic predators |
| Physiological stress on embryos while emergent | Embryonic development rate increased while emergent |
| Physiological stress on tending parent while emergent | Temporal and spatial refuge for adult spawners from aquatic predators |
| Tidal/seasonal migrants | |
| Time and energy expended, risk in tidal migration | |
| Increased risk of embryos to terrestrial and avian predators | Intertidal predation refuge for embryos from aquatic predators |
| Physiological stress on embryos while emergent | Embryonic development rate increased while emergent |
| Increased risk of adults to terrestrial and avian predators while spawning | Intertidal predation refuge for adult spawners from aquatic predators |

and atherinids. Taylor and DeMichele (1983) transplanted egg masses of *Fundulus heteroclitus,* spawned in empty mussel shells, to various heights within an intertidal Delaware marshland and observed nearly 100% mortality of continuously immersed eggs, apparently from suffocation during turbid ebb tides. Periodically emergent spawns in the low to high intertidal had nearly 100% survivorship. Kneib (1987) observed >95% greater recruitment of postlarval *Fundulus* spp. to predator exclosure cages in tidal marshland pools, compared to otherwise similar cages in which predators (including adult conspecifics) had been enclosed. In a complementary study, both the growth and survivorship of postlarval *Fundulus heteroclitus,* protected from predators within enclosures, were greater in low (versus high) intertidal pools (Kneib, 1993). Takita *et al.* (1984) observed that avian (egret) predators removed substantial numbers of adult *Menidia menidia* from spawning aggregations, but the magnitude of losses was not quantified. Middaugh (1981) provides analogous qualitative estimates of *M. menidia* embryo losses to 1 species of invertebrate plus 16 species of reptiles, fishes, and birds. Mammalian (ground squirrel) and avian (horned lark) predation on California grunion (*Leuresthes tenuis*) eggs buried in beach sand were noted by Olson (1950). Middaugh *et al.* (1983) evaluated the survivorship of natural spawns of *L. tenuis* and *M. menidia* laid on different substrates. Daily survivorship of *L. tenuis* embryos buried in beach sand was high (97% on average). Survivorship of *M. menidia* embryos averaged 76–94% for spawns laid within and at the entrance of abandoned fiddler crab burrows, on detrital mats, and on the stems and primary leaves of cordgrass but declined to less than 10% on exposed cordgrass surfaces where desiccation and temperature exceeded tolerances.

The survival of embryos of the osmerid, *Hypomesus pretiosus,* progressively decreased from natural spawning elevations high in the intertidal to nearly zero survival at elevations less than 2 m above mean low water (MLW), presumably due to predation (Loosanoff, 1937). The impact of invertebrate predation on the intertidal spawn of another osmerid, the capelin (*Mallotus villosus*), has been the subject of several more recent papers. Frank and Leggett (1981) observed that the mortality of capelin eggs was progressively greater at higher levels in the intertidal, and egg mortality was correlated with the density of maggots of a beach fly. DeBlois and Leggett (1993a,b) observed that predation by an intertidal amphipod consistently explained a large fraction of capelin egg mortality.

Several interesting case studies also exist for herring (*Clupea pallasi*). Taylor (1971) observed generally higher mortality of herring spawns developing at 9 and 18 m depths, relative to 0 m (MLW), presumably due to reduced water flow and poorer gas exchange at the greater depths. Jones (1972) on the other hand observed much greater (31% vs 13%) survivorship of embryos within herring spawns laid higher in the intertidal and attributed this to better gas exchange during aerial respiration while emergent for longer periods at higher tide elevations. Haegele (1993a,b) reviewed historical studies of predation mortality on Pacific herring spawns and, for one British Columbia stock, estimated average annual losses of 3.5 and 3.7% from predation by seabirds and benthic invertebrates, respectively.

One study of three species of sticklebacks (*Gasterosteus aculeatus, G. wheatlandi,* and *Pungitius pungitius*), nesting in tidal ponds of the St. Lawrence estuary, observed an estimated 30% loss in breeding adults (primarily males) due to predation by three species of birds (Whoriskey and FitzGerald, 1985).

Relatively few, exceptional studies to date have used experiments designed to test the concurrent effects of both physical stress and predation on intertidal spawn. Tewksbury and Conover (1987) placed spawns of the atherinid *Menidia menidia* in caged and uncaged artificial substrates in the high and middle intertidal and shallow subtidal (at 0.1 m below MLW) in Long Island, New York, salt marshes. The average survivorships of caged embryos were 36, 21, and 43% in the high and middle intertidal and subtidal; the survival of uncaged embryos was generally lower (1–15%) and poorest in the subtidal. Tewksbury and Conover (1987) observed little mortality due to physical stress and rightly concluded that most embryo mortality in their experiments was caused by aquatic predators (mainly adult *Fundulus*), rather than physiological stress while emergent.

In a recent study, MacDonald *et al.* (1995a,b) observed and manipulated the intertidal spawns of a newly discovered stickleback "type" within the *Gasterosteus aculeatus* species complex. At tidal ponds in Nova Scotia, Canada, males of the "rocky intertidal white stickleback" (a tidal/seasonal migrant) build and acquire spawns within nests of algal fragments in the unvegetated rocky intertidal, but then scatter the eggs over the barren rock substrate, where they develop in damp crevices without paternal care. By protecting fertilized eggs in cages placed at various tidal heights, MacDonald *et al.* (1995b) demonstrated that embryos in the low (mostly immersed) and high (mostly emergent) intertidal zone developed faster than embryos in the subtidal (continuously immersed). Embryo survivorship was indistinguishable among subtidal, low and high intertidal treatments and averaged 76% overall (MacDonald *et al.*, 1995b).

In a recent study of the puffer, *Takifugu niphobles* (a seasonal/tidal migrant), Yamahira (1996) outplanted artificially fertilized eggs to the high, mid, and low intertidal zones. During the early breeding season (when eggs were emergent during midday low tides), survivorship was lower in the high intertidal (25–85%) where puffer naturally spawn, compared to the mid and low intertidal (85–90%) where temperature and desiccation stresses were lower. Mortality rates due to physical stress were similar among tidal heights during the remainder of the spawning season, however, and predation mortality was indistinguishable between tide levels. Embryos developed 10 and 30% faster in the high compared to the mid and low intertidal, respectively (Yamahira, 1996).

Very few rigorous experimental studies of spawn survivorship have been conducted for resident intertidal fishes. Some studies of intertidal fishes have removed paternal egg guarders and subsequently observed large to complete losses of eggs to predators (usually fish of other, including tidal migrant, species: Almada *et al.*, 1992). Many analogous studies have observed moderate to large losses of embryos after removal of guardian males in subtidal spawners (e.g., Wirtz, 1978; DeMartini, 1987) or in subtidal species that spawn intertidally (DeMartini, 1978, 1991). In most of these studies, the documentation of predation as the cause of loss has been anecdotal or qualitative.

Only two quantitative studies of spawn survivorship exist for resident intertidal fishes [one partial (seasonal) resident, and a single year-round resident species]. Lindstrom and Ranta (1992) observed that gull and other avian predation affected the body size distribution of nesting males in the seasonally intertidal sand goby, *Pomatoschistus minutus*. Nests (bivalve shells) protected for 6 days by avian exclosures had larger sized male occupants than were present at the end of the experiment at otherwise similar nests that were unprotected from birds. The most straightforward interpretation was that cropping by birds at unprotected nests resulted in a greater turnover of occupants and less monopolization of these sites by large, dominant males (Lindstrom and Ranta, 1992).

Marliave (1981) removed algal (*Fucus distichus*) cover over high intertidal spawns of the cottid, *Clinocottus acuticeps*. Spawns incurred a 26% mortality (primarily due to desiccation, secondarily due to temperature stress) in plots with algae removed, compared to 0% loss of spawns in plots with protective algae intact (Marliave, 1981).

Experimental studies like those of Marliave (1981) and Lindstrom and Ranta (1992) are conspicuously rare for resident intertidal fishes. Additional studies are needed if we are to begin to understand how natural selection operates on the spawning behavior of resident species.

## C. Adaptations or Preadaptations?

Almada and Santos (1995) recently questioned whether several traits found in intertidal fishes (paternal care, courtship displays limited to the proximity of shelter) were true adaptations that evolved as species radiated into the intertidal zone from ancestral subtidal regions. The data reviewed by Alamada and Santos (1995), although largely restricted to Mediterranean blenniids, suggest that courtship behaviors may represent true adaptations, but that traits such as paternal care are "exaptations" (preadaptations) that evolved prior to ancestral colonization of the intertidal.

Additional traits of intertidally spawning fishes can benefit from an evaluation analogous to that of Almada and Santos (1995). Perhaps the most intriguing is intertidal spawning by subtidal species. Intertidal residents spawn intertidally because most, being small bodied and poor swimmers, are too susceptible to predation for subtidal spawning migrations (Gibson, 1986). Exceptions include species like the relatively large-bodied (to 18 cm: Qasim, 1957), seasonally resident blennioid *Pholis gunnellus,* which apparently migrates subtidally in the winter to spawn (Sawyer, 1967). A minority of partial intertidal residents, including small-bodied species like *Pomatoschistus minutus* (Gobiidae), migrate seasonally from the subtidal to spawn infralittorally (Hesthagen, 1977; Lindstrom and Ranta, 1992).

All this, of course, begs the question of why intertidal species evolved from subtidal ancestors in the first place. Was the intertidal primarily colonized so that adults could exploit untapped prey resources or gain a spatial and temporal refuge from predation? Was there a direct benefit to eggs spawned in the intertidal zone, and was offspring survivorship the major selective pressure (Robertson *et al.,* 1990)? If the latter, then both intertidal resident and subtidal fishes may have evolved intertidal spawning primarily to reduce offspring mortality. But why have many subtidal fishes evolved intertidal spawning when the physiological costs (for embryos) of doing so are many and perhaps greater than those experienced by (the adults of) some intertidal resident and amphibious species (Sayer and Davenport, 1991)? Especially informative would be comparisons of the relative reproductive success of individuals that spawn subtidally versus intertidally within populations of species in which both occur. Complementary studies of the trade-offs between foraging and refuging while adults are immersed versus emergent (Wyttenbach and Senn, 1993) also would be informative for resident species.

## D. Mixed Subtidal and Intertidal Spawners

Few data on reproductive success of such species exist. For some like the temperate wrasse *Crenilabrus (Symphurus) melops,* intertidal spawning may be less successful than subtidal spawning (Potts, 1984). Observations for *S. melops* indicate that recruit spawners and other small subordinate males are restricted to intertidal nests by large dominant males that monopolize subtidal nest sites (Potts, 1984). Analogous uses of the intertidal by subordinate individuals have been described for the sand goby (Lindstrom and Ranta, 1992) and for the "rocky intertidal white stickleback" (MacDonald *et al.,* 1995a,b). All of the three case studies to date thus indicate or suggest that intertidal spawning is inferior to subtidal spawning. Given this trivial sample size, however, other as yet unstudied cases may exist in which larger dominant males predominate at intertidal sites while smaller subordinate males are

growth relative to the seasonal spawning cycle of capelin (*Mallotus villosus*). *Can. J. Fish. Aquat. Sci.* **50,** 2581–2590.

de Jonge, J., de Ruiter, A. J. H., and van den Hurk, R. (1989). Testis-testicular gland complex of two *Tripterygion* species (Blennioidei, Teleostei): Differences between territorial and nonterritorial males. *J. Fish Biol.* **35,** 497–508.

DeMartini, E. E. (1978). Spatial aspects of reproduction in buffalo sculpin, *Enophrys bison. Environ. Biol. Fish.* **3,** 331–336.

DeMartini, E. E. (1986). Reproductive colorations, parental behavior and egg masses of kelp greenling, *Hexagrammos decagrammus,* and whitespotted greenling, *H. stelleri. Northwest Sci.* **60,** 32–35.

DeMartini, E. E. (1987). Paternal defence, cannibalism, and polygamy: Factors influencing the reproductive success of painted greenling (Pisces, Hexagrammidae). *Anim. Behav.* **35,** 1145–1158.

DeMartini, E. E. (1988a). Size-assortative courtship and competition in two embiotocid fishes. *Copeia* **1988,** 336–344.

DeMartini, E. E. (1988b). Spawning success of the male plainfin midshipman. I. Influences of male body size and area of spawning site. *J. Exp. Mar. Biol. Ecol.* **121,** 177–192.

DeMartini, E. E. (1990). Annual variations in fecundity, egg size and condition of the plainfin midshipman (*Porichthys notatus*). *Copeia* **1990,** 850–859.

DeMartini, E. E. (1991). Spawning success of the male plainfin midshipman. II. Substratum as a limiting spatial resource. *J. Exp. Mar. Biol. Ecol.* **146,** 235–251.

DeMartini, E. E., and Patten, B. G. (1979). Egg guarding and reproductive biology of the red Irish lord, *Hemilepidotus hemilepidotus* (Tilesius). *Syesis* **12,** 41–55.

Denny, M. W., and Shibata, M. F. (1989). Consequences of surf-zone turbulence for settlement and external fertilization. *Am. Nat.* **134,** 859–889.

Duarte, C. M., and Alcaraz, M. (1989). To produce many small or few large eggs: A size-dependent reproductive tactic of fish. *Oecologia (Berlin)* **80,** 401–404.

Elgar, M. A. (1990). Evolutionary compromise between a few large and many small eggs: Comparative evidence in teleost fish. *Oikos* **59,** 283–287.

Fishelson, L. (1976). Spawning and larval development of the blenniid fish *Meiacanthus nigrolineatus* from the Red Sea. *Copeia* **1976,** 798–800.

Fishelson, L. (1991). Comparative cytology and morphology of seminal vesicles in male gobiid fishes. *Japan. J. Ichthyol.* **38,** 17–30.

Frank, K. T., and Leggett, W. C. (1981). Subtidal occurrence of larval capelin (*Mallotus villosus villosus*) as related to the intertidal distribution of their eggs. *Rapp. Proces-Verbaux Reun. Cons. Perm. Explor. Mer* **178,** 179–180.

Fuhrman, F. A., Fuhrman, G. J., Dull, D. L., and Moser, H. S. (1969). Toxins from eggs of fishes and amphibia. *J. Agric. Food Chem.* **17,** 417–424.

Fuhrman, F. A., Fuhrman, G. J., and Rosen, J. S. (1970). Toxic effects produced by extracts of eggs of the cabezon *Scorpaenichthys marmoratus. Toxicon* **8,** 55–61.

Gibson, R. N. (1969). The biology and behaviour of littoral fish. *Oceanogr. Mar. Biol. Annu. Rev.* **7,** 367–410.

Gibson, R. N. (1982). Recent studies on the biology of intertidal fishes. *Oceanogr. Mar. Biol. Annu. Rev.* **20,** 363–414.

Gibson, R. N. (1986). Intertidal teleosts: Life in a fluctuating environment. *In* "The Behaviour of Teleost Fishes" (T. J. Pitcher, Ed.), pp. 388–408. Croom Helm, London.

Gibson, R. N. (1988). Patterns of movement in intertidal fishes. *In* "Behavioral Adaptation to Intertidal Life" (G. Chellazi and M. Vannini, Eds.), pp. 55–63. Plenum Press, Boulder, CO.

Gibson, R. N. (1992). Tidally-synchronized behaviour in marine fishes. *In* "Rhythms in Fishes" (M. A. Ali, Ed.), pp. 63–81. Plenum Press, New York.

Giorgi, A. E., and Congleton, J. L. (1984). Effects of current velocity on development and survival of lingcod, *Ophiodon elongtus,* embryos. *Environ. Biol. Fish.* **10,** 15–27.

Gladstone, W. (1987). The eggs and larvae of the sharpnose pufferfish *Canthigaster valentini* (Pisces: Tetraodontidae) are unpalatable to other reef fishes. *Copeia* **1987,** 227–230.

Gordon, C. D. (1965). "Aspects of the Life History of *Cymatogaster aggregata* Gibbons." M.Sc. thesis, University of British Columbia, Vancouver.

Gordon, M. S., Fisher, S., and Tarifeno, E. (1970). Aspects of the physiology of terrestrial life in amphibious fishes. II. The Chilean clingfish, *Sicyases sanguineus. J. Exp. Biol.* **53,** 559–572.

Goulet, D., and Green, J. M. (1988). Reproductive success of the male lumpfish (*Cyclopterus lumpus* L.) (Pisces: Cyclopteridae): Evidence against female mate choice. *Can. J. Zool.* **66,** 2513–2519.

Gross, M. R. (1996). Alternative reproductive strategies and tactics: Diversity within species. *Trends Ecol. Evol.* **11,** 92–98.

Gunn, J. S., and Thresher, R. E. (1991). Viviparity and the reproductive ecology of clinid fishes (Clinidae) from temperate Australian waters. *Environ. Biol. Fish.* **31,** 323–344.

Haegele, C. W. (1993a). Seabird predation of Pacific herring, *Clupea pallasi,* spawn in British Columbia. *Can. Fld.-Nat.* **107,** 73–82.

Haegele, C. W. (1993b). Epibenthic invertebrate predation of Pacific herring, *Clupea pallasi,* spawn in British Columbia. *Can. Fld.-Nat.* **107,** 83–91.

Hart, J. L. (1973). Fishes of the Pacific coast of Canada. *Fish. Res. Bd. Can. Bull.* **180,** 1–740.

Hashimoto, Y., Kawasaki, M., and Hatamo, M. (1976). Occurrence of a toxic phospholipid in cabezon roe. *Toxicon* **14,** 141–143.

Hastings, P. A. (1986). Habitat selection, sex ratio and sexual selection in *Coralliozetus angelica* (Blennioidea: Chaenopsidae). *In* "Indo-Pacific Fish Biology: Proc. 2nd Int. Conf. Indo-Pac. Fish." (T. Ueyeno, R. Arai, T. Taniuchi, and K. Matsuura, Eds.), pp. 785–793. Ichthyol. Soc. Japan, Tokyo.

Hatano, M., and Hashimoto, Y. (1974). Properties of a toxic phospholipid in northern blenny roe. *Toxicon* **12,** 231.

Hesthagen, I. H. (1977). Migrations, breeding, and growth in *Pomatoschistus minutus* (Pallas) (Pisces, Gobiidae) in Oslofjorden, Norway. *Sarsia* **63,** 17–26.

Hobson, E. S. (1972). Activity of Hawaiian reef fishes during the evening and morning transitions between daylight and darkness. *Fish. Bull. U.S.* **70,** 715–740.

Horn, M. H., and Gibson, R. N. (1988). Intertidal fishes. *Sci. Am.* **258**(1), 64–70.

Horn, M. H., and Riegle, K. C. (1981). Evaporative water loss and intertidal vertical distribution in relation to body size and morphology of stichaeoid fishes from California. *J. Exp. Mar. Biol. Ecol.* **50,** 273–288.

Hubbs, Carl L. (1921). The ecology and life history of *Amphigonopterus aurora* and other viviparous perches of California. *Biol. Bull.* **40,** 181–209.

Hubbs, Carl L., and Wick, A. N. (1951). Toxicity of the roe of the cabezon, *Scorpaenichthys marmoratus. Calif. Fish Game* **37,** 195–196.

Hubbs, Clark (1966). Fertilization, initiation of cleavage, and developmental temperature tolerance of the cottid fish, *Clinocottus analis. Copeia* **1966,** 29–42.

Hughes, G. W. (1986). Observations on the reproductive ecology of the crescent gunnel, *Pholis laeta,* from marine inshore waters of southern British Columbia. *Can. Fld.-Nat.* **100,** 367–370.

Ikebe, Y., and Oishi, T. (1996). Correlation between environmental parameters and behaviour during high tides in *Periophthalmus modestus. J. Fish Biol.* **49,** 139–147.

Ilich, I. P., and Kotrschal, K. (1990). Depth distribution and abundance of northern Adriatic rocky reef blennioid fishes (Blenniidae and *Tripterygion*). *P.S.Z.N.I. Mar. Ecol.* **11,** 277–289.

Jillett, J. B. (1968). The biology of *Acanthoclinus quadridactylus* (Bloch & Schneider) (Teleostei-Blennioidea). II. Breeding and development. *Austr. J. Mar. Freshw. Res.* **19,** 9–18.

Jones, B. C. (1972). Effect of intertidal exposure on survival and embryonic development of Pacific herring spawn. *J. Fish. Res. Bd. Can.* **29,** 1119–1124.

Kimura, S., Okazawa, T., and Mori, K. (1989). Reproductive biology of the tidepool gunnel *Pholis nebulosa* in Ago Bay, central Japan. *Bull. Japan. Soc. Sci. Fish.* **55,** 503–506.

Kneib, R. T. (1987). Predation risk and use of intertidal habitats by young fishes and shrimp. *Ecology* **68,** 379–386.

Kneib, R. T. (1993). Growth and mortality in successive cohorts of fish larvae within an estuarine nursery. *Mar. Ecol. Prog. Ser.* **94,** 115–127.

Kobayashi, T., Dotsu, Y., and Takita, T. (1971). Nest and nesting behavior of the mud skipper, *Periopthalmus cantonensis* in Ariake Sound. *Bull. Fac. Fish. Nagasaki Univ.* **32,** 27–40. [In Japan. with Engl. Abs.]

Kotrschal, K. (1988). Blennies and endolithic bivalves: Differential utilization of shelter in Adriatic Blenniidae (Pisces: Teleostei). *P.S.Z.N.I. Mar. Ecol.* **9,** 253–269.

Koya, Y., Munehara, H., and Takano, K. (1994). Reproductive cycle and spawning ecology in elkhorn sculpin, *Alcichthys alcicornis. Japan. J. Ichthyol.* **41,** 39–45.

Krejsa, R. J. (1964). Reproductive behaviour and sexual dimorphism in the manacled sculpin, *Synchirus gilli* Bean. *Copeia* **1964,** 448–450.

Laming, P. R., Funston, C. W., Roberts, D., and Armstrong, M. J. (1982). Behavioral, physiological and morphological adaptations of the shanny (*Blennius pholis*) to the intertidal habitat. *J. Mar. Biol. Assoc. U.K.* **62,** 329–338.

Lauth, R. R. (1989). Seasonal spawning cycles, spawning frequency, and batch fecundity of the cabezon, *Scorpaenichthys marmoratus,* in Puget Sound, Washington. *Fish. Bull. U.S.* **87,** 145–154.

Lewis, D. B. (1976). Studies of the biology of the lesser weaverfish *Trachinus vipera* Cuvier. I. Adaptations to a benthic habit. *J. Fish Biol.* **8,** 127–138.

Liem, A. H., and Scott, W. B. (1966). Fishes of the Atlantic coast of Canada. *Fish. Res. Bd. Can. Bull.* **155,** 1–485.

Lindquist, D. G. (1981). Reproduction of the onespot fringehead, *Neoclinus uninotatus,* in Monterey Harbor, California. *Bull. S. Calif. Acad. Sci.* **80,** 12–22.

Lindquist, D. G. (1985). Depth zonation, microhabitat, and morphology of three species of *Acanthemblemaria* (Pisces: Blennioidea) in the Gulf of California, Mexico. *P.S.Z.N.I. Mar. Ecol.* **6,** 329–344.

Lindstrom, K., and Ranta, E. (1992). Predation by birds affects population structure in breeding sand goby, *Pomatoschistus minutus,* males. *Oikos* **64,** 527–532.

Loosanoff, V. L. (1937). The spawning run of the Pacific surf smelt, *Hypomesus pretiosus* (Girard). *Int. Rev. Gesampten Hydrobiol. Hydrogr.* **36,** 170–187.

MacDonald, C. K. (1975). Notes on the family Gobiidae from Anaheim Bay. *Calif. Dep. Fish Game, Fish Bull.* **165,** 117–121.

MacDonald, J. F., Bekkers, J., MacIsaac, S. M., and Blouw, D. M. (1995a). Intertidal breeding and aerial development of embryos of a stickleback fish (*Gasterosteus*). *Behaviour* **132,** 1183–1206.

MacDonald, J. F., MacIsaac, S. M., Bekkers, J., and Blouw, D. M. (1995b). Experiments on embryo survivorship, habitat selection, and competitive ability of a stickleback fish (*Gasterosteus*) which nests in the rocky intertidal zone. *Behaviour* **132,** 1207–1221.

MacGinitie, G. E. (1939). The natural history of the blind goby, *Typhlogobius californiensis* Steindachner. *Am. Midl. Nat.* **21,** 481–505.

MacPherson, E. (1994). Substrate utilization in a Mediterranean littoral fish community. *Mar. Ecol. Prog. Ser.* **114,** 211–218.

Marliave, J. B. (1975). Seasonal shifts in the spawning site of a northeastern Pacific intertidal fish. *J. Fish. Res. Bd. Can.* **32,** 1687–1691.

Marliave, J. B. (1980). Spawn and larvae of the Pacific sandfish, *Trichodon trichodon. Fish. Bull. U.S.* **78,** 959–964.

Marliave, J. B. (1981). High intertidal spawning under rockweed, *Fucus distichus,* by the sharpnose sculpin, *Clinocottus acuticeps. Can. J. Zool.* **59,** 1122–1125.

Marliave, J. B. (1986). Lack of planktonic dispersal of rocky intertidal fish larvae. *Trans. Am. Fish. Soc.* **115,** 149–154.

Marliave, J. B., and DeMartini, E. E. (1977). Parental behaviour of intertidal fishes of the stichaeid genus *Xiphister. Can. J. Zool.* **55,** 60–63.

Marliave, J. B., and Peden, A. E. (1989). Larvae of *Liparis fucensis* and *Liparis callyodon:* Is the "cottoid bubblemorph" phylogenetically significant? *Fish. Bull. U.S.* **87,** 735–744.

Matarese, A. C., and Marliave, J. B. (1982). Larval development of laboratory-reared rosylip sculpin, *Ascelichthys rhodorus* (Cottidae). *Fish. Bull. U.S.* **80,** 345–355.

McDowall, R. M. (1968). *Galaxias maculatus* (Jenyns), the New Zealand whitebait. *N. Z. Fish. Res. Div. Fish. Res. Bull.* **2,** 1–84.

Middaugh, D. P. (1981). Reproductive ecology and spawning periodicity of the Atlantic silverside, *Menidia menidia* (Pisces: Atherinidae). *Copeia* **1981,** 766–776.

Middaugh, D. P., Domey, R. G., and Scott, G. I. (1984). Reproductive rhythmicity of the Atlantic silverside. *Trans. Am. Fish. Soc.* **113,** 472–478.

Middaugh, D. P., Scott, G. I., and Dean, J. M. (1981). Reproductive behavior of the Atlantic silverside, *Menidia menidia* (Pisces, Atherinidae). *Environ. Biol. Fish.* **6,** 269–276.

Middaugh, D. P., Kohl, H. W., and Burnett, L. E. (1983). Concurrent measurement of intertidal environmental variables and embryo survival for the California grunion, *Leuresthes tenuis* and the Atlantic silverside, *Menidia menidia* (Pisces: Atherinidae). *Calif. Fish Game* **69,** 89–96.

Mikulin, A. Y., and Soin, S. G. (1975). The functional significance of carotenoids in the embryonic development of teleosts. *J. Ichthyol.* **15,** 749–759.

Miller, P. J. (1979). Adaptiveness and implications of small size in teleosts. *Symp. Zool. Soc. London* **44,** 263–306.

Miller, P. J. (1984). The tokology of gobioid fishes. *In* "Fish Reproduction: Strategies and Tactics" (G. W. Potts and R. J. Wootton, Eds.), pp. 119–153. Academic Press, London.

Moffatt, N. M., and Thomson, D. A. (1978). Tidal influence on the evolution of egg size in the grunions (*Leuresthes,* Atherinidae). *Environ. Biol. Fish.* **3,** 267–273.

Morris, R. W. (1952). Spawning behavior of the cottid fish *Clinocottus recalvus* (Greeley). *Pac. Sci.* **6,** 256–258.

Morris, R. W. (1956). Clasping mechanism of the cottid fish *Oligocottus snyderi* Greeley. *Pac. Sci.* **10,** 314–317.

Munehara, H. (1991). Utilization and ecological benefits of a sponge as a spawning bed by the little dragon sculpin *Blepsias cirrhosus*. *Japan. J. Ichthyol.* **38,** 179–184.

Munehara, H. (1992). Utilization of polychaete tubes as spawning substrate by the sea raven *Hemitripterus villosus* (Scorpaeniformes). *Environ. Biol. Fish.* **33,** 395–398.

Munehara, H., Takano, K., and Koya, Y. (1989). Internal gametic association and external fertilization in the elkhorn sculpin, *Alcichthys alcicornis. Copeia* **1989,** 673–678.

Munehara, H., Takano, K., and Koya, Y. (1991). The little dragon sculpin *Blepsias cirrhosus,* another case of internal gametic association and external fertilization. *Japan. J. Ichthyol.* **37,** 391–394.

Nelson, J. S. (1994). "Fishes of the World," 3rd ed. Wiley, New York.

Olson, A. C., Jr. (1950). Ground squirrels and horned larks as predators upon grunion eggs. *Calif. Fish Game* **36,** 323–327.

Orton, G. L. (1955). Color variation in certain marine fish eggs. *Copeia* **1955,** 144–145.

Patzner, R. A. (1983). The reproduction of *Blennius pavo* (Teleostei, Blenniidae). I. Ovarial cycle, environmental factors and feeding. *Helgolander Meeresunters.* **36,** 105–114.

Patzner, R. A., Seiwald, M., Adlgasser, M., and Kaurin, G. (1986). The reproduction of *Blennius pavo* (Teleostei, Blenniidae). V. Reproductive behavior in natural environment. *Zool. Anz.* **216,** 338–350.

Peters, K. M. (1981). Reproductive biology and developmental osteology of the Florida blenny, *Chasmodes saburrae* (Perciformes: Blenniidae). *Northeast Gulf Sci.* **4,** 79–98.

Petersen, C. W. (1988). Male mating success, sexual size dimorphism, and site fidelity in two species of *Malacoctenus* (Labrisomidae). *Environ. Biol. Fish.* **21,** 173–183.

Pillsbury, R. W. (1957). Avoidance of poisonous eggs of the marine fish *Scorpaenichthys marmoratus* by predators. *Copeia* **1957,** 251–252.

Potts, G. W. (1984). Parental behaviour in temperate marine teleosts with special reference to the development of nest structures. *In* "Fish Reproduction: Strategies and Tactics" (G. W. Potts and R. J. Wootton, Eds.), pp. 223–244. Academic Press, London.

Potts, G. W. (1985). The nest structure of the corkwing wrasse, *Crenilabrus melops* (Labridae: Teleostei). *J. Mar. Biol. Assoc. U.K.* **65,** 531–546.

Prochazka, K. (1994). The reproductive biology of intertidal klipfish (Perciformes: Clinidae) in South Africa. *S. Afr. J. Zool.* **29,** 244–251.

Qasim, S. Z. (1957). The biology of *Centronotus gunnellus* (L.) (Teleostei). *J. Anim. Ecol.* **26,** 389–401.

Ragland, H. C., and Fischer, E. A. (1987). Internal fertilization and male parental care in the scalyhead sculpin, *Artedius harringtoni. Copeia* **1987,** 1059–1062.

Ralston, S. L., and Horn, M. H. (1986). High tide movements of the temperate-zone herbivorous fish *Cebidichthys violaceus* (Girard) as determined by ultrasonic telemetry. *J. Exp. Mar. Biol. Ecol.* **98,** 35–50.

Rasotto, M. B. (1995). Male reproductive apparatus of some Blennioidei (Pisces: Teleostei). *Copeia* **1995,** 907–914.

Robertson, R., Peterson, C. W., and Brawn, J. D. (1990). Lunar reproductive cycles of benthic brooding reef-fishes: Reflections of larval biology or adult biology? *Ecol. Monogr.* **30,** 311–329.

Ruchon, F., Laugier, T., and Quignard, J. P. (1995). Alternative male reproductive strategies in the peacock blenny. *J. Fish Biol.* **47,** 826–840.

Santos, R. S., and Almada, V. C. (1988). Intraspecific variations in reproductive tactics in males of the rocky intertidal fish *Blennius sanguinolentus* in the Azores. *In* "Behavioral Adaptation to Intertidal Life" (G. Chelazi and M. Vannini, Eds.), pp. 421–447. Plenum Press, Boulder, CO.

Sawyer, P. J. (1967). Intertidal life-history of the rock-gunnel, *Pholis gunnellus,* in the western Atlantic. *Copeia* **1967,** 55–61.

Sayer, M. D. J., and Davenport, J. (1991). Amphibious fish: Why do they leave water? *Rev. Fish Biol. Fish.* **1,** 159–181.

Schultz, L. P., and A. C. DeLacy. (1932). The eggs and nesting habits of the crested blenny, *Anoplarchus. Copeia* **1932,** 143–147.

Shiogaki, M., and Dotsu, Y. (1971). The life history of the clingfish, *Aspasma minima. Japan. J. Ichthyol.* **18,** 76–84. [In Japan. with Engl. Abs.]

Shiogaki, M., and Dotsu, Y. (1972a). Life history of the blennioid fish, *Dictyosoma burgeri. Bull. Fac. Fish. Nagasaki Univ.* **33,** 21–38. [In Japan. with Engl. Abs.]

Shiogaki, M., and Dotsu, Y. (1972b). The life history of the blenniid fish, *Neoclinus bryope. Bull. Fac. Fish. Nagasaki Univ.* **34,** 1–8. [In Japan. with Engl. Abs.]

Strathmann, R. R., and Chaffee, C. (1984). Constraints on egg masses. II. Effect of spacing, size, and number of eggs on ventilation of masses of embryos in jelly, adherent groups, or thin-walled capsules. *J. Exp. Mar. Biol. Ecol.* **84,** 85–93.

Sunobe, T., Ohta, T., and Nakazono, A. (1995). Mating system and spawning cycle in the blenny, *Istiblennius enosimae,* at Kagoshima, Japan. *Environ. Biol. Fish.* **43,** 195–199.

Suyama, M., and Uno, Y. (1957). Puffer toxin during the embryonic development of puffer, *Fugu* (*Fugu*) *niphobles* (J. et S.). *Bull. Japan. Soc. Sci. Fish.* **23,** 438–441. [In Japan. with Engl. Abs.]

Taborsky, M. (1994). Sneakers, satellites, and helpers: Parasites and cooperative behavior in fish reproduction. *Adv. Study Behav.* **23,** 1–100.

Takita, T., Middaugh, D. P., and Dean, J. M. (1984). Predation of a spawning atherinid fish *Menidia menidia* by avian and aquatic predators. *Japan. J. Ecol.* **34,** 431–437.

Taylor, F. H. C. (1971). Variation in hatching success in Pacific herring (*Clupea pallasii*) eggs with water depth, temperature, salinity, and egg mass thickness. *Rapp. Proces-Verbaux Reunions Cons. Perm. Int. Explor. Mer* **160,** 34–41.

Taylor, M. H. (1984). Lunar synchronization of fish reproduction. *Trans. Am. Fish. Soc.* **113,** 484–493.

Taylor, M. H. (1986). Environmental and endocrine influences on reproduction of *Fundulus heteroclitus. Am. Zool.* **26,** 159–171.

Taylor, M. H. (1990). Estuarine and intertidal teleosts. *In* "Reproductive Seasonality in Teleosts: Environmental Influences" (A. D. Munro, A. P. Scott, and T. J. Lam, Eds.), pp. 109–124. CRC Press, Boca Raton, FL.

Taylor, M. H., and DiMichele, L. (1983). Spawning site utilization in a Delaware population of *Fundulus heteroclitus* (Pisces: Cyprinodontidae). *Copeia* **1983,** 719–725.

Taylor, M. H., DiMichele, L., and Leach, G. J. (1977). Egg stranding in the life cycle of the mummichog, *Fundulus heteroclitus. Copeia* **1977,** 397–399.

Taylor, M. H., Leach, G. J., DiMichele, L., Levitan, W. M., and Jacob, W. F. (1979). Lunar spawning cycle in the mummichog, *Fundulus heteroclitus* (Pisces: Cyprinodontidae). *Copeia* **1979,** 291–297.

Templeman, W. (1948). The life history of the capelin (*Mallotus villosus*) in Newfoundland waters. *Bull. Newfoundland Govt. Lab. St. John's* **17,** 1–155.

Tewksbury, H. T., II, and Conover, D. O. (1987). Adaptive significance of intertidal egg deposition in the Atlantic silverside *Menidia menidia. Copeia* **1987,** 76–83.

Thomson, D. A., and Muench, K. A. (1976). Influence of tides and waves on the spawning behavior of the Gulf of California grunion, *Leuresthes sardina* (Jenkins and Evermann). *Bull. S. Calif. Acad. Sci.* **75,** 198–203.

Thresher, R. E. (1984). "Reproduction in Reef Fishes." T.F.H. Publ., Neptune City, NJ.

Veith, W. J. (1979). Reproduction in the live-bearing teleost *Clinus superciliosus. S. Afr. J. Zool.* **14,** 208–211.

Walker, B. W. (1952). A guide to the grunion. *Calif. Fish Game* **38,** 409–420.

Warner, R. R., and Harlan, R. K. (1982). Sperm competition and sperm storage as determinants of sexual dimorphism in the dwarf surfperch, *Micrometrus minimus. Evolution* **36,** 44–55.

Whoriskey, F. G., and FitzGerald, G. J. (1985). The effects of bird predation on an estuarine stickleback (Pisces: Gasterosteidae) community. *Can. J. Zool.* **63,** 301–307.

Wiebe, J. P. (1968). The reproductive cycle of the viviparous seaperch, *Cymatogaster aggregata* Gibbons. *Can. J. Zool.* **46,** 1221–1224.

Wirtz, P. (1978). The behaviour of the Mediterranean *Tripterygion* species (Pisces, Blennioidei). *Z. Tierpsychol.* **48,** 142–174.

Wootton, R. J. (1992). Constraints in the evolution of fish life histories. *Neth. J. Zool.* **42,** 291–303.

Wootton, R. J. (1994). Life histories as sampling devices: Optimum egg size in pelagic fishes. *J. Fish Biol.* **45,** 1067–1077.

Wourms, J. P., and Lombardi, J. (1992). Reflections on the evolution of piscine viviparity. *Am. Zool.* **32,** 276–293.

Wyttenbach, A., and Senn, D. G. (1993). Intertidal habitat: Does the shore level affect the nutritional condition of the shanny (*Lipophrys pholis,* Teleostei, Blenniidae)? *Experientia* **49,** 725–728.

Yamahira, K. (1994). Combined effects of tidal and diurnal cycles on spawning of the puffer, *Takifugu niphobles* (Tetraodontidae). *Environ. Biol. Fish.* **40,** 255–261.

Yamahira, K. (1996). The role of intertidal egg deposition on survival of the puffer, *Takifugu niphobles* (Jordan et Snyder), embryos. *J. Exp. Mar. Biol. Ecol.* **198,** 291–306.

Yonge, C. M. (1949). "The Sea Shore." Collins, London.

Yoshiyama, R. M., and Cech, J. J., Jr. (1994). Aerial respiration by rocky intertidal fishes of California and Oregon. *Copeia* **1994,** 153–158.

Yoshiyama, R. M., Valpey, C. J., Schalk, L. L., Oswald, N. M., Vaness, K. K., Lauritzen, D., and Limm, M. (1995). Differential propensities for aerial emergence in intertidal sculpins (Teleostei; Cottidae). *J. Exp. Mar. Biol. Ecol.* **191,** 195–207.

# 9

# Parental Care in Intertidal Fishes

Ronald M. Coleman

*Department of Integrative Biology, University of California, Berkeley, California*

## I. Introduction

Chapter 8 (DeMartini) provides a thorough review of intertidal spawning by fishes. Parental care is an important component of the reproductive biology of some of those fishes. My purpose in this chapter is to investigate parental care in intertidal fishes by addressing two main themes. First, is there something unique about parental care in this group? Second, can intertidal fishes help us understand the broader issues of parental care in fishes?

A rich history of research reports on the existence and form of parental care in particular fishes and a number of important synthetic works attempt to draw conclusions about patterns and processes. Breder and Rosen (1966) produced the definitive summary of the literature on reproduction in fishes up to the early 1960s. The sheer volume of information since then has prevented an update. Blumer (1979, 1982) summarized the literature on parental care in fishes and these works remain the principal references for determining which kind of care is seen in which family of fishes. Important synthetic and analytical works on parental care in fishes include Barlow (1974), Balon (1975), Loiselle and Barlow (1978), Perrone and Zaret (1979), Blumer (1979), Werren *et al.* (1980), Baylis (1981), Gittleman (1981), Gross and Shine (1981), Gross and Sargent (1985), Clutton-Brock (1991), and Smith and Wootton (1995).

Parental care is defined as "any investment by the parent in an individual offspring that increases the offspring's chance of surviving (and hence reproductive success) at the cost of the parent's ability to invest in other offspring" (p. 139, Trivers, 1972). In fish, parental care can take a variety of forms, including guarding, nest building and maintenance, substrate cleaning, fanning, internal gestation, removal of dead eggs, oral brooding, retrieval of eggs or fry that stray from the nest or school, cleaning of eggs or fry, external egg carrying, egg burying, moving eggs or young, coiling, ectodermal feeding, brood-pouch egg carrying, and splashing (Blumer, 1979, 1982). Gross and Sargent (1985) summarized much of the previous work to draw general conclusions about parental care in fishes. They concluded that parental care is performed in 21% of the families of bony fishes and is widely distributed phylogenetically. Guarding is the predominant form of care (it occurs in

*Intertidal Fishes: Life in Two Worlds*

95% of care-giving fishes). Uniparental care is most common (78% of families providing care have solitary male *or* solitary female care). Male care is more common than female parental care (61% of families showing care have male care). They also pointed out that for most fishes the morphology and physiology of the two parents do not seem to differ such that one would be better able to provide care, that is, a male fish could likely fan eggs as easily as a female fish could (cf., the presence of mammary glands in mammals).

## II. Patterns

### A. Forms of Care

How do intertidal fishes look when examined in light of these generalizations about fishes? For many intertidal fish, we have limited knowledge about the exact nature of parental care behavior because of the difficulties of observing the animals in the natural habitat without disturbing them. Frequently intertidal fishes spawn in complex habitats, often in concealed locations, for example, under rocks (Marliave and DeMartini, 1977), or macrophytes (Marliave, 1981a). In some parts of the world where intertidal fishes occur, lowest tides only occur at night during the breeding season, for example, Vancouver, British Columbia (Coleman, 1992), or observations are made more difficult by severe wave action. Underwater observations may be hampered by the disruptive optical properties of the halocline near rivermouths or by turbidity (personal observation). In a few cases, researchers have been able to get fish to spawn and provide parental care in the lab (e.g., Hughes, 1986; Shiogaki, 1987; Coleman, 1992) and in a very few, field manipulations have been performed (e.g., DeMartini, 1988). My first point is that we have relatively few data to draw upon when discussing parental care in intertidal fishes so quantitative comparisons are premature. But, even given the relatively little that we know about parental care in intertidal fishes, clearly, with a few important exceptions, intertidal fish are neither unusual nor restricted in the kinds of parental care they perform (Table 1).

The most important exception is the lack of posthatching care in all but one taxon. Except for the plainfin midshipman (*Porichthys notatus*) and possibly its unstudied relatives, no intertidal fish provides care past the point of hatching. To persons familiar with the extensive care of sunfishes (Centrarchidae) or cichlids (Cichlidae), this is surprising. But, it is not a phenomenon linked to the intertidal habitat; rather it is a characteristic of marine fish. There are no other temperate marine fish that provide posthatching care, and there are only two groups of tropical marine fish known to do so. *Acanthochromis polyacanthus,* a damselfish of the family Pomacentridae, provides biparental care of the free-swimming fry (Robertson, 1973; Thresher, 1985; Nakazono, 1993) and *Amphichthys cryptocentrus,* a tropical subtidal batrachoidid, apparently behaves much like the midshipman, guarding juveniles in its burrow (Hoffman and Robertson, 1983). Other examples may be found but clearly it remains a rare phenomenon. This fact precludes a number of the forms of care from being found in intertidal fish, for example, cleaning of fry, retrieval of fry, moving of young, and ectodermal feeding.

Of the other forms of care, most have been found in an intertidal fish, although this depends partly on how one defines intertidal fish. If one is concerned only with fish in which the eggs become emergent on a regular basis, the number of intertidal fishes decreases substantially and so too does the forms of care found. For example, external egg carrying and brood-pouch egg carrying is seen in the Syngnathidae, which includes the pipefishes and seahorses (Vincent *et al.,* 1995). These are shallow water fishes in general

**Table 1.** Parental Care in Selected Intertidal Fishes

| Family<br>*Species* | Summary of care | Location studied | Reference |
|---|---|---|---|
| Blenniidae | | | |
| Many | Male guards eggs laid in cavity | Mediterranean | Almada and Santos, 1995 |
| Hexagrammidae | | | |
| *Hexagrammos decagrammus* | Male guards egg mass | Washington State, California | DeMartini, 1986 |
| *Hexagrammos stelleri* | Male guards egg mass | Washington State, California | DeMartini, 1986 |
| Gobiesocidae | | | |
| *Gobiesox maendricus* | Male guards monolayer of eggs | British Columbia | Personal observation |
| *Gobiesox strumosus* | Male guards monolayer of eggs | Chesapeake Bay | Runyan, 1961 |
| Stichaeidae | | | |
| Many | Male *or* female coils around egg ball | British Columbia, Japan, NE Atlantic | see Table 2 |
| Pholidae | | | |
| *Pholis laeta* | Both parents coil around egg ball | British Columbia | Hughes, 1986 |
| *Pholis gunnellus* | Both parents coil around egg ball | British Isles | Qasim, 1957 |
| Batrachoididae | | | |
| *Porichthys notatus* | Male guards, fans, cleans monolayer of eggs *and* hatchlings | B.C., California, Washington | Crane, 1981; DeMartini, 1988 |
| Clinidae | | | |
| Several spp. | Viviparous | South Africa | Prochazka, 1994 |
| Gasterosteidae | | | |
| *Gasterosteus aculeatus* | Male builds nest, guards, fans eggs, then disperses them | Nova Scotia | MacDonald *et al.*, 1995a,b |
| Atherinidae | | | |
| *Leuresthes tenuis* | Female buries eggs in sand | California | Clark, 1938 |

*Note.* It should not be inferred from this table that members of a family, other than those studied, exhibit the stated behavior.

but a few, particularly various pipefishes such as the bay pipefish *Syngnathus leptorhynchus,* are often found in water measuring centimeters in depth, typically eelgrass beds or other aquatic vegetation. The males of this species carry the eggs in a brood pouch and occasionally males are left stranded in a tidepool or even in a clump of vegetation when the tide recedes (personal observation). I know of no example of a syngnathid that deliberately strands itself.

Internal gestation by females, or viviparity, is seen in several groups of intertidal fishes, including klipfishes (Family Clinidae) and surfperches (Family Embiotocidae)—again with the provision that these fish seldom remain behind when the tide recedes. There are 39 species of klipfishes endemic to South Africa and 25 of these occur in the intertidal zone. Prochazka (1994) reported on the reproductive biology of six of them: *Clinus acuminatus, C. agilis, C. cottoides, C. heterodon, C. superciliosus,* and *Muraenoclinus dorsalis.* These fish are all viviparous and have all evolved superembryonation, which means that a female contains broods of several different ages at the same time. Based on gonad indices, Prochazka estimated gestation periods of 4 to 6 months. Individual females contained as many as nine broods of eggs and embryos, though as with all viviparous fishes, total fecundity is small compared with similarly sized oviparous fishes.

The walleye surfperch *Hyperprosopon argenteum* (Family Embiotocidae) occurs from Vancouver Island to southern California, typically in the surf zone. Like all embiotocids, it is viviparous (Nelson, 1994). DeMartini *et al.* (1983) found that females produced broods averaging nine embryos after a 5- to 6-month gestation period. The extreme length of parental care produces young of such advanced size and developmental stage that nearly all males and 60% of the females attain sexual maturity in their first year.

Oral brooding, also called mouthbrooding, is one form of care that has not been found in an intertidal fish, despite being present in a wide diversity of fishes including four orders and eight families (Oppenheimer, 1970). The term mouthbrooding describes a parent keeping eggs or young in its mouth for a length of time, as distinct from a fish which picks up its offspring in its mouth to move them and keeps them in the mouth for only a matter of seconds. Mouthbrooders might retain the offspring for days or weeks (Oppenheimer, 1970). Research is ongoing with mouthbrooders to determine if the parent actually feeds itself during the period of oral brooding. In general, the parent does not feed, e.g., *Apogon doederleini,* a marine paternally mouthbrooding cardinalfish (Okuda and Yanagisawa, 1996), but exceptions exist in which the parent feeds solely to nourish the young such as *Tropheus moori* (Yanagisawa and Sato, 1990), or feeds both the young and itself such as *Microdontochromis* sp. (Yanagisawa *et al.,* 1996), both cichlids from Lake Tanganyika. Length of development—hence length of mouthbrooding and the cost of mouthbrooding if the parent does not feed itself—likely increases with both egg size and a decrease in temperature (Coleman, unpublished). Large eggs and low temperatures are a characteristic of many intertidal fishes so it may be that these factors make mouthbrooding prohibitively costly to the parent. If mouthbreeding is found in an intertidal fish, it will most likely be in a tropical species.

Splashing is a rare form of care known only in a few cases, most notably the spraying (or splashing) characin *Copeina arnoldi* (Krekorian, 1976; Krekorian and Dunham, 1972), a freshwater fish from South America. A pair spawns on leaves overhanging the water and male care consists of periodically splashing the leaves to keep the eggs moist. This type of care is rare in intertidal fishes (see midshipman, below) for several reasons. The churning surf and spray dwarf any effort a fish could make and the fact that the water recedes horizontally as well as vertically makes it difficult for a parent to remain in the water near an exposed egg mass for a lengthy period of time.

Egg burying is performed by a few intertidal spawners. The grunion (*Leuresthes tenuis*) is perhaps the most studied (Clark, 1938). During the summer in southern California, grunion breed at high tide on the beaches on nights just following the highest high tide (Walker, 1952). The fish arrive en masse and each female buries down into the sand. One or more males attempt to spawn with each female by releasing sperm into the sand above the buried female. Eggs are thus buried in the sand and the parents leave. The timing of spawning in relation to the tide cycle ensures that the eggs remain buried in the protection of the sand for 2 weeks until the next series of high tides, which will expose them and release the embryos.

Egg scattering has recently been described for the "white" stickleback, a subspecies or close relative of the threespine stickleback (*Gasterosteus aculeatus*) (MacDonald *et al.*, 1995a,b; Blouw, 1996). The typical reproductive cycle of the threespine stickleback is one of the best known behavioral sequences in biology (van Iersel, 1953). A male constructs a nest by gluing together plant material into a tube. He then courts a female and entices her into the nest, where she lays the eggs and he subsequently fertilizes them. The female departs and the male continues to court more females and to fan and guard the eggs. The eggs hatch and the male guards the fry for days or even weeks. The white stickleback has been found in the coastal waters of Nova Scotia, Canada (Blouw, 1996). It differs from other threespine sticklebacks in coloration, but more importantly in behavior. After a white male has acquired eggs, instead of caring for them, he takes mouthfuls of them and disperses them into the surrounding filamentous algae. In one particular study population, the males rapidly construct flimsy nests on bare rock in the subtidal or even intertidal. These males also disperse their eggs, and the eggs tend to end up in crevices between stones and may be exposed to air at low tide. Even so, such eggs typically survive to hatch (MacDonald *et al.,* 1995b). This species offers a wonderful opportunity to study the costs and benefits of guarding and the circumstances under which it can be lost.

Coiling is often attributed to intertidal fishes in particular. This is necessarily restricted to elongate fishes. Such a fish will wrap its body around the egg mass loosely forming part of a coil (e.g., some stichaeids), but in other cases forming more than one coil (e.g., wolf-eels, *Anarrhichthys ocellatus,* a very large subtidal fish with biparental care; personal observation). Coiling has been suggested to prevent desiccation of the eggs (Blumer, 1982), and this may be true to some extent, but some subtidal fish coil and some fish that coil in the intertidal also coil when fully submerged (Coleman, 1992). Coiling also likely serves a protective function, both to keep the eggs in place and possibly to deter predators (Coleman, 1992).

Guarding is the most common form of care by intertidal fishes, and fanning occurs in some cases, though it is harder to observe and study than guarding. Nest building or at least some degree of site alteration also happens in many species. The following accounts serve to provide typical examples of these common forms of parental care in intertidal fishes.

Cockscomb pricklebacks (*Anoplarchus purpurescens*) of the family Stichaeidae spawn high in the intertidal zone in late January and early February in shallow spaces under rocks (Coleman, 1992; Schultz and DeLacy, 1932; Peppar, 1965). The female alone provides parental care. The egg mass is only loosely attached to the substrate if at all, but the eggs adhere tightly to each other, forming a ball half the size of a golf ball. The female coils her body around the egg mass, which likely represents her total reproductive effort for the year (Figure 1). She both guards and fans the eggs for a period of 29 days. Much of the time she is submerged, but for a number of hours on low tides she and the egg mass are exposed to the air. She remains with the egg mass during these times. At the end of incubation the eggs hatch and the young immediately swim toward the surface. There is no further parental care.

Males were never found coiled around an egg mass and were not found in close association with guarding females (Coleman, 1992). No female was observed guarding more than one egg mass simultaneously. Undefended egg masses were never seen in the wild. In aquaria, males ignored females that were guarding eggs.

Some members of the closely related family Pholidae also spawn in the intertidal zone (Hughes, 1986; personal observation) but they are biparental, at least some of the time. Qasim (1957) reported on the rock gunnel (*Pholis gunnellus*) from the British Isles. He often found both parents guarding the eggs (8 cases), but found 11 cases with the female

**Figure 1.** Female cockscomb prickleback wrapped around her egg mass.

alone and 1 case with a solitary male. When I found the crescent gunnel *Pholis laeta* spawning at Cates Park, Vancouver, British Columbia, I located them under 0.5 to 1.0 m of rocks, at the interface between the rock jumble and the underlying sand/mud substrate (this was at approximately 0.9 m tide level). There I discovered 6 egg masses. In 2 cases the female was wrapped around the eggs and the male was close by (within 20 cm). In 4 other cases, no male was seen, although it would be easy to miss a fish under the circumstances, that is, under rocks and mud at midnight. Hughes (1986) reported 7 egg masses at Patricia Bay, on Vancouver Island. Of these, 4 had two adults coiled around them and 3 had one adult (sex unspecified). He also took 5 pairs into the lab and got them to breed. The egg mass was similar to that described above for the cockscomb prickleback, namely the eggs adhere to each other, but not to the substrate. Time budgets revealed that an egg mass was tended by one parent 23% of the time, by both parents 10% of the time, and was untended

67% of the time. The latter value may reflect the absence of other fish in the aquarium against which parents need to defend the egg mass.

The northern clingfish (*Gobiesox maeandricus*) lays its eggs in February in a monolayer about 10 cm in diameter on the underside of rocks in the intertidal zone near Vancouver, British Columbia (Figure 2). The male remains to guard them when the water recedes (Marliave and DeMartini, 1977) but there are no published accounts of the detailed behaviour of this fish. On the Atlantic coast, Runyan (1961) studied the subtidal *G. strumosus*. This fish spawns from April through August and its eggs are often found laid in oyster shells, in a monolayer. Some oyster shells contained eggs at different stages of development, suggesting a male may spawn with multiple females, or according to Runyan, with the same female repeatedly. As with its Pacific counterpart, the male is the nest guardian and Runyan reported fanning by males using anal and caudal fins.

As mentioned, the plainfin midshipman is unique among temperate marine fish, intertidal or otherwise, because it not only provides care for the eggs, but also cares for the hatchlings (Arora, 1948). Midshipman spawn in late spring and early summer along the Pacific Coast of North America. A male midshipman cleans a nest site underneath a rock, forming a cavity several times his body size (Figure 3). A female enters the nest and lays the eggs on the roof of the cave. The large bright yellow eggs are laid in patches and a male may acquire eggs from many females as evidenced by the various stages of development seen in some nests (Crane, 1981; DeMartini, 1988; Bass, 1996). The male guards and fans the eggs, cleans the nest site, and may splash water over the eggs when the eggs are drying out at low tide (Arora, 1948). The male does not appear to leave the nest and thus starves through the period of parental care. The eggs take weeks to hatch but when they do the hatchlings remain attached to the substrate for several more weeks via an adhesive disc. The hatchlings feed while still attached to the rock (Crane, 1981). Eventually they break

**Figure 2.** Male clingfish guarding his egg mass.

**Figure 3.** Male midshipman in his nest with eggs and juveniles.

free but may remain in the vicinity of the nest for longer still (Bass, 1996; personal observation). The detached young are unusual among marine fish because they are not larvae, they are juveniles, that is, they look like miniature adults (Arora, 1948).

Arora (1948) described some fascinating experiments in which he introduced two males each with their own nest into the same aquarium. The males fought after a few hours and one male was badly beaten. Implied in the performance of these experiments is that it is possible for researchers to take males, with their nests, and move them about without the male abandoning the nest. This opens up exciting avenues for manipulative experiments that surprisingly have not yet been pursued. Additionally, this remarkable fish also exhibits alternative male phenotypes: the large guarding males that dig nests, attract females, and provide parental care, but also smaller, fast-maturing "sneaker" males (Gross, 1984) that hang around the nest area and steal fertilizations (Bass, 1996; Brantley and Bass, 1994).

The red Irish lord *Hemilepidotus hemilepidotus* (Family Cottidae) presents an intriguing contrast to other known examples of parental care in intertidal fishes. Whereas in most species the roles of each sex are clearly defined, they appear much more flexible in the red Irish lord. DeMartini and Patten (1979) reported on 10 years of underwater and intertidal observations of this fish in which they observed both male and female guardians of the eggs. The eggs are laid between October and January on rock riprap and wooden dock pilings. The eggs are not in a monolayer, but rather a circular mass up to 10 eggs thick. The parent guards the eggs, but was not observed to fan. Unlike the previously described fish, red Irish lords do not remain with the eggs when the water recedes at low tide. Rather, the parent stays in the water but returns to the egg mass when the water returns. The authors distinguished between primary and secondary guardians. A fish was a primary guardian if it was the only guardian or if it was the closest fish to the egg mass. Other fish within 1 m of an egg mass were deemed secondary guardians.

This account highlights one of the recurring difficulties of parental care research, most noticeable when studying sedentary fish: Is the fish really providing parental care, or does it just happen to be in the area because the area offers feeding opportunities, camouflage, etc.? One test might be to threaten the eggs and see if the parent or parents respond.

Assuming that the presence of an adult represented guarding behavior, DeMartini and Patten (1979) found that the primary guardian was a female in 86% of the observations. In 35% of the observations, the primary guardian female was accompanied by one to four secondary guardians within 1.0 m of the egg mass. They also provided evidence that care may be biparental; removal of a female primary guardian resulted in guarding by a male. This system deserves much more attention by researchers.

Although blennies (Blenniidae) are often thought of as tropical reef fish, some species are important inhabitants of the rocky intertidal zone in parts of the Mediterranean (Kotrschal, 1988; Illich and Kotrschal, 1990; Almada and Santos, 1995). Blennies are cavity nesters, often using holes excavated by other organisms, or the hard remnants of other organisms including barnacles and bivalves. Females lay their eggs in the cavity and all species provide solitary male care of the eggs. Similarly, kelp and whitespotted greenling (*Hexagrammos decagrammus* and *H. stelleri,* family Hexagrammidae) provide solitary male care (DeMartini, 1986). Males were seen to guard up to seven egg masses at once.

Finally, many intertidal spawners provide no parental care (DeMartini, Chapter 8, this volume). For example, the sharpnose sculpin *Clinocottus acuticeps* lays its eggs high in the intertidal zone, at 3.0- to 3.7-m tide levels, that is, higher than most species that do provide parental care. And yet, after laying these eggs in a habitat that will ensure that they are exposed to the air for almost half their incubation period, the parents leave the eggs and do not return. Marliave (1981a) showed that most successful egg masses are laid under plants such as *Fucus distichus* and that removal of the plant material reduced egg survivorship. The plants likely provide protection from desiccation and temperature fluctuations, which a parent might do in other species.

In addition to cottids like *Clinocottus acuticeps,* there are various killifishes, smelts, herrings, etc., that spawn intertidally but do not provide parental care (DeMartini, Chapter 8, this volume). For this reason, I think it is an overstatement to characterize intertidal fish as having parental care (cf., Almada and Santos, 1995); many do, but many do not.

We have no data on parental care for many fishes known to utilize the intertidal zone and even spawn there. For example, perhaps the most intriguing intertidal spawner is *Trichodon trichodon,* the Pacific sandfish. Only a single egg mass of this species has been found (Marliave, 1981b). The mass contained very large eggs (3.52 mm in diameter) and was located between the 0.6- and 1.0-m tide levels in a surge channel on a rocky shoreline on

Vancouver Island, British Columbia. The egg mass was partially developed upon collection, but took an additional 8 months to hatch in the laboratory! This suggests an overall incubation period of more than 1 year. Does this species provide parental care? No one knows.

### B. Duration of Care

One aspect of parental care by some intertidal fishes stands out as unusual: It may last for a long period of time, for example, a month for cockscomb pricklebacks compared to a week for the freshwater bluegill sunfish *Lepomis macrochirus.* If we exclude posthatching care, the difference becomes even more extreme: bluegill sunfish eggs hatch in 2 to 3 days. Why is there such a difference in duration of care?

The explanation appears to stem from two facts (Coleman, in review). In general, intertidal fish showing extended care have large eggs and/or develop at low temperatures at temperate or boreal latitudes. Through manipulative experiments on cichlid fishes, Coleman showed that developmental time is a decreasing function of temperature and an increasing function of egg size. While the former result has long been known to fisheries biologists (e.g., Blaxter and Hempel, 1963), the extent of the effect has not been appreciated by parental care researchers. To a large degree, the extreme duration of parental care in some intertidal fishes is simply the product of low temperatures and large eggs.

This finding raises a new question: Why do some intertidal fishes produce large eggs? I suggest that large eggs give the emerging hatchlings a performance advantage over smaller hatchlings. Coleman and Galvani (1998) demonstrated in cichlid fishes that even over relatively small differences in egg size, larger eggs produce larger hatchlings. These small differences in offspring size create differences in the ability of offspring to swim against a current (Coleman, in preparation). We now know that the offspring of some intertidal fish do not passively enter the plankton, but rather appear to resist currents to maintain position nearshore (Marliave, 1986). Larger egg size, and hence larger offspring that are better swimmers, would allow these larvae more control over their position, either for feeding purposes or perhaps to control where they settle.

## III. Process

Why do some fishes, intertidal or otherwise, exhibit male care, others female care, others biparental care, and some no care at all? The origin of care states is a long-standing issue in evolutionary biology. Fishes in general offer unique opportunities to study the evolution of parental care. To examine the evolution of a trait, it is necessary to find variation in that trait. To study the origin of parental care states, we need clades that contain more than one care state so that we can examine the transition from one state to another. The models hinge on understanding these transitions. For this reason, several major groups of vertebrates, though interesting, can offer only limited insight into the origins of parental care.

Sargent and Gross (1993) summarized the distribution of care states among major groups of vertebrates. For example, parental care is highly developed in mammals—in fact all mammals exhibit parental care—making it impossible to discover the conditions under which care evolved. Similarly, almost all birds have biparental care (90%). Male-only care is found in a scant 2%. Teleost fishes have a more even distribution of care, with each state being well represented, though male care dominates (Blumer, 1982). Additionally, the

25,000 species of fish (Nelson, 1994) provide numerous examples for examination and experimentation.

Gross and Sargent (1985) presented a model for the evolution of male and female parental care in fishes. They argued that parental care has only one benefit, namely increased survivorship of the young. It has, however, three potential costs: a mating cost (an organism providing care cannot acquire additional matings), an adult survivorship cost (providing care reduces the resources a parent has available to maintain itself), and a future fertility cost (if fertility is a function of size, care reduces growth and hence fertility). The benefit of care is the same for either sex, as is the adult survivorship cost. How the other two costs compare for males versus females will determine which sex provides care.

A mating cost usually only applies to males because females are limited by egg production, not mating opportunities. Gross and Sargent (1985) argued that male territoriality can reduce or even remove this cost because females seek out a male and spawn in the male's territory.

The future fertility cost depends on how fertility changes with body size. With few exceptions, teleosts exhibit indeterminate growth, that is, they continue to grow after reaching sexual maturity. If fertility is a function of body size, as it is for many female fishes (i.e., the number of eggs a female lays increases with body size) and it often is for males (i.e., large males acquire more mates than small males) then this might affect the cost of parental care. Energy allocated to care, and hence not allocated to growth, will reduce the future fertility of a fish.

Gross and Sargent (1985) emphasized that the shape of the fertility versus body size curve is important. If the curve accelerates—large males do disproportionately well compared to small males—then a male losing growth is paying a larger cost than if the curve is linear. Conversely, if the curve decelerates—large males do better than small males, but not a great deal better—then a male losing growth is paying a smaller cost than if the curve is linear. This hypothesis explains the predominance of male care in teleost fishes. In general, female fecundity accelerates with body size while male fertility increases linearly or decelerates. Thus, it costs a male less to provide care than it would cost a female (Gross and Sargent, 1985). Testing the model has proved difficult.

Do intertidal fishes offer special opportunities for testing models of the evolution of parental care? Once again we must ask about variation. The nonintertidal sunfishes exhibit parental care, but it is always solitary male care—its origin remains a mystery. Similarly, solitary female care is found in over a dozen species yet, in many cases, these fish are viviparous (e.g., Poeciliidae), which introduces numerous confounding issues. Solitary care might evolve either from biparental care or from no care. Tropical freshwater cichlid fishes (family Cichlidae) provide the best opportunity for studying the transition from biparental care to solitary care since all three states (male, female, and biparental care) are found in the family and there are no cichlids without care. But what about the transition directly from no care to female care or no care to male care—How can we study that?

I suggest that the superfamily Stichaeoidea (the pricklebacks, gunnels, wolffishes, and quillfish) offers great promise for studies of the evolution of care states from no care (Table 2). Of the four families in the Stichaeoidea (Makushok, 1958), two have intertidal representatives (Stichaeidae, Pholidae) and one is subtidal (Anarhicadidae). The fourth family, the monotypic quillfish, Ptilichthyidae, is largely unknown.

The anarhichadids (wolffishes, 2 genera, 5 species) exhibit male-only or biparental care (Keats *et al.,* 1985; Marliave, 1987; personal observation). The pholids (gunnels, 4 genera, 14 species) exhibit biparental care, though possibly facultative female-only care (Hughes,

**Table 2.** Distribution of Parental Care in the Superfamily Stichaeoidea

| Family<br>Subfamily<br>*Species* | Parent | Multiple egg masses | Reference |
|---|---|---|---|
| Anarhichadidae | | | |
| *Anarhichas lupus* | Male | No | Keats *et al.*, 1985 |
| *Anarrhichthys ocellatus* | Biparental | No | Marliave, 1987; personal observation |
| Pholidae | | | |
| *Pholis laeta* | Biparental | No | Hughes, 1986 |
| *Pholis gunnellus* | Biparental | No | Qasim, 1957 |
| Ptilichthyidae | ? | | |
| Stichaeidae | | | |
| Stichaeinae | | | |
| *Ulvaria subbifurcata* | Male | Yes | LeDrew and Green, 1975<br>Green *et al.*, 1987 |
| Chirolophinae | | | |
| *Chirolophis japonicus* | Female | No | Shiogaki, 1983 |
| *Chirolophis ascanii* | Female | | Shiogaki, 1981* |
| *Chirolophis decoratus* | Female | No | D. Kent, personal communication |
| Lumpeninae | ? | | |
| Opisthocentrinae | | | |
| *Opisthocentrus tenuis* | Female | No | Shiogaki, 1981 |
| *Opisthocentrus ocellatus* | Female | No | Shiogaki, 1982 |
| Cebidichthyinae | | | |
| *Dictyosoma burgeri* | Male | | Shiogaki and Dotsu, 1972 |
| Xiphisterinae | | | |
| *Xiphister atropurpureus* | Male | Yes | Marliave and DeMartini, 1977 |
| *Xiphister mucosus* | Male | Yes | Marliave and DeMartini, 1977 |
| *Ernogrammus hexagrammus* | Male | | Shiogaki, 1981* |
| Alectriinae | | | |
| *Anoplarchus purpurescens* | Female | No | Schultz and DeLacy, 1932; Peppar, 1965; Coleman, 1992 |
| *Alectrias alectrolophus* | Female | | Shiogaki, 1981* |
| *Alectrias mutsuensis* | Female | | Shiogaki, 1985 |
| *Alectrias benjamini* | Female | | Shiogaki, 1987 |
| Azygopterinae | ? | | |
| Eulophinae | ? | | |

*Note.* A ? indicates unknown. Multiple egg masses refers to whether the parent guards more than one egg mass simultaneously. A * indicates that the information was taken from Shiogaki's (1981) summary.

1986; personal observation). The stichaeids (pricklebacks, 36 genera, 65 species) exhibit solitary male *or* solitary female care, but never biparental care (Coleman, 1992). This latter combination is intriguing: Why has female care evolved in some species and male care in others?

Fortunately, we have surprisingly good knowledge about parental care states in stichaeids, largely due to the efforts of Masaru Shiogaki (Table 2). Also, several researchers have worked on the phylogenetic systematics of the group (Makushok, 1958; Anderson, 1984; Stoddard, 1985) giving us a good idea of how the various species are related. More needs to be done, but we have a good start.

Parental care appears to follow phylogenetic lines, each subfamily exhibiting a characteristic care state (Table 2). In addition, Stoddard (1985) suggests that the clade containing *Xiphister* and the clade containing *Alectrias* and *Anoplarchus* are both monophyletic and together form a monophyletic group. If this is true, then male care and female care must have evolved multiple times within the family.

From my own observations, *Xiphister* and *Anoplarchus* are similar fishes. The two *Xiphister* species, *X. atropurpureus* and *X. mucosus,* both grow to a larger size than *Anoplarchus purpurescens,* but otherwise, all three species can be found in overlapping intertidal locations. And yet, the *Xiphister* spp. provide male care and *Anoplarchus purpurescens* provides female care. Why? We do not yet know why, but studying such similar, closely related fish doing opposite things at the same time in the same location seems to offer a far better prospect for understanding the evolution of parental care states than many other systems.

For example, Gross and Sargent (1985) argued that solitary female care will evolve only if (1) female fecundity does not accelerate with body size, (2) male fertility accelerates with body size, and/or (3) males are not territorial. In pricklebacks, the first condition appears to hold: female fecundity increases only linearly with body size (length or mass), likely because of the elongate body shape (Coleman, 1992).

The second and third conditions are harder to examine. In the stichaeids, male care is associated with males acquiring more than one egg mass to guard (Table 2). When only a single egg mass is guarded, it is the female that does so. These data indicate that under certain conditions males can obtain additional matings while already providing parental care. Perhaps this is because of territoriality. We know little about the detailed feeding habits of these species or how they use space because of the complex three-dimensional geometry of the rocky intertidal zone, but if the food resource of one species were monopolizable, such as macrophytes or small organisms living on macrophytes, then a male could have a feeding territory, acquire mates, guard eggs, and continue to acquire additional mates. Thus the mating cost would not exist for males of that species and male care might evolve. However, if a similar species had slightly different feeding habits that involved chasing mobile prey that could not be monopolized, males could not eliminate the mating cost of providing care and female care might evolve. Is this the explanation for care states in pricklebacks? Perhaps and perhaps not, but it definitely warrants investigation.

## IV. Conclusions and Recommendations

On the one hand, the answers to the first question asked in Section 1 are negative: that is, in general no particular aspect of parental care by intertidal fish seems to make them unique. Many species lack care and of those that provide it, care is usually by the male in keeping with the general pattern of teleost fishes. Examples of female care and biparental care also exist. Intertidal fishes exhibit a diversity of forms of care ranging from guarding and fanning through rare forms like coiling. The one noteworthy aspect of their parental care, namely the extended duration of care exhibited by some species, is likely the product of large egg size and temperature. However, this should not be discouraging. In fact, it may be particularly useful to researchers interested in parental care, because it means that there is nothing unique about intertidal fishes, which in turn means studying intertidal fish may provide information of general applicability to other fishes, intertidal or not.

Although in general parental care by intertidal fish is not unique, certain instances offer promise for future study. For example, the family Stichaeidae has solitary male and solitary female care and this may be a fabulous opportunity to test recent models of the evolution of parental care.

The study of parental care in intertidal fishes is still in its infancy. We have a growing list of fascinating fishes performing a range of behavior yet our knowledge has so many gaps that conclusions are premature. Sampling methods that utilize net tows or similar bulk collection methods are invaluable for determining which fish are present in the intertidal (e.g., Prochazka, 1996) but do not provide any information on parental care. More researchers need to get in the water, turn over rocks in the intertidal zone to observe these animals in the wild, and attempt to breed these fish in aquaria. Only then will we see the possibilities that this interesting group of fishes has to offer.

## References


Almada, V. C., and Santos, R. S. (1995). Parental care in the rocky intertidal: A case study of adaptation and exaptation in Mediterranean and Atlantic blennies. *Rev. Fish Biol. Fish.* **5,** 23–37.

Anderson, M. E. (1984). "On the Anatomy and Phylogeny of the Zoarcidae (Teleostei: Perciformes)." Ph.D. thesis, College of William and Mary, Virginia.

Arora, H. L. (1948). Observations on the habits and early life history of the batrachoid fish, *Porichthys notatus* Girard. *Copeia* **1948,** 89–93.

Balon, E. K. (1975). Reproductive guilds of fishes: A proposal and definition. *J. Fish. Res. Bd. Can.* **32,** 821–864.

Barlow, G. W. (1974). Contrasts in social behavior between Central American cichlid fishes and coral-reef surgeon fishes. *Am. Zool.* **14,** 9–34.

Bass, A. H. (1996). Shaping brain sexuality. *Am. Sci.* **84,** 352–363.

Baylis, J. R. (1981). The evolution of parental care in fishes, with reference to Darwin's rule of male sexual selection. *Environ. Biol. Fish.* **6,** 223–251.

Blaxter, J. H. S., and Hempel, G. (1963). The influence of egg size on herring larvae (*Clupea harengus* L.). *J. Cons. Perm. Int. Explor. Mer.* **28,** 211–240.

Blouw, D. M. (1996). Evolution of offspring desertion in a stickleback fish. *EcoScience* **3,** 18–24.

Blumer, L. S. (1979). Male parental care in the bony fishes. *Q. Rev. Biol.* **54,** 149–161.

Blumer, L. S. (1982). A bibliography and categorization of bony fishes exhibiting parental care. *Zool. J. Linn. Soc.* **76,** 1–22.

Brantley, R. K., and Bass, A. H. (1994). Alternative male spawning tactics and acoustic signals in the plainfin midshipman fish *Porichthys notatus* Girard (Teleostei, Batrachoididae). *Ethology* **96,** 213–232.

Breder, C. M., Jr., and Rosen, D. E. (1966). "Modes of Reproduction in Fishes." TFH Publications, Neptune City, NJ.

Clark, F. N. (1938). Grunion in southern California. *Calif. Fish Game* **24,** 49–54.

Clutton-Brock, T. H. (1991). "The Evolution of Parental Care." Princeton University Press, Princeton, NJ.

Coleman, R. M. (1992). Reproductive biology and female parental care in the cockscomb prickleback *Anoplarchus purpurescens* (Pisces: Stichaeidae). *Environ. Biol. Fish.* **35,** 177–186.

Coleman, R. M., and Galvani, A. P. (1998). Egg size determines offspring size in neotropical cichlid fishes (Teleostei: Cichlidae). *Copeia* **1998,** 209–213.

Crane, J. M., Jr. (1981). Feeding and growth by the sessile larvae of the teleost *Porichthys notatus. Copeia* **1981,** 895–897.

DeMartini, E. E. (1986). Reproductive colorations, paternal behavior and egg masses of kelp greenling, *Hexagrammos decagrammus,* and whitespotted greenling, *H. stelleri. Northwest Sci.* **60,** 32–35.

DeMartini, E. E. (1988). Spawning success of the male plainfin midshipman. I. Influences of male body size and area of spawning site. *J. Exp. Mar. Biol. Ecol.* **121,** 177–192.

DeMartini, E. E. (1990). Annual variation in fecundity, egg size and condition of the plainfin midshipman (*Porichthys notatus*). *Copeia* **1990,** 850–855.

DeMartini, E. E., Moore, T. O., and Plummer, K. M. (1983). Reproductive and growth dynamics of *Hyperprosopon argenteum* (Embiotocidae) near San Diego, California. *Environ. Biol. Fish.* **8,** 29–38.

DeMartini, E. E., and Patten, B. G. (1979). Egg guarding and reproductive biology of the red Irish lord, *Hemilepidotus hemilepidotus* (*tilesius*). *Syesis* **12,** 41–55.

Gittleman, J. L. (1981). The phylogeny of parental care in fishes. *Anim. Behav.* **29,** 936–941.

Green, J. M., Mathisen, A.-L., and Brown, J. A. (1987). Laboratory observations on the reproductive and agonistic behaviour of *Ulvaria subbifurcata* (Pisces: Stichaeidae). *Nat. Can.* **114,** 195–202.

Gross, M. R. (1984). Sunfish, salmon, and the evolution of alternative reproductive strategies and tactics in fishes. *In* "Fish Reproduction: Strategies and Tactics" (G. W. Potts and R. J. Wootton, Eds.), pp. 55–75. Academic Press, New York.

Gross, M. R., and Sargent, R. C. (1985). The evolution of male and female parental care in fishes. *Am. Zool.* **25,** 807–822.

Gross, M. R., and Shine, R. (1981). Parental care and mode of fertilization in ectothermic vertebrates. *Evolution* **35,** 775–793.

Hoffman, S. G., and Robertson, R. R. (1983). On the foraging and reproduction of Caribbean reef toadfishes (Batrachoididae). *Bull. Mar. Sci.* **33,** 919–927.

Hughes, G. W. (1986). Observations on the reproductive ecology of the crescent gunnel, *Pholis laeta,* from marine inshore waters of southern British Columbia. *Can. Field-Nat.* **100,** 367–370.

Illich, I. P., and Kotrschal, K. (1990). Depth distribution and abundance of northern Adriatic littoral rocky reef blennioid fishes (*Blenniidae* and *Tripterygion*). *Mar. Ecol.* **11,** 277–289.

Keats, D. W., South, G. R., and Steele, D. H. (1985). Reproduction and egg guarding by Atlantic wolffish (*Anarhichas lupus:* Anarhichidae) and ocean pout (*Macrozoarces americanus:* Zoarcidae) in Newfoundland waters. *Can. J. Zool.* **63,** 2565–2568.

Kotrschal, K. (1988). Blennies and endolithic bivalves: Differential utilization of shelter in Adriatic *Blenniidae* (Pisces: Teleostei). *Mar. Ecol.* **9,** 253–269.

Krekorian, C. O. (1976). Field observations in Guyana on the reproductive biology of the spraying characid, *Copeina arnoldi* Regan. *Am. Midl. Nat.* **96,** 88–97.

Krekorian, C. O., and Dunham, D. W. (1972). Preliminary observations on the reproductive and parental behavior of the spraying characid *Copeina arnoldi* Regan. *Z. Tierpsychol.* **31,** 419–437.

LeDrew, B. R., and Green, J. M. (1975). Biology of the radiated shanny *Ulvaria subbifurcata* Storer in Newfoundland (Pisces: Stichaeidae). *J. Fish Biol.* **7,** 485–495.

Loiselle, P. V., and Barlow, G. W. (1978). Do fishes lek like birds? *In* "Contrasts in Behaviour" (E. S. Reese and F. J. Lighter, Eds.), pp. 31–75. Wiley, New York.

MacDonald, J. F., Bekkers, J., MacIsaac, S. M., and Blouw, D. M. (1995a). Intertidal breeding and aerial development of embryos of a stickleback fish (*Gasterosteus*). *Behaviour* **132,** 1183–1206.

MacDonald, J. F., MacIsaac, S. M., Bekkers, J., and Blouw, D. M. (1995b). Experiments on embryo survivorship, habitat selection, and competitive ability of a stickleback fish (*Gasterosteus*) which nests in the rocky intertidal zone. *Behaviour* **132,** 1207–1221.

Makushok, V. M. (1958). The morphology and classification of the northern blennioid fishes (Stichaeoidae, Blennioidei, Pisces). *Proc. Zool. Inst. Acad. Sci. U.S.S.R.* **25,** 3–129. [In Russian; Engl. transl. by A. R. Gosline, *U.S. Fish Wildl. Serv.,* 1–105.]

Marliave, J. B. (1981a). High intertidal spawning under rockweed, *Fucus distichus,* by the sharpnose sculpin, *Clinocottus acuticeps. Can. J. Zool.* **59,** 1122–1125.

Marliave, J. B. (1981b). Spawn and larvae of the Pacific sandfish, *Trichodon trichodon. U.S. Natl. Mar. Fish Serv. Fish. Bull.* **78,** 959–964.

Marliave, J. B. (1986). Lack of planktonic dispersal of rocky intertidal fish larvae. *Trans. Am. Fish. Soc.* **115,** 149–154.

Marliave, J. B. (1987). The life history and captive reproduction of the wolf-eel *Anarrhichthys ocellatus* at the Vancouver Public Aquarium. *Int. Zoo Yb.* **26,** 70–81.

Marliave, J. B., and DeMartini, E. E. (1977). Parental behavior of intertidal fishes of the stichaeid genus *Xiphister. Can. J. Zool.* **55,** 60–63.

Nakazono, A. (1993). One-parent removal experiment in the brood-caring damselfish, *Acanthochromis polyacanthus,* with preliminary data on reproductive biology. *Aust. J. Mar. Freshw. Res.* **44,** 699–707.

Nelson, J. S. (1994). "Fishes of the World," 3rd ed. Wiley, Toronto.

Okuda, N., and Yanagisawa, Y. (1996). Filial cannibalism in a paternal mouthbrooding fish in relation to mate availability. *Anim. Behav.* **52,** 307–314.

Oppenheimer, J. R. (1970). Mouthbreeding in fishes. *Anim. Behav.* **18,** 493–503.

Peppar, J. L. (1965). "Some Features of the Life History of the Cockscomb Prickleback, *Anoplarchus purpurescens* Gill." M.Sc. thesis, University of British Columbia, Vancouver.

Perrone, M., and Zaret, T. M. (1979). Parental care patterns of fishes. *Am. Nat.* **113,** 351–361.

Prochazka, K. (1994). The reproductive biology of intertidal klipfish (Perciformes: Clinidae) in South Africa. *S. Afr. Tydskr. Dierk.* **29,** 244–251.

Prochazka, K. (1996). Seasonal patterns in a temperate intertidal fish community on the west coast of South Africa. *Environ. Biol. Fish.* **45,** 133–140.

Qasim, S. Z. (1957). The biology of *Centronotus gunnellus* (L.) (Teleostei). *J. Anim. Ecol.* **26,** 389–401.

Robertson, R. R. (1973). Field observations on the reproductive behaviour of a pomacentrid fish, *Acanthochromis polyacanthus. Z. Tierpsychol.* **32,** 319–324.

Runyan, S. (1961). Early development of the clingfish, *Gobiesox strumosus* Cope. *Chesapeake Sci.* **2,** 113–141.

Sargent, R. C., and Gross, M. R. (1993). Williams' principle: An explanation of parental care in teleost fishes. *In* "The Behaviour of Teleost Fishes" (T. J. Pitcher, Ed.), 2nd ed., pp. 333–361. Croom Helm, London.

Schultz, L. P., and DeLacy, A. C. (1932). The eggs and nesting habits of the crested blenny, *Anoplarchus. Copeia* **1932,** 143–147.

Shiogaki, M. (1981). Notes on the life history of the stichaeid fish *Opisthocentrus tenuis. Japan. J. Ichthyol.* **28,** 319–328.

Shiogaki, M. (1982). Life history of the stichaeid fish *Opisthocentrus ocellatus. Japan. J. Ichthyol.* **29,** 77–85.

Shiogaki, M. (1983). On the life history of the stichaeid fish *Chirolophis japonicus. Japan. J. Ichthyol.* **29,** 446–455.

Shiogaki, M. (1985). Life history of the stichaeid fish *Alectrias mutsuensis* in Mutsu Bay, northern Japan. *Sci. Rep. Aquacult. Cen. Aomori Pref.* **4,** 11–20.

Shiogaki, M. (1987). Life history of the stichaeid fish *Alectrias benjamini. Sci. Rep. Aquacult. Cen. Aomori Pref.* **5,** 9–20.

Shiogaki, M., and Dotsu, Y. (1972). Life history of the blennioid fish, *Dictyosoma burgeri. Contrib. Fish. Exp. Station Nagasaki Univ.* **34,** 21–38.

Smith, C., and Wootton, R. J. (1995). The costs of parental care in teleost fishes. *Rev. Fish Biol. Fish.* **5,** 7–22.

Stoddard, K. M. (1985). "A Phylogenetic Analysis of Some Prickleback Fishes (Teleostei, Stichaeidae, Xiphisterinae) from the North Pacific Ocean, with a Discussion of Their Biogeography." M.A. thesis, California State University, Fullerton.

Thresher, R. E. (1985). Distribution, abundance, and reproductive success in the coral reef fish, *Acanthochromis polyacanthus. Ecology* **66,** 1139–1150.

Trivers, R. I. (1972). Parental investment and sexual selection. *In* "Sexual Selection and the Descent of Man" (B. Campbell, Ed.), pp. 136–179. Aldine, Chicago.

van Iersel, J. J. A. (1953). An analysis of the parental behaviour of the male three-spined stickleback (*Gasterosteus aculaeatus* L.). *Behav. Suppl.* **3,** 1–159.

Vincent, A. C. J., Berglund, A., and Ahnesjo, I. (1995). Reproductive ecology of five pipefish species in one eelgrass meadow. *Environ. Biol. Fish.* **44,** 347–361.

Walker, B. W. (1952). A guide to the grunion. *Calif. Fish Game* **23,** 409–420.

Werren, J. H., Gross, M. R., and Shine, R. (1980). Paternity and the evolution of male parental care. *J. Theor. Biol.* **82,** 619–631.

Yanagisawa, Y., Ochi, H., and Rossiter, A. (1996). Intra-buccal feeding of young in an undescribed Tanganyikan cichlid *Microdontochromis* sp. *Environ. Biol. Fish.* **47,** 191–201.

Yanagisawa, Y., and Sato, T. (1990). Active browsing by mouthbrooding females of *Tropheus duboisi* and *Tropheus moorii* (Cichlidae) to feed the young and/or themselves. *Environ. Biol. Fish.* **27,** 43–50.

# 10

# Recruitment of Intertidal Fishes

Catherine A. Pfister

*Department of Ecology and Evolution, University of Chicago, Chicago, Illinois*

## I. Introduction

The resident intertidal fishes discussed in this volume have a diversity of life histories, morphologies, color patterns, and feeding habits. Nearly all, however, have one thing in common: they all have a period as larvae when they are planktonic. Following the planktonic phase, all fishes need to reestablish themselves in the nearshore population. I refer to the process by which larvae establish themselves in the benthic population as recruitment, and the objective of this chapter is to explore what we know about recruitment in intertidal fishes and the ecological arguments that suggest how recruitment processes can be integrated in population and assemblage level dynamics.

The study of larval ecology has a rich history in marine systems, and landmark studies by Thorson (1950) and Strathmann (1974) established an empirical and theoretical tradition of exploring the consequences of how marine propagules are packaged and distributed along the shoreline (Grosberg and Levitan, 1992). As marine ecologists have grappled with the questions of what processes are important to the distribution and abundance of organisms, the role that recruitment plays has become increasingly debated and analyzed (Gaines and Roughgarden, 1985, 1987; Sutherland, 1990; Menge, 1991), especially in fish populations (Victor, 1983; Shulman, 1985; Shulman and Ogden, 1987; Forrester, 1990; Doherty and Fowler, 1994; Pfister, 1996). For example, Is the supply of larvae to nearshore assemblages so bountiful that only interactions occurring after settlement (such as competition and predation) determine organism distribution and abundance? Or, alternatively, Are the number of individuals who survive successfully through the egg and larval stage, finally making it back to shore, so few that population growth is limited by this supply and populations are "recruitment limited"? These are questions that are only beginning to be answered for marine populations.

The study of recruitment has other implications for both population and genetic structure. The distance that marine propagules, including fish larvae, travel is unknown except for a handful of marine species that have been observed *in situ* (Marliave, 1986; Carlon and Olson, 1993). Since larval dispersal is capable of linking disparate populations, both demographically and genetically, understanding dispersal and the fate of recruitment events is

*Intertidal Fishes: Life in Two Worlds*

critically important to defining the geographic extent of populations. Thus, despite homing abilities and site fidelity in many species (Green, 1971, 1973; Yoshiyama *et al.,* 1992; Pfister, 1996), differentiation may be unlikely if larvae disperse some distance before joining the adult population.

In this chapter, I first summarize what we know about recruitment in intertidal fishes, including the timing of recruitment, the size of individuals when they recruit, and the patterns of recruitment that have been reported for some fishes. I also discuss the processes during the early life history stages that might affect the success of recruitment, such as competition, predation, and conspecific facilitation. Because patterns of recruitment in nature often indicate that recruitment is variable both temporally and spatially, I explore potential consequences of this variation for population dynamics. Finally, I suggest areas of future research.

## II. The Timing of Recruitment

The timing of recruitment in intertidal fishes is, of course, tightly related to the timing of spawning (see DeMartini, Chapter 8, this volume) and the time spent in the plankton and can be either seasonal or aseasonal. We know little about the time spent or distance traveled by intertidal fish larvae in the plankton. Only a handful of studies have estimated the time spent in the plankton. Cottids are estimated to spend a month or two (Washington *et al.,* 1984); *Oligocottus maculosus* persist for 2 months in the plankton at 10°C (Marliave, 1986). Similarly, clinids have an approximately 2-month planktonic period (Stepien *et al.,* 1991). A blenny (*Hypsoblennius gilberti*) is in the plankton for 2–3 months (Stephens *et al.,* 1970), while a goby (*Lepidogobius lepidus*) in the same region is thought to spend 3–4 months in the plankton (Grossman, 1979).

The seasonality often seen in spawning (DeMartini, Chapter 8, this volume) appears to result in a majority of intertidal fishes having a seasonal component to the timing of recruitment (Table 1), although this surely reflects a bias in the relatively high number of studies in temperate areas. For example, many tropical reef fishes are represented by species with both seasonal and continuous spawning and recruitment (e.g., Johannes, 1978; Victor, 1983). From Table 1, it is clear that the overwhelming majority of intertidal fish studies to date show a seasonal component to recruitment. The only exceptions to this are *L. lepidus, Diplodus sargus,* and the triplefin *Bellapiscis medius,* and a suggestion that spawning might occur year-round in *Eucyclogobius newberryi.*

Despite the number of species whose recruitment is thought to be seasonal (Table 1), we know very little about the environmental cues or selective pressures that may entrain seasonality in intertidal fishes. However, there are well-documented oceanographic phenomena that occur in many nearshore areas. Seasonal upwelling of cooler, nutrient-rich waters occurs from February and March through August in the northeast Pacific and Atlantic Oceans. Similarly, the spring and summer months in the Southern Hemisphere are periods of upwelling in the region of the Benguela Current in the southeastern Atlantic Ocean and the Humboldt current in the southeastern Pacific Ocean. Although these periods of seasonal upwelling may provide increased food resources for larval and juvenile fishes, they can also be associated with increased offshore Ekman transport (Parrish *et al.,* 1981) and may represent a trade-off for nearshore fishes between provision of maximal food resources and the risk of offshore dispersal.

This trade-off in the benefit and cost of timing recruitment with upwelling events may be the reason for conflicting data on the correlation between recruitment and upwelling

**Table 1.** The Timing of Recruitment for 37 Species Representing 10 Families Worldwide

| Family | Species | Locale | Time of recruitment | Reference |
|---|---|---|---|---|
| Blenniidae | *Hypsoblennius gilberti* | S. California | Aug–Oct | Stephens *et al.*, 1970 |
| | *Blennius pholis* | NW Europe | Aug–Sep | Qasim, 1957 |
| Clinidae | *Acanthoclinus fuscus* | Kaikoura, New Zealand | August | Berger and Mayr, 1992 |
| | *Clinus acuminatus* | Cape Peninsula, S. Africa | Dec–Feb | Prochazka, 1996 |
| | *Clinus agilis* | Cape Peninsula, S. Africa | Dec–Feb | Prochazka, 1996 |
| | *Clinus cottoides* | Cape Peninsula, S. Africa | Dec–Feb | Prochazka, 1996 |
| | *Clinus superciliosus* | E. Cape, S. Africa | Sep–Dec | Beckley, 1985 |
| | *Clinus superciliosus* | Cape Peninsula, S. Africa | Dec–Feb | Prochazka, 1996 |
| | *Mureanoclinus dorsalis* | Cape Peninsula, S. Africa | Mar–May | Prochazka, 1996 |
| Cottidae | *Artedius lateralis* | Washington outer coast | May–Aug | Personal observation |
| | *Clinocottus embryum* | Washington outer coast | Apr–Sep | Pfister, 1997 |
| | *Clinocottus globiceps* | N. California | Apr–Sep | Grossman, 1982 |
| | *Clinocottus globiceps* | Washington outer coast | Apr–Aug | Pfister, 1997 |
| | *Oligocottus maculosus* | Washington outer coast | May–Sep | Pfister, 1997 |
| | *Oligocottus snyderi* | N. California | Apr–Sep | Grossman, 1982; Freeman *et al.*, 1985 |
| | *Oligocottus snyderi* | Washington outer coast | Apr–Sep | Pfister, 1997 |
| Gobiesocidae | *Gastroscyphus hectoris* | South coast, New Zealand | Sep–Feb | Willis and Roberts, 1996 |
| | *Gobiesox meandricus* | Washington outer coast | May–Aug | Personal observation |
| | *Lepadogaster lepadogaster* | British Isles, NW Europe | Jul–Aug | Wheeler, 1969 |
| | *Trachelochismus pinnulatus* | South coast, New Zealand | Nov–Jun | Willis and Roberts, 1996 |
| Gobiidae | *Gobius cobitis* | British Isles, NW Europe | Aug–Oct | Wheeler, 1969 |
| | * *Eucyclogobius newberryi* | California | Year-round | Lafferty *et al.*, 1996 |
| | * *Lepidogobius lepidus* | Morro Bay, California | June, Nov, Dec | Grossman, 1979 |
| | *Pomatschistus microps* | British Isles, NW Europe | May–Aug | Wheeler, 1969 |
| Leiognathidae | *Leiognathus decorus* | Sulaibikhat Bay, Kuwait | May | Wright, 1989 |
| Pholidae | *Apodichthys flavidus* | N. California | Apr–Sep | Grossman, 1982 |
| | *Pholis ornata* | Yaquina Bay, Oregon | Mar–Sep | Barton, 1982 |
| | *Apodichthys fucorum* | San Luis Obispo County, California | Feb | Burgess, 1978 |
| Sparidae | *Diplodus cervinus* | South Africa | Aug–Nov | Christensen, 1978 |
| | *Sarpa salpa* | South Africa | Jun–Sep | Christensen, 1978 |
| | *Diplodus sargus* | E. Cape, S. Africa | Oct–Mar | Beckley, 1985 |
| | * *Diplodus sargus* | S. Africa | continuous | Christensen, 1978 |
| | *Sparodon durbanensis* | E. Cape, S. Africa | Oct–Mar | Beckley, 1985 |
| | *Diplodus cervinus* | E. Cape, S. Africa | Oct–Mar | Beckley, 1985 |
| Stichaeidae | *Anoplarchus purpurescens* | Washington outer coast | May–Aug | Personal observation |
| Tripterygiidae | * *Bellapiscis medius* | South coast, New Zealand | Continuous[a] | Willis and Robert, 1996 |
| | *Forsterygion lapillum* | South coast, New Zealand | Jan–May | Willis and Robert, 1996 |

*Note.* Most fishes appear to have seasonal appearance of recruits on the shore, except for those marked with an *.

[a] Although a distinct peak occurred in November.

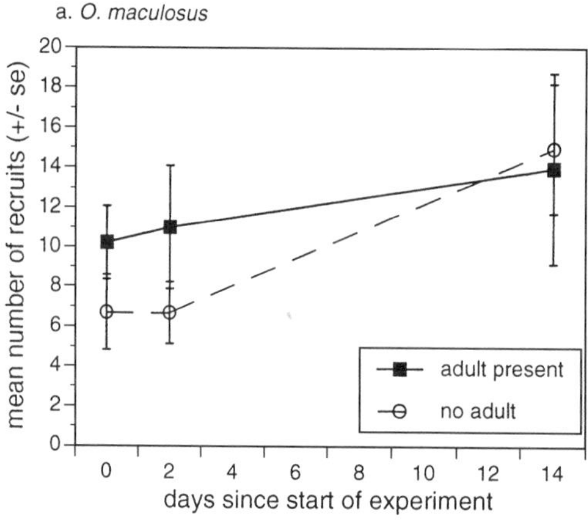


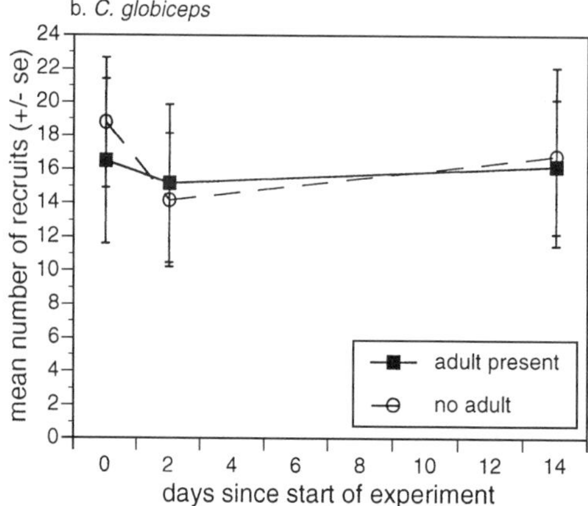


**Figure 1.** The number of recruits present in tidepools with a captive adult enclosed within a cage, versus control tidepools with the cage only (described in Section III.B). (a) The captive adult was *O. maculosus.* Repeated measures ANOVA indicated no effect of adult presence on the number of individuals recruiting ($P = 0.574$, $F_{1,8} = 0.344$); $n = 4$ pools with adult present, $n = 6$ pools without adults. Additionally, there was no evidence that the presence of adults altered the species composition of the dominant species recruiting (*C. globiceps:* $P = 0.532$, $t = 0.653$; *C. embryum:* $P = 0.816$, $t = 0.240$). (b) The captive adult was *C. globiceps.* Again, repeated measures ANOVA indicated no effect of adult presence on the number of individuals recruiting ($P = 0.703$, $F_{1,10} = 0.155$); $n = 5$ pools with adult present, $n = 6$ pools without adults. Again, there was no evidence that the presence of adults altered the species composition of the dominant species recruiting (*C. globiceps:* $P = 0.680$, $t = 0.427$; *C. embryum:* $P = 1.0$, $t = 0$).

between 16 and 17 recruits per tidepool in the *C. globiceps* experiment after 14 days. Additionally, there was no evidence that the presence of either species of adult altered the composition of the dominant species recruiting. *C. globiceps* recruits were equally abundant at the end of the experiment, whether an adult *O. maculosus* ($P = 0.532$, $t = 0.653$, unpaired $t$ test) or an adult *C. globiceps* ($P = 0.680$, $t = 0.427$) was present. Similarly, *C. embryum* recruited equally to control tidepools and those where an adult *O. maculosus* ($P = 0.816$, $t = 0.240$, unpaired $t$ test) or an adult *C. globiceps* ($P = 1.0$, $t = 0$) was present. Thus, to date, experimental manipulations and observations indicate that the presence of other fishes does not affect initial recruitment of tidepool sculpins, but does affect subsequent performance. However, more experimental work of the type described above is needed to know the generality of these results.

The effects of predation on either larval fishes or new recruits is relatively unexplored. I have found several species of intertidal sculpins to be cannibalistic, with adults preying on new recruits, but the incidence of this is low. In one case (*Clinocottus embryum*), cannibalism was documented in the laboratory only, while for another (*Cottus asper*), cannibalism was common in the field. In a tidal lagoon in Mendocino County, California, I found that the presence of adult *C. asper* decreased the survivorship of recruits by a factor of 3 and the growth rate by more than a factor of 2 relative to recruits that were separated from adults (Pfister, unpublished data).

Intertidal fishes have a diversity of predators, both from the sea and from the land. For example, I have found tidepool sculpins in the guts of a larger, subtidal species of sculpin (*Scorpaenichthys marmoratus,* the cabezon) and most locales have large, predatory fishes frequenting the intertidal zone during high tide. Additionally, seabirds are likely predators of intertidal fishes. Although few studies report prey items to the species level, there is evidence that pigeon guillemots, murres, cormorants, ducks, and herons consume fishes in the intertidal zone (Palmer, 1962; Drent, 1965; Ainley *et al.,* 1990). Perhaps because of the abundance and taxonomic diversity of intertidal fish predators, there are few studies that quantify the importance of these predators to the recruitment success or to adult abundance patterns of their prey.

Although the above summarizes some negative effects of competition, predation, and the effect of conspecifics, there may be positive effects of conspecifics (Sweatman, 1985). For marine invertebrates and fishes, there are a variety of experimental and observational studies that show differing responses of recruitment to conspecific residents (Caley *et al.,* 1996). Whether intertidal fishes show the same diversity of response has yet to be determined.

In addition to the effects that extraneous biotic factors have on recruitment success, recruitment may also be affected by factors intrinsic to the organism, such as the spawning type, larval duration in the plankton, and other life history attributes. For example, an analysis of tidepool sculpins revealed that although adult *C. globiceps* are no more abundant than adult *O. maculosus* at locales in the northeast Pacific, *C. globiceps* recruits nearly always outnumber *O. maculosus* recruits (Pfister, 1997). *O. maculosus* survive at a greater rate than *C. globiceps* and eventually surpass them in abundance with time. Thus, the life histories of these fishes differ such that high recruitment and subsequent higher mortality characterize *C. globiceps.* Although there are insufficient data on intertidal fishes to explore other correlates of recruitment, multispecies databases for marine fish populations have been used to reveal relationships among life history characteristics. For example, there is evidence that variation in recruitment is related positively to the change in length during the larval phase, a positive function of the duration of egg and larval phases (Pepin and Myers, 1991). However, variability in recruitment appears unrelated to the fecundity of a given species (Mertz and Myers, 1996).

## IV. Patterns of Recruitment

### A. The Size of Individuals at Recruitment

The size of an individual at recruitment is probably a function of many factors, such as egg size, the amount of yolk reserves, and the amount of time spent in the plankton. It is no surprise, then, that newly recruited intertidal fishes vary greatly in size (Table 3), ranging from 8 to 24 mm SL. Although there is variation among species within a family (e.g., Gobiidae, Cottidae) in the size of recruitment (Table 3), little is known about individual or population level differences within a species. Because there are so few studies that report the size of new recruits in addition to other important life history data, it is impossible at this time to explore whether there are correlates of size such as time spent in the plankton, size of adult, spawning mode, feeding type, or phylogenetic effects.

Information from individually marked sculpins indicate that linear growth rates following settlement can range from 6 to 16 mm per month, depending on the species. For example, tidepool sculpins in the northeast Pacific that are 20 mm SL show monthly growth rates as high as 6 mm for *O. maculosus* and *C. embryum,* 8 mm for *O. snyderi,* and 9 mm for *C. globiceps* (Pfister, 1997). Another cottid species that is often found in tidal, estuarine areas (*Cottus asper*) grew 16 mm/month in a lagoon in Northern California (Pfister, unpublished data). Growth rates of these fishes can vary among individuals within a population with the abundance of con- and heterospecifics (Pfister, 1995, unpublished data) and season (Freeman *et al.,* 1985; Pfister, 1997). Thus, the potential for rapid and variable early growth may result in a range of recruit sizes that diverge rapidly from those listed in Table 3.

### B. Spatial and Temporal Variability in Recruitment

Because the recruitment process usually involves successfully negotiating a planktonic phase and then finding suitable habitat for the juvenile and adult phases, it is not surprising that many benthic fish populations, including intertidal ones, experience temporal and spatial variability in the abundance of recruits. Although there has been a substantial amount published about variability in recruitment for both coral reef fishes and exploited fish populations (Hjort, 1914; Houde, 1987; Fogarty *et al.,* 1991; Doherty, 1991; Doherty and Fowler, 1994), few investigators have quantified this variability, both spatially and temporally, for intertidal fish populations (Grossman, 1982; Pfister, 1996, 1997).

Temporal variability in recruitment can have two components: within-year and among-year variability. Grossman (1982) quantified the abundance of intertidal fishes for 2 years in the northeast Pacific (Dillon Beach, CA), including the abundance of young-of-the-year (YOY) fishes. His data indicate that *O. snyderi* was the most abundant recruit in both years, although comparisons between the 2 years are hampered by different census schedules. Certainly 2 years may not be sufficient to capture the range of natural variability, although Yoshiyama *et al.* (1986) returned to the site 4 years later and reported a "virtual absence of YOY" individuals, suggesting that recruitment at this site may vary widely enough to include years of nearly total reproductive failure.

Pfister (1996) has censused tidepool sculpins in 11 tidepools in the northeast Pacific (coastal Washington state) annually for 9 years, and another group of 6 tidepools for 3 years. Tidepool sculpins recruit in a seasonal pulse from April to September, appearing in tidepools as 12–15 mm SL individuals that are still nearly transparent. They pigment and grow rapidly (see above). For the four most abundant species in these tidepools (all cottids), recruitment success could be variable among years, with the species that recruits

**Table 3.** Body Size at Recruitment for 23 Species Representing 8 Families Worldwide

| Family | Species | Locale | Size at recruitment | Reference |
|---|---|---|---|---|
| Blenniidae | *Hypsoblennius gilberti* | S. California | 18–21 mm SL | Stephens *et al.*, 1970 |
| Clinidae | *Clinus superciliosus* | E. Cape, S. Africa | 23 mm TL | Beckley, 1985 |
| Cottidae | *Artedius lateralis* | Washington outer coast | 15 mm SL | Personal observation |
| | *Clinocottus acuticeps* | Washington outer coast | 18 mm SL | Personal observation |
| | *Clinocottus embryum* | Washington outer coast | 13 mm SL | Pfister, 1997 |
| | *Clinocottus globiceps* | Washington outer coast | 13 mm SL | Pfister, 1997 |
| | *Clinocottus recalvus* | Laboratory | 10.8 mm TL | Morris, 1951 |
| | *Oligocottus maculosus* | Washington outer coast | 12 mm SL | Pfister, 1997 |
| | *Oligocottus snyderi* | Washington outer coast | 15 mm SL | Pfister, 1997 |
| | *Oligocottus snyderi* | Dillon Beach, California | 20 mm SL | Grossman and DeVlaming, 1984 |
| Gobiesocidae | *Gobiesox meandricus* | British Columbia, Canada | 14 mm TL | Marliave, 1977 |
| | *Gobiesox meandricus* | Washington outer coast | 13 mm SL | Personal observation |
| Gobiidae | *Chasmichthys dolichognathus* | Chiba Prefecture, Japan | 20 mm TL | Sasaki and Hattori, 1969 |
| | *Chasmichthys gulosus* | Chiba Prefecture, Japan | 21 mm TL | Sasaki and Hattori, 1969 |
| | *Gobius paganellus* | Isle of Man | 9.5–10.5 mm TL | Miller, 1961 |
| | *Lepidogobius lepidus* | Morro Bay, California | 18–20 mm TL | Grossman, 1979 |
| | *Typhlogobius californiensis* | Southern California | 20 mm TL | MacGinitie, 1939 |
| Pholidae | *Pholis ornata* | Yaquina Bay, Oregon | 18–20 mm SL | Barton, 1982 |
| | *Xererpes fucorum* | San Luis Obispo County, California | 24 mm SL | Burgess, 1978 |
| Sparidae | *Diplodus cervinus* | South Africa | 8 mm SL | Christensen, 1978 |
| | *Sarpa salpa* | South Africa | 9 mm SL | Christensen, 1978 |
| | *Diplodus sargus* | E. Cape, S. Africa | 9–10 mm SL | Christensen, 1978 |
| Stichaeidae | *Anoplarchus purpurescens* | Washington outer coast | 23 mm SL | Personal observation |

*Note.* Sizes are given in either total length (TL) or standard length (SL).

in the greatest abundance (*C. globiceps*) showing as much as a sixfold difference between years. Although *C. globiceps* was generally the most abundant recruit, in some years, at some sites, other species such as *O. snyderi* and *C. embryum* could dominate numerically.

In this guild of tidepool sculpins, recruitment among species also showed intriguing differences within a year when fishes were censused monthly for 3 years, with some species consistently recruiting earlier than others (Pfister, 1997). Despite the approximately constant ranking in time of appearance of each species, the exact timing of peak recruitment could differ from year to year. For example, *O. maculosus* recruited later than the other species, but its peak abundance in recruitment occurred in July, August, or September. Similarly, intertidal fishes on the southern coast of New Zealand also show interspecific differences in recruitment timing, with two species of each of the Gobiesocidae and Tripterygiidae showing different monthly peaks in the timing of recruitment (Willis and Roberts, 1996). Whether these interspecific recruitment differences reflect differences in the timing of spawning or differences in planktonic duration is unknown.

Thus, recruitment in this guild of tidepool fishes revealed within- and among-year, as well as spatial, variability in recruitment success, paralleling results from open ocean and coral reef fish populations (e.g., Houde, 1987; Doherty, 1991). Unfortunately, the generality of these results is unknown since so little published research is available on the early life history of intertidal fishes.

## V. Population Consequences of Recruitment Patterns

The motivation for quantifying the pattern of recruitment in fishes is that events occurring early in the life history may determine the abundance and distribution patterns of organisms as well as the genetic makeup of the population. There are relatively few studies that have investigated attributes that are relevant to the genetic structure of the population, but those that have done so report conflicting results. Marliave (1986) observed high frequencies of sculpin larvae at all stages in nearshore areas and suggested that offshore and longshore dispersal is restricted. Limited dispersal would suggest population differentiation. However, the studies of population genetic structure that I am aware of provide no evidence that populations are subdivided genetically (Yoshiyama and Sassaman, 1987; Waples, 1987). Yoshiyama and Sassaman (1987) found little evidence of genetic differentiation in allozymes for *Oligocottus snyderi* and *O. maculosus* among sites at Cape Arago, Oregon, and Pt. Pedras Blancas and Arena Cove, California, and for *Clinocottus analis* between Dillon Beach and Santa Cruz Island, California. Similarly, Waples (1987) found allele frequencies typical of those in populations with high gene flow for a California population of *Clinocottus analis.* These allozyme studies suggest that movement among sites is high enough to keep populations from being differentiated genetically by natural selection or drift. Given the very high site fidelity we have seen in tidepool sculpins once they have recruited (Green, 1971, 1973; Yoshiyama *et al.,* 1992; Pfister, 1996), movement among sites is presumably due to larval transport and the recruitment of individuals away from their source adult populations. Lack of genetic differentiation is not a feature only of cottid populations: *Anoplarchus purpurescens,* a stichaeid, is genetically undifferentiated over its entire range from Alaska to California (Yoshiyama and Sassaman, 1983).

Population and community ecologists have debated the extent to which populations are driven by early life history events, especially larval supply and recruitment (e.g., "supply-side ecology"; Lewin, 1986), versus postrecruitment processes. Of course, the debate is an artificially polarized one, since the abundance and dynamics of any organism are undoubtedly determined by a combination of both processes. The challenge is to determine what processes contribute the most to population growth and how variance in those processes affects population growth. For field populations this is a challenging task, since it requires an intimate knowledge of demography, including the myriad physical and biological interactions that may affect demography. Fortunately, there has also been some theoretical guidance in this area.

In exploited fish populations, "recruitment" is the phenomenon of a fish reaching harvestable size and can incorporate multiple years of life. In contrast, most ecologists refer to recruitment as the process by which new young are incorporated into the population, an event that can occur as early as the first days of an organism's life. It is important to note that how an investigator defines recruitment affects greatly the importance recruitment has on population dynamics, since, in the case of recruitment in fisheries, multiple years of survivorship are included in the interval prior to recruitment.

If recruitment were generally constant and a function of the local adult population, then simple, closed population models would be adequate. Thus, a simple birth and death model might suffice (e.g., Smith, 1954), or the Euler–Lotka model of estimating population growth could be used if age structure were known (Euler, 1970). If size- or stage-based information were available, a matrix population model could be used (e.g., Caswell, 1989; Getz and Haight, 1989) or a partial differential equation model of population projection (e.g., DeAngelis *et al.,* 1993). However, recruitment often is not constant through time (see above) and there is little evidence that the supply of new recruits is closely related to the number of local adults. For example, Grossman (1979) found no statistically significant correlation between benthic stock and YOY of *Lepidogobius lepidus* in California, and Grossman and DeVlaming (1984) found no correlation between the number of reproductive females and the number of recruits in *O. snyderi.* Additionally, my own studies of intertidal sculpins have never suggested a clear relationship between the number of adults and recruits (e.g., Table 2). Weak or nonexistent statistical relationships among adult stocks and recruitment also characterize many populations of exploited fishes (e.g., Hilborn and Walters, 1992). Thus, two complications enter into population models that seek to incorporate realistic recruitment patterns: a parameter for recruitment that can be variable and a population model that permits the influx of distant recruits to the established population (e.g., a population model that is "open").

The effects of adding variable recruitment to a population model can be difficult to determine, especially when other demographic variables vary, notably adult mortality (Warner and Hughes, 1988). Additionally, the type of variability that characterizes natural populations must be identified, and this requires long-term data on recruit abundance, information uncommon in the literature. From demographic data spanning 3 years on tidepool sculpins, I explored how variability in different demographic traits affected the variability in subsequent adult population size. I found that variability in adult survivorship was likely to impart much greater variability in adult population size than was variability in recruitment (Pfister, 1996). Studies that seek to understand the role of recruitment in the population dynamics of any fish species need to be examined in the context of the entire life history of the fish (e.g., Shulman and Ogden, 1987; Jones, 1990; Pfister, 1996). Additionally, there is great need for experimental manipulation of recruitment levels in natural populations, although this is difficult logistically (personal observation).

Most marine fishes can be characterized as open populations where larvae may be transported from the original adult population. Although there have been attempts to deal with populations that exchange individuals, including source-sink population models (Pulliam, 1988), metapopulation models (e.g., Hanski and Gilpin, 1996), and models designed specifically with recruitment processes in mind (Roughgarden *et al.,* 1985; Pfister and Bradbury, 1996), open population models represent unique challenges. For example, adding recruitment from outside the adult population by using a recruitment vector added to a matrix population model (Hughes, 1990; Pfister, 1996) forfeits the use of many tools of matrix models such as estimating the asymptotic population growth rate ($\lambda$) and the proportional contribution of different stages to population growth rate (elasticities). However, conventional methods of estimating the sensitivity of population growth rate to different parameters in the model remain possible.

There are an increasing number of analytical and simulation tools available to explore the effects of different recruitment patterns on the population trajectories of marine fishes, including intertidal ones. As more information becomes available on the natural history and demography of intertidal fishes, we can determine whether there are generalities about the role of recruitment in these fishes. The synthesis of extensive demographic data for terrestrial plants (e.g., Silvertown *et al.,* 1993) and a variety of other organisms (Pfister, 1998) as well as simulations of different life histories (Benton and Grant, 1996) indicate that events in the early life history may not have the greatest effect on population growth, especially for species with a lifespan of 2 or more years.

In addition to the effects that recruitment events have on single species dynamics, the pattern of recruitment among species may be important to the coexistence and persistence of assemblages. Theoretical work suggests that variability in recruitment sets the stage for a species to increase from low density during a strong episode of recruitment, while persisting through poor recruitment episodes with low adult mortality. Thus, as long as environmental variations result in different species having successful recruitment at different times, persistence is promoted (Chesson, 1985, 1990; Warner and Chesson, 1985). Other ecological theory has found a role for colonizing ability in facilitating the coexistence of several species. The Levins' metapopulation model (1970), modified by Tilman (1994) for multiple species, yields species coexistence when a trade-off exists between colonization ability and competition. Colonizing ability in Tilman's model is simply the probability that an individual successfully occupies a vacant site and is thus analogous to recruitment success. Thus, in addition to being a potentially important component of single species dynamics, differing recruitment success may contribute to coexistence in a multispecies assemblage. It is important to note that although intertidal fishes can reach high local densities, relatively little is known about the interactions among fishes and what effects these interactions might have on determining species abundance and diversity.

## VI. Future Research

Many areas of future research would help illuminate the early life history of intertidal fishes. First, studies of intertidal fishes would benefit greatly from increased attention to demography instead of simply occupancy patterns. Since the biogeography of intertidal fishes has become better known, it is timely to turn our attention to the determinants of abundance of these species, information that will be facilitated by intensive demographic analysis. Information about demography also provides a basis for life history comparisons among fish species. Using phylogenetic analyses, we may be able to understand how phylogenetic constraints interact with the environment to determine life history traits. For example, Is high recruitment variability associated with a longer period in the plankton or the size of the young at hatching? Additionally, Is the recruitment pattern related to longevity of a fish species, the number of competitors or predators it encounters, its spawning strategy, or the degree of genetic differentiation among different sites? Many of these questions will require long-term data on the biology of these fishes. Second, for intertidal fishes and many other marine organisms, we know relatively little about the transport processes that affect recruitment. Although the planktonic phase of the life cycle of intertidal fishes and other marine organisms has been a black box for biologists, increasing efforts are needed to unite nearshore physical oceanography with the life history of intertidal fishes.

## Acknowledgments

I thank the Miller Foundation for Basic Research in Science at the University of California at Berkeley and the Andrew Mellon Foundation for financial support during the writing of this manuscript. Comments from E. DeMartini, M. Chotkowski, and two anonymous reviewers improved an earlier version of this chapter.

## References

Ainley, D. G., Strong, C. S., Penniman, T. M., and Boekelheide, R. J. (1990). The feeding ecology of Farallon seabirds. *In* "Seabirds of the Farallon Islands" (D. G. Ainley and R. J. Boekelheide, Eds.), pp. 51–127. Stanford University Press, Stanford, CA.

Barton, M. (1982). Comparative distribution and habitat preferences of 2 species of stichaeoid fishes in Yaquina Bay, Oregon. *J. Exp. Mar. Biol. Ecol.* **59,** 77–87.

Beckley, L. E. (1985). Tide-pool fishes: Recolonization after experimental elimination. *J. Exp. Mar. Biol. Ecol.* **85,** 287–295.

Benton, T. G., and Grant, A. (1996). How to keep fit in the real world: Elasticity analyses and selection pressures on life histories in a variable environment. *Am. Nat.* **147,** 115–139.

Berger, A., and Mayr, M. (1992). Ecological studies on two intertidal New Zealand fishes, *Acanthoclinus fuscus* and *Forsterygion nigripenne robustum. N. Z. J. Mar. Freshw. Res.* **26,** 359–370.

Burgess, T. J. (1978). The comparative ecology of 2 sympatric polychromatic populations of *Xererpes fucorum* Jordan and Gilbert (Pisces: Pholididae) from the rocky intertidal zone of central California. *J. Exp. Mar. Biol. Ecol.* **35,** 43–58.

Caley, M. J., Carr, M. H., Hixon, M. A., Hughes, T. P., Jones, G. P., and Menge, B. A. (1996). Recruitment and the local dynamics of open marine populations. *Annu. Rev. Ecol. Systemat.* **27,** 477–500.

Carlon, D. B., and Olson, R. R. (1993). Larval dispersal distance as an explanation for adult spatial pattern in two Caribbean reef corals. *J. Exp. Mar. Biol. Ecol.* **173,** 247–263.

Caswell, H. (1989). "Matrix Population Models." Sinauer, Sunderland, MA.

Chesson, P. L. (1985). Coexistence of competitors in spatially and temporally varying environments: A look at the combined effects of different sorts of variability. *Theor. Pop. Biol.* **28,** 263–287.

Chesson, P. L. (1990). Geometry, heterogeneity and competition in variable environments. *Philos. Trans. R. Soc. London, Ser. B* **330,** 165–173.

Christensen, M. S. (1978). Trophic relationships in juveniles of three species of Sparid fishes in the South African marine littoral. *Fish. Bull.* **76,** 389–401.

DeAngelis, D. L., Rose, K. A., Crowder, L. B., Marschall, E. A., and Lika, D. (1993). Fish cohort dynamics: Application of complementary modeling approaches. *Am. Nat.* **142,** 604–622.

Doherty, P. J. (1991). Spatial and temporal patterns of recruitment. *In* "The Ecology of Fishes on Coral Reefs" (P. F. Sale, Ed.), pp. 261–293. Academic Press, San Diego.

Doherty, P. J., and Fowler, T. (1994). An empirical test of recruitment limitation in a coral reef fish. *Science* **263,** 935–939.

Drent, R. H. (1965). Breeding biology of the Pigeon Guillemot, *Cepphus columba. Ardea* **53,** 99–160.

Euler, L. (1970). A general investigation into the mortality and multiplication of the human species. *Theor. Pop. Biol.* **1,** 307–314.

Fogarty, M. J., Sissenwine, M. P., and Cohen, E. B. (1991). Recruitment variability and the dynamics of exploited marine populations. *Trends Ecol. Evol.* **6,** 241–246.

Forrester, G. E. (1990). Factors influencing the juvenile demography of a coral reef fish. *Ecology* **71,** 1666–1681.

Freeman, M. C., Neally, N., and Grossman, G. D. (1985). Aspects of the life history of the fluffy sculpin, *Oligocottus snyderi. Fish. Bull.* **83,** 645–655.

Gaines, S. D., and Roughgarden, J. (1985). Larval settlement rate: A leading determinant of structure in an ecological community of the marine intertidal zone. *Proc. Nat. Acad. Sci. USA* **82,** 3707–3711.

Gaines, S. D., and Roughgarden, J. (1987). Fish in offshore kelp forest affect recruitment to intertidal barnacle populations. *Science* **235,** 479–481.

Getz, W. M., and Haight, R. C. (1989). "Population Harvesting." Princeton University Press, Princeton, NJ.

Gosline, W. A. (1965). Vertical zonation of inshore fishes in the upper water layers of the Hawaiian Islands. *Ecology* **46,** 823–831.

Green, J. M. (1971). High tide movements and homing behavior of the tidepool sculpin *Oligocottus maculosus. J. Fish. Res. Bd. Can.* **28,** 383–389.

Green, J. M. (1973). Evidence for homing in the mosshead sculpin (*Clinocottus globiceps*). *J. Fish. Res. Bd. Can.* **30,** 129–130.

Grosberg, R. K., and Levitan, D. R. (1992). For adults only? Supply-side ecology and the history of larval biology. *Trends Ecol. Evol.* **7,** 130–133.

Grossman, G. D. (1979). Demographic characteristics of an intertidal bay goby (*Lepidogobius lepidus*). *Environ. Biol. Fish.* **4,** 207–218.

Grossman, G. D. (1982). Dynamics and organization of a rocky intertidal fish assemblage: The persistence and resilience of taxocene structure. *Am. Nat.* **119,** 611–637.

Grossman, G. D., and DeVlaming, V. (1984). Reproductive ecology of female *Oligocottus snyderi* Greeley: A North American intertidal sculpin. *J. Fish Biol.* **25,** 231–240.

Hanski, I. A., and Gilpin, M. E. (1996). "Metapopulation Biology: Ecology, Genetics and Evolution." Academic Press, San Diego.

Hilborn, R., and Walters, C. J. (1992). "Quantitative Fisheries Stock Assessment." Chapman and Hall, New York.

Hjort, J. (1914). Fluctuations in the great fisheries of northern Europe. *Rapp. P.-v Reun. Cons. Perm. Int. Explor. Mer.* **20,** 1–228.

Houde, R. D. (1987). Fish early life dynamics and recruitment variability. *Am. Fish. Soc. Symp.* **2,** 17–29.

Hughes, T. P. (1990). Recruitment limitation, mortality, and population regulation in open systems: A case study. *Ecology* **71,** 12–20.

Johannes, R. E. (1978). Reproductive strategies of coastal marine fishes in the tropics. *Environ. Biol. Fish.* **3,** 65–84.

Jones, G. P. (1990). The importance of recruitment to the dynamics of a coral reef fish population. *Ecology* **71,** 1691–1698.

Lafferty, K., Swenson, R. O., and Swift, C. C. (1996). Threatened fishes of the world: *Eucyclogobius newberryi* Girard, 1857 (Gobiidae). *Environ. Biol. Fish.* **46,** 254.

Levins, R. (1970). Extinction. *In* "Some Mathematical Questions in Biology. Lectures on Mathematics in the Life Sciences," Vol. 2, pp. 75–108. American Mathematical Society, Providence, RI.

Lewin, R. (1986). Supply-side ecology. *Science* **234,** 25–27.

MacGinitie, G. E. (1939). The natural history of the blind goby, *Typhlogobius californiensis* Steindachner. *Am. Midl. Nat.* **21,** 489–505.

Marliave, J. B. (1977). Substratum preferences of settling larvae of marine fishes reared in the laboratory. *J. Exp. Mar. Biol. Ecol.* **27,** 47–60.

Marliave, J. B. (1986). Lack of planktonic dispersal of rocky intertidal fish larvae. *Trans. Am. Fish. Soc.* **115,** 149–154.

Marsh, B., Crowe, T. M., and Siegfried, W. R. (1978). Species richness and abundance of clinid fish (Teleostei; Clinidae) in intertidal rock pools. *Zool. Afr.* **13,** 283–291.

Menge, B. M. (1991). Relative importance of recruitment and other causes of variation in rocky intertidal community structure. *J. Exp. Mar. Biol. Ecol.* **146,** 69–100.

Mertz, G., and Myers, R. A. (1996). Influence of fecundity on recruitment variability of marine fish. *Can. J. Fish. Aquat. Sci.* **53,** 1618–1625.

Miller, P. J. (1961). Age, growth and reproduction of the rock goby (*Gobius paganellus* L.) in the Isle of Man. *J. Mar. Biol. Assoc. U.K.* **41,** 737–769.

Morris, R. W. (1951). Early development of the cottid fish, *Clinocottus recalvus* (Greeley). *Calif. Fish Game* **37,** 281–300.

Nakamura, R. (1976). Temperature and vertical distribution of 2 tidepool fishes (*Oligocottus maculosus, O. snyderi*). *Copeia* **1976,** 143–152.

Palmer, R. S. (1962). "Handbook of North American Birds," Vol. I. Yale University Press, New Haven.

Parrish, R. H., Nelson, C. S., and Bakun, A. (1981). Transport mechanisms and reproductive success of fishes in the California Current. *Biol. Oceanogr.* **1,** 175–203.

Pepin, P., and Myers, R. A. (1991). Significance of egg and larval size to recruitment variability of temperate marine fish. *Can. J. Fish. Aquat. Sci.* **48,** 1820–1828.

Pfister, C. A. (1995). Estimating competition coefficients from census data: A test with field manipulations of tidepool fishes. *Am. Nat.* **146,** 271–291.

Pfister, C. A. (1996). The role and importance of recruitment variability to a guild of tide pool fishes. *Ecology* **77,** 1928–1941.

Pfister, C. A. (1997). Demographic consequences of within-year variation in recruitment. *Mar. Ecol. Prog. Ser.* **153,** 229–238.

Pfister, C. A. (1998). Patterns of variance in stage-structured populations: Evolutionary predictions and ecological implications. *Proc. Nat. Acad. Sci. USA* **95,** 213–218.

Pfister, C. A., and Bradbury, A. (1996). Harvesting red sea urchins: Recent effects and future predictions. *Ecol. Appl.* **6,** 298–310.

Prochazka, K. (1996). Seasonal patterns in a temperate intertidal fish community on the west coast of South Africa. *Environ. Biol. Fish.* **45,** 133–140.

Pulliam, R. (1988). Sources, sinks, and population regulation. *Am. Nat.* **132,** 652–661.

Qasim, S. Z. (1957). The biology of *Blennius pholis* L. (Teleostei). *Proc. Zool. Soc. London* **128,** 161–208.

Roughgarden, J., Iwasa, Y., and Baxter, C. (1985). Demographic theory for an open marine population with space-limited recruitment. *Ecology* **66,** 54–67.

Sasaki, T., and Hattori, J. (1969). Comparative ecology of 2 closely related sympatric gobiid fishes living in tide pools. *Japan. J. Ichthyol.* **15,** 143–145.

Sawyer, P. J. (1967). Intertidal life-history of the rock gunnel, *Pholis gunnellus,* in the western Atlantic. *Copeia* **1967,** 55–61.

Setran, A. C., and Behrens, D. W. (1990). New ecological information on *Scytalina cerdale* (Pisces: Scytalinidae) from a central rocky intertidal zone. *Environ. Biol. Fish.* **29,** 107–117.

Shafer, D. J. (1991). The brown goby, Bathygobius fuscus (Gobiidae): Responses to biological and physiochemical intertidal gradients. *Pacific Sci.* **45,** 102.

Shulman, M. J. (1985). Variability in recruitment of coral reef fishes. *J. Exp. Mar. Biol. Ecol.* **89,** 205–219.

Shulman, M. J., and Ogden, J. C. (1987). What controls tropical reef fish populations: Recruitment or benthic mortality? An example in Caribbean reef fish *Haemulon flavolineatum. Mar. Ecol. Prog. Ser.* **39,** 233–242.

Silvertown, J., Franco, M., Pisanty, I., and Mendoza, A. (1993). Comparative plant demography—Relative importance of life-cycle components to the finite rate of increase in woody and herbaceous perennials. *J. Ecol.* **81,** 465–476.

Smith, F. E. (1954). Quantitative aspects of population growth. *In* "Dynamics of Growth Processes" (E. J. Boell, Ed.), pp. 277–294. Princeton University Press, Princeton, NJ.

Stephens, J. S., Jr., Johnson, R. K., Key, G. S., and McCosker, J. E. (1970). The comparative ecology of 3 sympatric species of California blennies of the genus *Hypsoblennius* Gill (Teleostei, Blenniidae). *Ecol. Monogr.* **40,** 213–233.

Stepien, C. A., Phillips, H., Adler, J. A., and Mangold, P. J. (1991). Biogeographic relationships of a rocky intertidal fish assemblage in an area of cold water upwelling off Baja California, Mexico. *Pacific Sci.* **45,** 63–71.

Strathmann, R. (1974). The spread of sibling larvae of sedentary marine invertebrates. *Am. Nat.* **108,** 29–44.

Sutherland, J. P. (1990). Recruitment regulates demographic variation in a tropical intertidal barnacle. *Ecology* **71,** 955–972.

Sweatman, H. P. A. (1985). The influence of adults of some coral reef fish on larval recruitment. *Ecol. Monogr.* **71,** 1666–1681.

Thorson, G. (1950). Reproductive and larval ecology of marine invertebrates. *Biol. Rev.* **25,** 1–45.

Tilman, D. (1994). Competition and biodiversity in spatially structured habitats. *Ecology* **75,** 2–16.

Victor, B. J. (1983). Recruitment and population dynamics of a coral reef fish. *Science* **219,** 419–420.

Waples, R. S. (1987). A multispecies approach to the analysis of gene flow in marine shore fishes. *Evolution* **41,** 385–400.

Warner, R. R., and Chesson, P. L. (1985). Coexistence mediated by recruitment fluctuations: A field guide to the storage effect. *Am. Nat.* **125,** 769–787.

Warner, R. R., and Hughes, T. P. (1988). The population dynamics of reef fishes. *In* Proc. 6th Int. Coral Reef Symposium, Townsville, Australia.

Washington, B. B., Moser, H. G., Laroche, W. A., and Richards, W. J. (1984). Scorpaeniformes: Development. "Ontogeny and Systematics of Fishes," Special Publication 1. American Society of Ichthyologists and Herpetologists.

Wheeler, A. (1969). "The Fishes of the British Isles and North-West Europe." Michigan State University Press, East Lansing, MI.

Willis, T. J., and Roberts, C. D. (1996). Recolonisation and recruitment of fishes to intertidal rockpools at Wellington, New Zealand. *Environ. Biol. Fish.* **47,** 329–343.

Wright, J. M. (1989). Biology of *Leiognathus decorus* (Leiognathidae) in Sulaibikhat Bay, Kuwait. *Austr. J. Mar. Freshw. Res.* **40,** 179–185.

Yoshiyama, R. M., Gaylord, K. B., Philippart, M. T., Moore, T. R., Jordan, J. R., Coon, C. C., Schalk, L. L., Valpey, C. J., and Tosques, I. (1992). Homing behavior and site fidelity in intertidal sculpins (Pisces: Cottidae). *J. Exp. Mar. Biol. Ecol.* **160,** 115–130.

Yoshiyama, R. M., and Sassaman, C. (1983). Morphological and allozymic variation in the stichaeid fish *Anoplarchus purpurescens. Systemat. Zool.* **32,** 52–71.

Yoshiyama, R. M., and Sassaman, C. (1987). Geographic patterns of allozyme variation in three species of intertidal sculpins. *Environ. Biol. Fish.* **20,** 203–218.

Yoshiyama, R. C., Sassaman, C., and Lea, R. N. (1986). Rocky intertidal fish communities of California: Temporal and spatial variation. *Environ. Biol. Fish.* **17,** 23–40.

# 11

# Herbivory

Michael H. Horn
*Department of Biological Science, California State University, Fullerton, California*

F. Patricio Ojeda
*Departamento de Ecologia, Facultad de Ciencias Biologicas, Pontificia Universidad Catolica de Chile, Santiago, Chile*

## I. Introduction

The diets of coastal marine fishes represent a spectrum of food types from all animal material in carnivorous species, to a mixture of animal and plant material in omnivores, to virtually all plant matter in herbivorous species, at least in the adults. A wide variety of fish species are often found with some algal or seagrass matter in their guts, but whether these fishes are omnivores in the sense of obtaining energy and nutrients from plants as well animals remains largely unknown. Herbivores have been defined as those fishes with stomachs containing >50% plant material by volume, mass, or frequency of occurrence (Horn, 1989). This operational definition is followed here although the diets of most of the species discussed consist of far more than 50% plant matter.

Regardless of the exact definition, herbivore species richness is lower than that of either carnivorous or so-called omnivorous species in all aquatic habitats. The rocky intertidal zone illustrates this pattern, for as Gibson and Yoshiyama (Chapter 13, this volume) show, herbivores usually constitute no more than 10% of the species in the fish communities of this habitat. The small number of herbivorous fish species on rocky shores and other coastal habitats, especially in temperate and polar latitudes with high macroalgal biomass, continues to puzzle marine biologists and generate speculation as to cause. As exceptions to the rule of carnivory in the intertidal zone, herbivorous fishes command attention and prompt questions as to their phylogenetic position, distributional patterns, feeding specializations, dietary choice, digestive mechanisms, and ecological impacts on algal communities.

In this chapter, we attempt to answer these questions based on a general survey of the available information. Our emphasis is placed on the fishes of rocky shores that consume macroalgae, and to a lesser degree seagrasses, rather than diatoms or phytoplankton as a major part of their diets. The high incidence of apparent omnivory in rocky intertidal habitats (see Gibson and Yoshiyama, Chapter 13, this volume) is intriguing and provokes

*Intertidal Fishes: Life in Two Worlds*

uncertainty about the degree of specialization required to consume and digest plant material efficiently. Answers to questions surrounding the omnivorous mode of feeding and digestion rest on future research.

## II. Taxonomic Diversity and Latitudinal Patterns

### A. Family and Species Diversity

There are 426 recognized families of teleostean fishes, and 148 of these are placed in the acanthopterygian order Perciformes, a highly derived but polyphyletic assemblage of mostly marine shore fishes (Nelson, 1994). Against this backdrop of diversity, marine herbivorous fishes are concentrated in only about 19 families, and 15 of these are perciforms (Horn, 1989). Herbivorous species that occur in rocky intertidal habitats either as resident or periodic species represent almost the same set of families. Of the 15 or so families with intertidal representatives, all but 1 (Cottidae) are on the overall list of marine herbivorous fishes, and all but 3 (Cottidae, Mugilidae, Tetraodontidae) are perciform members (Table 1). Herbivory is virtually absent in several major groups of less highly derived teleosts such as the Anguilliformes and Gadiformes with long records in the ocean (see Nelson, 1994); this pattern is consistent with the view that carnivory is plesiomorphic and herbivory derived in ray-finned fish evolution (Winterbottom and McLennan, 1993).

At the species level, the total number of marine herbivorous fishes remains unknown (Horn, 1989), and the same can be said for the intertidal component. The reasons in both cases are mainly that the diets of too many species remain to be determined and that numerous species appear to be omnivorous, with the contribution of plant material to their nutritional requirements still unknown. Whatever the reasons, the actual number of herbivorous fish species seems certain to be small, probably accounting for even smaller proportions of teleosts and perciforms than at the family level. This statement, however, is not meant to imply that the 30 species in Table 1 represent a complete list of intertidal herbivorous fishes. Others are known from various habitats as can be seen, for example, in Chapters 3, 12, and 13 of this volume. Nevertheless, the premise of low herbivore species richness appears to be true and is supported by the fact that some intertidal community studies

**Table 1.** Names, Numbers, Proportions, Feeding Types, and Digestive Mechanisms of Herbivorous Fishes in 13 Intertidal Fish Communities

| Family<br>*Species* | Total species | Herbivores No. | Herbivores % | Feeding type | Digestive mechanism |
|---|---|---|---|---|---|
| Marshall Islands (11° N), Bussing (1972) | | | | | |
| | 33 | 8 | 24 | | |
| Acanthuridae | | | | | |
| *Acanthurus elongatus* | | | | ? | ? |
| *Acanthurus nigricans* | | | | ? | ? |
| *Acanthurus triostegus* | | | | Browser | I |
| Mugilidae | | | | | |
| *Crenimugil crenilabris* | | | | Grazer | II |
| *Mugil cephalus* | | | | Grazer | II |
| *Neomyxis chaptelii* | | | | Grazer | II |

**Table 1.** *Continued*

| Family<br>*Species* | Total species | Herbivores No. | Herbivores % | Feeding type | Digestive mechanism |
|---|---|---|---|---|---|
| Pomacanthidae | | | | | |
| *Centropyge flavissimus* | | | | Browser/grazer | I |
| Tetraodontidae | | | | | |
| *Canthigaster solandri* | | | | Grazer | I? |
| Barbados (13° N), Mahon and Mahon (1994) | | | | | |
| | 63 | >5 | >8 | | |
| Acanthuridae | | | | | |
| *Acanthurus bahianus* | | | | Grazer | II? |
| Blenniidae | | | | | |
| *Ophioblennius atlanticus* | | | | Grazer | I? |
| Pomacanthidae | | | | | |
| *Pomacanthus paru* | | | | Browser | I? |
| Scaridae | | | | | |
| *Sparisoma* spp. | | | | Browser? | III |
| Tetraodontidae | | | | | |
| *Canthigaster rostrata* | | | | Grazer | I? |
| Little Cayman (20° N), Potts (1980) | | | | | |
| | 15 | 3 | 20 | | |
| Acanthuridae | | | | | |
| *Acanthurus* sp. (*chirurgus*?) | | | | Grazer | II |
| Pomacanthidae | | | | | |
| *Pomacanthus paru* | | | | Browser | I? |
| Tetraodontidae | | | | | |
| *Canthigaster rostrata* | | | | Grazer | I? |
| Mexico, Gulf of California (31° N), Thomson and Lehner (1976) | | | | | |
| | 25 | 4 | 16 | | |
| Girellidae | | | | | |
| *Girella simplicidens* | | | | Browser/grazer | I |
| Kyphosidae | | | | | |
| *Hermosilla azurea* | | | | Browser/grazer | IV |
| Mugilidae | | | | | |
| *Mugil curema* | | | | Grazer | II |
| Pomacentridae | | | | | |
| *Eupomacentrus rectifraenum* | | | | Grazer | I? |
| Chile (32–33° S), Muñoz and Ojeda (1997) | | | | | |
| | 13 | 1 | 8 | | |
| Blenniidae | | | | | |
| *Scartichthys viridis* | | | | Grazer | I? |
| Chile (33° S), Stepien (1990) | | | | | |
| | 12 | 4 | 33 | | |
| Aplodactylidae | | | | | |
| *Aplodactylus punctatus* | | | | Browser | I |

*continues*

**Table 1.** *Continued*

| Family<br>*Species* | Total species | Herbivores No. | Herbivores % | Feeding type | Digestive mechanism |
|---|---|---|---|---|---|
| Blenniidae | | | | | |
| *Scartichthys viridis* | | | | Grazer | I? |
| Girellidae | | | | | |
| *Girella laevifrons* | | | | Browser/grazer | I |
| Gobiesocidae | | | | | |
| *Sicyases sanguineus* | | | | Grazer | I? |
| USA, California (33° N), Stepien (1990) | | | | | |
| | 9 | 1 | 11 | | |
| Girellidae | | | | | |
| *Girella nigricans* | | | | Browser/grazer | I |
| South Africa (34° S), Bennett *et al.* (1983) | | | | | |
| | 20 | 1 | 5 | | |
| Gobiidae | | | | | |
| *Caffrogobius caffer* | | | | Browser | I? |
| USA, California (36° N), L. G. Allen and M. H. Horn (unpublished data) | | | | | |
| | 28 | 2 | 7 | | |
| Stichaeidae | | | | | |
| *Cebidichthys violaceus* | | | | Browser | I |
| *Xiphister mucosus* | | | | Browser | I |
| USA, California (38° N), Grossman (1986) | | | | | |
| | 15 | 1 | 7 | | |
| Cottidae | | | | | |
| *Clinocottus globiceps* | | | | Browser | I? |
| France, Mediterranean (43° N), Gibson (1968) | | | | | |
| | 19 | 1 | 5 | | |
| Blenniidae | | | | | |
| *Parablennius (=Blennius) sanguinolentus* | | | | Grazer | I? |
| USA, Washington (48° N), Cross (1981) | | | | | |
| | 16 | 2 | 12 | | |
| Cottidae | | | | | |
| *Clinocottus globiceps* | | | | Browser | I? |
| Stichaeidae | | | | | |
| *Xiphister mucosus* | | | | Browser | I |
| USA, Alaska (60° N), Barber *et al.* (1995) | | | | | |
| | 19 | 2 | 11 | | |
| Cottidae | | | | | |
| *Clinocottus globiceps*[a] | | | | Browser | I? |
| Stichaeidae | | | | | |
| *Xiphister mucosus*[a] | | | | Browser | I |

*Note.* Arrangement is by increasing latitude of study location.

[a] These species are assumed to be herbivorous based on their diets in other localities; both were rare in the samples.

such as those on the Atlantic Coast of France (Gibson, 1972) and on the New Zealand coast (Willis and Roberts, 1996) have not recognized any species that depend primarily on a plant diet.

## B. Latitudinal Diversity

One of the most striking and still enigmatic distributional patterns in coastal marine waters is the higher species richness of herbivorous fishes in tropical latitudes than in temperate or polar latitudes despite large standing stocks of seaweeds in the latter regions. Horn's (1989) survey of 19 studies of shallow marine fish communities showed that the occurrence of herbivores is inversely correlated with latitude for both the absolute numbers and the proportions of herbivorous species in the community. This survey showed 15–28% herbivorous species in coral reef fish communities, 4–16% herbivores in fish assemblages in the 30°–40° latitudinal range, and no herbivorous species at and above 49°N or S. Intertidal communities, which form a subset of shallow marine communities as a whole, largely parallel the overall pattern as can be discerned from Table 1, and the tropical–temperate diversity gradient may be even sharper because dietary studies of intertidal fishes are scarce in the tropics where herbivorous species might be more prevalent (see Gibson and Yoshiyama, Chapter 13, this volume). Departures from a strict latitudinal pattern, however, have been noted in recent studies; for example, Table 1 shows that central Chile, based on at least one study (Stepien, 1990), has a higher proportion of herbivores than South Africa at a similar latitude and that an Alaskan community at 60°N contains two (11%) putative herbivores (Barber *et al.,* 1995).

Several hypotheses have been proposed to explain the rarity of herbivorous fishes at temperate and higher latitudes compared to tropical shores; however, most appear difficult to test, and none to date is backed by strong supportive evidence. Four of these hypotheses have been described by Horn (1989) and arguably apply to intertidal as well as subtidal species. The first is that insufficient time has elapsed to allow herbivores to expand from the tropics into temperate and polar seas. Mead (1970) speculated that the dramatic radiation of perciform fishes, the dominant fish group in the tropics, resulted from the evolution of the ability to consume and digest plant material. He argued that this radiation was relatively recent, too recent to allow the needed cold adaptation to occur for colonization of temperate waters. Choat (1991) counters this argument by asserting that most taxa have had almost all of the Tertiary period to expand into temperate regions. Mead's hypothesis seems untestable and also is weakened by the apparent polyphyletic origin (Johnson, 1993) of the order Perciformes.

The second hypothesis is that cold temperatures limit fish digestive physiology so that a primarily herbivorous diet is energetically infeasible. Gaines and Lubchenco (1982) suggested that if digestive efficiency is greatly reduced relative to energy demands in colder waters, a mainly herbivorous diet would not allow a fish to balance its energy budget. This hypothesis might be testable, and a resultant explanation for the diversity gradient could be that if the rate of passage of food through the gut is reduced and therefore food consumption rate also reduced at low temperatures, as has been shown by Horn and Gibson (1990), then energy intake from a plant diet could be insufficient to meet the fish's requirements. Both the kelp bed scorpaenid *Sebastes mystinus* (Hallacher and Roberts, 1985) and the intertidal stichaeid *Xiphister mucosus* (Cross, 1981) consume a higher proportion of macroalgae during the warmer compared to the cooler periods of the year. These two species might be candidates for testing this hypothesis and the third one below.

The third hypothesis is that suitable food for herbivorous fishes is seasonally unavailable at temperate and polar latitudes. If nonmigratory, strictly herbivorous fishes cannot persist without food as long as invertebrates can in seasonal environments, they might be excluded from these locations. This hypothesis is consistent with the summer highs and winter lows in macroalgal standing stocks in most mid-latitude habitats (see Murray and Horn, 1989) but is nevertheless weakened by the fact that some if not most temperate shores have large standing stocks of macroalgae throughout the year (Mann, 1973; Chapman and Lindley, 1980).

The fourth hypothesis is that increased algal toughness and chemical defenses and decreased nutritional quality restrict herbivory by fishes in temperate and higher latitudes. Little or no evidence appears to be available to support this hypothesis. Although toughness of large brown seaweeds may deter feeding by some fishes, several of the larger brown algae contribute importantly to the diets of temperate, mainly subtidal herbivores such as girellids, kyphosids, and odacids (Russell, 1983). The generally less tough red and green algae, which are major foods of temperate-zone herbivores (see Section III.C), do not appear to be depleted by these fishes in most cases (see Section VI). Moreover, structural and chemical traits are often well developed in tropical macroalgae, seemingly more so than in temperate species, and allow them to avoid, deter, or tolerate feeding by fishes (Hay, 1991). It is true that fucoid brown algae, which are rich in phlorotannins, deter feeding by northern temperate fish (Horn, 1989), but in Australian and New Zealand waters kyphosid and odacid fishes regularly consume fucoid algae (Russell, 1983) with even higher phlorotannin concentrations (Estes and Steinberg, 1988; Steinberg, 1988). Last, there seems to be no convincing evidence that temperate macroalgae are less nutritious than tropical forms.

More recently, Barry and Ehret (1993) and Hobson (1994), both with a focus on the northeastern Pacific, have proposed somewhat different explanations for the latitudinal gradient in herbivore species richness. The former authors maintain that a combination of factors has contributed to the scarcity of actively foraging herbivorous fishes in cooler waters. They argue that selection pressures for herbivory by fishes may have been reduced by the relatively greater productivity and seasonality in temperate systems such that the seasonal surpluses and shortages of food occur on time scales shorter than the generation time of the fishes. Shorter lived invertebrate herbivores (e.g., amphipods and worms) could increase their populations faster in reponse to enhanced food supplies. The resultant greater invertebrate biomass would reduce selection for herbivory among fishes and instead select for generalist carnivores and omnivores that can forage over algal turfs and rocky outcrops.

Hobson (1994) proposes that the reasons for the low diversity of temperate herbivores transcend herbivory and points out that closely related carnivores show a similar decline in species richness with latitude. The basis for the latitudinal gradient, according to Hobson, is not attributable to trophic relationships but to the effects of surface currents in the coastal upwelling system of the northeastern Pacific on the early life history stages of fishes whether of temperate or tropical derivation. More specifically, the diverse lines of warm water acanthopterygians, which contain the majority of herbivorous species, are limited in their distribution by conditions that place their pelagic eggs into currents of unfavorable surface transport. Although both Choat (1991) and Hobson (1994) argue that the diversity of all trophic groups decreases similarly with latitude, the assembled evidence for herbivorous fishes in general (Horn, 1989, Table VIII) indicates that the decline is proportionately greater for herbivores. Whether the same pattern holds for strictly intertidal herbivorous fishes is less clear (see Table 1); clarification requires increased knowledge of trophic positions among intertidal fish assemblages especially in the tropics.

## III. Feeding Mechanisms and Food Habits

Herbivorous fishes that feed in the rocky intertidal zone represent a partial array of the specialized behaviors and food-gathering mechanisms that characterize marine herbivores in general. Like subtidal feeders, intertidal herbivores must select from a broad diversity of available algal species to meet their energetic and nutritional requirements. How this is accomplished remains one of the fascinating and still inadequately answered questions in the study of fish herbivory, whatever the habitat. This section considers feeding behavior, tooth and jaw structure, food preferences, and factors influencing dietary choice found among intertidal herbivorous fishes. The relatively well-known herbivorous members of two families, the Blenniidae and Stichaeidae, are used to illustrate some of the specializations associated with an herbivorous lifestyle.

### A. Feeding Behavior

Algae-eating fishes, including intertidal species, can be classified as either grazers or browsers (Table 1; Figure 1; see Horn, 1989). Grazers ingest bits of inorganic material from the substratum as they feed by scraping or sucking, whereas browsers bite or tear pieces from more upright macroalgae and rarely consume any inorganic particles (Jones, 1968). This is an imperfect dichotomy because some fishes are difficult to place clearly into one category, and others, such as girellids (Orton, 1989; Table 1), employ both means to acquire plant material. Nevertheless, the classification serves an operational purpose and perhaps invites further investigation into the feeding behavior of different species.

Grazers, on average, can be expected to take smaller bites than browsers, not only as a function of the differences in size and stature of the algae they consume but also of differences in tooth and mouth structure of the fishes themselves. The prevailing notion is that grazers tend to feed nonselectively because their algal food is small and closely adherent to the substratum, whereas browsers feed on larger individual plants and therefore can and do feed more selectively (Lobel, 1981; Choat, 1982). Whether this distinction holds in most cases among intertidal fishes requires further study. Certainly there are exceptions. For example, the grazer *Parablennius sanguinolentus* in the western Mediterranean tends to feed selectively from an algal turf on the green alga *Ulva lactuca* and the few species of red algae in the turf but avoids an abundant brown alga (*Dictyota dichotoma*) in the same turf (Horn *et al.,* 1990; see Section IV.A). In contrast, the browser *Cebidichthys violaceus* on the California coast feeds both as juvenile (Horn *et al.,* 1982) and adult (Miller and Marshall, 1987) on a wide variety of red and green algae even though a smaller number of species predominate in the diets (see Section V.B).

Fish feeding methods have been divided into three types—biting, ram feeding, and suction feeding—based on ecomorphological analysis (Liem, 1980; Norton, 1995; Norton and Cook, Chapter 12, this volume). In biting, the oral jaws are used to remove part of a larger organism or tear attached prey from the substratum. In this context, most grazing and browsing herbivores would best be considered as biters rather than ram feeders or suction feeders. Once fish have grasped food in the oral jaws, they often use wrenching, thrashing, or spinning actions to tear small pieces from larger food masses (Norton and Cook, Chapter 12, this volume). Spin feeding, in which the food is grasped and the body rotated rapidly about the long axis to free a piece of food, is employed especially by elongate fishes such as anguillids (Helfman and Clark, 1986) and stichaeids (Miller and Marshall, 1987; Yoshiyama *et al.,* 1996; M.H.H., personal observation). The stichaeid observations include

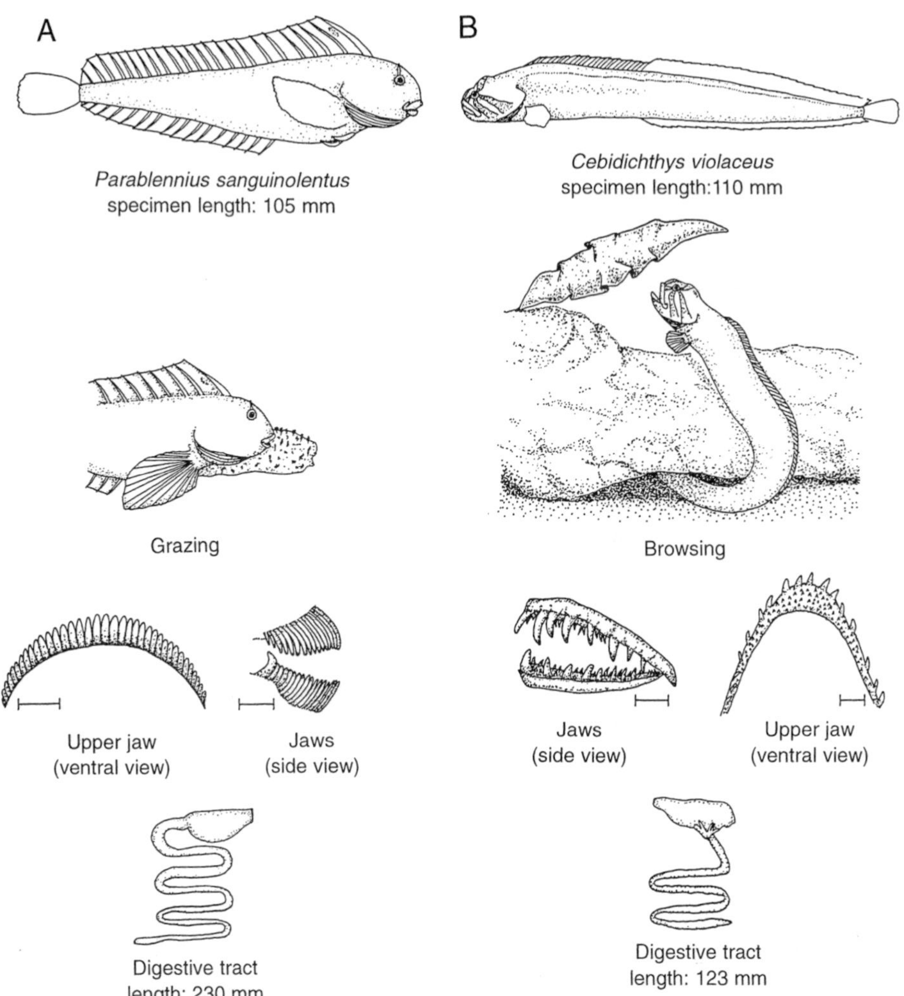


**Figure 1.** Comparison of the feeding behavior, jaw shape, tooth structure, and gut morphology of (A) *Parablennius sanguinolentus* (Blenniidae), a grazer, and (B) *Cebidichthys violaceus* (Stichaeidae), a browser. Digestive tract lengths, but not other dimensions, are drawn to the same scale as the fishes. *C. violaceus* has a true stomach; *P. sanguinolentus* is considered to have an intestinal swelling not a stomach. Scale bars = 1 mm.

those on the herbivorous pricklebacks *Cebidichthys violaceus* and *Xiphister mucosus,* indicating that spinning is an effective action for consuming algal as well as animal food material.

Another aspect of feeding behavior involves group foraging and territorial defense, which are frequently observed activities among herbivorous fishes in subtidal habitats and especially on tropical reefs. These behaviors are less common as established and predictable activities in intertidal environments. Thus, the roving, multispecies bands of parrotfishes (Scaridae) and surgeonfishes (Acanthuridae) (Ogden and Lobel, 1978) and the territorial damselfishes (Pomacentridae) (Hixon, 1983) so characteristic of subtidal tropical reefs are not common features of tidal pools, intertidal rock outcrops, or boulder fields. The most obvious reasons are the restricted space of the intertidal zone and the limited time for occupation by periodic intertidal fishes as a result of the ebb and flow of the tides, which create alternating conditions of exposure and submergence.

Exceptions to this pattern are seen in tropical latitudes where intertidal reef flats provide opportunities for feeding and even the development of territories. For example, Stephenson and Searles (1960) observed that rabbitfishes (Siganidae) and parrotfishes feed as mobile herbivores on intertidal beach rock at Heron Island on the Great Barrier Reef. Recently in American Samoa, Craig (1996) found that the browsing surgeonfish *Acanthurus lineatus* established aggressively defended feeding territories in the surf zone of the outer reef flat and reestablished them each morning and after displacement by rough surf or low tides. Territorial defense significantly reduced feeding rates, and rough surf greatly reduced the time that could be devoted to both feeding and defense.

## B. Tooth and Jaw Morphology

Herbivorous fishes seem to fall naturally under the biting category of feeding in Norton's (1995) ecomorphological analysis. According to Norton, biting performance is related to force generation and delivery at the expense of speed of movement and is improved by specializations that may include more robust jaws with cutting rather than "holding" teeth, enlarged adductor muscles, and more restricted jaw mobility as opposed to highly kinetic jaws. Biting fishes show an allometric increase in force generation, which should make tougher prey more vulnerable as these fish grow to adulthood (Norton and Cook, Chapter 12, this volume). Intertidal fishes with primarily macroalgal diets show these types of specializations to varying degrees. Typically, these species have short, blunt snouts with teeth that are closely set to form a cropping edge (Ogden and Lobel, 1978; Horn, 1989). Generally, the mouth of herbivorous fishes can be terminal in position as in browsing stichaeids (Figure 1B) or more ventrally positioned and downwardly directed as in grazing blenniids (Figure 1A).

Herbivorous members of both the Stichaeidae and Blenniidae have closely set teeth, but those of the blennies tend to be more numerous and uniform in size (Figures 1A, 1B). Kotrschal (1989) proposed an evolutionary scenario for the blennioid fishes in which the trend toward feeding on benthic organisms, including macroalgae, is accompanied by reduced jaw protrusibility and the formation of a single, closed row of incisiform teeth in both jaws. Within the Blenniidae, the tribe Blenniini are generalized, omnivorous browsers with the teeth borne on a semicircular jaw, whereas the Salariini are specialized grazers of fine epilithic algae (Kotrschal, 1989) as well as other small organisms on rock and coral surfaces (Springer, 1995). In the salariins, the numerous teeth are held loosely in a band of tissue (Springer, 1995) on a more flattened anterior jaw profile (Kotrschal, 1989).

Girellids (nibblers) including those occurring in the intertidal zone (Table 1) have spatulate, straight-edged or cuspate teeth, each of which is hinged with a ball-and-socket joint (Orton, 1989). These fishes are listed as both browsers and grazers in Table 1, and according to Orton, their tooth structure appears to allow a ripping action for effective browsing on foliose algae and a cropping action for efficient grazing on rock outcrops. Orton (1989) in his phylogenetic analysis of the Girellidae recognized two highly derived South Pacific members of the family without hinged jaw teeth, *Girella fimbriata* from the Kermadec Islands and *Girella nebulosa* from Easter Island. The teeth of these two species form a single row of heavy, chisel-shaped structures that appear almost like the fused teeth of parrotfishes when the jaws are closed. *G. fimbriata* and *G. nebulosa* are little studied but seem to be primarily subtidal species specialized for grazing rock surfaces (Orton, 1989).

Some fishes that become increasingly herbivorous with age also show ontogenetic changes in dentition. Porgies (family Sparidae) illustrate this pattern in that many of the omnivorous species undergo complex age-related changes in tooth morphology and food habits, usually with greater use of plant material as adults (e.g., Christensen, 1978; Stoner, 1980; Gerking, 1984). An intertidally occurring sparid exhibiting this pattern is *Sarpa salpa*. This species as an adult probably comes as close as any sparid to becoming a complete herbivore, and its canine teeth are replaced in older fish by incisors (Christensen, 1978), which are apparently better suited for taking bites of the algal material that becomes a larger part of its diet with age.

## C. Diet, Food Preferences, and Factors Influencing Preference

As a group, herbivorous fishes consume a great diversity of plant material, from diatoms (as in the Mugilidae) to large kelps (as in the Odacidae), to seagrasses (as in some Scaridae and Sparidae). A number of species of red and green algae, however, appear to predominate in the diets of most herbivorous fishes (Horn, 1989), including those occurring in the rocky intertidal zone. Evidence for the prevalence of these algae in the diets of intertidal herbivorous fishes includes that for *Clinocottus globiceps* (Cottidae) in California (Grossman, 1986; Yoshiyama *et al.*, 1996) and Washington (Cross, 1981); *Girella laevifrons* (Girellidae) in Chile (Stepien, 1990); *Girella nigricans* and *Hermosilla azurea* (Kyphosidae) in California (Barry and Ehret, 1993); *Sicyases sanguineus* (Gobiesocidae) in Chile (Paine and Palmer, 1978; Stepien, 1990); *Cebidichthys violaceus* and *Xiphister mucosus* (Stichaeidae) in California (Barton, 1982; Horn *et al.*, 1982); *Caffrogobius caffer* (Gobiidae) in South Africa (Berry *et al.*, 1982); *Siganus spinus* (Siganidae) on Guam (Bryan, 1975); and the blenniids *Alticus kirkii* in the Red Sea (Zander, 1967), *Blennius cristatus* in Puerto Rico and the Virgin Islands (Randall, 1967), *Istiblennius coronatus* in the Marshall Islands (Hiatt and Strasburg, 1960), *Scartichthys viridis* in Chile (Stepien, 1990; Muñoz and Ojeda, 1997), and *Parablennius sanguinolentus* in the Mediterranean Sea (Gibson, 1968; Horn *et al.*, 1990). Preference for red and green algae by herbivorous fishes thus crosses both taxonomic and geographic boundaries.

The red and green algae eaten by these fishes mostly are in the sheet-like, filamentous, or coarsely branched structural categories based on the functional-form model of Littler and Littler (1980). These algae tend to have thalli of high surface-to-mass ratios and a two-dimensional sheet-like structure (e.g., *Porphyra, Ulva*), a long ribbon-like form (e.g. *Enteromorpha*), or small, coarsely branched morphologies (e.g., *Gelidium, Microcladia, Acrosiphonia*) according to the model (also see Horn *et al.*, 1982). Moreover, some but not all of these algae are annuals rather than perennials, and they tend to be more palatable and higher in energy content than most perennial species (Littler and Littler, 1980). On temperate shores in California, however, perennial red algae such as *Mazaella* spp. (=*Iridaea* spp.) may compose a large portion of the diets of stichaeids during the winter when their preferred annual red algae (e.g., *Microcladia coulteri, Porphyra perforata, Smithora naiadum*) are rare or absent; the fishes are able to feed on the preferred algae only during the non-winter months when these algae are more available (Horn *et al.*, 1986).

Other types of algae in the functional-form model are less commonly eaten for reasons that may involve their availability in the intertidal habitat. As mentioned above, seasonal occurrence may prevent certain algae, apparently preferred for their relatively high protein content, from being eaten throughout the year. Other species appear to be available year-round and to be relatively nutritious, tender, and palatable, yet are not eaten by herbivorous fishes. For example, the turf-forming, highly branched red alga, *Endocladia muricata,* appears to be one of the best foods available on central California shores (Neighbors and Horn, 1991); however, it is not consumed by the herbivorous stichaeids living there (Horn *et al.,* 1982), possibly because the alga lives high in the intertidal habitat (Murray and Horn, 1989) and thus is largely unavailable to these relatively inactive fishes.

Intrinsic factors such as nutritional quality, structural defenses, chemical deterrents, or combinations of these also influence dietary choice. Fucoids (e.g., *Fucus gardneri, Hesperophycus californicus, Pelvetia compressa*), coarsely branched brown algae with relatively thick, tough thalli, are not, to our knowledge, eaten by any intertidal fish in north temperate waters even though these algae are often among the most abundant plants on the shore. Stichaeid fishes avoid fucoids perhaps in part because these algae usually occur high in the intertidal zone but more likely because they have tough thalli and produce phlorotannins (Steinberg, 1985) that appear to deter feeding (Irelan and Horn, 1991) and inhibit digestion (Horn *et al.,* 1985; Irelan and Horn, 1991; Boettcher and Targett, 1993) (see Section IV.B). The tough leathery category includes the laminarian brown algae (kelps), which also contain phlorotannins but in lower concentrations than the fucoids (Steinberg, 1985). Both of these groups of brown algae are not eaten by *C. violaceus,* perhaps in part because they have relatively indigestible $\beta$ linkages in their storage polysaccharides (see Montgomery and Gerking, 1980). Finally, the jointed calcareous functional-form group, which includes coralline red algae, is also rarely if ever consumed by intertidal herbivorous fishes (e.g., Neighbors and Horn, 1991), probably because they contain a high proportion of structural material and a relatively low amount of energy (Littler and Littler, 1980). Species in this group, such as *Corallina vancouveriensis,* may also be avoided by fishes because they produce (Ragan, 1981) bromophenolic secondary compounds.

In summary, studies on a limited number of intertidal fish species indicate that several factors may influence diets and food preferences. Temporal and spatial variation in availability of palatable algae affect dietary choice of herbivorous fishes. Some evidence exists to show that fishes select algal species that are relatively rich in protein and easy to digest (Horn and Neighbors, 1984). Other algae appear to be avoided because they are tough or calcareous or because they contain secondary compounds that reduce palatabilty or digestibility. In some cases, a combination of factors operates to cause algae either to be chosen or avoided.

## IV. Gut Morphology and Digestive Mechanisms

The challenges of herbivory appear to exert greater selective forces on digestion than on feeding. Choat (1991) stated this point well when he asserted that the critical evolutionary arena for herbivorous fishes lies within the alimentary canal where biochemical interactions are mediated by gut histology and microorganisms. He predicted that a variety of physiological and biochemical mechanisms for extracting nutrients and energy from algal cells would eventually be discovered among herbivorous fishes. Choat's comments appear to be valid despite instances of the digestive tracts of some herbivorous fishes showing no obvious specializations for handling a plant diet (e.g., herbivorous stichaeids). Even though knowledge of herbivore digestive physiology is still limited, several mechanisms for disrupting algal cell walls and digesting the cell contents have been described (Lobel, 1981; Horn, 1989). Cases of apparent absence of gut specializations in herbivorous species probably are a result of the lack of knowledge of gut structure and function in these fishes. In this section, emphasis is placed on the fundamental features of intestine length, gut transit time, and digestive mechanisms of fishes that feed on algae in the intertidal zone.

## A. Gut Length and Transit Times

Herbivorous fishes generally have longer guts relative to body size than omnivorous or carnivorous species (Horn, 1989, 1992). This pattern is characteristic of herbivorous taxa in different, unrelated families and also of those species that represent the only plant-eating taxa in otherwise carnivorous families. Herbivores usually start life as carnivores (White, 1993), and as they grow and become increasingly reliant on plant material their digestive tracts become longer relative to body length (Emery, 1973; Montgomery, 1977; Christensen, 1978; Barton, 1982; Lassuy, 1984). Relative gut length, however, does not always continue to increase with body size but may level off or even decline slightly once the fish reaches a certain length (Emery, 1973; Montgomery, 1977; Lassuy, 1984). In a frequently cited study, Al-Hussaini (1947) arranged 60 species of Red Sea fishes into four categories based on their food habits and relative gut lengths (defined as the esophagus–anus distance divided by standard body length). In his classification, plankton feeders had the shortest guts (0.5–0.7), carnivores somewhat longer guts (0.6–2.4), followed by omnivores (1.3–4.2), and then herbivores with the longest relative gut lengths (3.7–6.0). Caution is necessary, however, in using Al-Hussaini's values to predict the diets of other fishes without examining their actual food habits. One reason is simply that some herbivorous fishes have shorter guts than those in Al-Hussaini's herbivore category. For example, two intertidal species that probably come as close as any to being completely herbivorous are the stichaeid *Cebidichthys violaceus* and the blenniid *Parablennius sanguinolentus*. Nevertheless, the relative gut length of the former is 1.1 in small [~80–150 mm total length (TL)] specimens (Barton, 1982) to 1.7 in larger (~180–250 mm) fish (M.H.H. unpublished data), and the value for *P. sanguinolentus* is 2.4 (Goldschmid *et al.*, 1984). Certain other, subtidal fishes that consume large amounts of algae have relatively even shorter guts (Horn, 1989). A further caution against extrapolating trophic status from gut length is that the surface area of the intestinal mucosa per body weight or volume may decrease with size in some herbivorous fishes, including *Cebidichthys violaceus* (Montgomery, 1977) and others (Al-Hussaini, 1949; Gohar and Latif, 1959). These surface area reductions may be related to the lower metabolic rate of larger fish but further research is required before firmer conclusions can be made. A summary point is that relative gut lengths are most meaningful if the comparisons are made among related fish species with known different diets.

The importance of increased gut length in herbivorous fishes appears to translate into the value of a larger gut capacity for animals eating poor-quality foods. According to Sibly and Calow (1986), the net rate for obtaining energy is a product of (1) the weight of material in the digestive tract and (2) the amount of energy yielded from the material for a given retention period. The second quantity is lower for a poorer quality food; therefore, for a given nutritional requirement, the animal must carry a greater mass of digesta (the first quantity in the product above), which requires an increased gut capacity. Herbivorous fishes appear to meet the requirements of this model by distributing low-nutrient, high-fiber food along the extensive surface area of a long intestine to maximize digestion and assimilation (see Buddington *et al.*, 1987). These requirements are constrained by the size of the body cavity and the size and shape of the fish (Montgomery, 1977).

Small body size may be at least part of the reason why the smallest reef fishes (e.g., chaenopsids) are not herbivores (Kotrschal and Thomson, 1986) and why the young stages of herbivores are usually carnivorous (White, 1993). Also, larger body size seems to have evolved in herbivorous members of at least some teleostean families containing species with varied diets, a relationship between body size and food habits also seen in reptiles, birds, and mammals (see Horn, 1989; Hotton *et al.,* 1997). Among fishes, the largest members of the Odacidae from Australia and New Zealand (Gomon and Paxton, 1985), Pomacentridae from the Florida Keys (Emery, 1973), and Stichaeidae from the western coast of North America (Eschmeyer *et al.,* 1983) are herbivores (Horn, 1989). The largest stichaeids, *Cebidichthys violaceus* and *Xiphister mucosus,* live in intertidal to shallow subtidal habitats and are herbivores (see Section V.B). The frequent linkage between herbivory and large body size may be one of the reasons why herbivorous fishes are rare in the intertidal zone where a complex set of factors appear to select for small body size in most intertidal groups. A large body size is also considered to be an evolutionary response to nitrogen shortage (Mattson, 1980), and a long intestine, which can more easily occur in larger fishes, may also function to allow reabsorption of proteolytic enzymes, thereby conserving protein in herbivorous and omnivorous species (Hofer and Schiemer, 1981).

An often repeated scenario for food processing in herbivorous fishes is that they consume large amounts of plant material by more or less continuous feeding, pass the material quickly through the digestive tract, and assimilate a portion of it with moderate efficiency (Ogden and Lobel, 1978; Brett and Groves, 1979; Pandian and Vivekanandan, 1985; Buddington *et al.,* 1987). Gut transit or evacuation time is an important component of food processing by fishes because it influences assimilation efficiencies and consumption rates, the latter resulting from a return of appetite as the digestive tract empties (Fänge and Grove, 1979). Many factors including temperature, fish size, feeding method, food type, meal size, and feeding history of the fish affect gut transit time (or rate) (Fänge and Grove, 1979).

Most herbivores do appear to have shorter gut transit times than carnivores, but the data are sparse, particularly for intertidal species. In a review of herbivorous fish biology, Horn (1989) listed the gut transit times for only 13 species and, of these, only 5 occur in the intertidal zone. Three of the five were mullets (Mugilidae) and the fourth was the sparid *Sarpa salpa,* all with transit times of less than 10 h. The fifth species, *Cebidichthys violaceus,* was an exception to the scenario in requiring more than 50 h after a meal for undigested material to pass out of the gut (see Section V.B). Such variation, even though from a small number of species, suggests a diversity of feeding and digestive modes among marine herbivorous fishes and underscores the limited knowledge of these processes in intertidal species.

## B. Digestive Mechanisms

What are the mechanisms by which herbivorous fishes meet two major challenges of an algal diet, gaining access to the nutrients locked inside the algal cell wall and to the structural carbohydrates composing the wall itself? Lobel (1981) recognized that herbivorous fishes could achieve cell breakage by (I) lysis, resulting from gastric acidity; (II) mechanical action, by grinding in a gizzard-like stomach; or (III) mechanical action, by grinding in the pharyngeal jaws. Since then, it has become evident that a variety of herbivorous fishes harbor a hindgut microflora associated with digestion of algal polysaccharides to yield short chain fatty acids (Rimmer and Wiebe, 1987; Clements *et al.,* 1994; Kandel *et al.,* 1994; Clements and Choat, 1995). Furthermore, cellulolytic bacteria have been isolated from the intestinal tract of the sparid fish *Lagodon rhomboidalis,* a warm temperate omnivore of seagrass meadows and estuaries (Luczkovich and Stellwag, 1993). Thus, microbial fermentation has become recognized as a fourth mechanism (IV) by which herbivorous fishes gain access to algal cell contents and perhaps structural carbohydrates. These four digestive mechanisms have been described in previous reviews (Horn, 1989, 1992, 1998) and incorporated into a model of herbivorous fish guts using chemical reactor theory (Horn and Messer, 1992). Although the four mechanisms serve as a framework for investigations of herbivore digestive physiology, other mechanisms may be discovered as herbivorous fishes become more thoroughly studied.

All four digestive mechanisms are found among herbivorous fishes that reside in or periodically enter the rocky intertidal zone although the large number of question marks in Table 1 indicates that much remains to be learned about digestion in most fishes occurring in this habitat. The following accounts of the four mechanisms are based on Horn (1989, 1992).

Species with the Type I digestive mechanism are herbivores with thin-walled stomachs and no triturating mechanism other than the jaw bite; these species appear to use a highly acidic stomach environment to lyse algal cell walls and thereby gain access to the cell contents. These fishes typically have moderately long to long intestines but otherwise few apparent morphological specializations for herbivory. They are usually browsers and are predicted (Lobel, 1981) to feed mainly on large-celled green and red algae. Intertidal species with a Type I mechanism include those acanthurids with thin-walled, acidic stomachs, aplodactylids, girellids, and stichaeids (see Section V; Table 1). All these fishes are browsers or combination browsers/grazers. Other species that may belong in the Type I category include blenniids (see Section V.A), cottids, gobiesocids, and gobiids; however, too little is known about their digestive tracts including pH values to make a firm designation.

Herbivorous fishes with the Type II digestive mechanism use a thick-walled, muscular stomach to disrupt algal cell walls by grinding action. These species tend to have slightly acidic to slightly alkaline stomach fluids and a relatively long intestine. They are mostly grazers and therefore ingest quantities of sand and other sedimentary particles, which serve in the gizzard-like stomach to aid in the trituration of bacteria and various algal material consumed. Intertidal species with a Type II mechanism include mugilids and those acanthurids with thick-walled stomachs and nearly neutral gastric fluids (Table 1).

Those species with the Type III digestive mechanism also use physical means to rupture algal cell walls, in this case by trituration in pharyngeal jaws. Inorganic material may or may not be consumed with the food material; thus, the fishes are either grazers or browsers. The pharyngeal jaws serve either to grind or shred the algal material before it reaches the intestine. Fishes with this digestive mechanism lack a stomach and have a moderately long and slightly acidic to slightly alkaline intestine. Intertidal species with a Type III mechanism include mainly parrotfishes that periodically invade shallow waters on feeding excursions (Table 1).

Herbivorous fishes with the Type IV digestive mechanism contain diverse and abundant populations of microorganisms in their gut, especially in the distal regions. The hindgut may be enlarged into a chamber with or without blind ceca and each part set off from the rest by valves as in kyphosids, or it may not be obviously differentiated as in acanthurids or odacids. The overall gut is relatively long and a stomach may be present as in kyphosids and acanthurids or absent as in odacids. The action of this digestive mechanism occurs mainly in the distal intestine and appears to function in combination with others, for example, the Type I mechanism in kyphosids, or the Type III mechanism in odacids. An increasing variety of herbivorous fishes will likely be shown to obtain a portion of their energy from fermentative microbial digestion. One intertidal fish known to have a Type IV mechanism is *Hermosilla azurea,* a kyphosid that occurs in tidepools as a juvenile (Table 1).

## V. Case Studies of Herbivorous Taxa

The following two families each contain a relatively well-studied herbivorous fish species with abundant populations in rocky intertidal and shallow subtidal habitats. Knowledge of their feeding ecology and digestive physiology provides a comparative portrait of the herbivore biology of a grazer of algal turf (Figure 1A) and a browser of foliose macroalgae (Figure 1B).

### A. Blenniidae with Emphasis on *Parablennius sanguinolentus*

The Blenniidae (combtooth blennies) are bottom-dwelling fishes of shallow, mostly tropical to subtropical marine habitats; the family consists of several tribes and about 345 species (Nelson, 1994). Two of the tribes, Salariini and Parablenniini, are important in a discussion of marine herbivory. Salariin blennies make up about half the species in the family, and most feed by scraping algae and other organisms off the surface of rocks and coral using their small, close-set teeth (Springer, 1995; Figure 1A). Species in genera such as *Alticus, Andamia,* and *Salarias* occur in the supratidal zone and often feed out of water on algal turf (see Zander *et al.,* Chapter 3, this volume). A large salariin, *Scartichthys viridis,* grazes on algae growing on rock surfaces in the Chilean intertidal zone (Stepien, 1990; Muñoz and Ojeda, 1997), and its impact on the algal community is described in Section VI. Detailed information on feeding is available for only a few species in this large tribe, and digestion appears to be completely unstudied. Unpublished work by one of us (F.P.O.) shows that *S. viridis* feeds mainly on the green algae *Enteromorpha* and *Ulva* and a red algal turf consisting primarily of *Gelidium;* it lacks a stomach like other blenniids, and the slightly alkaline (pH 7.2–7.8) gut averages four times the body length. Given the apparently common feeding habit of scraping algae by its members, the tribe Salariini is certain to contain many as yet unstudied herbivorous species.

The Parablenniini, the second tribe of importance for herbivory, contains about 65 species (Nelson, 1994) with greatest species richness in and around the Mediterranean Sea (Springer, 1995). *Parablennius sanguinolentus* occurs mainly in shallow subtidal waters throughout the Mediterranean and in the intertidal zone of the Atlantic northward to southern France (Bath, 1973). This blenny is considered to be almost completely herbivorous and has been observed (Horn *et al.,* 1990) in shallow pools in the western Mediterranean to feed on the green alga *Ulva lactuca* or a red algal turf by taking a series of discrete bites involving a combined forward and lateral rotation of the body interspersed by periods of resting or watching or by retreats to cover. Gibson (1968) found 100% occurrence of algae in the stomachs of fish taken from the Banyuls region of France and Spain in the Mediterranean. He reasoned that the fish's small, close-set flexible teeth (Figure 1A) and presence of filamentous algae and sand in its stomach indicated that it grazes on fine algae growing on the walls and boulders of the pools where it lives. The infrequent occurrence of animal food in the fish's stomach was considered by Gibson to be an incidental result of algal grazing.

*P. sanguinolentus* shows an afternoon peak in feeding intensity that Taborsky and

Limberger (1980) interpreted as an adaptation to the daily accumulaton of photosynthetic products in its green macroalgal food, primarily *Ulva.* Subsequent research (Horn *et al.,* 1990) partially supported the Taborsky–Limberger hypothesis in that the *Ulva lactuca* upon which the fish was observed to feed attained its highest nutritional content in late afternoon, and, although large [~120 mm standard length (SL)] fish did not increase their feeding rate during the day, they did switch increasingly from feeding on a red algal turf to the more highly productive, higher quality *U. lactuca.* Smaller fish (~60 mm SL), however, did increase their feeding rate by afternoon but continued to feed on species of the red algal turf. Only larger fish, therefore, appeared to maximize their energy intake. The smaller fish may have continued to feed on the red algal turf because this turf is a more spatially and temporally predictable food supply, or they may have been prevented from switching to *Ulva* by the behavioral dominance of larger fish.

Herbivory based on diet in *P. sanguinolentus* is reflected in the fish's gut morphology and food processing capabilities. Its digestive tract was found to be the longest (2.41 × body length) and twice the relative length of the species with the next longest gut among 14 species of Adriatic blenniids (Goldschmid *et al.,* 1984). This blenny assimilated carbon and nitrogen from an *U. lactuca* diet at efficiencies within the reported ranges for other marine herbivorous fishes, indicating its ability to extract energy from an algal diet (Horn and Gibson, 1990). The type of digestive mechanism used by this and other herbivorous blenniids remains unknown although Horn (1989, 1992) speculated that they may belong in the Type I category. These fishes have long, coiled intestines and anterior intestinal swellings but lack a true stomach (Al-Hussaini, 1947; Goldschmid *et al.,* 1984; see Figure 1A). Gut pH apparently has not been determined in any herbivorous blenniids except *Scartichthys viridis,* as mentioned above.

## B. Stichaeidae with Emphasis on *Cebidichthys violaceus*

The Stichaeidae (pricklebacks) are elongate, bottom-dwelling fishes of cool temperate primarily inshore waters mainly of the North Pacific and also the North Atlantic; the family consists of several tribes and about 65 species (Nelson, 1994). Of the 23 stichaeid species recognized in the eastern North Pacific (Eschmeyer *et al.,* 1983), only the two largest, *Cebidichthys violaceus* and *Xiphister mucosus,* are considered to be herbivores (Barton, 1982; Horn *et al.,* 1982).

Both these pricklebacks begin life as carnivores (Setran and Behrens, 1993) and then shift to herbivory as small juveniles, coming to feed selectively as browsers in the rocky intertidal zone on red and green macroalgae and to avoid encrusting, calcareous and brown algae (Horn *et al.,* 1982). Compared to the small (44–227 mm) *C. violaceus* studied by Horn and co-workers, larger (300–710 mm) fish studied by Miller and Marshall (1987) fed with greater frequency on tough perennial plants especially the red algae *Mazzaella* spp. (=*Iridaea* spp.) and also appeared to forage lower in the intertidal zone. Miller and Marshall observed a 380-mm fish in an aquarium to undertake spin feeding in the process of grasping and swallowing an entire blade of *Mazzaella.* Based on laboratory feeding experiments, *C. violaceus* prefers three species of annual red algae with the highest protein contents among the eight species making up the bulk of the diet. This preference, accompanied by consistently high assimilation efficiencies, shows that the preferred species provide the greatest amount of protein per bite to the fish (Horn and Neighbors, 1984; Figure 2). Because these preferred algae are annuals, they are available mainly during the summer (Horn *et al.,* 1982). Diet choice from a seasonally fluctuating algal food supply

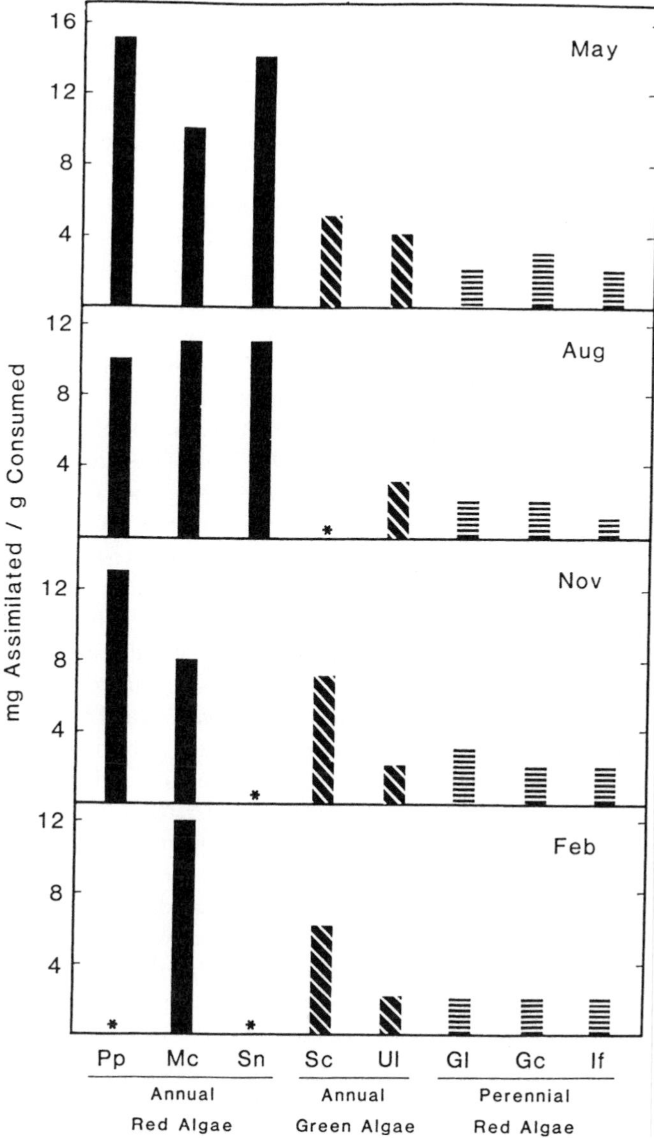


**Figure 2.** Protein assimilated (mg/g wet weight algae consumed) by *Cebidichthys violaceus* (Stichaeidae) from eight species of dietary algae during quarterly periods (May 1982–February 1983). Species arranged left to right by mean protein content for entire year. Asterisks (*) indicate species not found at study site. Pp, *Porphyra perforata;* Mc, *Microcladia coulteri;* Sn, *Smithora naiadum;* Sc, *Spongomorpha* (=*Acrosiphonia*) *coalita;* Ul, *Ulva lobata;* Gl, *Gigartina* (=*Mazzaella*) *leptorhynchos;* Gc, *Gigartina canaliculata* (=*Chondracanthus canaliculatus*); If, *Iridaea* (=*Mazzaella*) *flaccida.* From Horn and Neighbors (1984) with permission of the American Fisheries Society.

results in both *C. violaceus* and *X. mucosus* consuming carbohydrate-rich, protein-poor algae in the winter and protein-rich, carbohydrate-poor algae during the rest of the year (Horn *et al.,* 1986). Accordingly, growth rates of the two fishes predicted by Horn *et al.* (1986), using a growth curve based on the relative amounts of protein and energy assimilated from their diets, were at or just above maintenance levels in winter and higher during other times of the year.

Other factors influencing diet selection in addition to digestive efficiency in *C. violaceus* and *X. mucosus* include secondary metabolite concentrations, especially polyphenolic compounds (phlorotannins). They are found in laminarian and fucoid brown algae (Steinberg, 1985), which are avoided by these pricklebacks. In a force-feeding study, *C. violaceus* assimilated nitrogen but not carbon from the laminarian kelp *Macrocystis integrifolia* and neither element from the fucoid *Fucus gardneri (=F. distichus)* (Horn *et al.,* 1985). However, in another investigation in which polar and nonpolar extracts of the target algal species were coated on edible food strips, *M. integrifolia* did not deter the feeding nor affect the digestion of *C. violaceus* (Irelan and Horn, 1991). In the latter study, the polar, presumably phlorotannin-containing extract of *F. gardneri* applied to food strips reduced the feeding rate of *C. violaceus* and reduced nitrogen assimilation efficiency when they were force-fed to the fish. Further, Boettcher and Targett (1993) showed that assimilation efficiency in *X. mucosus* was little affected by phlorotannins of small molecular size (<5 kDa) but was significantly reduced by phlorotannins of larger molecular size (>10 kDa). The molecular size range of phlorotannins that reduced assimilation was also the size range of these substances most concentrated in the algae studied, including *F. gardneri.*

Herbivory based on diet and food preferences in *C. violaceus* and *X. mucosus* is reflected in gut morphology and pH, food processing capabilities, and growth responses of these two pricklebacks. Gut measurements expressed as a proportion of standard body length shows that both species in the herbivorous stage have only moderately long guts but nevertheless longer than two other intertidal stichaeids that remain carnivores into adulthood. Values for *C. violaceus* range from 1.1 in small (~80–150 mm) specimens (Barton, 1982) to 1.7 in larger (~180–250 mm) specimens (M.H.H., unpublished data); a value of 0.8 was obtained for *X. mucosus* (Barton, 1982). Gut lengths in the two carnivorous stichaeids are 0.45 for *Anoplarchus purpurescens* and 0.6 for *Xiphister atropurpureus* (Barton, 1982). Otherwise, the digestive tracts of *C. violaceus* and *X. mucosus* show no obvious structural specializations for herbivory. The gastric pH, however, of *C. violaceus* is highly acidic with values of 2.0–2.5 recorded by Urquhart (1984), who showed that acid lysis of the cell walls of the green alga *Ulva lobata* significantly increased the concentration of carbohydrate and protein in the medium compared to control solutions. This high gastric acidity along with the fish's thin-walled stomach and moderately long intestine combine to make its digestive mechanism fit the Type I category (see Section IV.B).

Assimilation efficiencies obtained in several studies show clearly that *C. violaceus* (and in fewer cases, *X. mucosus*) can digest and absorb carbohydrate, lipid, and protein from consumed algal material (Edwards and Horn, 1982; Horn and Neighbors, 1984; Horn *et al.,* 1986). Protein assimilation efficiencies as high as 95% for *C. violaceus* and 98% for *X. mucosus* have been obtained when these pricklebacks were fed their preferred dietary algae (Horn *et al.,* 1986). These efficiencies match the high values recorded for carnivorous fishes (Kapoor *et al.,* 1975). Even more convincing evidence for assimilation of algal compounds by *C. violaceus* comes from data showing that radioactively labeled ($^{14}C$) algal material largely free of epibiotic bacteria and diatoms is absorbed from the digestive tract into the fish's body (Horn *et al.*, 1985). The most compelling evidence for the ability of *C. violaceus* to survive and grow entirely on a plant diet and a wide range of protein levels has been obtained in experiments in which the fish were fed diets of differing protein content. The fish grew not only on a pelleted all-algal food representing its natural summer diet (10% protein) but also on this food modified to make a diet as low as 7% protein (Fris and Horn, 1993) and as high as 50% protein (Horn *et al.*, 1995). *C. violaceus* grew fastest on diets of intermediate protein content (Fris and Horn, 1993; Horn *et al.*, 1995) and converted food protein to body protein (net protein utilization) at levels that are above average for vertebrates in general (Fris and Horn, 1993).

The digestive tract of *C. violaceus* and other marine herbivorous fishes has been modeled as a chemical reactor (Horn and Messer, 1992) based on the Penry and Jumars (1987) application of chemical reactor theory to animal guts. As a browser, *C. violaceus* has a relatively high-quality diet because little inorganic material is ingested. The alimentary canal can be modeled as continuous-flow, stirred-tank reactor (acidic stomach) followed by a plug-flow reactor (intestine). The stomach acts as a batch reactor because the fish appears to feed intermittently (Ralston and Horn, 1986) under the constraints of an intertidal habitat and perhaps the cost of digestion. The model predicts that the reaction rate will be low for a fish with a very low throughput rate and a moderately long intestine. A low reaction rate signals a limited energy intake per unit time. The model, therefore, is in accord with the observed characteristics of this fish: sluggish activity and movements based on tracking with ultrasonic telemetry (Ralston and Horn, 1986), a relatively unspecialized digestive tract (see above), and a very slow food passage time (53 h) through its gut (Urquhart, 1984).

## VI. Ecological Impacts of Intertidal Herbivorous Fishes

One of the most active and successful areas of research on marine herbivorous fishes in recent decades has been that focused on the impacts of their grazing and browsing on shallow water benthic communities. Most of this work has been conducted in subtidal habitats in tropical waters, especially on coral reefs, and shows how parrotfishes, by their scrapings and excavations of hard substrata, contribute to erosional and sedimentary processes (e.g., Randall, 1967; Ogden, 1977). Moreover, their intensive cropping of algal stands changes the abundance, composition, and spatial patterns of these algal populations (e.g., John and Pople, 1973; Lewis, 1986). At the same time, territorial herbivores, especially damselfishes, protect some algal stands from the roving herbivores and create a mosaic of luxuriant algal growth inside the territories and heavily cropped zones outside the territories (e.g., Montgomery, 1980; Hixon and Brostoff, 1983).

In contrast, few studies have been conducted on the impacts of herbivorous fishes in strictly intertidal habitats although some of the work on coral reefs or other hard substrata in the tropics has included intertidal portions of the habitat. Two examples of such work were experimental investigations, one conducted on intertidal beach rock at Heron Island on the Great Barrier Reef (Stephenson and Searles, 1960) and the other on a breakwater system in the Gulf of Guinea (John and Pople, 1973). At Heron Island, exclosures demonstrated that several species of grazing and browsing fishes, especially rabbitfishes, were responsible for the paucity of beach rock algae. In the Gulf of Guinea, group foraging by roving surgeonfishes and parrotfishes appeared to be the main factor responsible for the low algal diversity and absence of large algae. Browsing by these fishes on algae suspended

endeavors. Knowledge of the ecological impacts of herbivorous fishes residing in or periodically invading the intertidal zone is meager, and additions to the few completed experimental studies could prove to be rewarding and beneficial.

## Acknowledgments

We thank Steven Murray for information on algal taxonomy and Janine Kido for drawing Figures 1 and 3A. We are grateful to the U.S. National Science Foundation (M.H.H.) and the Chilean FONDECYT and FONDAP (F.P.O.) for support of our research on herbivorous fishes.

## References

Al-Hussaini, A. H. (1947). The feeding habits and the morphology of the alimentary tract of some teleosts living in the neighborhood of the Marine Biological Station, Ghardaqa, Red Sea. *Pub. Mar. Biol. Stn., Ghardaqa, Red Sea* **5,** 1–61.

Al-Hussaini, A. H. (1949). On the functional morphology of the alimentary tract of some fish in relation to differences in their feeding habits: Anatomy and histology. *Q. J. Microsc. Sci.* **90,** 109–139.

Bakus, G. J. (1969). Energetics and feeding in shallow marine waters. *Int. Rev. Gen. Exp. Zool.* **4,** 275–369.

Barber, W. E., McDonald, L. L., Erickson, W. P., and Vallarino, M. (1995). Effect of the *Exxon Valdez* oil spill on intertidal fish: A field study. *Trans. Am. Fish. Soc.* **124,** 461–476.

Barry, J. P., and Ehret, M. J. (1993). Diet, food preference, and algal availability for fishes and crabs on intertidal reef communities in southern California. *Environ. Biol. Fish.* **37,** 75–95.

Barton, M. G. (1982). Intertidal vertical distribution and diets of five species of central California stichaeoid fishes. *Calif. Fish Game* **68,** 174–182.

Bath, H. (1973). Blenniidae. *In* "Check-list of the Fishes of the North-eastern Atlantic and of the Mediterranean" (J. C. Hureau and Th. Monod, Eds.), Vol. I, pp. 519–527. UNESCO, Paris.

Bennett, B., Griffiths, C. L., and Penrith, M.-L. (1983). The diets of littoral fish from the Cape Peninsula. *S. Afr. J. Zool.* **18,** 343–352.

Berry, P. F., van der Elst, Hanekom, P., Joubert, C. S. W., and Smale, M. J. (1982). Density and biomass of the ichthyofauna of a Natal littoral reef. *Mar. Ecol. Prog. Ser.* **10,** 49–55.

Boettcher, A. A., and Targett, N. M. (1993). Role of polyphenolic molecular size in reduction of assimilation efficiency in *Xiphister mucosus. Ecology* **74,** 891–903.

Brett, J. R., and Groves, T. D. D. (1979). Physiological energetics. *In* "Fish Physiology," Vol. VIII, "Bioenergetics and Growth" (W. S. Hoar, D. J. Randall, and J. R. Brett, Eds.), pp. 279–352. Academic Press, New York.

Bryan, P. G. (1975). Food habits, functional digestive morphology, and assimilation efficiency of the rabbitfish *Siganus spinus* (Pisces, Siganidae) on Guam. *Pacific Sci.* **29,** 269–277.

Buddington, R. K., Chen, J. W., and Diamond, J. (1987). Genetic and phenotypic adaptations of intestinal nutrient transport to diet in fish. *J. Physiol.* **393,** 261–281.

Bussing, W. A. (1972). Recolonization of a population of supratidal fishes at Eniwetok Atoll, Marshall Islands. *Atoll Res. Bull.* **154,** 1–7.

Chapman, A. R. O., and Lindley, J. E. (1980). Seasonal growth of *Laminaria solidungula* in the Canadian High Arctic in relation to irradiance and dissolved nutrient concentrations. *Mar. Biol.* **57,** 1–5.

Choat, J. H. (1982). Fish feeding and the structure of benthic communities in temperate waters. *Annu. Rev. Ecol. Syst.* **13,** 423–449.

Choat, J. H. (1991). The biology of herbivorous fishes on coral reefs. *In* "The Ecology of Fishes on Coral Reefs" (P. F. Sale, Ed.), pp. 120–155. Academic Press, San Diego.

Christensen, M. S. (1978). Trophic relationships in juveniles of three species of sparid fishes in the South African marine littoral. *Fish. Bull. U.S.* **76,** 389–401.

Clements, K. D., and Choat, J. H. (1995). Fermentation in tropical marine herbivorous fishes. *Physiol. Zool.* **68,** 355–378.

Clements, K. D., Gleeson, V. P., and Slaytor, M. (1994). Short-chain fatty acid metabolism in temperate marine herbivorous fish. *J. Comp. Physiol. B* **164,** 372–377.

Craig, P. (1996). Intertidal territoriality and time-budget of the surgeonfish, *Acanthurus lineatus,* in American Samoa. *Environ. Biol. Fish.* **46,** 27–36.

Cross, J. N. (1981). "Structure of a Rocky Intertidal Fish Assemblage." Ph.D. dissertation, University of Washington, Seattle.

Ebeling, A. W., and Hixon, M. A. (1991). Tropical and temperate reef fishes: Comparison of community structures. *In* "The Ecology of Fishes on Coral Reefs" (P. F. Sale, Ed.), pp. 509–563. Academic Press, San Diego.

Edwards, T. W., and Horn, M. H. (1982). Assimilation efficiency of a temperate-zone intertidal fish (*Cebidichthys violaceus*) fed diets of macroalgae. *Mar. Biol.* **67,** 247–253.

Emery, A. R. (1973). Comparative ecology and functional osteology of fourteen species of damselfish (Pisces: Pomacentridae) at Alligator Reef, Florida Keys. *Bull. Mar. Sci.* **23,** 649–770.

Eschmeyer, W. N., Herald, E. S., and Hammann, H. (1983). "A Field Guide to Pacific Coast Fishes of North America." Houghton Mifflin, Boston.

Estes, J. A., and Steinberg, P. D. (1988). Predation, herbivory and kelp evolution. *Paleobiology* **14,** 19–36.

Fänge, R., and Grove, D. (1979). Digestion. *In* "Fish Physiology," Vol. VIII, "Bioenergetics and Growth" (W. S. Hoar, D. J. Randall, and J. R. Brett, Eds.), pp. 161–260. Academic Press, New York.

Fris, M. B., and Horn, M. H. (1993). Effects of diets of different protein content on food consumption, gut retention, protein conversion,and growth of *Cebidichthys violaceus* (Girard), an herbivorous fish of temperate zone marine waters. *J. Exp. Mar. Biol. Ecol.* **166,** 185–202.

Gaines, S. D., and Lubchenco, J. (1982). A unified approach to marine plant–herbivore interactions. II. Biogeography. *Annu. Rev. Ecol. Syst.* **13,** 111–138.

Gerking, S. D. (1984). Fish as primary consumers: Assimilation and maintenance ration of an herbivorous fish, *Sarpa salpa,* feeding on a green alga. *Trans. Am. Fish. Soc.* **113,** 378–387.

Gibson, R. N. (1968). The food and feeding relationships of littoral fish in the Banyuls region. *Vie Mil. Ser. A* **19,** 447–456.

Gibson, R. N. (1972). The vertical distribution and feeding relationships of intertidal fish on the Atlantic coast of France. *J. Anim. Ecol.* **41,** 189–207.

Gibson, R. N. (1982). Recent studies on the biology of intertidal fishes. *Oceanogr. Mar. Biol. Annu. Rev.* **20,** 363–414.

Gohar, H. A. F., and Latif, A. F. A. (1959). Morphological studies on the gut of some scarid and labrid fishes. *Pub. Mar. Biol. Stn. Ghardaqa, Red Sea* **10,** 145–190.

Goldschmid, A., Kotrschal, K., and Wirtz, P. (1984). Food and gut length of 14 Adriatic blenniid fish (Blenniidae: Percomorpha; Teleostei). *Zool. Anz.* **213,** 145–150.

Gomon, M. F., and Paxton, J. R. (1985). A revision of the Odacidae, a temperate Australian-New Zealand labroid fish family. *Indo-Pacific Fish.* **8.**

Grossman, G. D. (1986). Food resource partitioning in a rocky intertidal fish assemblage. *J. Zool. London B* **1,** 317–355.

Hallacher, L. E., and Roberts, D. A. (1985). Differential utilization of space and food by inshore rockfishes (Scorpaenidae: *Sebastes*) of Carmel Bay, California. *Environ. Biol. Fish.* **12,** 91–110.

Harris, L. G., Ebeling, A. W., Laur, D. R., and Rowley, R. J. (1984). Community recovery after storm damage: A case of facilitation in primary succession. *Science* **224,** 1336–1338.

Hay, M. E. (1986). Associational plant defenses and the maintenance of species diversity: Turning competitors into accomplices. *Am. Nat.* **128,** 617–641.

Hay, M. E. (1991). Fish-seaweed interactions on coral reefs: Effects of herbivorous fishes and adaptations of their prey. *In* "The Ecology of Fishes on Coral Reefs" (P. F. Sale, Ed.), pp. 96–119. Academic Press, San Diego.

Helfman, G. S., and Clark, J. B. (1986). Rotational feeding: Overcoming gape-limited foraging in anguillid eels. *Copeia* **1986,** 679–685.

Hiatt, R. W., and Strasburg, D. W. (1960). Ecological relationships of the fish fauna on coral reefs of the Marshall Islands. *Ecol. Monogr.* **30,** 65–127.

Hixon, M. A. (1983). Fish grazing and community structure of reef corals and algae: A synthesis of recent studies. *In* "The Ecology of Deep and Shallow Coral Reefs" (M. L. Reaka, Ed.), NOAA Symp. Ser. Undersea Res., Vol. 1, pp. 79–87. Natl. Oceanic Atmos. Adm., Rockville, MD.

Hixon, M. A. (1986). Fish predation and local prey diversity. *In* "Contemporary Studies on Fish Feeding" (C. A. Simenstad and G. M. Cailliet, Eds.), pp. 235–257. Junk, Dordrecht, The Netherlands.

Hixon, M. A., and Brostoff, W. N. (1983). Damselfish as keystone predators in reverse: Intermediate disturbance and diversity of reef algae. *Science* **220,** 511–513.

Hobson, E. S. (1994). Ecological relations in the evolution of acanthopterygian fishes in warm-temperate communities of the northeastern Pacific. *Environ. Biol. Fish.* **40,** 49–90.

Hofer, R., and Schiemer, F. (1981). Proteolytic activity in the digestive tract of several species of fish with different feeding habits. *Oecologia (Berlin)* **48,** 342–345.

Horn, M. H. (1989). Biology of marine herbivorous fishes. *Oceanogr. Mar. Biol. Annu. Rev.* **27,** 167–272.

Horn, M. H. (1992). Herbivorous fishes: Feeding and digestive mechanisms. *In* "Plant-Animal Interactions in the Marine Benthos" (D. M. John, S. J. Hawkins, and J. H. Price, Eds.), pp. 339–362. Clarendon Press, Oxford.

Horn, M. H. (1998). Feeding and digestion. *In* "The Physiology of Fishes" (D. H. Evans, Ed.), pp. 43–63. CRC Press, Boca Raton, FL.

Horn, M. H., and Gibson, R. N. (1990). Effects of temperature on the food processing of three species of seaweed-eating fishes from European coastal waters. *J. Fish Biol.* **37,** 237–247.

Horn, M. H., Mailhiot, K. F., Fris, M. B., and McClanahan, L. L. (1995). Growth, consumption, assimilation and excretion in the marine herbivorous fish *Cebidichthys violaceus* (Girard) fed natural and high protein diets. *J. Exp. Mar. Biol. Ecol.* **190,** 97–108.

Horn, M. H., and Messer, K. S. (1992). Fish guts as chemical reactors: A model of the alimentary canals of marine herbivorous fishes. *Mar. Biol.* **113,** 527–535.

Horn, M. H., Murray, S. N., and Edwards, T. W. (1982). Dietary selectivity in the field and food preferences in the laboratory for two herbivorous fishes (*Cebidichthys violaceus* and *Xiphister mucosus*) from a temperate intertidal zone. *Mar. Biol.* **67,** 237–246.

Horn, M. H., Murray, S. N., Fris, M. B., and Irelan, C. D. (1990). Diurnal feeding periodicity of an herbivorous blenniid fish, *Parablennius sanguinolentus,* in the western Mediterranean. *In* "Trophic Relationships in the Marine Environment" (M. Barnes and R. N. Gibson, Eds.), Proc. 24th Eur. Mar. Biol. Symp., pp. 170–182. Aberdeen University Press, Aberdeen, Scotland.

Horn, M. H., and Neighbors, M. A. (1984). Protein and nitrogen assimilation as a factor in predicting the seasonal macroalgal diet of the monkeyface prickleback. *Trans. Am. Fish. Soc.* **113,** 388–398.

Horn, M. H., Neighbors, M. A., and Murray, S. N. (1986). Herbivore responses to a seasonally fluctuating food supply: Growth potential of two temperate intertidal fishes based on the protein and energy assimilated from their macroalgal diets. *J. Exp. Mar. Biol. Ecol.* **103,** 217–234.

Horn, M. H., Neighbors, M. A., Rosenberg, M. J., and Murray, S. N. (1985). Assimilation of carbon from dietary and nondietary macroalgae by a temperate-zone intertidal fish, *Cebidichthys violaceus* (Girard) (Teleostei: Stichaeidae). *J. Exp. Mar. Biol. Ecol.* **86,** 241–253.

Hotton, N., III, Olson, E. C., and Beerbower, R. (1997). Amniote origins and the discovery of herbivory. *In* "Amniote Origins: Completing the Transition to Land" (S. Sumida and K. L. M. Martin, Eds.), pp. 207–264. Academic Press, San Diego.

Irelan, C. D., and Horn, M. H. (1991). Effects of macrophyte secondary chemicals on food choice and digestive efficiency of *Cebidichthys violaceus* (Girard), an herbivorous fish of temperate marine waters. *J. Exp. Mar. Biol. Ecol.* **153,** 179–194.

John, D. M., and Pople, W. (1973). The fish grazing of rocky shore algae in the Gulf of Guinea. *J. Exp. Mar. Biol. Ecol.* **11,** 81–90.

Johnson, G. D. (1993). Percomorph phylogeny: Progress and problems. *Bull. Mar. Sci.* **52,** 3–28.

Jones, R. S. (1968). Ecological relationships in Hawaiian and Johnston Island Acanthuridae (surgeonfishes). *Micronesica* **4,** 309–361.

Kandel, J. S., Horn, M. H., and Van Antwerp, W. (1994). Volatile fatty acids in the hindguts of herbivorous fishes from temperate and tropical marine waters. *J. Fish Biol.* **45,** 527–529.

Kapoor, B. G., Smit, H., and Verighina, I. A. (1975). The alimentary canal and digestion in teleosts. *Adv. Mar. Biol.* **13,** 109–239.

Kotrschal, K. (1989). Trophic ecomorphology in eastern Pacific blennioid fishes: Character transformation of oral jaws and associated change of their biological roles. *Environ. Biol. Fish.* **24,** 199–218.

Kotrschal, K., and Thomson, D. A. (1986). Feeding patterns in eastern tropical Pacific blennioid fishes (Teleostei: Tripterygiidae, Labrisomidae, Chaenopsidae, Blenniidae). *Oecologia (Berlin)* **70,** 367–378.

Lassuy, D. R. (1984). Diet, intestinal morphology, and nitrogen assimilation efficiency in the damselfish, *Stegastes lividus,* in Guam. *Environ. Biol. Fish.* **10,** 183–193.

Lewis, S. M. (1986). The role of herbivorous fishes in the organization of a Caribbean reef community. *Ecol. Monogr.* **56,** 183–200.

Liem, K. (1980). Acquisition of energy by teleosts: Adaptive mechanisms and evolutionary patterns. *In* "Environmental Physiology of Fishes" (M. A. Ali, Ed.), pp. 299–334. Plenum Press, New York.

Littler, M. M., and Littler, D. S. (1980). The evolution of thallus form and survival strategies in benthic marine macroalgae: Field and laboratory tests of a functional form model. *Am. Nat.* **116,** 25–44.

Lobel, P. S. (1981). Trophic biology of herbivorous reef fishes: Alimentary pH and digestive capabilities. *J. Fish Biol.* **19,** 365–397.

Lubchenco, J., and Gaines, S. D. (1981). A unified approach to marine plant-herbivore interactions. I. Populations and communities. *Annu. Rev. Ecol. Syst.* **12,** 405–437.

Luczkovich, J. J., and Stellwag, E. J. (1993). Isolation of cellulolytic microbes from the intestinal tract of the pinfish, *Lagodon rhomboides:* Size-related changes in diet and microbial abundance. *Mar. Biol.* **116,** 381–388.

Mahon, R., and Mahon, S. D. (1994). Structure and resilience of a tidepool fish assemblage at Barbados. *Environ. Biol. Fish.* **41,** 171–190.

Mann, K. H. (1973). Seaweeds: Their productivity and strategy for growth. *Science* **182,** 975–981.

Mattson, W. J., Jr. (1980). Herbivory in relation to plant nitrogen content. *Annu. Rev. Ecol. Syst.* **11,** 119–161.

Mead, G. W. (1970). A history of South Pacific fishes. *In* "Scientific Exploration of the South Pacific" (W. S. Wooster, Ed.), pp. 236–251. National Academy of Sciences, Washington, DC.

Miller, K. A., and Marshall, W. H. (1987). Food habits of large monkeyface prickleback, *Cebidichthys violaceus. Calif. Fish Game* **73,** 37–44.

Montgomery, W. L. (1977). Diet and gut morphology in fishes, with special reference to the monkeyface prickleback, *Cebidichthys violaceus* (Stichaeidae: Blennioidei). *Copeia* **1977,** 178–182.

Montgomery, W. L. (1980). The impact of non-selectve grazing by the giant blue damselfish, *Microspathodon dorsalis,* on algal communities in the Gulf of California, Mexico. *Bull. Mar. Sci.* **30,** 290–303.

Montgomery, W. L., and Gerking, S. D. (1980). Marine macroalgae as foods for fishes: An evaluation of potential food quality. *Environ. Biol. Fish.* **5,** 143–153.

Muñoz, A. A., and Ojeda, F. P. (1997). Feeding guild structure of a rocky intertidal fish assemblage in central Chile. *Environ. Biol. Fish.* **49,** 471–479.

Murray, S. N., and Horn, M. H. (1989). Seasonal dynamics of macrophyte populations from an eastern North Pacific rocky-intertidal habitat. *Bot. Mar.* **32,** 457–473.

Neighbors, M. A., and Horn, M. H. (1991). Nutritional quality of macrophytes eaten and not eaten by two temperate-zone herbivorous fishes: A multivariate comparison. *Mar. Biol.* **108,** 471–476.

Nelson, J. S. (1994). "Fishes of the World," 3rd ed. Wiley, New York.

Norton, S. F. (1995). A functional approach to ecomorphological patterns of feeding in cottid fishes. *Environ. Biol. Fish.* **44,** 61–78.

Ogden, J. C. (1977). Carbonate sediment production by parrotfish and sea urchins on Caribbean reefs. *Stud. Geol. (Tulsa, OK)* **4,** 281–288.

Ogden, J. C., and Lobel, P. S. (1978). The role of herbivorous fishes and urchins in coral reef communities. *Environ. Biol. Fish.* **3,** 49–63.

Orton, R. D. (1989). "The Evolution of Dental Morphology in the Girellidae (Acanthopterygii: Perciformes), with a Systematic Revision of the Girellidae." Ph.D. dissertation, University of California, Los Angeles.

Paine, R. T., and Palmer, A. R. (1978). *Sicyases sanguineus:* A unique trophic generalist from the Chilean intertidal zone. *Copeia* **1978,** 75–81.

Pandian, T. J., and Vivekanandan, E. (1985). Energetics of feeding and digestion. *In* "Fish Energetics: New Perspectives" (P. Tytler and P. Calow, Eds.), pp. 99–124. Johns Hopkins University Press, Baltimore.

Penry, D. L., and Jumars, P. A. (1987). Modeling animals guts as chemical reactors. *Am. Nat.* **129,** 69–96.

Potts, G. W. (1980). The littoral fishes of Little Cayman (West Indies). *Atoll Res. Bull.* **241,** 43–52.

Ragan, M. A. (1981). Chemical constituents of seaweeds. *In* "The Biology of Seaweeds" (C. S. Lobban and M. J. Wynne, Eds.), pp. 589–626. University of California Press, Berkeley.

Ralston, S. L., and Horn, M. H. (1986). High tide movements of the temperate-zone herbivorous fish *Cebidichthys violaceus* (Girard) as determined by ultrasonic telemetry. *J. Exp. Mar. Biol. Ecol.* **98,** 35–50.

Randall, J. E. (1967). Food habits of reef fishes of the West Indies. *Stud. Trop. Oceanogr.* **5,** 665–847.

Rimmer, D. W., and Wiebe, W. J. (1987). Fermentative microbial digestion in herbivorous fishes. *J. Fish Biol.* **31,** 229–236.

Russell, B. C. (1983). The food and feeding habits of rocky reef fish of northeastern New Zealand. *N. Z. J. Mar. Freshw. Res.* **17,** 121–145.

Setran, A. C., and Behrens, D. W. (1993). Transitional ecological requirements for early juveniles of two sympatric stichaeid fishes, *Cebidichthys violaceus* and *Xiphister mucosus. Environ. Biol. Fish.* **37,** 381–395.

Sibly, R. M., and Calow, P. (1986). "Physiological Ecology of Animals, an Evolutionary Approach." Blackwell, Oxford.

Springer, V. G. (1995). Blennies. *In* "Encyclopedia of Fishes" (J. R. Paxton and W. N. Eschmeyer, Consultant Eds.), pp. 216–219. Academic Press, San Diego.

Steinberg, P. D. (1985). Feeding preferences of *Tegula funebralis* and chemical defenses of marine brown algae. *Ecol. Monogr.* **55,** 333–349.

Steinberg, P. D. (1988). Effects of quantitative and qualitative variations in phenolic compounds on feeding in three species of marine invertebrate herbivores. *J. Exp. Mar. Biol. Ecol.* **120,** 221–237.

Stepien, C. A. (1990). Population structure, diets and biogeographic relationships of a rocky intertidal fish assemblage in central Chile: High levels of herbivory in a temperate system. *Bull. Mar. Sci.* **47,** 598–612.

Stephenson, W., and Searles, R. B. (1960). Experimental studies on the ecology of intertidal environments at Heron Island. I. Exclusion of fish from beach rock. *Austr. J. Mar. Freshw. Res.* **2,** 241–267.

Stoner, A. W. (1980). The feeding ecology of *Lagodon rhomboides* (Pisces: Sparidae): Variation and functional responses. *Fish. Bull. U.S.* **78,** 337–352.

Taborsky, M., and Limberger, D. (1980). The activity rhythm of *Blennius sanguinolentus* Pallas, an adaptation to its food source? *P.S.Z.N.I. Mar. Ecol.* **1,** 143–153.

Thomson, D. A., and Lehner, C. E. (1976). Resilience of a rocky intertidal fish community in a physically unstable environment. *J. Exp. Mar. Biol. Ecol.* **22,** 1–29.

Urquhart, K. A. F. (1984). "Macroalgal Digestion by *Cebidichthys violaceus,* a Temperate Marine Fish with Highly Acidic Stomach Fluids." M.A. thesis, California State University, Fullerton.

White, T. C. R. (1993). "The Inadequate Environment: Nitrogen and the Abundance of Animals." Springer-Verlag, Berlin.

Willis, T. J., and Roberts, C. D. (1996). Recolonisation and recruitment of fishes to intertidal rockpools at Wellington, New Zealand. *Environ. Biol. Fish.* **47,** 329–343.

Winterbottom, R., and McLennan, D. A. (1993). Cladogram versatility: Evolution and biogeography of acanthuroid fishes. *Evolution* **47,** 1557–1571.

Yoshiyama, R. M., Wallace, W. D., Burns, J. L., Knowlton, A. L., and Welter, J. R. (1996). Laboratory food choice by the mosshead sculpin, *Clinocottus globiceps* (Girard) (Teleostei; Cottidae), a predator of sea anemones. *J. Exp. Mar. Biol. Ecol.* **204,** 23–42.

Zander, C. D. (1967). Beiträge zur ökologie und biologie litoralbewohnender Salariidae und Gobiidae (Pisces) aus dem Roten Meer. *Meteor Forsch. Ergebn. Reihe D* **2,** 69–84.

# 12

# Predation by Fishes in the Intertidal

Stephen F. Norton

*Department of Biology, East Carolina University, Greenville, North Carolina*

Amy E. Cook

*Department of Ecology and Evolutionary Biology, University of California, Irvine, California*

## I. Introduction

The intertidal environment presents unique challenges to its inhabitants. Daily and seasonal fluctuations in water coverage, temperature, turbulence, and water chemistry impose special pressures on intertidal organisms, including fishes (e.g., Denny, 1988; Mathieson and Nienhuis, 1991; Raffaelli and Hawkins, 1996). In response, fishes and their potential prey have evolved many unique adaptations to intertidal life, as summarized in other chapters in this book. The interaction of abiotic factors, biological processes, and intrinsic adaptations not only determines the distributional limits of individual species of intertidal fishes, but also drives variation in the distribution and abundance of their potential prey, leading to patchiness on both short (e.g., meters and hours) and long (e.g., thousands of kilometers and months) scales.

Two groups of fishes use the intertidal as a foraging arena; for some intertidal fishes, the residents, their entire juvenile and adult lives are spent interacting with other intertidal organisms (Gibson, 1969). A second group, the transients, includes species for which the intertidal provides a nursery area for juveniles that then migrate into the subtidal as they mature. Other species in the transient group are those primarily subtidal fishes that forage occasionally in the intertidal. Regardless of the group, the dietary patterns that we observe for fishes foraging in the intertidal depend not only on the spatial and temporal distribution of predators and prey, but also on the functional and energetic interactions between predators and prey. In this chapter, we outline the relative importance of the various intertidal habitats as foraging areas for both resident and transient intertidal fishes and we summarize some of the dietary patterns exhibited by fishes that live in these habitats. We also discuss the role of functional and ontogenetic considerations in producing interspecific and intraspecific differences in diet. Finally, we address the potential for predation pressure by fishes to impact the distribution and abundance of other intertidal organisms. Our contribution

*Intertidal Fishes: Life in Two Worlds*

**Table 1.** Recent Studies (Approximately Since 1980) of the Feeding Ecology of Intertidal Fishes

| Study | Predator | Onto. | Habitat | Sites | Seasons | Years | Data | Prey | Other |
|---|---|---|---|---|---|---|---|---|---|
| South Africa | | | | | | | | | |
| Stobbs, 1980 | 1 | Q | TP | Ml | ?? | Ml | FOO, %N | N | |
| Bennett *et al.*, 1983 | A | D | TP | ?? | ?? | ?? | FOO, %V | N | |
| Lasiak, 1986 | A | D | SB | 1 | Ml | 1 | FOO, %W | N | |
| Coetzee, 1986 | 1 | N | TP,SB | 1 | Ml | Ml | FOO, OPS | N | |
| Romer and McLachlan, 1986 | 1 | Q | SB | 1 | Ml | Ml | FOO | Y | 24 h |
| Lasiak and McLachlan, 1987 | A | N | SB | 1 | 1 | 1 | FOO, %W | N | 24 h |
| Harrison, 1991 | 1 | Q | MG | 1 | Ms | 1 | FOO, %N, OPS | Y | |
| Chile | | | | | | | | | |
| Cancino and Castilla, 1988 | 1 | Q | TP | Ml | Ml | Ml | FOO, W, N | N | prey size |
| Stepien, 1990 | A | D | TP | 1 | 1 | 1 | %V (semi) | N | |
| Varas and Ojeda, 1990 | A | N | TP | Ml | Ml | 1 | %W, N | N | |
| Muñoz and Ojeda, 1997 | A | N | TP | Ml | Ml | Ml | %W, N | N | |
| Australia | | | | | | | | | |
| Beumer, 1978 | S | N | MG | 1 | Ms | 1 | FOO | N | |
| Robertson, 1984 | A | D | EG,MF | 1 | Ml | Ml | %V | Y | |
| Robertson and Lenanton, 1984 | A | Q | SB | Ms | Ml | 1 | OPS | N | day/night |
| Robertson and White, 1986 | S | N | EG | 1 | 1 | 1 | %V | N | 24 h |
| Morton *et al.*, 1987 | A | N | SM | 1 | Ms | 1 | %V | N | day/night |
| Morton *et al.*, 1988 | S | N | SM | 1 | Ms | 1 | %V | N | day/night |
| Brewer and Warburton, 1992 | 1 | Q | MG,EG | 1 | Ms | 1 | OPS | Y | |
| California & Baja California | | | | | | | | | |
| Crane, 1981 | 1 | Q | TP | 1 | Ms | 1 | FOO | N | |
| Barton, 1982 | T | N | TP | Ml | 1 | 1 | FOO | N | |
| Horn *et al.*, 1982 | T | N | TP | 1 | Ms | 1 | W | Y | |
| Horn, 1983 | T | N | TP | 1 | Ms | 1 | %W | Y | |
| Freeman *et al.*, 1985 | 1 | Q | TP | 1 | Ms | Ms | FOO, %W | N | sex |
| Grossman, 1986 | A | Q | TP,CB | 1 | Ms | Ml | FOO, %W | N | |
| Wells, 1986 | 1 | N | TP | 1 | Ml | 1 | FOO, N | N | |
| Miller and Marshall, 1987 | 1 | N | TP | Ml | Ml | Ml | FOO | D | |
| Stebbins, 1988 | A | N | TP | 1 | ?? | ?? | FOO | D | |
| Stepien *et al.*, 1988 | 1 | N | TP | 1 | ?? | ?? | %N | N | |
| Ruiz-Campos and Hammann, 1991 | 1 | Q | TP | 1 | Ml | 1 | %N, %V, FOO, IRI | N | |
| Barry and Ehret, 1993 | S | N | TP | Ml | Ml | Ml | %V | D | |
| Setran and Behrens, 1993 | T | D | TP | Ml | Ms | 1 | IRI | D | |

| | | | | | | | | | |
|---|---|---|---|---|---|---|---|---|---|
| Washington/British Columbia | | | | | | | | | |
| Cross, 1981 | A | Q | TP,CB | Ms | Ms | Ml | %N, %W | D | |
| Birtwell *et al.*, 1983 | A | N | MF,SM | Ml | 1 | 1 | N | Q | |
| Hughes, 1985 | T | Q | EG,MF,TP | Ml | Ml | 1 | IRI, %N, %V | N | |
| Armstrong *et al.*, 1995 | 1 | Q | MF,EG | 1 | Ms | 1 | IRI, %N, %W, FOO | Q | |
| Eastern Canada and Gulf of Maine | | | | | | | | | |
| Worgan and Fitzgerald, 1981 | T | N | SM | 1 | 1 | 1 | OPS | N | sex |
| Dutil and Fortin, 1983 | A | Q | SM | Ml | 1 | 1 | N, FOO | N | |
| Delbeek and Williams, 1988 | T | Q | SM | Ms | Ms | 1 | %N | Q | |
| Moring, 1988 | 1 | Q | TP | Ms | ?? | ?? | FOO, %W | N | |
| Poulin and Fitzgerald, 1989 | T | N | SM | 1 | 1 | 1 | %N, FOO | N | |
| Black and Miller, 1991 | A | N | TP | Ms | 1 | 1 | D | Q | |
| Ojeda and Dearborn, 1991 | A | N | TP | 1 | 1 | 1 | FOO, %N | N | |
| Rangeley and Kramer, 1995 | 1 | N | TP | Ml | Ms | Ml | FOO | N | |
| Western North Atlantic (Cape Cod to Florida) | | | | | | | | | |
| Kneib and Stiven, 1978 | 1 | Q | SM | 1 | Ms | 1 | FOO | N | |
| McDermott, 1983 | A | N | SB | 1 | Ml | Ml | %W | Y | |
| Kleypas and Dean, 1983 | 1 | N | SM | 1 | 1 | 1 | FOO, %N | N | |
| Feller *et al.*, 1990 | 1 | N | SM | 1 | 1 | Ms | %N | Y | 24 h |
| Feller and Coull, 1995 | 1 | N | SM | 1 | 1 | Ml | N | Y | 24 h |
| Gulf of Mexico | | | | | | | | | |
| Modde and Ross, 1983 | A | Q | SB | Ml | Ml | 1 | %V | N | |
| McMichael and Ross, 1987 | T | Q | SB | Ml | Ml | Ml | %V, %N, FOO | N | |
| Thayer *et al.*, 1987 | 1 | Q | MG/EG | Ml | Ml | 1 | FOO | N | |
| Rozas and LaSalle, 1990 | 1 | N | SM | Ml | Ms | 1 | N, %V, FOO | N | flood/ebb |
| Ley *et al.*, 1994 | A | N | MG | Ms | Ms | 1 | OPS | N | |
| Eastern North Atlantic (Portugal to Norway) | | | | | | | | | |
| Collins, 1981 | 1 | N | TP | Ms | Ms | 1 | OPS, FOO, %N | N | sex |
| Ansell and Gibson, 1990 | T | Q | SB | 1 | 1 | 1 | %N, V | Y | |
| O'Farrell and Fives, 1990 | S | Q | TP | Ms | Ms | 1 | FOO, N, OPS | N | |
| Deady and Fives, 1995 | 1 | Q | TP | Ms | Ml | Ml | FOO, OPS | N | |
| Antarctic and Sub-Antarctic | | | | | | | | | |
| Duarte and Moreno, 1981 | 1 | N | TP | Ms | 1 | 1 | FOO, %W | Y | |
| Blankley, 1982 | A | D | TP | Ml | Ml | 1 | %W, FOO, N | N | |
| Blankley and Grindley, 1985 | 1 | N | TP | ?? | ?? | 1 | %W | Y | |

*continues*

**Table 1.** *Continued*

| Study | Predator | Onto. | Habitat | Sites | Seasons | Years | Data | Prey | Other |
|---|---|---|---|---|---|---|---|---|---|
| Indian Ocean | | | | | | | | | |
| Pillai *et al.*, 1992 | A | N | CR | 1 | 1 | 1 | D | N | |
| Persian Gulf | | | | | | | | | |
| Wright, 1988 | A | Q | MF | 1 | Ml | 1 | FOO, OPS | N | |
| Wright, 1989 | 1 | Q | MF | 1 | Ml | 1 | OPS | N | |
| Ismail and Clayton, 1990 | 1 | D | TP | Ml | Ms | 1 | FOO | N | |
| Barak *et al.*, 1994 | 1 | Q | MF | 1 | Ml | 1 | OPS, FOO | N | |
| Northwest Pacific | | | | | | | | | |
| Tsurpalo, 1993 | T | N | TP | Ml | 1 | 1 | FOO.OPS | N | |
| Noichi *et al.*, 1993 | A | N | SB | 1 | 1 | Ml | %N, FOO, OPS | N | |
| Malaysia | | | | | | | | | |
| Sasekumar *et al.*, 1984 | A | N | MG | 1 | 1 | Ms | %V | N | |
| Ong and Sasekumar, 1984 | A | N | MG | Ml | Ml | 1 | %V | N | |
| Tropical Atlantic | | | | | | | | | |
| Tararam and Wakabara, 1982 | 1 | N | TP | 1 | Ml | 1 | FOO, %V | N | |

*Note.* Predator = types of fish sampled: A = all, S = subset, T = taxonomic group, 1 = single predator; Onto. = Ontogenetic data: N = none, D = descriptive, Q = quantitative; Habitat = Habitat coverage: TP = tidepool and rocky intertidal, CB = cobble beach, SB = sand beach, EG = eelgrass, MF = mudflat, CR = coral rubble, SM = salt marsh, MG = mangrove creek/forest; Sites = areas sampled: 1 = one site, Ms = multiple sites with separated data, Ml = multiple sites with lumped data; Seasons = seasons sampled: 1 = one season, Ms = multiple seasons with separated data, Ml = multiple seasons with lumped data; Years = years of sampling: 1 = one year, Ms = multiple years with separated data, Ml = multiple years with lumped data; Data = type of diet data: D = descriptive, FOO = frequency of occurrence, W = gravimetric importance, %W = relative gravimetric importance, V = volumetric importance, %V = relative volumetric importance, N = number of prey, %N = relative numeric importance, IRI = index of relative importance, OPS = other point schemes; Prey = estimates of prey abundance: N = none, D = descriptive, Q = quantitative; Other = other features.

updates and extends earlier summaries of predation by intertidal fishes, including Gibson (1969, 1982) and Horn and Gibson (1988).

## II. Recent Studies of the Diets of Intertidal Fishes

Continued investigation over the past 20 years into the foraging biology of intertidal fishes has added tremendously to the body of knowledge that was summarized and synthesized by Gibson (1969, 1982). In particular, these recent studies have expanded coverage of different habitats and geographic regions and have provided additional support for the prevalence of spatial, seasonal, and ontogenetic shifts in the diets of intertidal fishes (Table 1). In several cases, attempts have been made to examine some of the underlying forces responsible for these shifts by sampling the prey base available to intertidal predators or to estimate the potential impact of fish predation on prey populations. We have listed many of these recent studies as well as some of their attributes (geographic location, types of fish species included, major habitat sampled, consideration of spatial, temporal, and ontogenetic variation, and method of quantifying diet and prey) in Table 1. The references in Table 1 emphasize studies published since 1980 and, in conjunction with the references contained in Gibson (1969, 1982), provide an extensive overview of the published studies of the diets of intertidal fishes.

In the past 2 decades no region of the world has seen greater advancement in our understanding of diets of intertidal fishes than has the Southern Hemisphere, especially for fish communities in temperate waters. Several studies have described the diets of the rich intertidal fish fauna of South Africa from sandy beaches (Lasiak, 1983, 1986; Lasiak and McLachlan, 1987) and rocky intertidal/nearshore subtidal habitats (Stobbs, 1980; Butler, 1982; Bennett *et al.,* 1983; Coetzee, 1986). Research in Australia has provided new data on the diets of fish communities in salt marsh and mangrove creeks (Beumer, 1978; Morton *et al.,* 1987, 1988; Brewer and Warburton, 1992), eelgrass meadows (Robertson, 1980, 1984; Robertson and White, 1986), and sand beaches (Robertson and Lenanton, 1984). Several studies have focused on diets of the rocky intertidal fishes in central Chile (Cancino and Castilla, 1988; Stepien, 1990; Varas and Ojeda, 1990; Muñoz and Ojeda, 1997). Some dietary information exists even for the intertidal ichthyofauna of the sub-Antarctic (Duarte and Moreno, 1981; Blankley, 1982; Blankley and Grindley, 1985). In contrast to progress in these regions, the diets of intertidal fishes from the southern Atlantic (e.g., Argentina, Brazil, Namibia) have not been well described (but see Tararam and Wakabara, 1982).

In the Northern Hemisphere much of the new information on the diets of intertidal fishes has come from the northeast Pacific. These studies have added detail (larger sample sizes, broader geographic coverage, examination of seasonal variation, etc.) to the earlier studies summarized by Gibson (1969, 1982). Special attention has been given to the rocky intertidal of California (Crane, 1981; Barton, 1982; Horn *et al.,* 1982; Horn, 1983; Freeman *et al.,* 1985; Grossman, 1986; Wells, 1986; Stebbins, 1988; Stepien *et al.,* 1988; Barry and Ehret, 1993; Setran and Behrens, 1993) and of Baja California (Ruiz-Campos and Hammann, 1991) and to rocky intertidal and cobble beaches (Cross, 1981, 1982) and mudflats and eelgrass beds (Birtwell *et al.,* 1983; Hughes, 1985; Armstrong *et al.,* 1995) in Washington. The diets of intertidal fishes farther north in Alaska and British Columbia have not been well characterized, nor is there much information on the diets of intertidal fishes from the northwest Pacific (i.e., Kamchatka Peninsula, Japan, Korea, China, but see Noichi *et al.,* 1993; Tsurpalo, 1993).

Quantitative descriptions of the diets of fishes from the temperate North Atlantic have existed for quite a long time (see Gibson, 1969, 1982). Recent studies in eastern Canada and the Gulf of Maine have added information on the diets of fishes from salt marshes (Worgan and Fitzgerald, 1981; Dutil and Fortin, 1983; Delbeek and Williams, 1988; Poulin and Fitzgerald, 1989) and the rocky intertidal (Moring, 1988; Black and Miller, 1991; Ojeda and Dearborn, 1991; Rangely and Kramer, 1995). Other studies have expanded our knowledge of the diets of the sandy beach fishes of the central United States and Gulf of Mexico (e.g., Kleypas and Dean, 1983; McDermott, 1983; Modde and Ross, 1983; McMichael and Ross, 1987; Feller and Coull, 1995). Because the diets of the fishes of the eastern North Atlantic had been well characterized in studies before 1980, only a few new quantitative studies (e.g., Collins, 1981; Ansell and Gibson, 1990; O'Farrell and Fives, 1990; Deady and Fives, 1995) have focused on these areas in the past 15 years; instead, attention has been devoted to more experimental studies (see Section VII).

The diets of the intertidal fishes from the tropics are very poorly known. Much of the attention of tropical ichthyologists has focused on coral reef fishes, primarily on subtidal populations. Many studies have documented the diets of subtidal populations of coral reef fishes (e.g., Randall, 1967; Williams and Hatcher, 1983; Thresher and Colin, 1986; Wainwright, 1988), but the diets of the rich intertidal ichthyofauna of coral reefs, which includes both resident species and the transient juveniles of nominally subtidal species (e.g., Mahon and Mahon, 1994), have not yet been widely investigated. Two recent contributions that provide some indications of the likely diets of intertidal coral reef fishes include Sano *et al.* (1984) and Pillai *et al.* (1992). Sano *et al.* (1984) described the diets of over 188 fish species from shallow (1–2.5 m) coral reefs off Okinawa and classified them into trophic categories. Pillai *et al.* (1992) included qualitative descriptions of the diet of intertidal fishes from a coral atoll in the Indian Ocean. One surprising conclusion from both of these studies is the lack of planktivores among the intertidal fishes, even though this is a common foraging strategy among subtidal populations.

Other tropical habitats have also been poorly sampled.

While mangals or mangrove forests dominate 60–75% of the shoreline in the tropics (Walsh, 1974), only a few studies have reported on the diets of fishes from these areas (e.g., Beumer, 1978; Sasekumar *et al.,* 1984; Thayer *et al.,* 1987; Brewer and Warburton, 1992; Ley *et al.,* 1994) with the notable exception of mudskippers (see reviews in Gibson, 1969; Clayton, 1993). Coverage of the diets of fishes from soft sediment habitats (mud and sand) in the tropics has also been sparse. Some initial samplings from these habitats exist for the Persian Gulf (Wright, 1988, 1989; Ismail and Clayton, 1990; Barak *et al.,* 1994), the Indian Ocean (Pillai *et al.,* 1992), and the tropical Atlantic Ocean (Sanusi, 1980). Large underexplored areas of the tropical Atlantic Ocean, the Indo-Pacific region, the Indian Ocean, and the Red Sea present opportunities for future studies.

## III. Influence of Habitat on the Diets of Intertidal Fishes

The intertidal is not a homogeneous environment and major intertidal habitat types are shaped by their physical characteristics (e.g., substrate type, wave energy) and by dominant organisms (e.g., seagrasses, mangroves). Most discussions of the predominant intertidal habitat categories include the rocky intertidal, sandy beaches, mudflats, seagrass beds, mangals, and salt marshes. The distribution of an individual fish species is typically confined to only one or two of these major habitats and the distribution of a single individual may be even more restricted. Prey distributions are also delimited by environmental characteristics; thus the overlap of the distributions of potential prey and the potential foraging areas of predators defines the widest possible spectrum of prey items from which only a small subset are found in the final diet.

Within each major habitat category, gradients of biotic and abiotic factors produce consistent patterns in the distribution and abundance of intertidal organisms that can be used to characterize distinct zones (reviewed by geographic region in contributions in Mathieson and Nienhuis, 1991). Zonation within a habitat provides a finer scale modification on the prey suite available to foraging fish. While a single intertidal prey species may be restricted biogeographically, functional groups (sometimes with tight taxonomic affinities, but sometimes as a result of convergence) appear consistently within the same major zones worldwide. For example, mussels (*Mytilus* spp. in the temperate Northern Hemisphere, *Perna perna* in the tropical and subtropical Atlantic coast of Africa, *Choromytilus meridionalis* in southern Africa, and *Perumytilus purpuratus* in Chile) and oysters (*Chama* spp. and *Ostrea* spp. in the tropics) often form bands in the lower eulittoral areas of the rocky intertidal, but can be replaced by analogous organisms like tunicates (*Pyura* spp.) in parts of southern Africa and Chile (Mathieson and Nienhuis, 1991). Thus it is not surprising to find convergence or similarities in the diets of fishes that forage in similar intertidal zones across the globe.

The richness of an intertidal habitat as a foraging arena for fishes depends not only on the primary productivity and food web processes intrinsic to that intertidal area, but also on productivity brought into the intertidal via waves and currents from the nearshore subtidal and pelagic zones and to a lesser extent on productivity from terrestrial sources, primarily via insects (Mathieson and Nienhuis, 1991; Raffaelli and Hawkins, 1996). Among and within the major intertidal habitats, the relative importance of these sources of productivity will vary spatially and temporally. Spatial gradients in productivity may occur not only along the shore, but also as a function of height above the low-tide line. Temporal variation in productivity can be measured not only daily and seasonally, but also interannually, often in response to differences in the intensity of physical factors (e.g., wave action, nutrient availability, temperature, light levels) impacting not only the intertidal, but also offshore areas (e.g., upwelling). In the following sections we will describe the major sources of productivity, outline the major potential prey groups, and present some examples of fish diets for each of the six major intertidal habitats.

## A. Rocky Intertidal

Over the years investigations of the factors that influence the distribution and abundance of rocky intertidal organisms have played a key role in the development of many fundamental principles in marine ecology, such as competitive dominance (Connell, 1961), intermediate disturbance hypothesis (Sousa, 1979), keystone predators (Paine, 1974), and supply-side ecology (Gaines and Roughgarden, 1985) (see also contributions in Moore and Seed, 1986; Mathieson and Nienhuis, 1991; and Raffaelli and Hawkins, 1996). These studies have typically focused on the major occupiers of space in the rocky intertidal, including macroalgae and surfgrasses and large sessile invertebrates (e.g., barnacles, mussels, tube-dwelling polychaetes, tunicates). Even when not directly consumed by fishes, these sessile invertebrates, algae, and plants provide habitat and food for a wide variety of mobile invertebrates, including gammarid and caprellid amphipods, isopods, crabs, shrimp, errant polychates, gastropods, and chitons, that are major components of the diets of rocky intertidal fishes (see below). Among the major intertidal habitats, the rocky intertidal probably presents the greatest taxonomic breadth and functional diversity of potential prey for intertidal fishes.

Predictable patterns in the distribution and abundance of intertidal organisms have lead to the description of analogous zones that are typical of rocky intertidal habitats across the globe (see Mathieson and Nienhuis, 1991; Raffaelli and Hawkins, 1996). Although several schemes for defining these zones exist (see discussions in Russell, 1991), the general order and character is reasonably consistent worldwide and we use the system proposed by Stephenson and Stephenson (1972). Moving from upland terrestrial habitats to the subtidal, the first defined zone is the supralittoral fringe with encrusting lichens and blue-green algae, semiterrestrial gammarid amphipods, and littorine snails. The next zone is the midlittoral zone with bands of barnacles (e.g., *Balanus* spp., *Chthamalus* spp., *Tetraclita* spp.) and then mussels (*Mytilus* spp. and related genera). The lowest zone in the intertidal is the infralittoral fringe, which is dominated by diverse macroalgae in most areas and may include extensive populations of tunicates in the Southern Hemisphere. Tidepools within the rocky intertidal provide refuges for many fishes at low tide, while others find refuge under algae or boulders (Martin, 1995). Rocky intertidal fishes may forage over all these zones, or may be restricted due to tide level or low vagility to a subset of intertidal zones and therefore a restricted spectrum of potential prey.

Primary producers in the rocky intertidal include both the macroscopic forms of surfgrasses, and those of red, green, and brown algae as well as the microscopic stages of the three algal groups, plus diatoms, dinoflagellates, and cyanobacteria. Factors affecting the productivity and biomass of intertidal algae and seagrasses (e.g., light levels, nutrients, hydrodynamic conditions) have a strong influence over the size of the energetic pie available for higher trophic levels (Raffaelli and Hawkins, 1996). Fluctuations in productivity and biomass of these primary producers affect fishes directly (i.e., herbivorous fishes) and indirectly through impacts on energy flow, especially through detritally based food webs (i.e., carnivorous fishes). Seasonal changes in productivity have been linked to seasonal changes in fish diets (e.g., Bennett *et al.,* 1983; Grossman, 1986). Direct utilization of this primary productivity by intertidal fishes is far more common in tropical and warm temperate habitats than in cold temperate and boreal areas, in spite of the abundance of algae and surfgrasses at these higher latitudes. Consideration of the underlying forces leading to this strong latitudinal cline in the intensity and richness of herbivory by intertidal fishes has generated considerable discussion (see Horn, 1989; Horn and Ojeda, Chapter 11, this volume).

Other dominant space-occupiers in the rocky intertidal are sessile macroinvertebrates such as mussels, barnacles, and tube-building polychaetes. These organisms not only represent potential food sources for intertidal fishes but, along with macroalgae and eelgrasses, provide three-dimensional structure that forms critical microhabitats for some intertidal fishes and for other invertebrates. Both barnacles and mussels possess adaptations such as strong physical attachment to the substrate and tough outer shells that enable them to withstand the intense hydrodynamic forces that often occur in the rocky intertidal (Denny, 1988); these adaptations also serve to restrict the list of potential predators (including fishes) to those that have specific adaptations to remove these prey from the substrate and then process them (see Section IV).

Small, mobile crustaceans are important components of the diets of many rocky intertidal fishes. The populations of these mobile crustaceans fall or rise in conjunction with changes in primary production and especially in response to fluctuations in detrital food webs (Laur and Ebeling, 1983). The short generation time and direct development of many crustaceans (e.g., amphipods and isopods) allows populations to respond quickly to local fluctuations in productivity, such as upwelling or deposition of drift algae (Griffiths *et al.*, 1983; Field and Griffiths, 1991). While copepods have planktonic dispersal, their short generation times also allow their populations to respond quickly to productivity fluctuations. The relative abundance of some characteristic crustacean groups can provide insight into the importance of intrinsic versus extrinsic sources of productivity in the rocky intertidal. For example, comparisons of the relative abundance of harpacticoid copepods, which are benthically oriented and with many species restricted to the intertidal (Coull and Feller, 1988; Feller *et al.*, 1990), versus the abundance of calanoid copepods, which are more characteristic of pelagic habitats, provide indications as to the importance of intrinsic versus offshore production in intertidal habitats. While decapod crustaceans have year-round recruitment patterns in tropical waters (albeit with seasonal peaks), in temperate waters there is usually a distinct settlement period that coincides with the yearly peak in productivity; this annual reproductive cycle limits numerical responses to short-term or local variations in productivity, but does make their populations sensitive to interannual variation.

Chitons and gastropods, including herbivorous limpets and littorine snails and predatory whelks, are also important components of the mobile fauna in the rocky intertidal. Temporal and spatial variability in the distribution and abundance of these prey may reflect a mix of local and regional phenomena influencing both adults and larvae. Like many crustaceans, adult gastropods and chitons often demonstrate limited mobility (Wells, 1980; Garrity and Levings, 1983) and post-settlement populations may show significant variability due to differences in local living conditions (food, predators, etc.) (Faller-Fritsch and Emson, 1986; Jardine, 1986). This may be reinforced for species with direct larval development that act more as a closed population. In species with planktonic larvae, local recruitment may depend less on the reproductive output of local adults and more on the factors effecting reproductive output, larval survival, and supply in the metapopulation (Gaines and Lafferty, 1995; Havenhand, 1995).

Small crustaceans are clearly the predominant prey items (Figures 1a–1c) in the diets of rocky intertidal fishes from several sites around the world: Chile (Varas and Ojeda, 1990), South Africa (Bennett *et al.*, 1983), and California (Grossman, 1986). Depending on the schemes used in individual studies to categorize prey (i.e., a combined decapod category versus separate categories for crabs and shrimp), either gammarid amphipods or decapods are the top-ranked prey for the communities as a whole and for the numerically dominant predators in each community. Differences among these sites also exist. Isopods are a highly ranked prey category for most South African species, but are a less important, albeit widespread, diet item of intertidal fishes in California and Chile. Macroalgae are important food items for a few predatory species in both South Africa and California, but are exploited more generally in the rocky intertidal of Chile (see also Stepien, 1990). In spite of their abundance, few intertidal fishes feed on barnacles or cnidarians, although specialists may be found in these communities (e.g., Yoshiyama *et al.*, 1996a,b). Piscivores are rare in all three temperate rocky intertidal communities. Molluscivores are also rare in these temperate systems (except limpet specialists like clingfish), but may be more common in the rocky intertidal of the tropics (Vermeij, 1978; Bertness *et al.*, 1981; Menge *et al.*, 1986; but see Ortega, 1986). These overall patterns are similar to those described by Cross (1982) who compared the diets of rocky intertidal fishes from southern California, Washington, and the

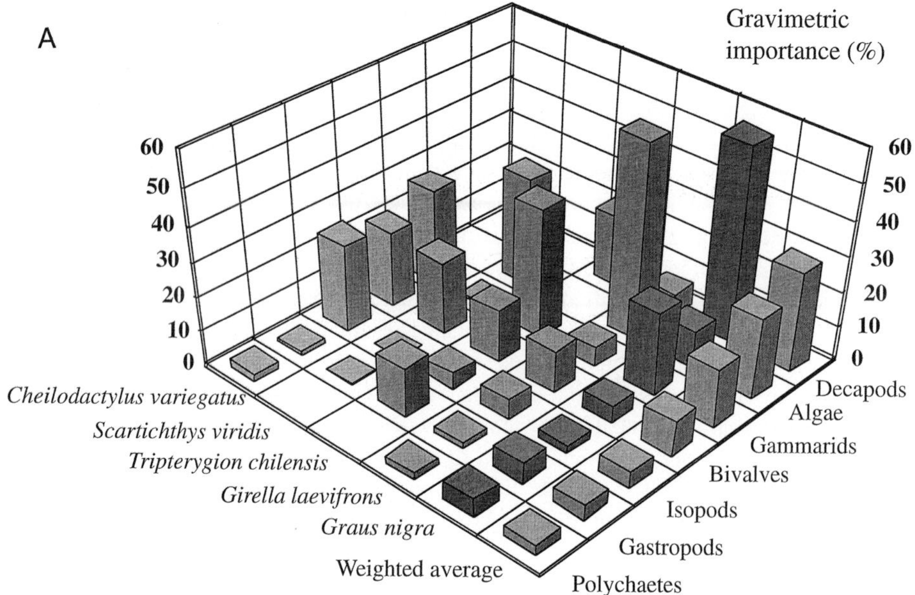


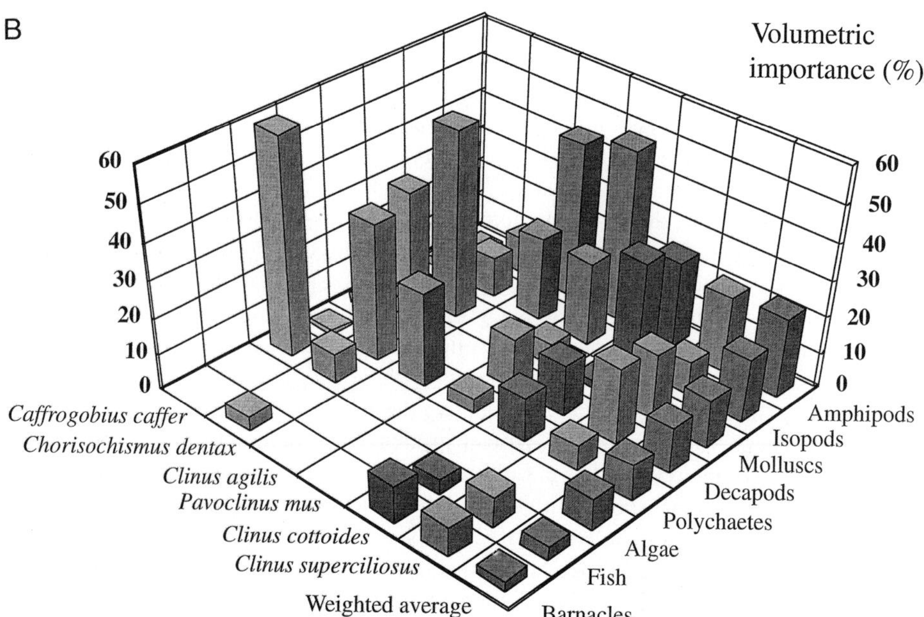


**Figure 1.** Diets of rocky intertidal fishes from (a) Chile (data from Varas and Ojeda, 1990), (b) South Africa (data from Bennett *et al.,* 1983), and (c) central California (data from Grossman, 1986). Included with each is an average composite diet, calculated by weighting the diets of each predator species by the number of individuals included in the sample. Values from this weighted average were used to rank the importance of the various prey categories.

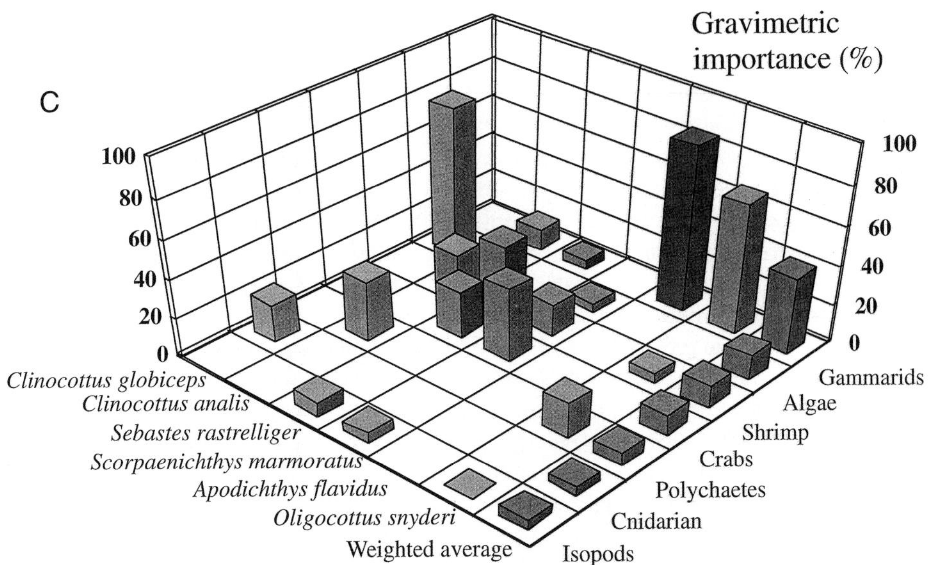


**Figure 1.** *Continued*

Atlantic coast of France. In rocky intertidal communities, most of the resources utilized by fishes are part of *in situ* food webs and are not prey (e.g., pelagic mysids, calanoid copepods) carried into the intertidal from offshore sources.

## B. Sandy Beaches

The distribution of potential prey available to fishes in sandy beach habitats has also been well studied (e.g., McLachlan and Erasmus, 1983; Brown and McLachlan, 1990; contributions in Mathieson and Nienhuis, 1991). As in the rocky intertidal, interactions among the intrinsic physiological limits of individual species, spatial and temporal variation in physical conditions, and the presence and abundance of other members of the sandy beach fauna play important roles in determining the distribution and abundance of potential prey communities. Several schemes have been proposed to describe patterns of zonations on sandy beaches in which particle size and degree of wave exposure define the gradients (see Raffaelli and Hawkins, 1996). Productivity on sandy beaches may range from domination by intrinsic sources (e.g., production of diatoms) on dissipative beaches to domination by imported material from offshore areas either as phyto- and zooplankton or as drift macroalgae on reflective beaches (Brown and McLachlan, 1990; Jaramillo *et al.,* 1993). This variation leads to differences in the types and abundance of prey in the intertidal and to differences in the diets of intertidal fishes.

On sandy beaches, many potential prey are infaunal, with most of the biomass buried in the sediments. As outlined by Brown and McLachlan (1990) the dominant macroscopic taxa in this category include burrowing bivalves (e.g., *Donax* and *Tellina* spp.), tube-dwelling polychaetes (e.g., *Arenicola, Scolelepis, Spio, Glycera* spp.), and burrowing shrimp (*Callianassa* spp.). In high-energy habitats sand crabs (i.e., *Emerita* and *Hippa* spp.) may often provide a significant fraction of the infaunal biomass. Other potentially important macrofauna include sand dollars, holothurians, and ophiuroids. Meiofauna (harpacticoid copepods, cumaceans, tanaids, nematodes, etc.) are also important members of this infaunal community. On the surface, mobile macrocrustaceans, such as benthic mysids, shrimp, crabs, isopods, and gammarid amphipods, may provide significant energy sources for fishes. Gastropods do not usually form an important component of the fauna of sandy beaches. If inputs of offshore productivity are high, then prey characteristic of more pelagic habitats, such as calanoid copepods, invertebrate larvae, and pelagic mysids, become available to fishes foraging in the intertidal. On some sandy beaches drift algae collect at the intertidal/subtidal margin. Significant spatial and temporal variation in the production and deposition of this material makes this an ephemeral, but valuable resource (Robertson and Lenanton, 1984). Where deposition occurs decomposition of macrophytes overwhelms the usual preeminence of microscopic algae as the primary energy source supporting prey communities on sandy beaches. These deposits of algae can attract not only large numbers of semiterrestrial amphipods, isopods, and crabs, but also a variety of terrestrial insects, including beetles, wrack flies, and crickets, which may then be available to intertidal fishes (Stenton-Dozey and Griffiths, 1983; Brown and McLachlan, 1990).

On sandy shores, predators may specialize either on small planktonic prey (e.g., calanoid copepods, mysids) or on larger benthic prey (e.g., gammarid amphipods, bivalves, sand crabs, polychaetes) (Lasiak, 1983; McDermott, 1983; Modde and Ross, 1983; Robertson and Lenanton, 1984). This can be seen clearly in Figures 2a and 2b by comparing the top two prey types in diets of Gulf of Mexico fishes (calanoid copepods and pelagic mysids) with those of New Jersey fishes (polychaetes and gammarid amphipods). Opportunistic foraging on planktonic resources by fishes that normally feed on benthic prey has also been reported from South Africa (Lasiak and McLachlan, 1987). In areas of sandy beach that accumulate large quantities of detached macrophytes, a detrital-based food web develops and gammarid amphipods and errant polychates from the macrophytes are the major ele-

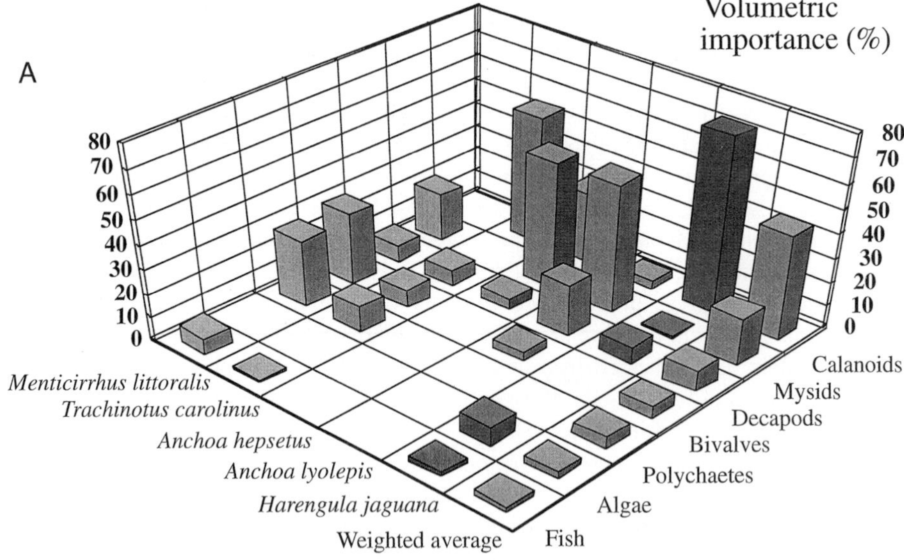


**Figure 2.** Diets of sandy beach fishes from (a) the Gulf of Mexico (data from Modde and Ross, 1983), (b) New Jersey (data from McDermott, 1983), and (c) Australia (data from Robertson and Lenanton, 1984). Included with each is an average composite diet, calculated by weighting the diets of each predator species by the number of individuals included in the sample. Values from this weighted average were used to rank the importance of the various prey categories.

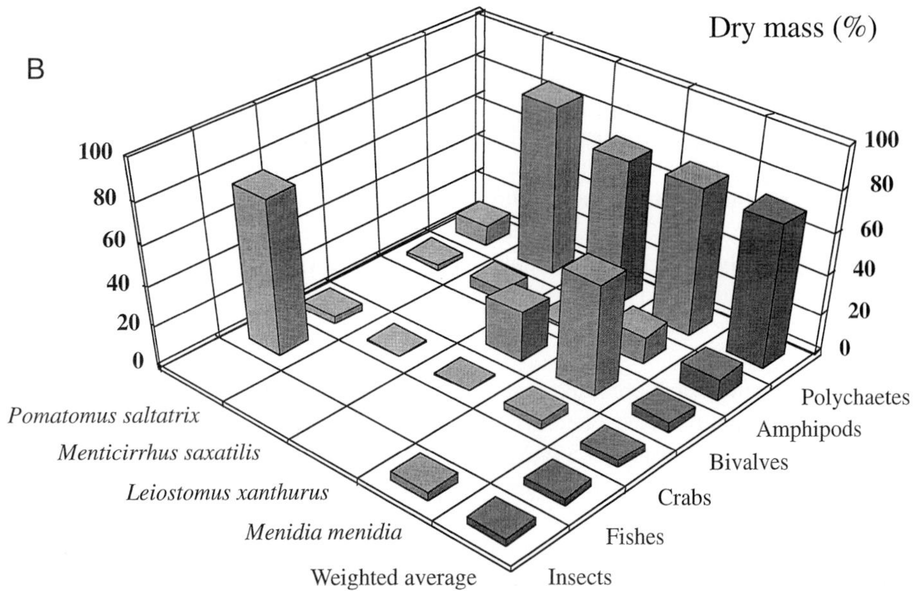


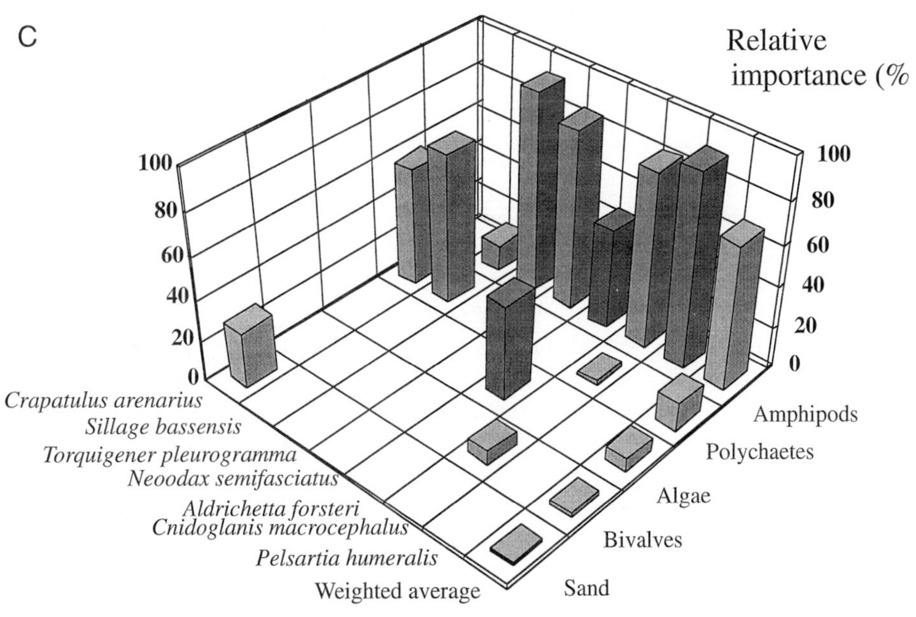


**Figure 2.** *Continued*

ment in the diet of fishes that forage in this area. For example, while polychaetes are the dominant prey of *Sillago bassensis* collected in these drift areas (Figure 2c), bivalves are the dominant prey of individuals collected from sandy beaches (Robertson and Lenanton, 1984). Herbivorous fishes from sandy beaches include some species (e.g., *Aldrichetta forsteri* from Australia, Figure 2c) that harvest macroscopic algae and others (e.g., mullet) that ingest diatoms (Romer and McLachlan, 1986).

## C. Mudflats

Intrinsic sources of primary productivity in mudflats include microscopic sources (i.e., diatoms, cyanobacteria, archaebacteria) and macroalgae that may form thick layers over the mud surface (e.g., *Gracilaria* spp., *Ulva* spp.). Chemoautotrophic archaebacteria found below the redox potential discontinuity provide a source of organic molecules unique to mudflats among the major intertidal habitats (Fenchel and Blackburn, 1979). As in sandy beach habitats, the mudflat fauna includes both epifauna species (mobile crustaceans, gastropods, errant polychaetes) and infaunal species (tube-dwelling polychaetes, burrowing amphipods and callianassid shrimp, bivalves). In the absence of sharp physical gradients, biological zones are not as distinct as in rocky intertidal or sandy beach habitats (e.g., Woodin, 1974).

Infaunal prey are even more important in the diets of fishes collected from mudflats than from other intertidal habitats. Flounders, one of the best studied of the fishes associated with mudflats, feed extensively on burrowing polychaetes and bivalves, often nipping off exposed siphons, tentacles, or hind parts (Macer, 1967; Kuipers, 1977; deVlas, 1979a). Other important prey in the diet of flounders include gammarid amphipods and crangonid shrimp. These crustaceans also dominate the diets of groups of fishes collected from mudflats off Wales (Macer, 1967). Meiofauna, especially harpacticoid copepods and nematodes, are a major component of the diets of some mudflat fishes (e.g., Feller *et al.,* 1990; Feller and Coull 1995; Coull *et al.,* 1995).

## D. Seagrass Beds

Although most seagrass/eelgrass beds are sublittoral, they often extend into adjoining intertidal areas. Most seagrass beds are found in soft sediments, including sand and mud substrates, but surfgrasses may also be present in lower rocky intertidal areas. Seagrass beds alter the hydrodynamic environment, trapping fine particles and organic materials. Not only are seagrasses themselves a potential source of food for herbivorous fishes (e.g., Horn, 1989; Luczkovich and Stellwag, 1993; Luczkovich *et al.,* 1995) and the base of a rich detrital food chain, but also they provide an attachment surface for a variety of epiphytic/epizootic organisms, such as red and green algae, diatoms, bryozoans, and hydroids (McRoy and Helfferich, 1977; Zieman, 1982). They also provide shelter among their roots and blades for shrimp, gammarid and caprellid amphipods, harpacticoid copepods, gastropods, young scallops, and juvenile fishes. Other organisms in seagrass beds are more characteristic of the muddy or sandy sediments in which seagrasses are rooted.

As in the other intertidal habitats, mobile crustaceans are the central elements in the diet of seagrass fishes (Figure 3a). In particular, shrimp have a key place in the diets of these fishes from Australia (Robertson, 1984, 1988). Polychaetes from the sediments and seagrass roots also are important prey items. Some seagrass fishes are specialized as herbivores on seagrass blades and algae. In some seagrass beds, production from terrestrial or freshwater habitats may be transferred to marine intertidal fishes through ingestion of adult and larval insects.

## E. Mangals

Mangals occur in sheltered, soft sediment, intertidal areas of the tropics and subtropics. Determining the relative contributions of mangrove plants, microscopic algae, and macroscopic algae to the overall primary production of the marine component of mangals is

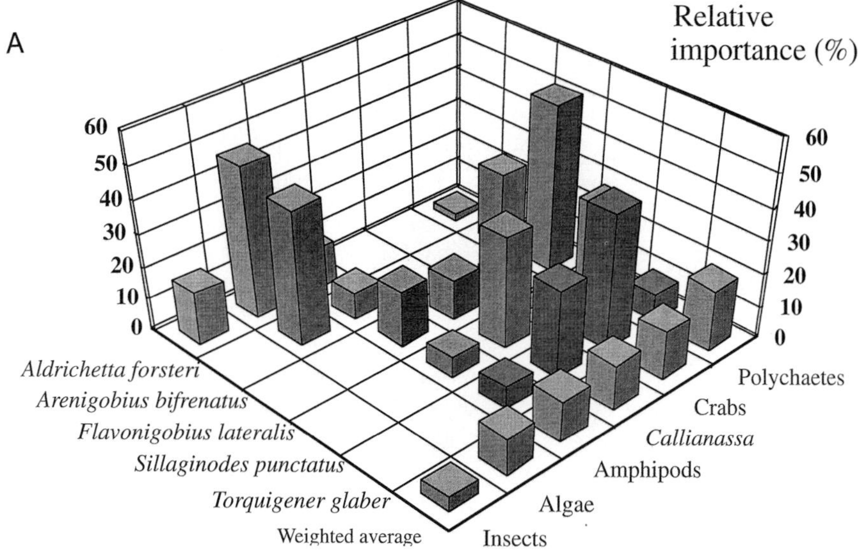


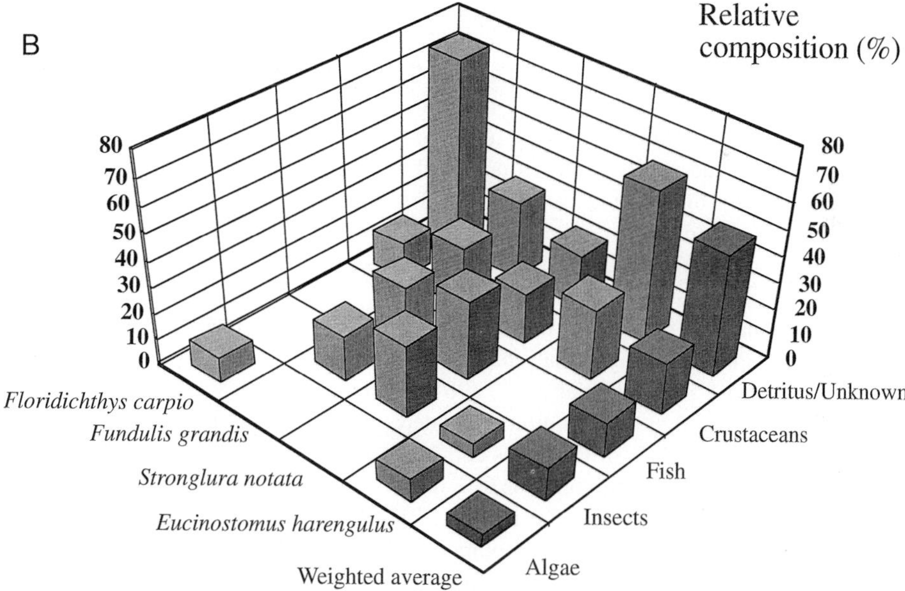


**Figure 3.** Diets of fishes from (a) eelgrass and mudflats from Australia (data from Robertson 1984), (b) mangrove forests of Florida Bay (data from Ley *et al.*, 1994), and (c) an Australian salt marsh (data from Morton *et al.*, 1987). Included with each is an average composite diet, calculated by weighting the diets of each predator species by the number of individuals included in the sample. Values from this weighted average were used to rank the importance of the various prey categories.

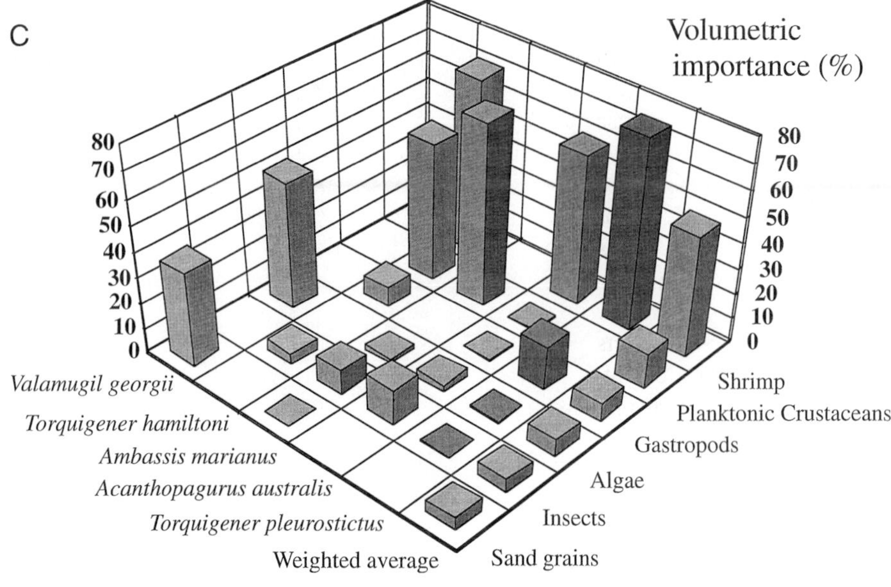


**Figure 3.** *Continued*

complicated because large amounts of energy and biomass production are primarily linked to terrestrial food webs (Teas, 1983; Gearing *et al.,* 1984; Por and Dor, 1984). Marine organisms in mangals can be separated into those that use the hard substrates created by the supporting roots of the mangrove plants and those that inhabit the underlying mud. For example, oysters, common elements in mangal ecosystems, attach to roots, while other bivalves are buried in the mud. Other important mangal organisms include gastropods living on the roots and mud surface, isopods that bore into mangrove roots, epizootic barnacles and tunicates on the roots, semiterrestrial and marine crabs, burrowing shrimp and polychaetes in the mud, and even juvenile spiny lobsters. Mangals can be significant nursery areas for commercially important shrimp and fishes (e.g., Sasekumar *et al.,* 1992), but are rapidly being destroyed to create shrimp ponds.

Not only do mangals support the usual mix of carnivorous, omnivorous, and herbivorous fishes found in other intertidal habitats, but detritivores are especially important (Figure 3b; see also Beumer, 1978; Sasekumar *et al.,* 1984; Ley *et al.,* 1994). Material in the detritus includes microscopic algae, foraminifera, and decaying plant material. In a study by Ong and Sasekumar (1984), 29% of the fishes collected subtidally off mangrove shores in Malaysia were classified as detritivores (>20% detritus in diet). Some mangal species also consume macroalgae (e.g., Figure 3b; Beumer, 1978). Macroinvertebrate prey in the diets of mangal fishes includes gammarid amphipods, tanaids, crabs, shrimp, and terrestrial insects. One unusual prey extracted from the mangal mud in Malaysia are sipunculids, which are important in the diets of marine catfishes (*Arius sagor*), mudskippers (*Periophthalmodon schlosseri*), and even small jacks (*Trachyurus maculatus*) (Sasekumar *et al.,* 1984). We caution, however, that in light of the large area of the tropics and subtropics that are covered by mangals and of the relatively few studies done on this habitat, these studies provide just a first indication of food web relationships in this important habitat.

### F. Salt Marshes

Salt marshes include areas of emergent vegetation and of tidal creeks carrying water into and out of the marsh. The emergent macrophytes include grasses, herbs, and shrubs that are rooted in mud. These vascular plants are the major source of primary productivity in salt marshes, but the pathway to higher trophic levels is not through consumption of plant material by herbivores, but through decomposition by saprophytes (Day *et al.,* 1989). Important contributions to primary productivity are also made by macroscopic and microscopic algae (Zedler, 1982). Associated invertebrates include aquatic and semiterrestrial crabs, gammarid amphipods, mussels and gastropods, errant and tube-dwelling polychaetes, and oligochaetes. One important potential source of prey for salt marsh fishes are the larvae and adults of insects, especially oligohaline mosquitoes and dipterans.

Potential foraging areas of salt marsh fishes include the areas around the salt marsh vegetation during periods of inundation and the connecting channels. For the Australian salt marsh community described by Morton *et al.* (1987), the dominant prey of these fishes was shrimp, including juveniles of commercially important species (Figure 3c). The importance of both channels and vegetated sites can be seen in the importance not only of surface prey, such as insects and gastropods, but also of planktonic crustaceans (see also Joyce and Weisberg, 1986). Meiofauna may also constitute a large component of the diet of salt marsh fishes (e.g., Dutil and Fortin, 1983; Walters *et al.,* 1996).

## IV. A Functional Approach to Understanding Fish Diets

Spatial and temporal variability in the distribution and abundance of predators and their potential prey play an important role in establishing broad limits on the types and numbers of prey that a predator might encounter and therefore eat. However, at any particular place and time, a foraging fish will be exposed to and have to choose from a tremendous variety of sizes and types of potential prey. One recurring observation from diet studies is that individuals of two species foraging in the same area are likely to have dramatically different diets, both qualitatively and quantitatively, in spite of the presence of the same spectrum of prey (e.g., Figures 1–3; Norton, 1991b). The same will often be true of two individuals of the same species that vary in size. These intra- and interspecific differences may be due to fine-scale habitat partitioning (e.g., Nakamura, 1971) or to stochastic events in prey encounter frequency, but they may also reflect the interplay of morphological, behavioral, and physiological characteristics of predators and potential prey. These include the interactions between predator capture ability and prey escape ability and between predator digestion and prey resistance among others.

Viewed another way, while two populations of a single species that live in different habitats or sites may be exposed to different suites of potential prey and therefore exploit different prey taxa, similarities in their prey capture, prey processing, and digestive abilities will often lead to similarities in the functional aspects of prey in the diet. For example, the same morphological adaptations that allow *Clinocottus globiceps* from Washington to graze on intertidal algae (Cross, 1981) also allow this species to bite pieces from intertidal anemones in Oregon and California (Grossman, 1986; Yoshiyama *et al.,* 1996a, b). Similarly, while seasonal variation in diet may parallel seasonal changes in the prey community, the functional aspects of the prey in the diet may remain the same, even if the taxa do not. For example, while seasonal changes in availability may shift the relative abundance of shrimp and small fishes in the diet, both prey types are elusive and are vulnerable to the same attack strategy. Therefore, our ability to predict the diet of particular predators or to understand the underlying mechanisms leading to differences in the diet of two predators requires that we know not only the prey spectrum available to predators, but also the relative functional abilities of predators and prey. This will determine the outcome of a particular predator–prey encounter.

Several studies have examined some of the morphological and physiological differences among intertidal fishes that play important roles in determining interspecific differences in diet (e.g., deGroot, 1969; Beumer, 1978; Grossman, 1986; Horn, 1989). However, these studies have focused primarily on the functional attributes of the predators and the other half of the interaction, the functional attributes of the prey, has received less attention.

## A. Prey Capture

The major prey capture modes of fishes have been organized into three categories: biting, ram feeding, and suction feeding (Liem, 1980). In biting, the oral jaws of the predator are used to cut a piece from a larger prey or to remove a prey organism from the substrate. In ram feeding, the predator captures its prey by swimming at and then engulfing the prey with its mouth open, often aided by rapid protrusion of the fish's premaxilla (Figure 4). In suction feeding, explosive expansion of the buccal cavity of the predator draws a stream of water into the mouth and the prey is dragged in with this stream (Figure 5).

These descriptions require two caveats. First, although we have presented these three feeding modes as unique and distinct, in practice they better identify ends of a continuum (Norton, 1995). For example, an individual strike may combine a ram component (movement of the predator toward the prey) and a suction component (movement of the prey toward the predator) (Norton and Brainerd, 1993). Second, any fish should be capable of performing all these functions to some extent. Biting at a minimum requires a predator to close its jaws. To ram feed a predator need only open its mouth and swim forward. Suction feeding has many similarities to respiration. However, even if an individual fish can, in

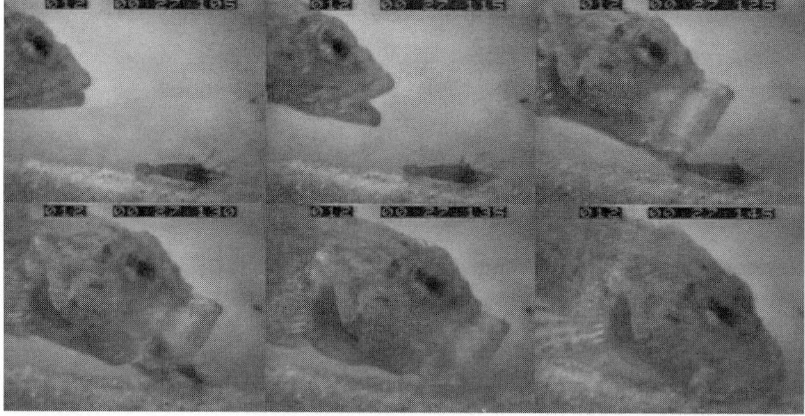

**Figure 4.** Successful capture of a hippolytid shrimp by *Artedius lateralis*. The images were taken with a NAC200 at 200 fields $s^{-1}$. The duration of the sequence is 45 milliseconds (ms) and the time between fields is either 10 ms (first and second fields on top row, second and third fields on bottom row) or 5 ms (all other fields).

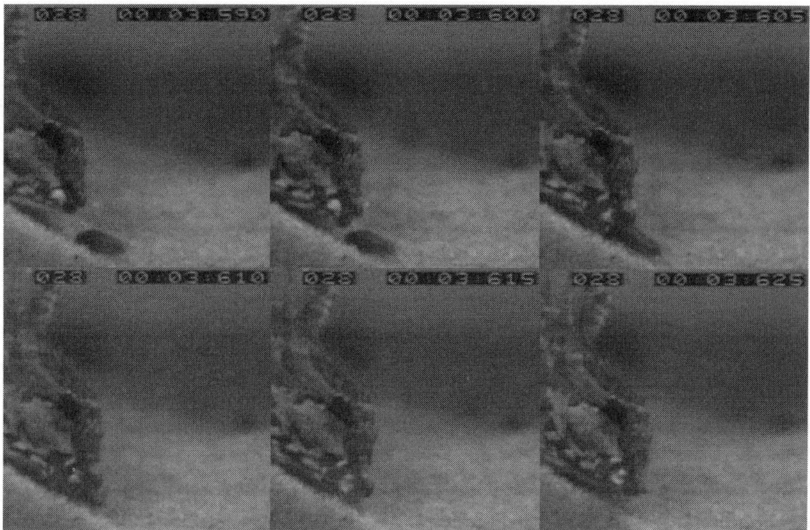

**Figure 5.** Successful capture of a gammarid amphipod by *Oligocottus snyderi.* The images were taken with a NAC200 at 200 fields $s^{-1}$. The duration of the sequence is 45 ms and the time between fields is either 10 ms (first and second fields on top row, second and third fields on bottom row) or 5 ms (all other fields).

theory, perform all three feeding modes, it will not do them all equally well. Suites of specific morphological, behavioral, and physiological specializations produce clear functional differences in feeding ability that are often associated with interspecific differences in diet (Wainwright, 1988; Norton, 1991a; Norton, 1995). In most intertidal fish communities we can identify specific species that are specialized for biting, ram feeding, or suction feeding by these specializations and use this information to predict differences in their utilization of the available prey.

There are clear functional contrasts in several behavioral and morphological characteristics between biters and ram/suction feeders (e.g., Kotrschal, 1988, 1989; Barel *et al.,* 1989; Liem, 1993; Norton, 1995; Wainwright and Richard, 1995). These differences are centered on trade-offs between characteristics affecting force generation as the jaws close, the key to success for biting fishes, and characteristics affecting speed of jaw movement, the key for ram and suction feeding fishes (Norton, 1991a; Wainwright and Richard, 1995). In vertebrates, force generation by muscle fibers increases in proportion to increases in cross-sectional area (Altringham and Johnston, 1982; Johnston and Salamanski, 1984), while contractile speed is correlated with muscle length. The critical muscle for biting fishes is the adductor mandibulae complex, responsible for retraction of the upper and lower jaws. Qualitative comparisons among intertidal cottid fishes from the Pacific Northwest indicate that biters (defined by their feeding behavior) have the largest relative cross-sectional area in the adductor musculature (Norton, unpublished) and should be capable of greater force generation than ram or suction feeding specialists. The jaw lever systems of biters should be specialized for force transmission, while those of ram feeders and suction feeders should be specialized for speed of jaw opening and closing (Wainwright and Richard, 1995). In addition to differences in force generation ability, biters have robust jaws with cutting teeth and tight ligamentous connections between jaw elements that allow efficient force transmission, but limit jaw kinesis (Purcell and Bellwood, 1993). For a few biting species, force can also come from the axial musculature during a variation of biting called rotational feeding (Helfman and Clark, 1984; Horn and Ojeda, Chapter 11, this volume). In ram and suction feeders lighter jaws with more flexible connections permit faster jaw movements; the teeth are either cardiform or caniform to hold prey in the jaws prior to swallowing.

Ram and suction feeding specialists also have clear behavioral and morphological differences. The key to success of suction feeders lies in the velocity of the water flow, generated by the predator, that passes by the prey because the drag force experienced by the prey is a function of the water velocity squared (Denny, 1988; Norton, 1995). This flow originates from the pressure differential (up to 50 kPa) created during rapid expansion (often in less than 30 ms) of the volume of the buccal cavity (Lauder, 1980). For a set change in buccal volume water velocity at the mouth opening should be inversely correlated with mouth area; suction feeding specialists typically have relatively small mouths (Norton, 1991a, 1995; Norton and Brainerd, 1993). Ram feeders are not so constrained, and a larger mouth area may increase capture success (Norton, 1991a). Attack behaviors are also different. To minimize attenuation of the velocity of water passing by the prey, suction feeders would have more success in strikes initiated very close to the prey. For ram feeders, attacks initiated at some distance from the prey allow predators enough time and distance to be at peak velocity as the predator reaches the prey.

These contrasts in morphology and attack and strike behavior can be clearly seen when comparing the feeding strategy of *Artedius lateralis,* an intertidal cottid fish whose diet consists primarily of shrimp and other fishes, with that of *Oligocottus snyderi,* a syntopic intertidal cottid whose diet consists primarily of gammarid amphipods (e.g., Cross, 1981; Grossman, 1986). *A. lateralis* (Figure 4) begins its attack (start of forward motion) and strike (initiation of mouth opening) much farther away from the prey, accelerates to a greater relative velocity, and presents a much larger mouth area to the prey than does *O. snyderi* (Figure 5). In the strike of *O. snyderi,* the prey is drawn into the mouth of the predator, while *A. lateralis* closes its jaws around the prey. These morphological and behavioral differences among cottids lead to dramatic differences in capture success when faced with difficult prey such as shrimp (Norton, 1991a, 1995).

## B. Prey Processing

After prey items are captured, the oral jaws (mandible and premaxilla), the pharyngeal jaws, or both may participate in prey processing. In generalized teleosts that utilize ram or suction feeding, both the oral jaws and pharyngeal jaws typically lack dental or muscular specializations for elaborate prey processing. For these species, prey processing may consist only of holding the prey prior to swallowing. In some cases, the prey processing phase may be used to separate food items from nonnutritive debris (e.g., oral winnowing; Laur and Ebeling, 1983; Drucker and Jensen, 1991). For other fishes, especially species specialized for biting, prey are shredded or crushed during processing as an initial phase of mechanical digestion (Brett, 1979; Palmer, 1979; Wainwright, 1988). For some biters (e.g., Tetradontiformes, Sparidae), the oral jaws are used both to capture and to crush or shred prey. In the Labroidei (Liem and Greenwood, 1981), which includes intertidal species from the Labridae, Scaridae, Pomacentridae, Odacidae, and Embiotocidae, the oral jaws are specialized for prey capture alone, while osteological and myological specializations in the pharyngeal jaws are responsible for a well-developed prey processing ability. In these families, the lower pharyngeal jaws are fused, the upper pharyngeal jaws form a stable joint with the neurocranium, and pharyngeal muscles extend as a muscular sling from the neurocranium and pectoral girdle to the lower pharyngeal jaws (Liem, 1986; Liem and Sanderson, 1986; Wainwright, 1988; Drucker and Jensen, 1991). This arrangement allows the generation of powerful forces by the pharyngeal jaws during prey processing.

## C. Prey Digestion

A number of morphological specializations (e.g., gut length, presence and number of pyloric cecae, gizzard-like stomach) and physiological specializations (e.g., pH, enzyme types and concentrations, symbiotic microbes) of the digestive tract can be linked to dietary specializations in fishes (Suyehiro, 1942; Zihler, 1982; Horn, 1989; Verigina, 1991). In particular, plant material appears to represent a very difficult challenge for fishes (reviewed in Horn, 1989; Horn and Ojeda, Chapter 11, this volume). There is very strong evidence for a positive correlation between gut length and the importance of plant material in the diet. In a few cases, symbiotic microbes that provide cellulases to assist in plant digestion have been isolated (Rimmer and Weibe, 1987; Horn, 1989; Luczkovich and Stellwag, 1993). In other species (e.g., mullets) mechanical trituration of food items with sand in a muscular stomach assists in digestion (Horn, 1989). Levels of other specific enzymes, such as carbohydrases and chitinases (derived from symbiotic bacteria), may also vary as a function of diet and ontogeny in fishes (Goodrich and Morita, 1977; Clark *et al.,* 1984; Danulat, 1986; Seiderer *et al.,* 1987; Sabapathy and Teo, 1993), but few studies have addressed these issues specifically for intertidal fishes.

## D. Functional Interactions between Intertidal Fish and Major Prey Taxa

Prey are not passive elements in a predator–prey encounter. While some prey taxa (e.g., gammarid amphipods, harpacticoid copepods) have relatively few defenses once encountered, other prey taxa have well-developed morphological, chemical, and/or behavioral defenses that only specialized predators can easily overcome (Norton, 1995). For example, a heterogeneous collection of prey, including mysids, calanoid copepods, shrimp, and other fish (i.e., elusive prey), can rapidly accelerate away from an attacking predator. This prey group is most vulnerable to ram feeding tactics (Norton, 1991a). Even within these groups escape potential may be quite variable (Daniel and Meyerhofer, 1989); for example, caridean shrimp have twice the escape ability of hippolytid shrimp (Norton, 1991a), and these shrimp taxa have better escape abilities than crangonid and panaeid shrimps, which rely more on crypsis to avoid predators (Nemeth, 1997). In the next few paragraphs we discuss some antipredator attributes of major prey taxa of fishes and some of the ways that intertidal fishes can overcome them.

Polychaetes and other marine worms (e.g., nematodes) can be captured efficiently by either biting or suction feeding, but probably not by ram feeding unless they are in the water column, such as during reproductive swarms. For many species the presence of chemical and mechanical defenses (e.g., setae) discourage attack or lead to subsequent rejection. Errant polychaetes may be ingested whole, but tube-dwelling polychaetes often suffer partial predation, including loss of feeding structures (i.e., palps, feeding crowns) or tail segments in the case of the deposit feeder, *Arenicola* (De Vlas, 1979b). The ability of many tube-dwellers to retract quickly to the safe haven of the tube requires a stealthy, precise approach by a predator as exhibited by many biters and suction feeders. Digestion of marine worms, especially nematodes, occurs quickly (Scholz *et al.*, 1991; Coull *et al.*, 1995) and this leads to chronic underestimation of these prey types by all diet metrics, except perhaps frequency of occurrence.

The capture, processing, and digestion of molluscs by fishes are highly variable. The most common method of capturing bivalves is probably by biting (Brett, 1979), less often by suction feeding. The tenacity with which mussels (Witman and Suchanek, 1984; Denny, 1988) and oysters adhere to the substrate may present an especially difficult removal problem for a generalized fish. Biting may be used to pull whole animals from the substrate or to bite off the exposed siphons (De Vlas, 1979a; Peterson and Quammen, 1982). Whole bivalves are then crushed by the oral or pharyngeal jaws (e.g., Brett, 1979). Without this processing, bivalves that have shells that fit together snugly (e.g., *Clinocardium* spp.) may survive the digestive process and emerge alive in the feces (Norton, 1988).

Most gastropods can be captured easily by either biting or suction, but the tenacity of limpets (also Polyplacophora or chitons) prevents easy removal from the substrate (Branch and Marsh, 1978; Smith, 1991). Clingfishes (Gobieosocidae) have developed several unique strategies to dislodge limpets (Paine and Palmer, 1978; Stobbs, 1980; Stadler, 1988), some of which take advantage of the stability provided by their ventral suction disk. Other fishes, such as *Clinocottus analis,* capture these prey using highly-developed suction feeding strategies (Norton, unpublished). Nonoperculate gastropods are often ingested whole, while most predators that feed regularly on operculate gastropods crush them with either their oral or pharyngeal jaws (e.g., Palmer, 1979). In cases in which the shells are not crushed, the shell and operculum prevent digestive enzymes from accessing the soft tissues and these operculate gastropods (including most prosobranchs, but few opistobranchs or pulmonates) may survive ingestion and emerge alive (Cancino and Castilla, 1988; Norton, 1988). For example, we have observed live operculate gastropods in the feces of both *C. analis* and *Gobiesox meandricus.* This feature may preclude broader exploitation of operculate gastropods and many bivalve species by typical intertidal fishes that lack the morphological specializations needed to break through the shell (Norton, 1988).

As we illustrated earlier, crustaceans are typically the most important prey in the diets of intertidal fishes, but they also present a variety of challenges. Many groups of crustaceans that are important parts of the intertidal fauna (e.g., caridean shrimp, calanoid copepods, mysids) are elusive and present strong challenges to capture due to a combination of high sensitivity to approaching predators and the ability to achieve rapid escape velocities. Fishes that typically feed on these elusive prey tend to use ram feeding tactics (Coughlin and Strickler, 1990; Norton, 1991a, 1995). Mechanical defenses are seen in many crab larvae that swarm in nearshore waters and that, prior to settling, have long spines that inhibit swallowing. Still other crustaceans (e.g., gammarid amphipods, harpacticoid copepods) have no strong defenses against predation once discovered and may be captured by any of the three feeding strategies. Sessile and sedentary crustaceans, such as barnacles and caprellid amphipods, may be removed completely or in pieces (e.g., cirri) by biting. Crustaceans are typically swallowed without extensive prey processing, except for some hard-shelled crabs that may be crushed prior to swallowing. While most crustaceans do not present a great digestive challenge to fishes, the shells of ostracods may provide a digestive barrier analogous to that presented by the shells of bivalves and the shell/operculum combination of many gastropods, as we have observed live ostracods in the feces of several intertidal and subtidal fish species.

The other potential animal taxa also present a diverse set of challenges. For example, other arthropods that occur in the diets of fishes, such as insects and aquatic mites, are probably not difficult to capture. However, many mite species are chemically defended, and others can survive ingestion by a still unknown mechanism (Norton, personal observation). Although piscivores are not very common in the intertidal, those species that are specialized to feed on other members of the ichthyofauna tend to use ram feeding tactics and prey are typically ingested whole.

Plant and algal material present special problems for fishes, not only in capture, but also in digestion. Microscopic algae, such as diatoms, are brought into the buccal cavity by suction feeding if on the benthos or by ram feeding if suspended in the water column and collected on mucus pads associated with long gill rakers and the pharyngeal apparatus (Sanderson and Wassersug, 1990; Sanderson *et al.,* 1991; Cech and Massingill, 1995). Capture of macroalgae and plant material typically requires well-developed biting ability to cut pieces from larger items or to scrape algae from the substrate. Following capture, herbivorous fishes employ any of several diverse strategies for digestion, including the activities of symbiotic bacteria that produce cellulases (Horn, 1989; Horn and Ojeda, Chapter 11, this volume).

## V. Ontogenetic Changes in the Diets of Intertidal Fishes

Most intertidal fishes leave the intertidal for a period as pelagic larvae but eventually metamorphose and return to settle in the intertidal as juveniles. As they grow through adulthood, they may experience up to a 15-fold increase in length, although 2- to 4-fold increases are more common (Cross, 1981; Bennett *et al.,* 1983; Modde and Ross, 1983; Lasiak, 1986; McMicheal and Ross, 1987). An observation common to many diet studies is that a change in diet accompanies the growth of a predator (see references in Table 2). These changes may be both qualitative (i.e., presence/absence of prey types) and quantitative (i.e., shifts in the relative importance of prey types, shifts in prey size) and reflect changes in the predator's nutritional requirements and functional abilities and in the nutritional content and functional attributes of the available prey.

If we look at a variety of studies that have examined patterns of ontogenetic changes in diet, we see reasonably consistent patterns of change in the importance of specific prey types during the ontogeny of most intertidal fishes (Table 2). Typical of these patterns are the ontogenetic changes that we observed in the diet of *Clinocottus analis,* collected from tidepools in February and March 1991 off Santa Barbara, California (Norton, unpublished). Harpacticoid copepods are found in the diets of almost all of the smallest fish [<34 mm standard length (SL)], but they drop out of the diet entirely by 40 mm SL (Figure 6). Gammarid amphipods are also important in the diets of the smallest fish, but they also disappear from the diets of larger fish. In contrast, crabs and molluscs (limpets and chitons primarily) first enter in the diets of fish over 40 mm SL and appear more frequently in larger fish. Algae (primarily filamentous green algae) also increase in frequency in larger fish. A few prey types, such as isopods and polychaetes, remain relatively constant elements of the diet of *C. analis.*

These results are mirrored in other studies on a variety of intertidal fishes. Often harpacticoid copepods are the dominant prey by number and amphipods (primarily gammarids) are the gravimetric or volumetric dominant in the diets of the smallest intertidal fishes, but

**Table 2.** Changes in Diet, Indicating Which Taxonomic Groups Are Added, Dropped, or Remain in the Diet Throughout the Ontogeny of a Species

| | Frequency of occurrence | | |
|---|---|---|---|
| Prey type | Added | Dropped | Remains |
| Sponges | 1 | 0 | 0 |
| Hydrozoans | 2 | 2 | 1 |
| Anthozoans | 1 | 0 | 0 |
| Polychaetes | 3 | 9 | 18 |
| Ostracods | 0 | 1 | 1 |
| Copepods | 2 | 26 | 11 |
| Barnacles | 3 | 4 | 1 |
| Mysids | 2 | 2 | 5 |
| Cumaceans | 2 | 3 | 1 |
| Isopods | 7 | 3 | 16 |
| Gammarids | 6 | 10 | 32 |
| Caprellids | 1 | 0 | 1 |
| Shrimps | 7 | 4 | 6 |
| Ghost shrimps | 0 | 2 | 2 |
| Sand crabs | 6 | 0 | 0 |
| Brachyuran crabs | 11 | 0 | 8 |
| Insects | 0 | 4 | 5 |
| Limpets | 2 | 0 | 1 |
| Snails | 1 | 2 | 3 |
| Bivalves | 6 | 1 | 3 |
| Algae | 10 | 0 | 7 |
| Echinoderms | 1 | 0 | 0 |
| Fish | 12 | 0 | 3 |

*Note.* Each number is a sum of the occurrence of these changes for all the species in the studies examined (Johnson, 1968; Bellinger and Avault, 1971; Montgomery, 1977; Christensen, 1978; Grossman, 1980; Cross, 1981; Bennett *et al.*, 1983; Modde and Ross, 1983; Robertson, 1984; Robertson and Lenanton, 1984; Freeman *et al.*, 1985; Hughes, 1985; Grossman, 1986; Kneib, 1987; McMicheal and Ross, 1987; Cancino and Castilla, 1988; Poulin and Fitzgerald, 1989; Armstrong *et al.*, 1995; Toepfer and Fleeger, 1995).

they both drop out of the diet during ontogeny (Christensen, 1978; Cross, 1981; Freeman *et al.*, 1985; Grossman, 1986; Setran and Behrens, 1993). As copepods and gammarids disappear from the diets of larger individuals, most intertidal fishes demonstrate a net increase in prey richness during their ontogeny as new taxonomic groups are added to the diet. The prey taxa most frequently added are crabs, fishes, and algae (Table 2). Exceptions are found for those fish species (e.g., engraulids) that are specialists on small prey or that demonstrate relatively small increases in size during ontogeny and retain the diet types and diversity of their earliest juveniles throughout their ontogeny (Grossman, 1980; Kneib,

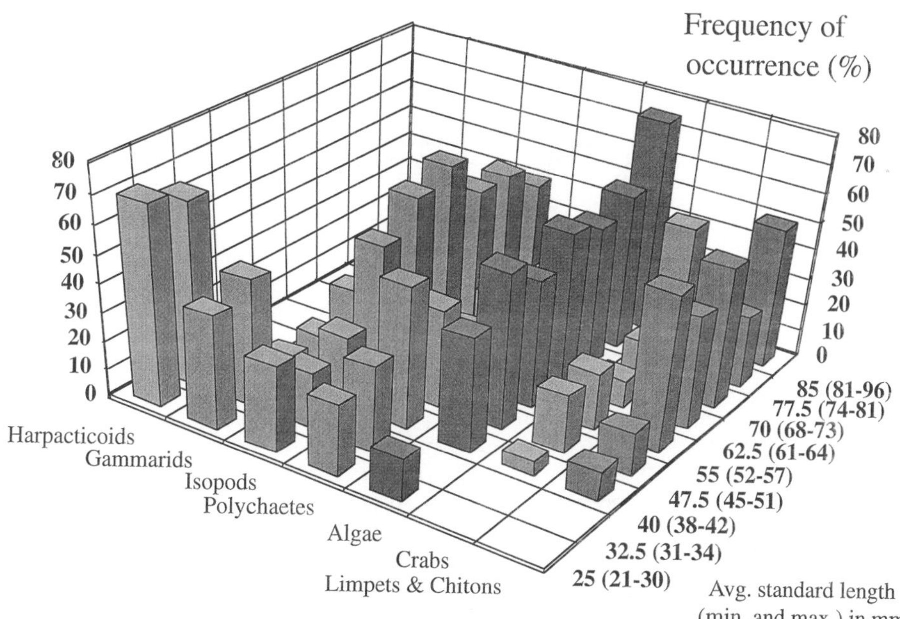


**Figure 6.** Ontogenetic changes in the diet of *Clinocottus analis* collected from tidepools at Coal Oil Pt., California, in February and March 1991. Individual diets were determined by examining the feces of each individual held for 48 h in separate containers in the laboratory. To determine the patterns of ontogenetic change, we ordered the fish from smallest to largest and calculated the frequency of occurrence of prey taxa for running blocks of 20 fish. We present the frequency of occurrence data for nine nonoverlapping blocks at 7.5-mm intervals and indicate the average (minimum and maximum) standard lengths (mm) for each block on the right axis.

1978; Toepfer and Fleeger, 1995). At the same time that larger fish show increases in diet richness, there is also a tendency toward greater specialization.

Several authors have hypothesized that the limited range of small prey types available in many intertidal environments should lead to greater dietary overlap and therefore more competition among younger/smaller fishes than among larger individuals (Gibson, 1972; Crane, 1981; Bennett *et al.,* 1983). Only two studies have measured overlap between life history stages. Hughes (1985) found that both intraspecific and interspecific dietary overlaps in pholids were greater between small size classes than between the largest size classes. In contrast, Setran and Behrens (1993) found no size-related differences in interspecific overlap between two stichaeid species. To address this issue in more depth, we calculated overlap values (Morisita's index; Morisita, 1959) from diet studies in sandy beaches and rocky intertidal habitats (Johnson, 1968; Bellinger and Avault, 1971; Christensen, 1978; Modde and Ross, 1983; Hughes, 1985; Grossman, 1986; McMichael and Ross, 1987). In our intraspecific comparison we calculated overlap values between the two smallest size classes, between the two largest size classes, and between the smallest and largest size classes for each of the species in these studies. We found that average overlap dropped, but not significantly (paired $t$ test, $t = 1.525$, $P = 0.146$), from 0.73 between the two smallest size classes to 0.63 between the two largest size classes (Figure 7). As expected from the general progression of prey types during ontogeny, diet overlap values calculated between the largest and smallest individuals of a species were even lower.

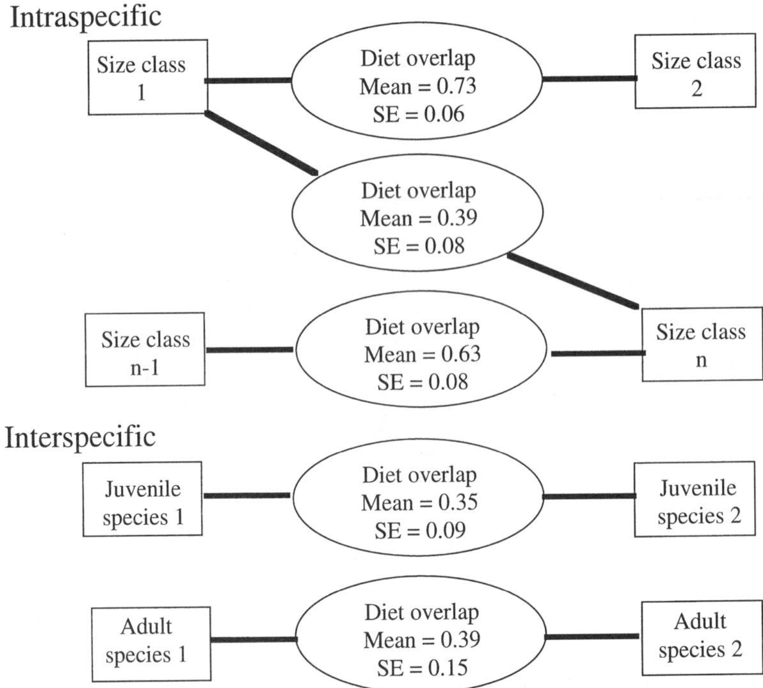


**Figure 7.** Ontogenetic patterns of intraspecific and interspecific diet overlap. For the intraspecific comparisons we determined diet overlap values between the two smallest size classes and the two largest size classes, and between the largest and smallest size classes in several studies (see text). In the interspecific comparisons we calculated overlap values between the smallest size classes of species pairs and for the largest size classes of species pairs. The data shown are the means and SE of these overlap values.

In our interspecific comparisons we calculated diet overlap between the smallest size classes and between the largest size classes for pairs of species. Overlap between the smallest groups was not significantly different from that between the largest groups ($t$ test, $t = 0.230$, $P = 0.824$). Overlap between the two smallest size classes of a single species was significantly greater than interspecific overlap between small size classes of different species ($t$ test, $t = 2.688$, $P < 0.05$). However, for the larger size classes, there was no significant difference between intraspecific overlap of consecutive size classes and interspecific overlap between species of similar sizes ($t$ test, $t = 0.919$, $P = 0.382$). These comparisons are limited by the small sample sizes available for calculating for interspecific overlap and the fact that interspecific diet data were available only for sandy beaches. These observations would indicate that while differences in diet between two individuals (even of the same species) may be influenced by differences due to size, intrinsic characteristics (physiology, morphology, behavior) that differ between species may be even more important in determining overall dietary patterns, particularly in smaller size classes.

As always, interpreting the ecological relevance of these overlap values to the levels of inter- and intraspecific competition among species can be problematic. High overlap values could indicate severe competition for limited resources or they may simply reflect convergence on an abundant resource. Similarly, low overlap values may be the product of strong competition in the past that has lead to niche complimentarity or may simply reflect autecological adaptations of each species to a diverse suite of potential prey. There is some evidence for niche complementarity among small juvenile fishes in sandy beach habitats (Lasiak, 1986; McMichael and Ross, 1987). However, in general there is insufficient information on the abundance of resources available to these fishes, daily ration, or population size to postulate the level of food competition, past or present, among small fishes in the intertidal. Very few experiments addressing competition among intertidal fishes have been done.

Changes in the types of prey in the diet are not the only ontogenetic changes that occur. The dramatic increases in body size demonstrated by intertidal fishes must be accompanied by increases in food intake. This could be accomplished by eating more prey individuals or by taking larger prey. A general observation from most studies is that intertidal fishes do not increase the number of prey consumed, but instead take fewer, larger prey during their ontogeny (Bellinger and Avault, 1971; Robertson, 1980; Hughes, 1985; Cancino and Castilla, 1988; Mullaney and Gale, 1996). For some fishes, this is accomplished by the addition of larger prey taxa to the diet (Hughes, 1985), for others by consuming larger individuals within the same prey taxa (Robertson, 1980; Cancino and Castilla, 1988) or both (Bellinger and Avault, 1971; Mullaney and Gale, 1996). This increase in maximum prey size is typically accompanied by an increase in minimum prey size. Two exceptions are seen in *Sillaginodes punctatus* (Robertson, 1980) and *Sicyases sanguineus* (Cancino and Castilla, 1988) for whom the maximum prey size increases during ontogeny but the minimum prey size does not change.

## A. Factors Driving Ontogenetic Changes in Diet

These consistent ontogenetic changes in diet appear to be driven by the straightforward relationships between gape changes by predators and the size spectrum of prey and by more complex relationships between the developing functional characteristics of predators and of prey as they grow. These two areas, gape limitation and functional constraints, may interact; for example, because very small fishes cannot generate enough force (see below) to bite pieces from larger prey, they are limited to those prey that fit within their gape. The gape (measured as mouth diameter or radius) of newly settled intertidal fishes is quite small (Table 3) and typically increases isometrically during ontogeny (e.g., Norton, 1995; Cook, 1996). There are very few intertidal prey types that are small enough (<2 mm) to be consumed by newly settled juveniles. In estuaries, prey types that fall into this group are restricted to copepods, crab zoea, barnacle nauplii, ostracods, and cumaceans (Grossman, 1980; Houser and Allen, 1996). The larvae of other important prey, such as shrimp and crabs, settle into the intertidal from the plankton at sizes larger than 2 mm (Cook, unpublished).

Not only does gape limitation serve to restrict the sizes and therefore the types of prey in the diet of small fish, but changes may also occur during ontogeny that affect the ability of an individual to capture a prey item by affecting the predator's ability to execute a ram, suction, or biting attack on a prey item. For example, *Clinocottus analis* appears to use a suction feeding attack strategy as an adult (Norton, 1991a) and a more ram feeding strategy as an early juvenile (Cook, 1996). Although comprehensive studies on the scaling of attack kinematics in fishes are lacking, we can draw some predictions based on functional morphological studies.

We can distill the critical elements of each attack strategy to its primary functional component; for ram feeding it is speed, for biting it is force generation, and for suction feeding it is subambient pressure generation. We then can make some predictions on the influence

**Table 3.** Minimum Body Size and Gape of Intertidal Fishes

| Species | Smallest size in intertidal (mm) | Gape (mm) | Reference |
|---|---|---|---|
| *Cebidichthys violaceus* | 17 | 1.5 | Setran and Behrens, 1993 |
| *Xiphister mucosus* | 17 | 1.7 | Setran and Behrens, 1993 |
| *Clinus acuminatus* | 34 | 4.1 | Bennett *et al.*, 1983 |
| *Clinus agilis* | 32 | 3.5 | Bennett *et al.*, 1983 |
| *Clinus aguillaris* | 56 | 4.5 | Bennett *et al.*, 1983 |
| *Clinus capensis* | 56 | 6.2 | Bennett *et al.*, 1983 |
| *Clinus cottoides* | 33 | 4.3 | Bennett *et al.*, 1983 |
| *Clinus dorsali* | 36 | 3.2 | Bennett *et al.*, 1983 |
| *Clinus heterodon* | 32 | 3.8 | Bennett *et al.*, 1983 |
| *Clinus superciliosus* | 25 | 2.8 | Bennett *et al.*, 1983 |
| *Clinus venustris* | 32 | 3.2 | Bennett *et al.*, 1983 |
| *Pavoclinus mus* | 30 | 2.1 | Bennett *et al.*, 1983 |
| *Blennioclinus brachycephalus* | 30 | 2.4 | Bennett *et al.*, 1983 |
| *Caffrogobius caffer* | 27 | 3.0 | Bennett *et al.*, 1983 |
| *Chorisochismus dentex* | 15 | 2.4 | Bennett *et al.*, 1983 |
| *Halidesmus scapularis* | 63 | 3.2 | Bennett *et al.*, 1983 |
| *Lagodon rhomboides* | 16 | 1.4 | Kjelson and Johnson, 1976 |
| *Leiostomus xanthurus* | 16 | 1.7 | Kjelson and Johnson, 1976 |
| *Oligocottus maculosus* | 12 | 1.3 | personal observation |
| *Leptocottus armatus* | 16 | 1.9 | personal observation |

of ontogeny on attack strategy. For ram feeding, functional morphological studies have shown that while larger fish swim at absolutely higher speeds than smaller fish, the relative burst speed remains constant (Beamish, 1978; Richard and Wainwright, 1995). This means that as an individual grows, its ability to overtake a fixed-sized prey should improve, leading to increasing capture success on elusive prey like shrimp and other fishes. For biting fishes, because the maximum force generated by a vertebrate muscle is a function of the cross-sectional area of that muscle, a doubling of fish length should lead to a fourfold increase in cross-sectional area of muscle and in force generation. This allometric increase in force generation should make tougher prey more susceptible to biting predators as a direct function of growth. This prediction is consistent with the observation that fishes that are biters as adults initially consume prey whole, then begin to bite soft prey parts (barnacle cirri, polychaetes parts, etc.), and finally feed on prey than can only be excised (bivalve siphons, seagrasses) or crushed (mussels, whole barnacles, gastropods) by high forces (e.g., Luczkovich *et al.,* 1995). These changes in muscle area are also correlated with changes in dentition that improve biting ability (e.g., Stoner and Livingston, 1984). Unfortunately, at the present time we can make no predictions on the ontogenetic changes that might be seen in suction feeding ability as there are no data on the ontogenetic scaling of subambient pressures during feeding.

## VI. Other Factors Influencing Diet Choice in Fishes

The spectrum of prey that ultimately ends up in the diet of an intertidal fish depends on more than overlap in the distributions of predators and prey and the ability of predators to capture, handle, and digest prey items successfully. Diets are also influenced by the ability of fishes to detect suitable prey and by the relative rewards that different prey provide. Fishes use a variety of sensory modalities to detect prey, including vision (e.g., Hobson, 1991), chemoreception (e.g., Luczkovich, 1988), and mechanoreception (e.g., Montgomery and Coombs, 1998). Prey differ in the degree to which they may generate signals, and predators differ in their ability to perceive these signals. Very little research on these types of interaction between predators and prey has been conducted on intertidal fishes. Holmes and Gibson (1983) demonstrated that differences in seaching and hunting tactics by various species of flatfish played a major role in creating interspecific differences in diet. Other experiments have ranked the relative importance of prey characteristics such as movement, size, color, and shape on prey selection by 15-spined sticklebacks (Kislalioglu and Gibson, 1976a) and of prey activity and size for pinfish (Luczkovich, 1988). Much further research remains to be done in this area, and the tractability of intertidal fishes to laboratory manipulations or even field experiments (e.g., in tidepools) should provide opportunities for generating important insights into the role of prey detection on fish diets.

Another important element likely to shape prey selection by intertidal fishes is the relative energetic or nutritional gain provided by particular prey items. These cost/benefit considerations typically fall under the rubric of optimal foraging. Unfortunately, very little work has been done on profitability of different prey. Experiments on the foraging of catadromous, littoral sticklebacks and intertidal stichaeids are notable exceptions to this pattern. Both 15-spined sticklebacks (Kislalioglu and Gibson, 1976b) and three-spined sticklebacks (Ibrahim and Huntingford, 1989) selected prey types or sizes that should yield greater energetic benefit, measured as grams or joules of prey per second of handling time. Research by Horn and his colleagues has demonstrated that the algae selected by two species of stichaeids were higher in energetic value (Neighbors and Horn, 1991) and assimilation efficiency (Horn and Neighbors, 1984) than underrepresented taxa, but also that nutritional constraints and seasonal availability play an important role in determining diets in the field (Horn, 1983; Horn and Neighbors, 1984). This is another area that merits additional attention.

## VII. Impact of Fish Predation on Intertidal Communities

Studies of community organization, especially of invertebrate communities, have documented the key roles that variation in recruitment, competitive interactions, physical disturbance, and predation have in regulating the diversity and abundance of intertidal organisms (e.g., Connell, 1961; Paine, 1974; Lewis, 1977; Lubchenco, 1978; Sousa, 1979; Connell, 1985; Gaines and Roughgarden, 1985). Those studies that focused on the impacts of predation in the intertidal have emphasized the role of invertebrates or birds as predators (e.g., Paine, 1974; Underwood and Jernakoff, 1981; Feare and Summer, 1986; Marsh, 1986; Lindberg *et al.*, 1987; Raffaelli *et al.*, 1989). Several previous reviews of the biology of intertidal communities (Gibson, 1969, 1982; Horn and Gibson, 1988; Raffaelli and Hawkins, 1996) have downplayed the importance of fish predation on community structure. However, we feel that an increasing body of knowledge is emerging that indicates that predation by intertidal fishes can have dramatic effects on their prey. Evidence for the impact of fish predation on intertidal communities rests on three types of information: elaborate antipredator adaptations of intertidal organisms, comparison between estimates of consumption by fishes with estimates of production of intertidal prey, and experimental manipulations of predators and/or potential prey in intertidal communities.

Predators can inflict direct and indirect costs on prey populations. Direct costs include complete mortality of individuals or partial predation, often of exposed feeding structures (cirri of barnacles, feeding tentacles of polychaetes, or siphons of bivalves). Partial predation may not only drain energetic resources from the prey to regenerate these structures, but also lead to a decline in near-term productivity because of lost foraging time and lowered foraging efficiency during regeneration. Indirect costs on prey include the energy expenditures for defensive compounds or structures and costs due to spatial or temporal shifts to suboptimal habitats to avoid predators (e.g., Wells, 1980; Kunz and Connor, 1986). Additional impacts from fish predators may occur through control of critical invertebrate predators (e.g., Posey, 1986) or by mediating intra- and interspecific competitive interactions among prey populations.

## A. Antipredator Adaptations of Intertidal Organisms

Antipredator adaptations of invertebrates, algae, and macrophytes to reduce predation represent an evolutionary response to the impact of past predation pressure (Vermeij, 1987). The difficulties with consumption of algae and seagrasses by fishes have been well summarized by Horn (1989) and Horn and Ojeda (Chapter 11, this volume). Antipredator adaptations of invertebrates against predators are also well developed and include crypsis (e.g., Reimchen, 1989), mechanical defenses against capture and ingestion (e.g., Vermeij, 1978; Palmer, 1979; Lowell, 1987), behavioral avoidance of predators (e.g., Wells, 1980; Garrity and Levings, 1983; Garrity *et al.,* 1986; Kunz and Connor, 1986), chemical defenses against ingestion and digestion (Yoshiyama and Darling, 1982; Pawlick *et al.,* 1986), and physical barriers against digestion (see above; Norton, 1988). While some adaptations are effective against a wide range of invertebrate and vertebrate predators, others appear directed at only a subset of potential intertidal predators. For example, color mimicry of tubeworms by a littorine snail is only effective against visual predators like fishes and other vertebrates (Reimchen, 1989). Similarly, chemical defenses found in the foot of the limpet *Collisella limatula* are effective against intertidal fish and crabs, but do not deter predation by starfish, octopi, or sea gulls (Pawlik *et al.,* 1986). The effectiveness of these defenses in reducing vulnerability is borne out in both laboratory and field experiments. In laboratory feeding trials, effective use of fast tail flips by hippolytid shrimps resulted in <25% capture success rate by suction feeding cottids, and even ram feeding species were successful in less than half their attacks (Norton, 1991a). Similarly, laboratory experiments by Kunz and Connor (1986) documented a sixfold greater mortality of intertidal limpets without home scars than limpets with home scars. In field manipulations in Panama, 26% of the limpets denied access to home scars in the intertidal were lost to predators in less than 4 weeks (Garrity and Levings, 1983).

## B. Production versus Consumption

Two observational approaches have been used to gauge the impact of fish predation on intertidal organisms: estimates of energy or biomass transfer between multispecies compartments in food webs, and production versus loss estimates for single prey species. Comparisons of the magnitude of the consumption by fish of intertidal organisms have been made with the estimates of production or standing stock for whole trophic levels (e.g.,

Baird and Milne, 1981; Bennett *et al.,* 1983, Robertson, 1984; Gee, 1987). The consumption estimates are derived from diet analysis, estimates of gut clearance time, and quantification of the abundance of the predators (e.g., De Vlas, 1979a; Robertson, 1984; Grossman, 1986). These estimates alone are not sufficient to demonstrate that intertidal fishes exert predation pressure on invertebrate or algal communities; they must represent a significant proportion of intertidal production by intertidal prey. Food webs and energy flow have been studied in all the major intertidal habitats (see examples in Raffaelli and Hawkins, 1996). Estimates of the effects of fish consumption on production by lower food web categories varies widely. For example, fishes consume approximately 54% of the yearly production by benthic invertebrates and zooplankton in the surf zone of South Africa (McLachlan and Romer, 1990). In the Ythan Estuary of Scotland, two fish species, *Pomatoschistus minutus,* the sand goby, and *Platichthys flesus,* the flounder, together consume over 17% of the 47.4 gC $m^{-2}$ $year^{-1}$ that is produced by invertebrate suspension feeders and deposit feeders (Baird and Milne, 1981). In Australia, benthic fishes consume only 6.5% of the macroinvertebrate production in thick seagrass beds, but over 38% of macroinvertebrate production in bare and lightly grassed areas (Robertson, 1984). These food web analyses indicate that the fish community can consume significant amounts of the productivity of their major prey.

The second approach is to compare mortality or biomass losses to the total population or to production estimates for a single prey species, especially if losses can be directly tied to specific predators. From this approach, it is clear that predation by fishes can exert major impacts on the populations of specific invertebrates. Calculations derived from this approach can produce impressive estimates of consumption rates by intertidal fishes; Kelso (1979) estimated that populations of *Fundulus heteroclitus* remove over 546,000 *Mya arenaria* (softshell clam) per kilometer of shoreline per day during peak predation periods. Total consumption of benthic invertebrates by plaice alone averaged 5 g $m^{-2}$ $year^{-1}$ along intertidal flats in the Wadden Sea (De Vlas, 1979a). This loss can represent a significant fraction of the prey population or biomass. Parry (1982) calculated that foraging in the intertidal by wrasses resulted in a 50% loss of the population of the limpet *Patella peroni* and had significant impacts on the populations of two other limpet species in Australia. Similarly, Blankley and Branch (1985) estimated that the Antarctic cod, *Notothenia coriiceps,* was responsible for between 15 and 30% of the mortality experienced by the sub-Antarctic limpet, *Nacella delesserti,* and that this fish, along with a starfish and a gull, has a major impact on younger limpets (<5 years old) for whom mortality rates range from 75% for first-year individuals to 38.8% for 3- to 5-year-old individuals. Twelve percent of the mortality of *Asellops intermedia,* the dominant harpacticoid in the top 5 mm of sediments in the Exe estuary, is due to consumption by plaice, shrimp, and gobies (Gee, 1987). From loss rates of marked hermit crabs, Bertness (1981) estimated that yearly predation rates by intertidal fishes exceed 300% of the standing stock of individuals in +1-m tidepools and 35% in +4-m pools. These would not appear to be trivial impacts.

Partial predation can also exert a significant toll on energy allocation of benthic invertebrates. Experiments by Peterson and Quammen (1982) demonstrated that while direct mortality due to siphon cropping by fishes was low, frequent siphon cropping resulted in a 60% reduction in growth rate by *Protothaca staminea* in clean-sand habitats. Similar results were obtained by De Vlas and Beukema who estimated that 22% of the diet of plaice in the Dutch Wadden Sea consisted of the tail tips of *Arenicola marina* (De Vlas, 1979a); that predation by plaice and flounder consumed 26% of the mean annual biomass, roughly equivalent to the biomass invested by these worms on reproduction (Beukema and De Vlas,

1979); and that this partial predation resulted in a reduction in body size for individuals of between 16 and 44% (De Vlas, 1979b). Finally, estimates for the magnitude of indirect costs (e.g., production of defensive spines, lost foraging time) have not been measured for the prey of intertidal fishes.

### C. Experimental Manipulations of Predators or Prey

The strongest evidence for the impact of fish predation on intertidal organisms is derived from experimental manipulations, focused either on specific fish species–prey species interactions or as part of a more general examination of predator effects that include not only fish but also crabs and birds. Typically, these involve either caging predators in or caging predators out of an area. These types of studies are difficult logistically, and separating cage artifacts from treatment effects is a constant problem (e.g., Hurlberg and Oliver, 1980; Raffaelli and Hawkins, 1996). In many cases, those excluded include not only fishes, but also large invertebrate predators and shorebirds, making assignment of relative importance problematic. The results of these experiments have been mixed; several careful studies have concluded that fishes may have minimal impacts on invertebrate communities (e.g., Quammen, 1984; Gee *et al.,* 1985; Kneib, 1987; Gibbons, 1988; Jaquet and Raffaelli, 1989; Machado *et al.,* 1996). Yet, others have demonstrated significant direct and indirect impacts of fish predation on invertebrates and algae (e.g., Reise, 1977; Dethier, 1980; Bertness *et al.,* 1981; Wiltse *et al.,* 1984; Joyce and Weisberg, 1986; Menge *et al.,* 1986; Posey, 1986; Kneib, 1987; Thrush *et al.,* 1991; Robles and Robb, 1993; Walters *et al.,* 1996). For example, Posey (1986) demonstrated that exclusion of *Leptocottus armatus* allowed the burrowing ghost shrimp, *Callianassa californiensis,* to expand from its normal zone in the mid-intertidal into the lower intertidal, with major potential impacts on the whole infaunal community. Similarly, Kneib (1987) documented higher abundances of benthic invertebrates in inclusion cages containing large *Fundulus heteroclitus* because predation by *Fundulus* depresses the abundance of another important salt marsh predator, *Palaemonetes pugio.* Other manipulative experiments have demonstrated the role of fish predation in restricting the distribution of a number of other intertidal invertebrates, including harpacticoid copepods (Dethier, 1980) and a pulmonate snail (Joyce and Weisberg, 1986). In the rocky intertidal of the Bay of Panama, predation appears to dominate overall community structure as consumption by herbivorous and carnivorous fishes and crabs leads to a complex cascade of direct negative and positive effects on sessile invertebrates and algae (Menge *et al.,* 1986). Even in cases in which fish predation does not appear to impact the numbers of invertebrate prey, size-selective predation by fishes may mediate intraspecific interactions, especially by improving the recruitment of small individuals (Kent and Day, 1983; Wilson, 1989). At present, there are too few experimental studies of the impacts of fish predators on intertidal communities to elicit more general patterns (e.g., tropical vs. temperate, rocky intertidal vs. mud), and this remains a fruitful area for future research.

## VIII. Summary

In conclusion, our understanding of dietary patterns has progressed on several fronts over the past 20 years. We have more information on broad geographic patterns of prey use and patterns of local spatial and temporal variability in diet. We are beginning to develop a better picture of the distribution of potential prey available to fishes in the various intertidal habitats. Insights from functional morphology and biomechanics have provided a predictive framework for exploring the important role of functional interactions between predator and prey in determining diet, but the relative importance of energetic considerations during prey selection (e.g., optimal foraging theory) has largely been unexplored for intertidal fishes. For most intertidal fishes there are clear ontogenetic changes in diet that reflect gape limitations and functional constraints on predators. Finally, several lines of evidence indicate that for some intertidal organisms and in some areas fishes do exert considerable predation pressure on their prey and can play a major role in structuring prey communities.

## Acknowledgments

We thank the editors, Mike Horn, Karen Martin, and Mike Chotkowski, for inviting us to contribute to this volume. Jen Zamon and Lara Ferry-Graham provided insightful comments on the manuscript. Dr. Karel Liem allowed us to borrow the NAC200 high-speed videocamera. Dr. Elizabeth Brainerd assisted during the videotaping. Jim McCullough and Shane Anderson at U.C.S.B. helped to collect the *Clinocottus analis* at Coal Oil Pt., California. Pat Guyette and the Interlibrary Loan Office at E.C.U. did a great job in providing many of the references in Table 1.

## References

Altringham, J. D., and Johnston, I. A. (1982). The pCa-tension and force-velocity characteristics of skinned fibers isolated from fish fast and slow muscle. *J. Physiol.* **333,** 421–449.

Ansell, A. D., and Gibson, R. N. (1990). Patterns of feeding and movement of juvenile flatfishes on an open sandy beach. *In* "Trophic Relationships in the Marine Environment" (M. Barnes and R. N. Gibson, Eds.), pp. 191–207. Aberdeen University Press, Aberdeen.

Armstrong, J. L., Armstrong, D. A., and Matthews, S. B. (1995). Food habits of estuarine staghorn sculpins, *Leptocottus armatus,* with focus on consumption of juvenile Dungeness crab, *Cancer magister. Fish. Bull.* **93,** 456–470.

Baird, D., and Milne, H. (1981). Energy flow in the Ythan Estuary, Aberdeenshire, Scotland. *Est. Coast. Shelf Sci.* **13,** 455–472.

Barak, N. A. E., Salman, N. A., and Ahmad, S. N. (1994). The piscivorous feeding of mudskipper *Periophthalmus waltoni* Koumaus from Khor Al-Zubair, Northwest Arabian Gulf. *Pakistan J. Zool.* **26,** 280–283.

Barel, C. D. N., Anker, G. Ch., Witte, F., Hoogerhoud, R. J. C., and Goldschmidt, T. (1989). Constructional contraints and its ecomorphological implications. *Acta Morphol. Neerl. Scand.* **27,** 23–109.

Barry, J. P., and Ehret, M. J. (1993). Diet, food preference, and algal availability for fishes and crabs on intertidal reef communities in southern California. *Environ. Biol. Fish.* **37,** 75–95.

Barton, M. G. (1982). Intertidal vertical distribution and diets of five species of central California stichaeoid fishes. *Calif. Fish Game* **68,** 174–182.

Beamish, F. W. H. (1978). Swimming capacity. *In* "Fish Physiology" (W. S. Hoar and D. J. Randall, Eds.), Vol. 7, pp. 101–187. Academic Press, New York.

Bellinger, J. W., and Avault, J. W., Jr. (1971) Food habits of juvenile pompano, *Trachinotus carolinus,* in Louisiana. *Trans. Am. Fish. Soc.* **100,** 486–494.

Bennett, B., Griffiths, C. L., and Penrith, M. (1983). The diets of littoral fish from the Cape Peninsula. *S. Afr. J. Zool.* **18,** 343–352.

Bertness, M. D. (1981). Predation, physical stress, and the organization of a tropical rocky intertidal hermit crab community. *Ecology* **62,** 411–425.

Bertness, M. D., Garrity, S. D., and Levings, S. C. (1981). Predation pressure and gastropod foraging: A tropical-temperate comparison. *Evolution* **35,** 995–1007.

Beukema, J. J., and De Vlas, D. (1979). Population parameters of the lugworm, *Arenicola marina,* living on tidal flats in the Dutch Wadden Sea. *Neth. J. Sea Res.* **13,** 331–353.

Beumer, J. P. (1978). Feeding ecology of four fishes from a mangrove creek in north Queensland, Australia. *J. Fish Biol.* **12,** 475–490.

Birtwell, I. K., Wood, M., and Gordon, D. K. (1983). Data report on fish diets and faunal diversity in the Somass River Estuary, Port Alberni, British Columbia. *Can. Data Rep. Fish. Aquat. Sci.* **396.**

Black, R., and Miller, R. J. (1991). Use of the intertidal by fish in Nova Scotia. *Environ. Biol. Fish.* **31,** 109–121.

Blankley, W. O. (1982). Feeding ecology of three inshore fish species at Marion Island (Southern Ocean). *S. Afr. J. Zool.* **17,** 164–170.

Blankley, W. O., and Branch, G. M. (1985). Ecology of the limpet *Nacella delesserti* (Philippi) at Marion Island in the sub-Antarctic Southern Ocean. *J. Exp. Mar. Biol. Ecol.* **92,** 259–281.

Blankley, W. O., and Grindley, J. R. (1985). The intertidal and shallow subtidal food web at Marion Island. *In* "Antarctic Nutrient Cycles and Food Webs" (W. R. Siegfried, P. R. Condy, and R. M. Laws, Eds.), pp. 630–636. Springer-Verlag, Berlin.

Branch, G. M., and Marsh, A. C. (1978). Tenacity and shell shape in six *Patella* species: adaptive features. *J. Exp. Mar. Biol. Ecol.* **34,** 111–130.

Brett, J. R. (1979). Some morphological and behavioral adaptations of pile perch (*Rhacochilus vacca*) feeding on mussels (*Mytilus edulus*). *Can. J. Zool.* **57,** 658–664.

Brewer, D. T., and Warburton, K. (1992). Selection of prey from a seagrass/mangrove environment by golden lined whiting, *Sillago analis* (Whitley). *J. Fish Biol.* **40,** 257–271.

Brown, A. C., and McLachlan, A. (1990). "Ecology of Sandy Shores." Elsevier, Amsterdam.

Butler, G. S. (1982). Daily feeding periodicity of the intertidal goby *Caffrogobius caffer. S. Afr. J. Zool.* **17,** 182–189.

Cancino, J. M., and Castilla, J. C. (1988). Emersion behavior and foraging ecology of the common Chilean clingfish, *Sicyases sanguineus* (Pisces: Gobiesocidae). *J. Nat. Hist.* **22,** 249–261.

Carr, W. E. S., and Adams, C. A. (1973). Food habits of juvenile marine fishes occupying seagrass beds in the estuarine zone near Crystal River, Florida. *Trans. Am. Fish. Soc.* **102,** 511–540.

Cech, J. J., Jr., and Massingill, M. J. (1995). Tradeoffs between respiration and feeding in Sacramento blackfish, *Orthodon microlepidotus. Environ. Biol. Fish.* **44,** 157–163.

Christensen, M. S. (1978). Trophic relationships in juveniles of three species of sparid fishes in the South African marine littoral. *Fish. Bull.* **76,** 389–401.

Clark, J., McNaughton, J. E., and Stark, J. R. (1984). Metabolism in marine flatfish. 1. Carbohydrate digestion in Dover sole (*Solea solea* L.). *Comp. Biochem. Physiol. B* **77,** 821–827.

Clayton, D. A. (1993). Mudskippers. *Oceanogr. Mar. Biol. Annu. Rev.* **31,** 507–577.

Coetzee, P. S. (1986). Diet composition and breeding cycle of blacktail, *Diplodus sargus capensis* (Pisces: Sparidae), caught off St. Croix Island, Algoa Bay, South Africa. *S. Afr. J. Zool.* **21,** 237–243.

Collins, S. P. (1981). Littoral and benthic investigations on the west coast of Ireland. XIII. The biology of *Gobiusculus flavescens* (Fabricius) on the Connemara coast. *Proc. R. Ir. Acad.* **81b,** 63–87.

Connell, J. H. (1961). The influence of intra-specific competition and other factors on the distribution of the barnacle *Chthamalus stellatus. Ecology* **42,** 710–723.

Connell, J. H. (1985). The consequences of variation in initial settlement vs. post-settlement mortality in rocky intertidal communities. *J. Exp. Mar. Biol. Ecol.* **93,** 11–45.

Cook, A. E. (1996). Ontogeny of feeding morphology and kinematics in juvenile fishes: A case study of the cottid fish *Clinocottus analis. J. Exp. Biol.* **199,** 1961–1971.

Coughlin, D. J., and Strickler, J. R. (1990). Zooplankton capture by a coral reef fish: An adaptive response to evasive prey. *Environ. Biol. Fish.* **29,** 35–42.

Coull, B. C., and Feller, R. J. (1988). Site-to-site variability in abundance of meiobenthic copepods along a tidal gradient over 24 hours. *Hydrobiologia* **167,** 477–483.

Coull, B. C., Greenwood, J. G., Fielder, D. R., and Coull, B. A. (1995). Subtropical Australian juvenile fish eat meiofauna: Experiments with winter whiting *Sillago maculata* and observations of other species. *Mar. Ecol. Prog. Ser.* **125,** 13–19.

Crane, J. M., Jr. (1981). Feeding and growth by the sessile larvae of the teleost *Porichthys notatus. Copeia* **1981,** 895–897.

Cross, J. N. (1981). "Structure of a Rocky Intertidal Fish Assemblage." Ph.D. thesis, University of Washington.

Cross, J. N. (1982). Resource partitioning in three rocky intertidal fish assemblages. *In* "Gutshop '81" (G. M. Cailliet and C. A. Simenstad, Eds.), pp. 142–150. Washington Sea Grant Press, Seattle, WA.

Daniel, T. L., and Meyerhofer, E. (1989). Size limits in escape locomotion of carridean shrimp. *J. Exp. Biol.* **143,** 245–265.

Danulat, E. (1986). The effect of various diets on chitinase and beta-glucosidase activities and the condition of cod, *Gadus morhua* (L.). *J. Fish Biol.* **28,** 191–197.

Day, J. W., Jr., Hall, C. A. S., Kemp, W. M., and Yanes-Arancibia, A. (1989). "Estuarine Ecology." Wiley, New York.

Deady, S., and Fives, J. M. (1995). The diet of corkwing wrasse, *Crenilabrus melops,* in Galway Bay, Ireland and in Dinard, France. *J. Mar. Biol. Assoc. U.K.* **75,** 635–649.

de Groot, S. J. (1969). Digestive system and sensorial factors in relation to the feeding behavior of flatfish (Pleuronectiformes). *J. Cons. Int. Explor. Mer.* **32,** 385–395.

Delbeek, J. C., and Williams, D. D. (1988). Feeding selectivity of four species of sympatric stickleback in brackish-water habitats in eastern Canada. *J. Fish Biol.* **32,** 41–62.

Denny, M. W. (1988). "Biology and Mechanics of the Wave-Swept Environment." Princeton University Press, Princeton, NJ.

Dethier, M. N. (1980). Tidepools are refuges: Predation and the limits of the harpacticoid copepod *Tigriopus californicus* (Baker). *J. Exp. Mar. Biol. Ecol.* **42,** 99–111.

De Vlas, J. (1979a). Annual food intake by plaice and flounder in a tidal flat area in the Dutch Wadden Sea, with special reference to consumption of regenerating parts of macrobenthic prey. *Neth. J. Sea Res.* **13,** 117–153.

De Vlas, J. (1979b). Secondary production by tail regeneration in a tidal flat population of lugworms (*Arenicola marina*), cropped by flatfish. *Neth. J. Sea Res.* **13,** 362–393.

Drucker, E. G., and Jensen, J. S. (1991). Functional analysis of a specialized prey processing behavior: Winnowing by surfperches (Teleostei: Embiotocidae). *J. Morphol.* **210,** 267–287.

Duarte, W. E., and Moreno, C. A. (1981). The specialized diet of *Harpagifer bispinis. Hydrobiologia* **80,** 241–250.

Dutil, J. D., and Fortin, M. (1983). La communaufe de poissons d'un mafecage intertidal de l'estuaraire du Saint-Laurent. *Nat. Can.* **110,** 397–410.

Faller-Fritsch, R. J., and Emson, R. H. (1986). Causes and patterns of mortality in *Littorina rudis* (Maton) in relation to intraspecific variation: A review. *In* "The Ecology of Rocky Coasts" (P. G. Moore and R. Seed, Eds.), pp. 157–177. Columbia University Press, New York.

Feare, C. S., and Summers, R. W. (1986). Birds as predators on rocky shores. *In* "The Ecology of Rocky Coasts" (P. G. Moore and R. Seed, Eds.), pp. 249–264. Columbia University Press, New York.

Feller, R. J., and Coull, B. C. (1995). Non-selective ingestion of meiobenthos by juvenile spot (*Leiostomus xanthurus*) (Pisces) and their daily ration. *Vie Mil.* **45,** 49–59.

Feller, R. J., Coull, B. C., and Hentschel, B. T. (1990). Meiobenthic copepods: Tracers of where juvenile *Leiostomus xanthurus* (Pisces) feed? *Can. J. Fish. Aquat. Sci.* **47,** 1913–1919.

Fenchell, T., and Blackburn, H. (1979). "Bacteria and Mineral Cycling." Springer-Verlag, Berlin.

Field, J. G., and Griffiths, C. L. (1991). Littoral and sublittoral ecosystems of Southern Africa. *In* "Intertidal and Littoral Ecosystems (Ecosystems of the World 24)" (A. C. Mathieson and P. H. Nienhuis, Eds.), pp. 323–346. Elsevier, London.

Freeman, M. C., Neally, N., and Grossman, G. D. (1985). Aspects of the life history of the fluffy sculpin, *Oligocottus snyderi. Fish. Bull.* **83,** 645–655.

Gaines, S. D., and Roughgarden, J. (1985). Larval settlement rate, a leading determinant of structure in an ecological community of the intertidal zone. *Proc. Natl. Acad. Sci. USA* **82,** 3707–3711.

Gaines, S. D., and Lafferty, K. D. (1995). Modeling the dynamics of marine species: The importance of incorporating larval dispersal. *In* "Ecology of Marine Invertebrate Larvae" (L. R. McEdwards, Eds.), pp. 389–412. CRC Press, Boca Raton, FL.

Garrity, S. D., and Levings, S. C. (1983). Homing to scars as a defense against predators in the pulmonate limpet *Siphonaria gigas* (Gastropoda). *Mar. Biol.* **72,** 319–324.

Garrity, S. D., Levings, S. C., and Caffey, H. M. (1986). Spatial and temporal varitation in shell crushing by fishes on rocky shores of Pacific Panama. *J. Exp. Mar. Biol. Ecol.* **103,** 131–142.

Gearing, J., Gearing, P., Rodelli, M., Marshall, N., and Sasejumar, A. (1984). Initial findings from stable carbon isotope ratios in west coast mangrove areas of peninsular Malaysia. *In* "Proceedings of the Asian Symposium on the Mangrove Environment—Research and Management" (E. Saepadmo, A. N. Rao, and D. J. Macintosh, Eds.), pp. 488–495. UNESCO, Kuala Lumpur.

Gee, J. M. (1987). Impact of epibenthic predation on estuarine intertidal harpacticoid copepod populations. *Mar. Biol.* **96,** 497–510.

Gee, J. M., Warwick, R. M., Carey, J. T., and George, C. L. (1985). Field experiments on the role of epibenthic predators in determining prey densities in an estuarine mudflat. *Est. Coast. Shelf Sci.* **21,** 429–448.

Gibbons, M. J. (1988). Impact of predation by juvenile *Clinus superciliosus* on phytal meiofauna: Are fish important predators. *Mar. Ecol. Prog. Ser.* **45,** 13–22.

Gibson, R. N. (1969). The biology and behavior of littoral fish. *Oceanogr. Mar. Biol. Annu. Rev.* **7,** 367–410.

Gibson, R. N. (1972). The vertical distribution and feeding relationships of intertidal fish on the Atlantic coast of France. *J. Anim. Ecol.* **41,** 189–207.

Gibson, R. N. (1982). Recent studies on the biology of intertidal fishes. *Oceanogr. Mar. Biol. Annu. Rev.* **20,** 363–414.

Goodrich, T. D., and Morita, R. Y. (1977). Bacterial chitinase in the stomachs of marine fishes from Yaquina Bay, Oregon, USA. *Mar. Biol.* **41,** 355–360.

Griffiths, C. L., Stenton-Dozey, J., and Koop, K. (1983). Kelp wrack and energy flow through a sandy beach. *In* "Sandy Beaches as Ecosystems" (A. McLachlan and T. Erasmus, Eds.), pp. 547–556. Dr. W. Junk, The Hague.

Grossman, G. D. (1980). Ecological aspects of ontogenetic niche shifts in prey size utilization in the bay goby (Pisces: Gobiidae). *Oecologia* **47,** 233–238.

Grossman G. D. (1986). Food resource partitioning in a rocky intertidal fish assemblage. *J. Zool. London (B)* **1,** 317–355.

Harrison, T. D. (1991). A note on the diet and feeding selectivity of juvenile riverbream *Acanthopagrus berda* (Forskal, 1775) in a subtropical mangrove creek. *S. Afr. J. Zool.* **26,** 36–42.

Havenhand, J. N. (1995). Evolutionary ecology of larval types. *In* "Ecology of Marine Invertebrate Larvae" (L. R. McEdward, Ed.), pp. 79–122. CRC Press, Boca Raton, FL.

Hobson, E. S. (1991). Trophic relationships of fishes specialized to feed on zooplankters above coral reefs. *In* "The Ecology of Fishes on Coral Reefs" (P. F. Sale, Eds), pp. 65–95. Academic Press, New York.

Holmes, R. A., and Gibson, R. N. (1983). A comparison of predatory behavior in flatfish. *Anim. Behav.* **31,** 1244–1255.

Horn, M. H. (1983). Optimal diets in complex environments: Feeding strategies of two herbivorous fishes from a temperate rocky intertidal zone. *Oecologia* **58,** 345–350.

Horn, M. H. (1989). Biology of marine herbivorous fishes. *Oceanogr. Mar. Biol. Annu. Rev.* **27,** 167–272.

Horn, M. H., and Gibson, R. N. (1988). Intertidal fishes. *Sci. Am.* **258,** 64–70.

Horn, M. H., Murray, S. N., and Edwards, T. W. (1982). Dietary selectivity in the field and food preferences in the laboratory for two herbivorous fishes (*Cebidichthys violaceus* and *Xiphister mucosus*) from a temperate intertidal zone. *Mar. Biol.* **67,** 237–246.

Horn, M. H., and Neighbors, M. A. (1984). Protein and nitrogen assimilation as a factor in predicting the seasonal macroalgal diet of the monkeyface prickleback. *Trans. Am. Fish. Soc.* **113,** 388–396.

Houser, D. S., and Allen, D. M. (1996). Zooplankton dynamics in an interidal salt-marsh basin. *Estuaries* **19,** 659–673.

Hughes, G. W. (1985). The comparative ecology and evidence for resource partitioning in two pholidid fishes (Pisces: Pholididae) from southern British Columbia eelgrass beds. *Can. J. Zool.* **63,** 76–85.

Hurlberg, L. W., and Oliver, J. S. (1980). Caging manipulations in marine soft-bottom communities: Importance of animal interactions or sedimentary habitat modifications. *Can. J. Fish. Aquat. Sci.* **37,** 1130–1139.

Ibrahim, A. A., and Huntingford, F. A. (1989). Laboratory and field studies on diet choice in three-spined sticklebacks *Gasterosteus aculeatus* L., in relation to profitability and visual features of prey. *J. Fish Biol.* **2,** 245–258.

Ismail, W. A., and Clayton, D. A. (1990). Biology of *Omobranchus punctatus* (Blenniidae) on rocky shores in Kuwait. *Cybium* **14,** 285–293.

Jaquet, N., and Raffaelli, D. (1989). The ecological importance of the sand goby *Pomatoschistus minutus* (Pallas). *J. Exp. Mar. Biol. Ecol.* **128,** 147–156.

Jaramillo, E., McLachlan, A., and Coetzee, P. (1993). Intertidal zonation patterns of macroinfauna over a range of exposed sandy beaches in south-central Chile. *Mar. Ecol. Prog. Ser.* **101,** 105–118.

Jardine, I. W. (1986). Height on the shore as a factor influencing growth rate and reproduction of the top-shell *Gibbula cineraria* (L.). *In* "The Ecology of Rocky Coasts" (P. G. Moore and R. Seed, Eds.), pp. 117–135. Columbia University Press, New York.

Johnson, C. R. (1968). Food of the the buffalo sculpin, *Enophrys bison. J. Fish. Res. Bd. Can.* **25,** 807–811.

Johnston, I. A., and Salamanski, J. (1984). Power output and force-velocity relationships of red and white muscle fibers from the Pacific blue marlin (*Makria nigricans*). *J. Exp. Biol.* **111,** 171–177.

Joyce, A. A., and Weisberg, S. B. (1986). The effects of predation by the mummichog, *Fundulus heteroclitus* (L.) on the abundance and distribution of the salt marsh snail, *Melampus bidentatus* (Say). *J. Exp. Mar. Biol. Ecol.* **100,** 295–306

Kelso, W. E. (1979). Predation on soft-shell clams, *Mya arenaria,* by the common mummichog, *Fundulus heteroclitus. Estuaries* **2,** 249–254.

Kent, A. C., and Day, R. W. (1983). Population dynamics of an infaunal polychaete: The effect of predators and an adult-recruit interaction. *J. Exp. Mar. Biol. Ecol.* **73,** 185–203.

Kislalioglu, M., and Gibson, R. N. (1976a). Some factors governing prey selection by the 15-spined stickleback, *Spinachia spinachia* (L.). *J. Exp. Mar. Biol. Ecol.* **25,** 159–169.

Kislalioglu, M., and Gibson, R. N. (1976b). Prey 'handling time' and its importance in food selection by the 15-spined stickleback, *Spinachia spinachia* (L.). *J. Exp. Mar. Biol. Ecol.* **25,** 151–158.

Kjelson, M. A., and Johnson, G. N. (1976). Further observations of the feeding ecology of postlarval pinfish, *Lagodon rhomboides,* and spot, *Leiostmus xanthurus. Fish. Bull.* **74,** 423–432.

Kleypas, J., and Dean, J. M. (1983). Migration and feeding of the predatory fish, *Bairdiella chrysura* Lacepede, in an intertidal creek. *J. Exp. Mar. Biol. Ecol.* **72,** 199–209.

Kneib, R. T. (1978). Habitat, diet, reproduction, and growth of the spotfin killifish, *Fundulus luciae,* from a North Carolina salt marsh. *Copeia* **1978,** 164–168.

Kneib, R. T. (1986). The role of *Fundulus heteroclitus* in salt marsh trophic dynamics. *Am. Zool.* **26,** 259–269.

Kneib, R. T. (1987). Predation risk and use of intertidal habitats by young fishes and shrimp. *Ecology* **68,** 379–386.

Kneib, R. T., and Stiven, A. E. (1978). Growth, reproduction, and feeding of *Fundulus heteroclitus* (L.) on a North Carolina salt marsh. *J. Exp. Mar. Biol. Ecol.* **31,** 121–140.

Kotrschal, K. (1988). Evolutionary patterns in tropical marine reef fish feeding. *Z. Zool. Syst. Evol.-forsch.* **26,** 51–64.

Kotrschal, K. (1989). Trophic ecomorphology in eastern Pacific blennioid fishes: Character transformation of oral jaws and associated change of their biological roles. *Environ. Biol. Fish.* **24,** 199–218.

Kuipers, B. R. (1977). On the ecology of juvenile plaice on a tidal flat in the Wadden Sea. *Neth. J. Sea Res.* **11,** 56–91.

Kunz, C., and Connor, V. (1986). Roles of the home scar of *Collisella scabra* (Gould). *Veliger* **29,** 25–30.

Lasiak, T. A. (1983). The impact of surf-zone fish communities on faunal assemblages associated with sandy beaches. *In* "Sandy Beaches as Ecosystems" (A. McLachlan and T. Erasmus, Eds.), pp. 501–506. Dr. W. Junk, The Hague.

Lasiak, T. A. (1986). Juveniles, food and the surf zone habitat: Implications for teleost nursery areas. *S. Afr. J. Zool.* **21,** 51–56.

Lasiak, T., and McLachlan, A. (1987). Opportunistic utilization of mysid shoals by surf-zone teleosts. *Mar. Ecol. Prog. Ser.* **37,** 1–7.

Lauder, G. V. (1980). The suction feeding mechanism in sunfishes (*Lepomis*): An experimental analysis. *J. Exp. Biol.* **88,** 49–72.

Laur, D. R., and Ebeling, A. W. (1983). Predator-prey relationships in surfperches. *Environ. Biol. Fish.* **8,** 217–229.

Lewis, J. R. (1977). The role of physical and biological factors in the distribution and stability of rocky shore communities. *In* "Biology of Benthic Organisms" (B. F. Keegan, P. O'Ceidigh, and P. J. S. Boaden, Eds.), pp. 417–424. Pergamon Press, Oxford.

Ley, J. A., Montague, C. L., and McIvor, C. C. (1994). Food habits of mangrove fishes: A comparison along estuarine gradients in northeastern Florida Bay. *Bull. Mar. Sci.* **54,** 881–899.

Liem, K. F. (1980). Adaptive significance of intra- and interspecific differences in the feeding repertoire of cichlid fishes. *Am. Zool.* **20,** 295–314.

Liem, K. F. (1986). The pharyngeal jaw apparatus of the Embiotocidae (Teleostei): A functional and evolutionary perspective. *Copeia* **1986,** 311–323.

Liem, K. F. (1993). Ecomorphology of the teleost skull. *In* "The Skull," Vol. 3, "Functional and Evolutionary Mechanisms" (J. Hanken and B. K. Hall, Eds.). pp. 422–452. University of Chicago Press, Chicago.

Liem, K. F., and Greenwood, P. H. (1981). A functional approach to the phylogeny of the pharyngognath teleosts. *Am. Zool.* **15,** 427–454.

Liem, K. F., and Sanderson, S. L. (1986). The pharyngeal jaw apparatus of labrid fishes: A functional morphological perspective. *J. Morphol.* **187,** 143–158.

Lindberg, D. R., Warheit, K. I., and Estes, J. S. (1987). Prey preference and seasonal predation by oystercatchers on limpets at San Nicholas Island, California, USA. *Mar. Ecol. Prog. Ser.* **39,** 105–113.

Lowell, R. B. (1987). Safety factors of tropical versus temperate limpet shells: Multiple selection pressures on a single structure. *Evolution* **41,** 638–650.

Lubchenco, J. (1978). Plant species diversity in a marine intertidal community, importance of herbivore food preference and algal competitive abilities. *Am. Nat.* **112,** 23–39.

Luczkovich, J. J. (1988). The role of prey detection in the selection of prey by pinfish *Lagodon rhomboides* (Linnaeus). *J. Exp. Mar. Biol. Ecol.* **123,** 15–30.

Luczkovich, J. J., Norton, S. F., and Gilmore, R. G., Jr. (1995). The influence of oral anatomy on prey selection during ontogeny of two percoid fishes, *Lagodon rhomboides* and *Centropomus unidecimalis. Environ. Biol. Fish.* **44,** 79–95.

Luczkovich, J. J., and Stellwag, E. J. (1993). Isolation of cellulolytic microbes from the intestinal tract of the pinfish, *Lagodon rhomboides:* Size-related changes in diet and microbial abundance. *Mar. Biol.* **116,** 389–379.

Macer, C. T. (1967). The food web in Red Wharf Bay (North Wales) with particular reference to young plaice (*Pleuronectes platessa*). *Helgolander. Wiss. Meeresunters.* **15,** 560–573.

Machado, K. R. S., Chapman, A. R. O., and Coutinho, R. (1996). Consumer species have limited and variable roles in community organization on a tropical intertidal shore. *Mar. Ecol. Prog. Ser.* **134,** 73–83.

Mahon, R., and Mahon, S. D. (1994). Structure and resilience of a tidepool fish assemblage at Barbados. *Environ. Biol. Fish.* **41,** 171–190.

Marsh, C. P. (1986). Rocky intertidal community organization: The impact of avian predators on mussel recruitment. *Ecology* **67,** 771–786.

Martin, K. L. M. (1995). Time and tide wait for no fish: Intertidal fishes out of water. *Environ. Biol. Fish.* **44,** 165–181.

Mathieson, A. C., and Nienhuis, P. H. (Eds.) (1991). "Intertidal and Littoral Ecosystems (Ecosystems of the World 24)." Elsevier, London.

McDermott, J. J. (1983). Food web in the surf zone of an exposed sandy beach along the mid-Atlantic coast of the United States. *In* "Sandy Beaches as Ecosystems" (A. McLachlan and T. Erasmus, Eds.), pp. 529–538. Dr. W. Junk, The Hague.

McLachlan, A., and Erasmus, T. (1983). "Sandy Beaches as Ecosystems." Dr. W. Junk, The Hague.

McLachlan, A., and Romer, G. (1990). Trophic relationships in a high energy beach and surf-zone ecosystem. *In* "Trophic Relationships in the Marine Environment" (M. Barnes and R. N. Gibson, Eds.), pp. 356–371. Aberdeen University Press, Aberdeen.

McMichael, R. H., Jr., and Ross, S. T. (1987). The relative abundance and feeding habits of juvenile kingfish (Sciaenidae: *Menticirrhus*) in a Gulf of Mexico surf zone. *Northeast Gulf Sci.* **9,** 109–123.

McRoy, C. P., and Helfferich, C. (1977). "Seagrass Ecosystems." Dekker, New York.

Menge, B. A., Lubchenco, J., Ashkenas, L. R., and Ramsey, F. (1986). Experimental separation of effects of consumers on sessile prey in the low zone of a rocky shore in the Bay of Panama: Direct and indirect consequences of food web complexity. *J. Exp. Mar. Biol. Ecol.* **100,** 225–269.

Miller, K. A., and Marshall, W. H. (1987). Food habits of large monkeyface prickleback *Cebidichthys violaceus. Calif. Fish Game* **73,** 37–44.

(Modde, T., and Ross, S. T. (1983). Trophic relationships of fishes occurring within a surf zone habitat in the northern Gulf of Mexico. *Northeast Gulf Sci.* **6,** 109–120.

Montgomery, J. C., and Coombs, S. (1998). Peripheral encoding of moving sources by the lateral line system of a sit-and-wait predator. *J. Exp. Biol.* **201,** 91–102.

Montgomery, W. M. (1977). Diet and gut morphology in fishes, with special reference to the monkeyface prickleback, *Cebidichthys violaceus* (Stichaeidae: Blennoidei). *Copeia* **1977,** 178–182.

Moore, P. G., and Seed, R. (1986). "The Ecology of Rocky Coasts." Columbia University Press, New York.

Moring, J. R. (1988). Food habits and algal associations of juvenile lumpfish, *Cyclopterus lumpus* L., in intertidal waters. *Fish. Bull.* **87,** 233–237.

Morisita, M. (1959). Measuring the dispersion of individuals and analysis of distributional patterns. *Mem. Fac. Sci. Kyushu Univ. Ser. E (Biol.)* **2,** 215–235.

Morton, R. M., Beumer, J. P., and Pollock, B. R. (1988). Fishes of a subtropical Australian saltmarsh and their predation upon mosquitoes. *Environ. Biol. Fish.* **21,** 185–194.

Morton, R. M., Pollock, B. R., and Beumer, J. P. (1987). The occurrence and diet of fishes in a tidal inlet to a saltmarsh in southern Moreton Bay, Queensland. *Austr. J. Ecol.* **12,** 217–237.

Mullaney, M. D., Jr., and Gale, L. D. (1996). Morphological relationships in ontogeny: Anatomy and diet in gag, *Mycteroperca microlepis* (Pisces: Serranidae). *Copeia* **1996,** 167–180.

Muñoz, A. A., and Ojeda, F. P. (1997). Feeding guild structure of a rocky intertidal fish assemblage in central Chile. *Environ. Biol. Fish.* **49,** 471–479.

Nakamura, R. (1971). Food of two cohabiting tidepool Cottidae. *J. Fish. Res. Bd. Can.* **28,** 928–932.

Neighbors, M. A., and Horn, M. H. (1991). Nutritional quality of macrophytes eaten and not eaten by two temperate-zone herbivorous fishes: A multivariate comparison. *Mar. Biol.* **108,** 471–476.

Nemeth, D. H. (1997). Modulation of buccal pressure during prey capture in *Hexagrammos decagrammos* (Teleostei: Hexagrammidae). *J. Exp. Biol.* **200,** 2145–2154.

Noichi, T., Kusano, M., Ueki, D., and Senta, T. (1993). Feeding habits of fishes eating settled larval and juvenile Japanese flounder (*Paralichthys olivaceus*) at Yanagihama Beach, Nagasaki Prefecture. *Bull. Fac. Fish. Nagasaki Univ. Chodai Suikenpo* **73,** 1–6.

Norton, S. F. (1988). The role of the gastropod shell and operculum in inhibiting predation by fishes. *Science* **241,** 92–94.

Norton, S. F. (1991a). Capture success and diet of cottid fishes: The role of predator morphology and attack kinematics. *Ecology* **72,** 1807–1819.

Norton, S. F. (1991b). Habitat use and community structure in an assemblage of cottid fishes. *Ecology* **72,** 2181–2192.

Norton, S. F. (1995). A functional approach to ecomorphological patterns of feeding in cottid fishes. *Environ. Biol. Fish.* **44,** 61–78.

Norton, S. F., and Brainerd, E. L. (1993). Convergence in the feeding mechanics of ecomorphologically similar species in the Centrarchidae and Cichlidae. *J. Exp. Biol.* **176,** 11–29.

O'Farrell, M. M., and Fives, J. M. (1990). The feeding relationships of the shanny, *Lipophrys pholis* (L.) and Montagu's blenny, *Coryphoblennius galerita* (L.) (Teleostei: Blenniidae). *Ir. Fish. Invest. Ser. B* **36,** 3–16.

Ojeda, F. P., and Dearborn, J. H. (1991). Feeding ecology of benthic mobile predators: Experimental analyses of their influence in rocky subtidal communities of the Gulf of Maine. *J. Exp. Mar. Biol. Ecol.* **149,** 13–44.

Ong, T. L., and Sasekumar, A. (1984). The trophic relationship of fishes in the shallow waters adjoining a mangrove shore. *In* "Proceedings of the Asian Symposium on the Mangrove Environment—Research and Management" (E. Saepadmo, A. N. Rao, and D. J. Macintosh, Eds.), pp. 453–469. UNESCO, Kuala Lumpur.

Ortega, S. (1986). Fish predation on gastropods on the Pacific coast of Costa Rica. *J. Exp. Mar. Biol. Ecol.* **97,** 181–191.

Paine, R. T. (1974). Intertidal community structure, experimental studies on the relationship between a dominant competitor and its principal predator. *Oecologia* **15,** 93–120.

Paine, R. T., and Palmer, A. R. (1978). *Sicyases sanguineus:* A unique trophic generalist from the Chilean intertidal zone. *Copeia* **1978,** 75–80.

Palmer, A. R. (1979). Fish predation and the evolution of gastropod shell structure: Experimental and geographic evidence. *Evolution* **33,** 697–713.

Parry, G. D. (1982). The evolution of the life histories of four species of intertidal limpets. *Ecol. Monogr.* **52,** 65–91.

Pawlick, J. R., Albizati, K. F., and Faulkner, D. J. (1986). Evidence of a defensive role for limatulone, a novel triterpene from the intertidal limpet *Collisella limatula. Mar. Ecol. Prog. Ser.* **30,** 251–260.

Peterson, C. H., and Quammen, M. L. (1982). Siphon nipping: Its importance to small fishes and its impact on growth of the bivalve *Protothaca staminea* (Conrad). *J. Exp. Mar. Biol. Ecol.* **63,** 249–268.

Pillai, C. S. G., Gopakumar, G., and Mohan, M. (1992). Ichthyofauna of the intertidal reef flats of Minicoy Atoll, Lakshadweep: An analysis of its structure, relative abundance and food. *J. Mar. Biol. Assoc. India* **34,** 74–83.

Por, F. D., and Dor, I. Eds. (1984). "Hydrobiology of the Mangal." Dr. W. Junk, The Hague.

Posey, M. H. (1986). Predation on a burrowing shrimp: Distribution and community consequences. *J. Exp. Mar. Biol. Ecol.* **103,** 143–161.

Poulin, R., and Fitzgerald, G. J. (1989). Early life histories of three sympatric sticklebacks in a salt-marsh. *J. Fish Biol.* **34,** 207–221.

Purcell, S. W., and Bellwood, D. R. (1993). A functional analysis of food procurement in two surgeonfish species, *Acanthurus nigrofuscus* and *Ctenochaetus striatus* (Acanthuridae). *Environ. Biol. Fish.* **37,** 139–159.

Quammen, M. L. (1984). Predation by shorebirds, fish, and crabs on invertebrates in intertidal mudflats: An experimental test. *Ecology* **65,** 529–537.

Raffaelli, D., Conacher, A., McLachlan, H., and Emes, C. (1989). The role of epibenthic crustacean predators in an estuarine food web. *Estuar. Coast. Shelf Sci.* **28,** 149–160.

Raffaelli, D., and Hawkins, S. (1996). "Intertidal Ecology." Chapman and Hall, London.

Randall, J. E. (1967). Food habits of reef fishes of the West Indies. *Stud. Trop. Oceanogr.* **5,** 665–847.

Rangeley, R. W., and Kramer, D. L. (1995). Use of rocky intertidal habitats by juvenile pollock, *Pollachius virens. Mar. Ecol. Prog. Ser.* **126,** 9–17.

Reimchen, R. E. (1989). Shell color ontogeny and tubeworm mimicry in a marine gastropod, *Littorina mariae. Biol. J. Linn. Soc.* **36,** 97–109.

Reise, K. (1977). Predator exclusion experiments in an intertidal mud flat. *Helgolander Wiss. Meeresunters* **30,** 263–271.

Richard, B. A., and Wainwright, P. C. (1995). Scaling the feeding mechanism of largemouth bass (*Micropterus salmoides*): Kinematics of prey capture. *J. Exp. Biol.* **198,** 419–433.

Rimmer, D. W., and Wiebe, W. J. (1987). Fermentive microbial digestion in herbivorous fishes. *J. Fish Biol.* **31,** 229–236.

Robertson, A. I. (1980). The structure and organization of an eelgrass fish fauna. *Oecologia* **47,** 76–82.

Robertson, A. I. (1984). Trophic interactions between the fish fauna and macrobenthos of an eelgrass community in Western Port, Victoria. *Aquat. Bot.* **18,** 135–153.

Robertson, A. I. (1988). Abundance, diet and predators of juvenile banana prawns, *Penaeus merguiensis,* in a tropical mangrove estuary. *Austr. J. Mar. Freshw. Res.* **39,** 467–478.

Robertson, A. I., and Lenanton, R. C. J. (1984). Fish community structure and food chain dynamics in the surf-zone of sandy beaches: The role of detached macrophyte detritus. *J. Exp. Mar. Biol. Ecol.* **84,** 265–283.

Robertson, C. H., and White, R. W. G. (1986). Feeding patterns of *Nesogobius* sp., *Gymnapristes marmoratus, Neoodax balteeatus,* and *Acanthaluteres spilomelanurus* from a Tasmanian seagrass meadow. *Austr. J. Mar. Freshw. Res.* **37,** 481–489.

Robles, C., and Robb, J. (1993). Varied carnivore effects and the prevalence of intertidal algal turfs. *J. Exp. Mar. Biol. Ecol.* **166,** 65–91.

Romer, G. S., and McLachlan, A. (1986). Mullet grazing on surf diatom accumulations. *J. Fish Biol.* **28,** 93–104.

Rozas, L. P., and LaSalle, M. W. (1990). A comparison of the diets of Gulf killifish, *Fundulus grandis* Baird and Girard, entering and leaving a Mississippi brackish marsh. *Estuaries* **13,** 332–336.

Ruiz-Campos, G., and Hammann, M. G. (1991). Feeding of the woolly sculpin, *Clinocottus analis* (Pisces: Cottidae), in Todos Santos Bay, Baja California, Mexico. *Southwest. Nat.* **36,** 348–353.

Russell, G. (1991). Vertical distribution. *In* "Intertidal and Littoral Ecosystems (Ecosystems of the World 24)" (A. C. Mathieson and P. H. Nienhuis, Eds.), pp. 43–66. Elsevier, The Hague.

Sabapathy, U., and Teo, L. H. (1993). A quantitative study of some digestive enzymes in the rabbitfish, *Siganus canaliculatus,* and the sea bass, *Lates calcarifer. J. Fish. Biol.* **42,** 595–602.

Sanderson, S. L., Cech, J. J., Jr., and Patterson, M. R. (1991). Fluid dynamics in suspension-feeding blackfish. *Science* **251,** 1346–1348.

Sanderson, S. L., and Wassersug, R. (1990). Suspension feeding vertebrates. *Sci. Am.* **262,** 96–101.

Sano, M., Shimizu, M., and Nose, Y. (1984). "Food Habits of Teleostean Reef Fishes in Okinawa Island, Southern Japan." University of Tokyo Press, Tokyo.

Sanusi, S. S. (1980). "A Study on Grazing as a Factor Influencing the Distribution of Benthic Littoral Algae." M.Sc. thesis, University of Ghana.

Sasekumar, A., Chong, V. C., Leh, U., and Cruz, R. D. (1992). Mangroves as a habitat for fish and prawns. *Hydrobiologia* **247,** 195–207.

Sasekumar, A., Ong, T. L., and Thong, K. L. (1984). Predation of mangrove fauna by marine fishes. *In* "Proceedings of the Asian Symposium on the Mangrove Environment—Research and Management" (E. Saepadmo, A. N. Rao, and D. J. Macintosh, Eds.), pp. 378–384. UNESCO, Kuala Lumpur.

Scholz, D. S., Matthews, L. A., and Feller, R. J. (1991). Detecting selective digestion of meiobenthic prey by juvenile spot *Leiostomus xanthurus* (Pisces) using immunoassays. *Mar. Ecol. Prog. Ser.* **72,** 59–67.

Seiderer, L. J., Davis, C. L., Robb, F. T., and Newell, R. C. (1987). Digestive enzymes of the anchovy *Engraulis capensis* in relation to diet. *Mar. Ecol. Prog. Ser.* **35,** 15–23.

Setran, A. C., and Behrens, D. W. (1993). Transitional ecological requirements for early juveniles of two sympatric stichaeid fishes, *Cebidichthys violaceus* and *Xiphister mucosus. Environ. Biol. Fish.* **37,** 381–395.

Smith, A. M. (1991). The role of suction in the adhesion of limpets. *J. Exp. Biol.* **161,** 151–169.

Sousa, W. (1979). Experimental investigations of disturbance and ecological succession in a rocky intertidal algal community. *Ecol. Monogr.* **49,** 227–254.

Stadler, J. H. (1988). "Feeding Biology of the Northern Clingfish, *Gobiesox maeandricus;* Diet, Morphology, and Behavior." Unpublished masters thesis, College of Oceans and Fisheries Science, University of Washington, Seattle.

Stebbins, T. D. (1988). The role of sea urchins in mediating fish predation on a commensal isopod (Crustacea: Isopoda). *J. Exp. Mar. Biol. Ecol.* **124,** 97–113.

Stenton-Dozey, J. M. E., and Griffiths, C. L. (1983). The fauna associated with kelp stranded on a sandy beach. *In* "Sandy Beaches as Ecosystems" (A. McLachlan and T. Erasmus, Eds.), pp. 557–568. Dr. W. Junk, The Hague.

Stephenson, T. A., and Stephenson, A. (1972). "Life between Tidemarks on Rocky Shores." W. H. Freeman, San Francisco.

Stepien, C. A. (1990). Population structure, diets and biogeographic relationships of a rocky intertidal fish assemblage in central Chile: High levels of herbivory in a temperate system. *Bull. Mar. Sci.* **47,** 598–612.

Stepien, C. A., Glattke, M., and Fink, K. M. (1988). Regulation and significance of color patterns of the spotted kelpfish, *Gibbonsia elegans* Cooper, 1864 (Blennioidei: Clinidae). *Copeia* **1988,** 7–15.

Stobbs, R. E. (1980). Feeding habits of the giant clingfish *Chorisochismus dentex* (Pisces: Gobiesocidae). *S. Afr. J. Zool.* **15,** 146–149

Stoner, A. W., and Livingston, R. J. (1984). Ontogenetic patterns in diet and feedng morphology in sympatric sparid fishes from seagrass meadows. *Copeia* **1984,** 174–187.

Suyehiro, Y. (1942). A study of the digestive system and feeding habits of fish. *Japan. J. Zool.* **10,** 1–303.

Tararam, A. S., and Wakabara, Y. (1982). Notes on the feeding of *Blennius cristatus* Linneaus from a rocky pool of Itanhéim, São Paulo State. *Bol. Int. Oceanogr.* **31,** 1–3.

Teas, H. J. (Ed.) (1983). "Biology and Ecology of Mangroves." Dr. W. Junk, The Hague.

Thayer, G. W., Colby, D. R., and Hettler, W. F., Jr. (1987). Utilization of the red mangrove prop root habitat by fishes in south Florida. *Mar. Ecol. Prog. Ser.* **35,** 25–38.

Thresher, R. E., and Colin, P. L. (1986). Trophic structure, diversity, and abundance of fishes of the deep reef (30–300m) at Enewetak, Marshall Islands. *Bull. Mar. Sci.* **38,** 253–272.

Thrush, S. F., Pridmore, R. D., Hewitt, J. E., and Cummings, V. J. (1991). Impact of ray feeding disturbances on sandflat macrobenthos: Do communities dominated by polychaetes or shellfish respond differently? *Mar. Ecol. Prog. Ser.* **69,** 245–252.

Toepfer, C. S., and Fleeger, J. W. (1995). Diet of juvenile fishes *Citharichthys spilopterus, Symphurus plagiusa* and *Gobionellus boleosoma. Bull. Mar. Sci.* **56,** 238–249.

Tsurpalo, A. P. (1993). Feeding characteristics of littoral fish, *Alectrias alectrolophus alectrolophus* and *Stichaeopsis nana* (Stichaeidae) of Shikotan Island (Kuril Islands). *J. Ichthyol.* **33,** 139–144.

Underwood, A. J., and Jernakoff, P. (1981). Effects of interactions between algae and grazing gastropods on the structure of a low shore intertidal algal community. *Oecologia* **48,** 221–233.

Varas, E., and Ojeda, F. P. (1990). Intertidal fish assembages of the central Chilean coast: Diversity, abundance and trophic patterns. *Rev. Biol. Mar. Valparaiso* **25,** 59–70.

Verigina, I. A. (1991). Basic adaptations of the digestive system in bony fishes as a function of diet. *J. Ichthyol.* **31,** 8–20.

Vermeij, G. J. (1978). "Biogeography and Adaptation: Patterns of Marine Life." Harvard University Press, Cambridge, MA.

Vermeij, G. J. (1987). "Evolution and Escalation." Princeton University Press, Princeton, NJ.

Wainwright, P. C. (1988). Morphology and ecology: Functional basis of feeding constraints in Caribbean labrid fishes. *Ecology* **69,** 635–645.

Wainwright, P. C., and Richard, B. A. (1995). Predicting patterns of prey use from morphology of fishes. *Environ. Biol. Fish.* **44,** 97–113.

Walsh, G. E. (1974). Mangroves, a review. *In* "Ecology of Halophytes" (R. J. Reimold and W. J. Queen, Eds.), pp. 51–174. Academic Press, New York.

Walters, K., Jones, E., and Etherington, L. (1996). Experimental studies of predation on metazoans inhabiting *Spartina alterniflora* stems. *J. Exp. Mar. Biol. Ecol.* **195,** 251–265.

Wells, A. W. (1986). Aspects of the ecology and life history of the wooly sculpin, *Clinocottus analis,* from southern California. *Calif. Fish Game* **72,** 213–226.

Wells, R. A. (1980). Activity pattern as a mechanism of predator avoidance in two species of acmeid limpet. *J. Exp. Mar. Biol. Ecol.* **48,** 151–168.

Williams, D. McB., and Hatcher, A. I. (1983). Structure of fish communities on outer slopes of inshore, mid-shelf, and outer shelf reefs of the Great Barrier Reef. *Mar. Ecol. Prog. Ser.* **10,** 239–250.

Wilson, W. H., Jr. (1989). Predation and the mediation of intraspecific competition in an infaunal community in the Bay of Fundy. *J. Exp. Mar. Biol. Ecol.* **132,** 221–245.

Wiltse, W. I., Foreman, K. H., Teal, J. M., and Valiela, I. (1984). Effects of predators and food resources on the macrobenthos of salt marsh creeks. *J. Mar. Res.* **42,** 923–942.

Witman, J. D., and Suchanek, T. H. (1984). Mussels in flow: Drag and dislodgment by epizoans. *Mar. Ecol. Prog. Ser.* **16,** 259–268.

Woodin, S. A. (1974). Polychaete abudance patterns in a marine soft-sediment environment: The importance of biological interactions. *Ecol. Monogr.* **44,** 171–187.

Worgan, J. P., and Fitzgerald, G. J. (1981). Diet activity and diet of three sympatric sticklebacks in tidal salt marsh pools. *Can. J. Zool.* **59,** 2375–2379.

Wright, J. M. (1988). Recruitment patterns and trophic relationships of fish in Sulaibikhat Bay, Kuwait. *J. Fish Biol.* **33,** 671–687.

Wright, J. M. (1989). Biology of *Leiognathus decorus* (Leiognathidae) in Sulaibikhat Bay, Kuwait. *Austr. J. Mar. Freshw. Res.* **40,** 179–185.

Yoshiyama, R. M., and Darling, J. D. S. (1982). Grazing by the intertidal fish *Anoplarchus purpurescens* upon a distasteful polychaete worm. *Environ. Biol. Fish.* **7,** 39–45.

Yoshiyama, R. M., Knowlton, A. L., Welter, J. R., Comfort, S., Hopka, B. J., and Wallace, W. D. (1996a). Laboratory behavior of mosshead sculpins *Clinocottus globiceps* toward their sea anemone prey. *J. Mar. Biol. Assoc. U.K.* **76,** 793–809.

Yoshiyama, R. M., Wallace, W. D., Burns, J. L., Knowlton, A. L., and Welter, J. R. (1996b). Laboratory food choice by the mosshead sculpin, *Clinocottus globiceps* (Girard) (Teleostei; Cottidae), a predator of sea anemones. *J. Exp. Mar. Biol. Ecol.* **204,** 23–42.

Zedler, J. B. (1982). "The Ecology of Southern California Coastal Salt Marshes: A Community Profile." U.S. Fish & Wildlife Service Biological Services Program. FWS/OBS-81/54.

Zieman, J. C. (1982). "The Ecology of the Seagrasses of South Florida: A Community Profile." U.S. Fish & Wildlife Service Biological Services Program. FWS/OBS-82/25.

Zihler, F. (1982). Gross morphology and configuration of digestive tracts of Cichlidae (Teleostei, Perciformes): Phylogenetic and functional signficance. *Neth. J. Zool.* **32,** 544–571.

# 13

# Intertidal Fish Communities

R. N. Gibson

*Centre for Coastal and Marine Sciences, Dunstaffnage Marine Laboratory, Oban, Argyll, Scotland*

R. M. Yoshiyama

*Department of Wildlife, Fish and Conservation Biology, University of California, Davis, California*

## I. Introduction

Intertidal fishes are generally considered to be those that live their postlarval lives in the intertidal zone and possess particular morphological, physiological, and behavioral adaptations that enable them to do so. The combined populations of such species can be considered to be the intertidal fish community, taking the definition of community as "an assemblage of species populations which occur together in space and time" (Begon *et al.,* 1990). Such a view, however, is incomplete because it is based on the situation at low tide. At high tide, numerous other species move into previously emersed areas and some of them may interact with members of the resident community. On this basis, both groups of species constitute the total community of fishes inhabiting the intertidal zone. Generally speaking, the residents have received the most attention because only rarely have both sections of any one community been described in equivalent detail. The reasons for this difference are probably twofold. First, the resident species, because of their adaptations to intertidal life, are a particularly fascinating group to study. In contrast, those fishes that move into the intertidal zone at high tide are essentially extensions of the wider inshore ichthyofauna and show no special adaptation to intertidal life. Second, sampling at low tide is considerably easier than at high tide, particularly on rocky shores. Consequently, most intertidal observations and collections are made when the tide is out.

Historically, the study of intertidal fish communities has been geographically patchy and until recently most studies have been carried out on the coasts of the northeast Pacific (British Columbia to southern California) or western Europe (Great Britain to the western Mediterranean). As a result, most of the current information originates from these areas, particularly from North America. The results of studies in the latter area can therefore be



*Intertidal Fishes: Life in Two Worlds*

used as a framework with which results from other locations can be compared. This chapter concentrates on rocky shore communities and first considers community structure in terms of its functional components. It then describes how structure varies over a range of spatial and temporal scales, includes a brief discussion of the few known instances of species interactions within particular communities and assesses their importance in structuring such communities.

## II. Functional Components of Community Structure

### A. Duration of Occupancy (Residents and Visitors)

Fishes that use the intertidal zone as a habitat do so for greatly different lengths of time or proportions of their life history. At one extreme are the species that live there for almost all their lives; at the other are those that enter for only brief periods during high tide. Recognition of these differences in duration of intertidal residence have led to various attempts at classifying intertidal fishes on the basis of their duration of occupancy of the zone. Several schemes have been proposed (Breder, 1948; Gibson, 1969, 1988; Potts, 1980; Thomson and Lehner, 1976; Grossman, 1982; Moring, 1986; Mahon and Mahon, 1994) and all make the basic distinction between the permanent inhabitants (the residents) and others that visit the intertidal zone for varying lengths of time during their life (the visitors or transients). There are, however, no clear boundaries between the various categories that have been proposed and there is a continuum of types between the two extremes. The major factor that determines the duration of intertidal residence is the nature of the substratum. In intertidal areas that are devoid of refuges, such as sandy beaches or mudflats, the great majority of fishes are present only at high tide and the community consists almost solely of transients. The main exceptions to this rule are the mudskippers, tropical gobiid fishes that construct their own refuges in the form of burrows in the mud (Clayton, 1993). Several other gobies (e.g., *Lepidogobius lepidus* as studied by Grossman, 1979) occupy intertidal invertebrate burrows. On rocky shores, the "typical" (Breder, 1948) resident species are small, cryptic forms that show numerous morphological, physiological, and behavioral adaptations to intertidal life. They settle intertidally as larvae and grow, reproduce, and die there. Other species may settle intertidally but once they have reached a particular size or maturity stage they move into deeper water where they complete their life cycle. Many of these, variously called "partial residents" (Gibson, 1969), "seasonal transients" (Potts, 1980) or "secondary residents" (Thomson and Lehner, 1976), show no particular adaptations to intertidal life. Their duration of residence may vary from a few weeks to a few years and they are usually the juveniles of more mobile forms that have their main distribution in subtidal areas. As adults they may use the intertidal area as a foraging or breeding ground on a tidal, diel, or seasonal basis. Finally, there are the "casual" or "accidental" species, usually pelagic planktivores that feed over the intertidal zone at high tide and occasionally become trapped in pools at low tide. Such species probably play little part in the ecology of the intertidal zone. Except in a general sense, the usefulness of such a classification system for comparative purposes is rather restricted for several reasons (Mahon and Mahon, 1994). First, the relative proportions of residents and transients may vary seasonally so that the time of year at which collections are made must be taken into account (Bennett, 1987). Second, community structure depends on the size and level of the area sampled, as well as the overall size of the sample itself. Large tidepools lower on the shore,

**Table 1.** Comparison of Community Structure on Rocky Shores with Respect to Residency Status of Members

| Location | PR S | PR I | SR S | SR I | T S | T I | Species | Individuals | Reference |
|---|---|---|---|---|---|---|---|---|---|
| Azores | 69 | | | | 31 | | 13 | ? | Santos *et al.*, 1994 |
| Chile | 55 | 6 | 45 | 94 | | | 11 | 597 | Varas and Ojeda, 1990 |
| Chile | 67 | 83 | 25 | 17 | 8 | 1 | 12 | 649 | Stepien, 1990 |
| Little Cayman | 7 | 33 | 93 | 68 | | | 15 | 120 | Potts, 1980 |
| Barbados | 54 | 61 | 46 | 39 | | | 63 | 3432 | Mahon and Mahon, 1994 |
| South Africa | 39 | | 61 | | | | 18 | 1014 | Christensen and Winterbottom, 1981 |
| South Africa | 62 | | 38 | | | | 21 | 2595 | Bennett, 1987 |
| South Africa | 100 | 100 | | | | | 20 | 5049 | Prochazka, 1996 |
| Taiwan | 18 | 33 | 76 | 51 | 6 | 17 | 194 | 8595 | Chang *et al.*, 1977 |
| Taiwan | 23 | 20 | 74 | 75 | 3 | 4 | 177 | 4162 | Lee, 1980 |
| USA, Maine | 0 | | 100 | | | | 22 | 1135 | Moring, 1990 |
| USA, Washington | 26 | 83 | 21 | 16 | 53 | 1 | 34 | 14799 | Cross, 1981 |
| USA, California | 31 | 79 | 35 | | 35 | | 29 | 2857 | Grossman, 1982 |
| USA, California | 42 | | 45 | | 13 | | 31 | 1599+ | Moring, 1986 |
| USA, California | 54 | 90 | 33 | 9 | 13 | 1 | 24 | 3703 | Yoshiyama, 1981 |
| Gulf of California | 68 | 99 | | | 32 | 1 | 52 | 13680 | Thomson and Lehner, 1976 |

*Note.* PR = primary residents; SR = secondary residents; T = transients (Thomson and Lehner, 1976). The entries in the table represent the relative number of species (S) and individuals (I) in each residential category as a percentage of the total number of species (Species) and individuals (Individuals).

for example, are likely to contain more species (Green, 1971a; Gibson, 1972; Prochazka and Griffiths, 1992; Mahon and Mahon, 1994) and a higher proportion of transients than small pools at a higher level. Third, where the description of community structure in such terms has been attempted, authors have rarely used directly equivalent categories so that comparisons between such studies are difficult. Nevertheless, some broad differences between areas are detectable (Table 1). Prochazka (1996) notes the remarkable absence of transient species on the west coast of South Africa in comparison with other regions. In Maine, where fish are absent from tidepools in winter (Moring, 1990), there are no resident species. Overall, it seems likely that both resident and transient species contribute significantly to community structure but the extent to which they do so varies both seasonally and geographically.

## B. Age Structure of Populations and the Function of the Intertidal Zone as a Nursery Ground

If resident species are defined as those that live all or most of their lives in the intertidal zone then the age structure of the population will reflect the longevity of the species concerned. Examination of available life history data indicates that most resident species are relatively short lived, probably 5–6 years or fewer (Gibson, 1969; Stepien, 1990), but some species reach 10 years or more. In the latter case the older, larger individuals usually occur at the lowest levels on the shore or even subtidally. Studies of populations on the west coast of the United States provide a good example of the variety in age structure and distribution that can be encountered. *Oligocottus maculosus* populations in northern California, Puget

Sound (Washington) and Vancouver Island consist mainly of 0- and 1-year age classes, although small numbers of 2-year and perhaps older fish are present (Atkinson, 1939; Green, 1971a; Moring, 1979). There are two age classes of *O. snyderi* at Dillon Beach, California, judging from the frequency distribution of fish lengths in that population (Freeman *et al.*, 1985). Chadwick (1976), in analyzing vertebral annuli, showed that there were as many as five age groups (beyond the 0-year class) of *O. maculosus* and four to six age groups of *Clinocottus globiceps* present at northern California and Vancouver Island localities, although the older age groups were substantially less abundant. Mgaya (1992) similarly identified five age groups for *C. globiceps* (including 0-year fish) and four age groups for *O. maculosus* in a different Vancouver Island population. These may well be the upper age limits for the relatively small species that are restricted to the intertidal habitat, although the largest (and oldest) individuals of some of those species (e.g., *C. analis, C. globiceps, C. recalvus*) tend to occur in the lowest tidepools and at the interface with the shallow subtidal zone (Williams, 1957; Wells, 1986; R.M.Y. personal observation). The larger species that occur intertidally as juveniles and later shift to subtidal areas attain more advanced ages, for example, 13+ years for *Scorpaenichthys marmoratus* (O'Connell, 1953) and up to 18 years for *Cebidichthys violaceus* (Marshall and Echeverria, 1992). In addition to these spatial differences in age structure, seasonal differences are also likely in those areas where reproduction is seasonal with young-of-the-year being dominant immediately after settlement.

The observation that tidepools sometimes contain a large number of juveniles of species that are predominantly subtidal has led to the suggestion that the intertidal zone acts as a nursery ground for those species. Such a suggestion is based on the assumption that the area is a temporal and spatial refuge from predators, which are considered to be larger and more numerous in the subtidal zone. Juveniles in pools, although protected from subtidal predators over the low-tide period, may be subject to the alternative risk of predation from birds (Mahon and Mahon, 1994; Prochazka, 1996). The relative risks from these two types of predator have not been evaluated. Whether pools are of greater significance than subtidal regions for the growth and survival of juveniles seems to vary with location, and the perception of the relatively greater importance of pools for subtidal species may have arisen in the absence of information on the presence of juveniles in the subtidal. Where such comparisons between intertidal and subtidal populations have been made (Bennett, 1987; Smale and Buxton, 1989), intertidal juvenile densities are no higher than in the subtidal zone. In addition, it has been noted that for some areas the majority of juveniles occurring intertidally belong to resident species and that the juveniles of deeper water species are comparatively rare (Prochazka, 1996; Willis and Roberts, 1996). In central Chilean intertidal habitats the juveniles of some species were never seen subtidally (Varas and Ojeda, 1990) and on the East Cape, South Africa, pools are considered to be important nursery areas for some transient species (Beckley, 1985a). Yet, at other locales the intertidal habitat is evidently of importance to subtidal species. The Patagonian intertidal ichthyofauna, for example, is dominated by juvenile cod icefishes (Nototheniidae) whose adults live in deeper water (Pequeño and Lamilla, 1995). Similarly, in Barbados many partial residents of intertidal pools are juveniles of reef species and may represent an overflow from the deeper reef areas (Mahon and Mahon, 1994). This intertidal reservoir of juveniles may play a role in replenishing populations in subtidal reef habitats. In New South Wales all 25 dominant species (98% of the total) in one large tidepool occurred as juveniles and only 10 species were also represented as adults (Lardner *et al.*, 1993). In this case, however, the pool had a limited connection to the subtidal region. Pools on muddy and sandy beaches are also known to act as nursery areas where the absence of predators is considered to enhance juvenile survival (Crabtree and Dean, 1982; van der Veer and Bergman, 1986).

## C. Trophic Structure of Communities

In rocky intertidal areas the majority of species are carnivorous and feed entirely on benthic invertebrates. Other species include variable proportions of algae in their diet (omnivores) and relatively few feed predominantly or solely on algae (herbivores). Such a classification into these three basic feeding categories must be qualified by recognizing that carnivores and herbivores represent the extremes of a spectrum of dietary composition and that in some species a shift from carnivory to herbivory takes place as fish grow as, for example, in the giant goby, *Gobius cobitis* (Gibson, 1970), the stichaeids *Cebidichthys violaceus* and *Xiphister mucosus* (Montgomery, 1977; Barton, 1982; Horn *et al.,* 1982; Horn, 1983; Miller and Marshall, 1987), and the sculpin *Clinocottus globiceps* (Grossman, 1986a; R.M.Y., unpublished data). Consequently the trophic structure of the community may depend on the age structure of its component species. Nevertheless, species that rely on algae for their main energy source seem to be in the minority in all the communities so far investigated (Table 2), although few studies have been done in the tropics where herbivory might be expected to be more common. Within the carnivores and omnivores the most common prey are small Crustacea, particularly amphipods, decapods, and copepods. The last group is especially common in the diets of smaller individuals. Other prey groups such as polychaetes, molluscs, and echinoderms feature relatively rarely in the diets of most species. Few resident species include fish in their diets, suggesting that at low tide at least, tidepools are relatively free of predators. One exception in the Gulf of California is the piscivorous *Paralabrax maculofasciatus,* which may act as a keystone predator (Thomson and Lehner, 1976) in regulating species diversity of tidepool communities.

Concentration on relatively few prey groups means that diet overlap can sometimes be high but differences in the relative proportions of the main prey types are usually recognizable, suggesting that available prey resources are being partitioned. In the African Cape Peninsula community, many species feed on the same prey species in roughly equivalent amounts but relative specialists feeding on particular groups such as barnacles and ophiuroids are also present (Bennett *et al.,* 1983). In an extensive study in California, Grossman (1986a) found dietary overlap to be low. Six basic groups of crustacean feeders could be distinguished together with a specialist algae/sea anemone feeder. In this location, food resource partitioning was considered to be responsible for community organization because other likely factors such as predation, competition for space, and environmental disturbance seemed to have little effect.

## D. Community Structure in Terms of Behavior

Observations of fish in tidepools over the low-tide period, or when they are more widely dispersed at high tide, suggest that they fall essentially into two categories; midwater schooling species and cryptic benthic species (Thomson and Lehner, 1976; Potts, 1980; Christensen and Winterbottom, 1981; Lardner *et al.,* 1993; Mahon and Mahon, 1994). These two categories also roughly correspond to the transients and residents, respectively. The midwater schooling species are usually the juveniles of subtidal forms and often make up more than half of the community in terms of numbers of individuals but rather less in terms of species. The cryptic benthic species category includes a variety of behavioral types

**Table 2.** Relative Abundances of Species in Different Trophic Categories in Some Rocky Shore Intertidal Fish Communities

| Location | Carnivores | Omnivores | Herbivores | Species in sample | Basis of assessment | Reference |
|---|---|---|---|---|---|---|
| Taiwan | 16 | 76 | 8 | 37 | % occurrence | Chang and Lee, 1969 |
| France, Atlantic | 46 | 64 | 0 | 13 | % occurrence | Gibson, 1972 |
| Chile | 60 | 0 | 40 | 10 | % volume | Stepien, 1990 |
| Marshall Islands | 60 | 28 | 12 | 33 | % of total weight of sample | Bussing, 1972 |
| Chile | 69 | 23 | 8 | 13 | % total food weight | Muñoz and Ojeda, 1997 |
| Gulf of California | 70 | 30 | 0 | 10 | % volume, 10 dominant species only | Thomson and Lehner, 1976 |
| France, Mediterranean | 74 | 21 | 5 | 19 | % occurrence | Gibson, 1968b |
| South Africa | 85 | 15 | 0 | 20 | % occurrence | Bennett *et al.*, 1983 |
| USA, California | 86 | 7 | 7 | 15 | % food weight | Grossman, 1986b |

*Note*. The numbers in the categories are percentages of the total number of species at each location. Carnivores are arbitrarily classified as species with $<5\%$ plant material in the diet, omnivores those with 5–69%, and herbivores those with $\geq 70\%$.

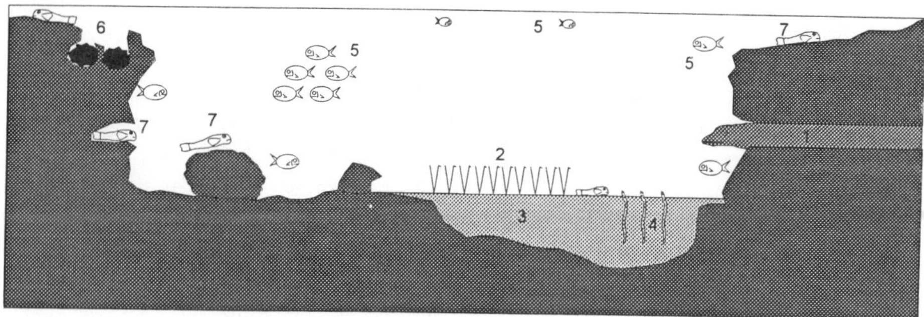


**Figure 1.** Diagrammatic representation of the distribution of fishes in a Barbados tidepool. (1) Eel, (2) turtlegrass, (3) sand, (4) spaghetti eels, (5) partial resident juvenile reef fishes, (6) the sea urchin *Echinometra lucunter* in burrows that are inhabited by nineline goby and red clingfish, (7) cryptic/benthic true resident species. Reproduced from Mahon and Mahon (1994) with kind permission from Kluwer Academic Publishers.

that are usually solitary and well camouflaged and may be territorial. They are generally small and secretive in their behavior and may inhabit weed, holes, open areas, or very specific microhabitats (Figure 1). The activity patterns of most species in the wild are unknown but, because most are likely to be visual feeders, the majority are probably diurnal (Thomson and Lehner, 1976). Feeding has also rarely been observed, particularly in cryptic species, but in the absence of direct evidence, analyses of diet and morphology can be used to deduce feeding behavior. Within the three main categories of carnivores, omnivores, and herbivores there is likely to be a wide range of feeding behavior employed ranging from the planktivory of the midwater schooling transients, through browsing of algae and sessile invertebrates, to specialized hunting methods of those species with eel-like body forms that can penetrate small spaces inaccessible to wider bodied species.

# III. Spatial Variation in Community Structure

Apart from the functional aspects of community structure, several levels of spatial organization can also be discerned, depending on the scale over which communities are viewed. Such scales range from the global to the individual.

## A. Global Patterns: Variation among Continents and with Latitude

Despite the numerous species of fishes found in the intertidal zone, it is remarkable that the majority of the truly resident species belong to a relatively small number of families, particularly the Blenniidae (blennies), Gobiidae (gobies), Cottidae (sculpins), Clinidae (clinids), Gobiesocidae (clingfishes), Tripterygiidae (triplefin blennies), and Stichaeidae (pricklebacks). Most of these families have a wide distribution but intertidal fish communities in different parts of the world tend to be dominated by members of one or a few of these principal families; the clinids in South Africa, the triplefins in New Zealand, the sculpins on the west coast of North America, and the blennies and gobies in the eastern Atlantic and Mediterranean. Other parts of the world have been less well studied but are likely to be similar.

In addition to these continental taxonomic differences in community structure, latitudinal patterns can also be discerned, with the species and family richness of communities increasing toward the equator (Table 3). These relationships are statistically significant (Spearman's rank correlation, $P < 0.0002$) but the data are greatly biased toward the Northern Hemisphere and probably also by the fact that the actual number of species and families recorded at any one location is dependent on the length of the study and the size and number of the pools sampled (Ibáñez *et al.*, 1989; Mahon and Mahon, 1994). Nevertheless, certain major groups are usually consistently abundant in intertidal communities at different points within any one region (Figure 2A), although the particular species that are present vary from locality to locality, generally in a latititudinal pattern of species replacement as, for example, in the cottids of the Pacific coast of North America (Figure 2B) (cf. Green, 1971a; Cross, 1981; Grossman, 1982; Yoshiyama *et al.*, 1986). Within these patterns of species replacement there are often particularly noticeable breakpoints; Point Conception, California (Hubbs, 1948; Briggs, 1974; Horn and Allen, 1978; Horn *et al.*, 1983) and the Basque coast of northern Spain (Ibáñez *et al.*, 1989) are two examples. A longitudinal pattern of familial replacement has been described for the coasts of South Africa where the Clinidae and Gobiesocidae are dominant on the west coast and are gradually replaced in an easterly direction by the Sparidae, Gobiidae, Blenniidae, Mugilidae, and Cheilodactylidae (Prochazka and Griffiths, 1992).

## B. Regional Variation in Community Structure

At the most obvious level, rocky intertidal fish communities are geographically patchily distributed because rocky habitats occur along the coast in patches, often separated by extensive stretches of sandy beach. The species composition of the communities inhabiting such distinct habitats have rarely been compared but there seems to be little interchange between them. Where such interchange occurs, it is the sandy beach species that occur as visitors to rocky areas rather than the reverse. In the Azores, for example, of the 13 species found on rocky shores only 3 are found on adjacent sandy beaches and all 3 are juveniles of pelagic species (Santos *et al.*, 1994). In the Gulf of California, also, many of the transient species are represented by juveniles of sandy beach forms (Thomson and Lehner, 1976) and such observations probably apply to many other parts of the world.

What this spatial separation means for the underlying dynamic structure of such taxonomically and ecologically distinct communities is not immediately obvious. Does the habitat patchiness mean that spatially separated intertidal fish communities, and the populations that constitute them, exist as isolated and independent units, or are they somehow connected? If they are connected, is it by dispersal and mutual recruitment of individuals during vagile life stages or, in the case of rocky shore populations, by "diffusion" of populations along subtidal rocky expanses or narrow reef corridors that connect intertidal rocky patches? A question related to the preceding ones, but at a smaller spatial scale (10s to 100s of meters, instead of kilometers) is whether, at any given locality, the resident intertidal fish fauna is distinct from the subtidal fauna or simply an extension of a much larger but essentially similar community of fishes that is centered in the subtidal areas. For this last question it is reasonable to consider the transient species as peripheral components of subtidal communities, but for residents the situation is not as clear cut. In a summary of previous sampling surveys at two California localities and in northern British Columbia–southeast Alaska, Yoshiyama *et al.* (1987) showed that the species compositions of the respective intertidal and subtidal fish faunas were essentially distinct and that relatively few

**Table 3.** Taxonomic Structure of Rocky Shore Intertidal Fish Communities Based on Collections Made at Low Water

| Location | Lat. | Long. | Families | Species | Dominant family by species | Dominant family by individuals | Reference |
|---|---|---|---|---|---|---|---|
| USA, Alaska | 60° N | 148° W | 6 | 18 | Cottidae, Stichaeidae, Pholidae | Stichaeidae, Pholidae, Liparidae, Cottidae | Barber *et al.*, 1995 |
| Scotland | 56° N | 5° W | 9 | 11 | Blenniidae, Gadidae | Blenniidae, Pholidae, Cottidae | Gibson, unpublished |
| Kamchatka | 55° N | 165° E | 8 | 15 | Cottidae, Stichaeidae, Hexagrammidae | Cottidae, Hexagrammidae, Stichaeidae | Matyushin, 1982 |
| Komandorskie Islands | 55° N | 166° E | 9 | 20 | Cottidae, Liparidae, Stichaeidae | Cottidae, Liparidae, Stichaeidae | Pinchuk, 1976a |
| Paramushir Island | 50° N | 156° E | 5 | 9 | Cottidae, Pholidae | Cottidae | Pinchuk, 1976a |
| England, Guernsey | 49° N | 2° W | 10 | 15 | Gobiidae, Blenniidae | Gobiidae, Atherinidae, Blenniidae | Wheeler, 1970 |
| France, Atlantic | 49° N | 4° W | 9 | 13 | Blenniidae, Gobiidae, Gobiesocidae, Gadidae | Gobiidae, Blenniidae, Syngnathidae, Gadidae | Gibson, 1972 |
| USA, Washington | 47° N | 122° W | 4 | 9 | Cottidae, Pholidae | Cottidae, Gobiesocidae | Armstrong *et al.*, 1976 |
| Iterup Island | 45° N | 148° E | 6 | 14 | Stichaeidae, Cottidae | Cottidae, Stichaeidae | Pinchuk, 1976b |
| Kunashir Island | 44° N | 146° E | 6 | 10 | Pholidae, Stichaeidae, Cottidae | Cottidae, Pholidae, Stichaeidae | Pinchuk, 1976a |
| Shikotan Island | 44° N | 147° E | 4 | 10 | Sticheidae, Cottidae | Cottidae, Stichaeidae | Pinchuk, 1976a |
| USA, Maine | 44° N | 68° W | 17 | 22 | Gasterosteidae, Cottidae | Cyclopteridae, Gasterosteidae, Pholidae, Cottidae | Moring, 1990 |
| France, Mediterranean | 43° N | 3° E | 8 | 19 | Blenniidae, Gobiidae, Gobiesocidae | Gobiidae, Blenniidae, Gobiesocidae | Gibson, 1968 |
| Spain, Mediterranean | 42° N | 3° E | 3 | 17 | Blenniidae | Blenniidae, Gobiidae | Macpherson, 1994 |
| USA, New England | 42° N | 71° W | 12 | 13 | Gasterosteidae | Labridae, Pholidae, Cottidae | Collette, 1986 |
| USA, California | 41° N | 124° W | 8 | 20 | Cottidae | ? | Moring, 1986 |
| Spain, Atlantic | 41° N | 3° W | 12 | 31 | Blenniidae | Blenniidae, Gobiesocidae, Gobiidae | Ibanez *et al.*, 1989 |
| Portugal | 39° N | 9° W | 4 | 7 | Blenniidae | Blenniidae | Arruda, 1979 |
| USA, California | 38° N | 123° W | 11 | 29 | Cottidae | Cottidae, Scorpaenidae, Pholidae | Grossman, 1982 |
| Portugal | 38° N | 9° W | 9 | 15 | Blenniidae, Gobiidae | Gobiidae, Blenniidae | Beja, 1995 |
| USA, California | 37° N | 122° W | 9 | 24 | Cottidae, Stichaeidae | Cottidae, Stichaeidae | Yoshiyama, 1981 |
| Azores | 37° N | 28° W | 8 | 13 | Blenniidae, Mugilidae | Blenniidae, Mugilidae | Santos *et al.*, 1994 |
| USA, Gulf of California | 31° N | 113° W | >16 | >52 | Labrisomidae, Gobiidae, Pomacentridae | Pomacentridae, Labrisomidae, Kyphosidae, Gobiidae | Thomson and Lehner, 1976 |
| Bahamas | 26° N | 79° W | 10 | 11 | Pomacentridae | ? | Breder, 1948 |

| | | | | | | | |
|---|---|---|---|---|---|---|---|
| Oman | 24° N | 58° E | 11 | 29 | Blenniidae, Gobiidae, Pomacentridae | Blenniidae, Pomacentridae, Gobiidae | Gibson, unpublished |
| Taiwan[a] | 22° N | 121° E | 46 | 177 | Labridae, Blenniidae, Muraenidae, Pomacentridae | Pomacentridae, Blenniidae, Labridae, Acanthuridae | Lee, 1980 |
| Taiwan[a] | 22° N | 121° E | 53 | 192 | Labridae, Blenniidae, Pomacentridae, Gobiidae | Gobiidae, Labridae, Atherinidae | Chang *et al.*, 1977 |
| Little Cayman Island | 21° N | 80° W | 11 | 15 | Pomacentridae, Gerridae | Gobiidae, Pomacentridae, Labridae | Potts, 1980 |
| Barbados | 13° N | 60° W | 28 | 63 | Labrisomidae, Gobiidae, Labridae, Muraenidae | Labrisomidae, Acanthuridae, Labridae, Pomacentridae | Mahon and Mahon, 1994 |
| Marshall Islands | 11° N | 162° E | 16 | 33 | Acanthuridae, Blenniidae, Gobiidae, Holocentridae | Pomacentridae, Blenniidae, Acanthuridae | Bussing, 1972 |
| Costa Rica | 10° N | 86° W | 22 | 46 | Clinidae, Labridae, Pomacentridae, Muraenidae | Clinidae, Pomacentridae, Gobiidae | Weaver, 1970 |
| Australia, NSW[a] | 32° S | 152° E | 42 | 99 | Pomacentridae, Serranidae, Labridae, Clinidae | Tetraodontidae, Pomacentridae, Mugilidae, Blenniidae | Lardner *et al.*, 1993 |
| South Africa | 33° S | 27° E | 19 | 36 | Blenniidae, Sparidae, Serranidae | Sparidae, Mugilidae | Christensen and Winterbottom, 1981 |
| Chile | 33° S | 71° W | 9 | 12 | Blenniidae, Kyphosidae, Labrisomidae | Tripterygiidae, Clinidae, Blenniidae, Kyphosidae | Stepien, 1990 |
| Chile | 33° S | 71° W | 8 | 11 | Blenniidae, Kyphosidae, Tripterygiidae | Kyphosidae, Batrachoididae, Tripterygiidae | Varas and Ojeda, 1990 |
| South Africa | 34° S | 18° E | 5 | 20 | Clinidae, Gobiesocidae | Clinidae, Gobiesocidae | Prochazka, 1996 |
| South Africa | 34° S | 20° E | 9 | 21 | Clinidae | Clinidae, Sparidae | Bennett, 1987 |
| South Africa | 34° S | 26° E | 15 | 35 | Clinidae, Sparidae, Blenniidae | Clinidae, Sparidae, Blenniidae | Beckley, 1984 |
| New Zealand | 41° S | 174° E | 9 | 26 | Tripterygiidae, Gobiesocidae | Tripterygiidae, Gobiesocidae, Plesiopidae | Willis and Roberts, 1996 |
| Patagonia | 42° S | 73° W | 21 | 28 | Nototheniidae, Ophidiiidae, Labrisomidae, Clinidae | Nototheniidae, Galaxiidae, Normanichthyidae, Labrisomidae | Pequeño and Lamilla, 1995 |
| Patagonia | 53–55° S | 68–70° W | 9 | 20 | Nototheniidae, Zoarcidae, Harpagiferidae, Atherinidae | ? | Pequeño *et al.*, 1995 |

*Note.* Dominant families are those with the most species and individuals listed in order of abundance. Latitudes and longitudes are approximate. The data are sorted by latitude from North to South.

[a]Collections made from pools with connections to the subtidal and that may not, therefore, be directly comparable with the other data.

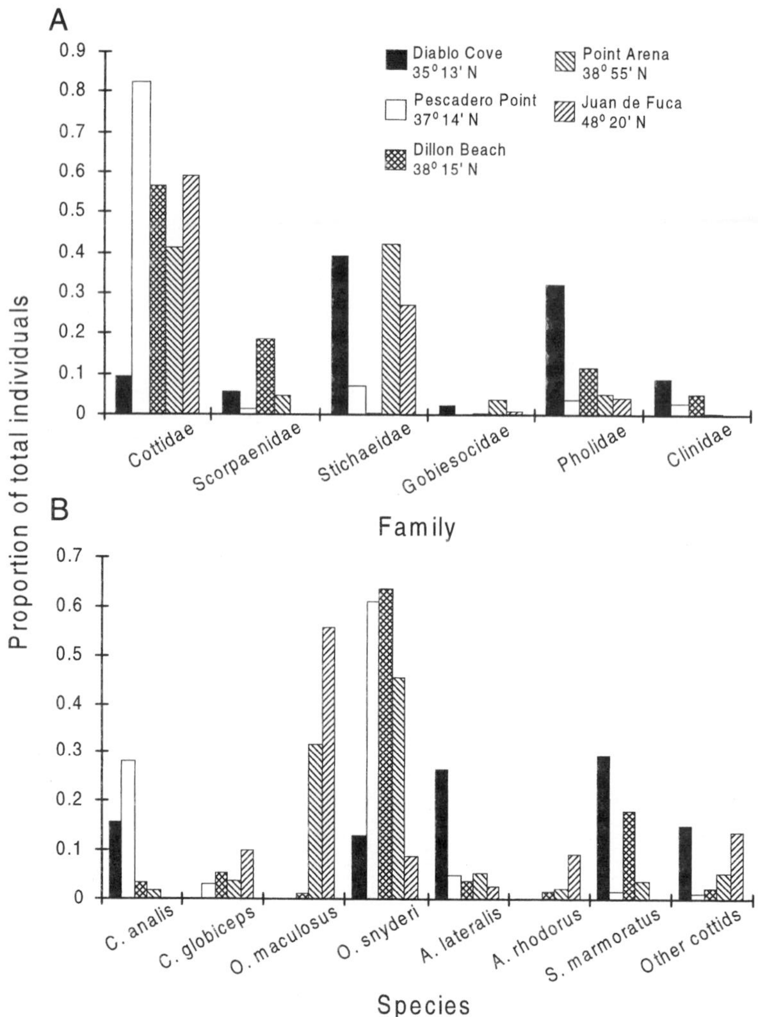


**Figure 2.** Proportional abundances of intertidal fish families (A) and species of Cottidae (B) at five localities on the North American Pacific coast. The names and latitudes of each locality are given in A. In B the genera represented by the major species are *Clinocottus, Oligocottus, Artedius* (*A. lateralis*), *Ascelichthys* (*A. rhodorus*), and *Scorpaenichthys.* Data are from Cross (1981), Grossman (1982), and Yoshiyama *et al.* (1986, 1987).

species were abundant in both intertidal and subtidal zones. The indication, therefore, is that the intertidal fish community is a fairly well-defined entity, structurally distinct from that of subtidal waters. This observation further implies that intertidal fishes probably do not, in this area at least, use subtidal rocky areas to any significant extent as corridors or "stepping stones" for dispersal from one intertidal patch to another. In contrast, Mahon and Mahon's (1994) observations of the continuously changing relationship between tidepool size and body size, assemblage structure and particularly the proportion of partial residents, led them to suggest that in Barbados "the tidepool fish assemblage may grade into the subtidal fish assemblage in which the partial residents of tidepools predominate."

In general, relatively few species are restricted in their distribution to the area between the tidemarks whereas many others never enter the intertidal zone at all.

Between these extremes there are some species that are equally abundant within a restricted depth range of which the intertidal zone represents a significant part and others whose occupation of the intertidal zone represents only the shallow fringe of their overall bathymetric distribution. Low-water mark, in any case, is not a rigid boundary and the inhabitants of a low-level rockpool could be categorized as intertidal when the pool is exposed on spring tides but as subtidal when it is continually submerged on neap tides.

## C. Species Dispersal and the Connectedness of Geographically Disjunct Communities

Yet, spatially separated populations of at least some intertidal fish species maintain a certain degree of connection, judging from population genetic studies that indicate gene flow is maintained between population patches within fairly broad geographical areas. For several intertidal cottid species from central California (north of Point Conception) to southern Oregon, Yoshiyama and Sassaman (1987) estimated that dispersal (gene flow) rates were high enough to maintain genetic continuity between population patches. In this study, population patches corresponded to geographically discrete areas of rocky intertidal habitat. However, the estimated numbers of dispersants, as represented by the migration rate mN (average number of immigrants per generation per population patch), did not seem high enough to affect the local demographics of population patches to any significant degree; for example, mN values were estimated to be ~143 for *Oligocottus maculosus* and within the range 34–88 for *O. snyderi.* A numerical estimate of gene flow could not be made for *Clinocottus analis,* but it is probably at least as high as the rates for the other two species (Yoshiyama and Sassaman, 1987). Waples (1987) estimated relatively low gene flow rates (mN ~ 6–8) for *C. analis* populations on the southern California–northern Baja California coast and Isla de Guadelupe (>275 km offshore), but even those values are not low enough to indicate genetic isolation of the population patches in that region. Waples' gene flow estimates were correspondingly higher for other inshore species that have greater dispersal capabilities, as would be expected. In addition, the two clinid species *Gibbonsia montereyensis* and *G. metzi,* which are primarily constituents of the cold-temperate Oregon fauna north of Point Conception but are also found at scattered localities within the warm-temperate California region (southern and Baja California), show little genetic divergence of isolated southern populations from the much larger populations in northern areas (Stepien and Rosenblatt, 1991). These clinids possibly have wide dispersal during the planktonic larval stage (Stepien *et al.,* 1991), as is true for the cottid *Scorpaenichthys marmoratus,* which has been commonly taken in inshore plankton collections (O'Connell, 1953). Thus, at least a number of intertidal fishes evidently possess a metapopulation structure in the broad sense (Meffe and Carroll, 1994), that is, discrete subpopulations separated by various distances along the coast and connected by the dispersal of variable numbers of individuals during some life stage.

Most of the resident fishes in the intertidal habitat (e.g., blenniids, cottids, stichaeids, gobiesocids) typically lack swim bladders and are probably incapable of sustained, long-distance travel as adults. The eggs of intertidal fishes are generally deposited on demersal substrata (e.g., rocks, crevices, pilings; Marliave and DeMartini, 1977; DeMartini, 1978; DeMartini and Patten, 1979; Marliave, 1981) and are not subject to transport. Also, a number of species are viviparous, as are the marginally intertidal embiotocids on the western

North American coast (Hubbs, 1921; DeMartini, 1969) and the many (18 species) intertidal clinids of South Africa (Veith and Cornish, 1986; Prochazka, 1994). Others are pouch brooders that carry their eggs and hatchlings, such as the syngnathid pipefishes worldwide (e.g., Herald, 1941; Vincent *et al.,* 1994). These "live-bearing" species have no egg or larval transport. For the egg-depositing species, dispersal of individuals between intertidal patches, to the extent that it occurs, must be accomplished during the planktonic larval stage and mediated by currents close to the shore. However, direct underwater observations and sampling of larval stages of rocky intertidal fishes in the water column at different points along and away from (seaward) shore indicate that in the protected inner coastal waters of Vancouver Island, British Columbia, intertidal fish larvae do not occur very far offshore (no more than 20 m from the shoreline) or laterally along the shore away from rocky areas (Marliave, 1986). The larvae, in fact, are active swimmers that show daily patterns of schooling and dispersion, maintain schools near topographic prominences, and can resist displacement by currents (Marliave, 1986). Hence, these intertidal fish larvae evidently are not routinely dispersing en masse between localities. Cross (1981) accordingly observed only small numbers of juveniles of several species that presumably had been dispersed via planktonic transport between disjunct localities on the Washington inner coast region (Strait of Juan de Fuca and Puget Sound). Perhaps, then, any substantial dispersal that occurs between populations is episodic, for example, coinciding with storm events that wash the larvae away from or along the shore to distant locations, although Marliave's (1986) results indicated that even storm-generated turbulence prior to one sampling period did not cause significant alongshore dispersion of larvae. Presumably, the free-swimming young of these rocky intertidal species behave in like manner at exposed open coast localities, but there the more turbulent conditions most likely result in greater interlocality dispersion. The possibility must also be considered that the association of the larval and juvenile forms of some species with clumps of floating weed may lead to their dispersal (Kingsford, 1992; Davenport and Rees, 1993).

## D. Spatial Patterns between and within Localities

Spatial organization of intertidal fishes is also seen on a smaller spatial scale along a particular stretch of coastline or within a single intertidal locality. Patchiness of the substratum is a major consideration and the differences between sediment and rocky shores have already been referred to in relation to the taxonomic and resident/transient structure of communities. In the former context, it is notable that two of the major resident intertidal families, the Blenniidae and Gobiidae, occupy rocky and sedimentary shores to different extents. Members of the Blenniidae are rare outside rocky areas, whereas numerous goby species are found on both sandy and muddy substrata (cf. Zander, 1967). Even within rocky areas, differences in community structure are often evident. In California, for example, areas comprising extensive boulder fields that are regularly exposed to air during low-tide periods are often dominated by the eel-shaped stichaeids and pholids, with the clingfish *Gobiesox maeandricus* occasionally abundant. In contrast, areas containing numerous tidepools with heavy vegetation harbor large numbers of cottid species and a variety of other cryptic forms such as kelpfish (*Gibbonsia* spp.) and juvenile greenlings (*Hexagrammos* spp.), as well as stichaeids and pholids (Yoshiyama, 1981; Yoshiyama *et al.,* 1986). In the Mediterranean, several species of blennies, gobies, and triplefins have a clear preference for different substrata (Macpherson, 1994) and even particular types of rock structure can affect the species composition of the fish communities that live there (Beja, 1995). The degree of exposure to wave action is also known to affect distribution and abundance, as on the Atlantic coast of Spain where blennies are more common in exposed sites and gobies in more sheltered ones (Ibáñez *et al.,* 1989).

Studies at different localities have shown that intertidal fish species have characteristic distributions with respect to level on the shore (e.g., Weaver, 1970; Green, 1971a; Gibson, 1972; Horn and Riegle, 1981; Yoshiyama, 1981; Barton, 1982); that is, they show patterns of vertical zonation equivalent to those of intertidal algae and invertebrates (Stephenson and Stephenson, 1972; Ricketts *et al.,* 1985). Certain fishes are more tolerant of prolonged confinement to tidepools or even of exposure to air and, therefore, occur higher in the intertidal zone. The cottids *Oligocottus maculosus* and *O. snyderi,* for example, at localities on Vancouver Island show somewhat different vertical distributions in addition to having different microhabitat preferences. *O. maculosus* tends to occur in open areas of tidepools, and *O. snyderi* among heavy vegetation (Nakamura, 1976). Cover in the form of rocks and boulders is of great significance in determining the abundance of many species (Gibson, 1972; Marsh *et al.,* 1978; Prochazka and Griffiths, 1992). The presence of weed seems less important because fish biomass and the number of species and individuals were not correlated with weed cover on shores in the Cape Peninsula (Bennett and Griffiths, 1984) and removal of weed cover from a shore in Nova Scotia had no effect on fish abundance (Black and Miller, 1991). A final factor that may have overriding importance on rocky shore community structure, particularly in terms of abundance and size structure, is the area, volume, and rugosity of tidepools at low tide (Gibson, 1972; Bennett and Griffiths, 1984; Prochazka and Griffiths, 1992). In an extensive study on a Barbados shore with a very shallow slope such that vertical height between pools was negligible, Mahon and Mahon (1994) were able to demonstrate positive significant correlations between pool volume and the number of species (Figure 3), number of individuals, total fish biomass, and the weight of 10 individual species for which there were sufficient data. Furthermore, the proportion of partial residents increased with increasing pool size. Such correlations with pool size are not unexpected because larger pools provide a greater range of microhabitats and, if fish move out of pools at high tide, a greater "collection area" as the tide ebbs. Also, large pool size

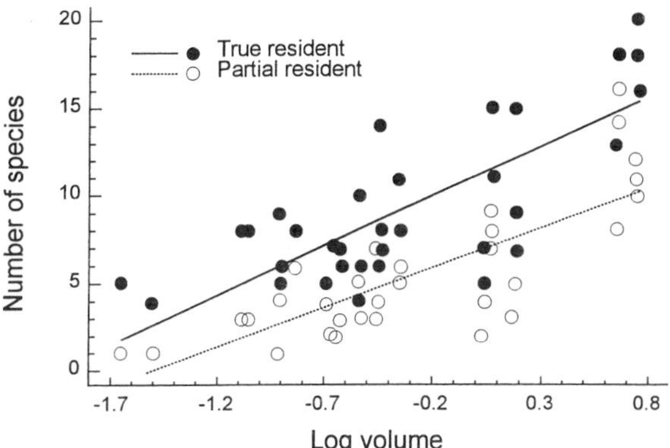


**Figure 3.** The relationship between the numbers of species and pool volume for true and partial resident species of Barbados tidepools. Reproduced from Mahon and Mahon (1994) with kind permission from Kluwer Academic Publishers.

confers greater chemical and physical stability on the water and allows a wider range of species to survive there at low tide.

In addition to those fishes that have size- or age-related depth distributions, certain other inshore species show differences in depth preferences between the sexes. Stepien and Rosenblatt (1991) state that the North American Clinidae show segregation by depth, with males usually occurring deeper than females and juveniles. Accordingly, in both *Gibbonsia montereyensis* and *G. elegans,* females predominate in intertidal areas and males are proportionately more common at increasing depths (Williams, 1954; Stepien and Rosenblatt, 1991). The reasons underlying such sex-related segregation by depth are not known. In these clinid species, the females migrate seasonally to the deeper areas to lay their eggs in algal nests that are guarded by the males (Stepien and Rosenblatt, 1991). Perhaps the females and juveniles experience more favorable growth and survival conditions at the shallower depths and therefore concentrate there, whereas the males must remain in deeper waters where, for some unknown reason, the nesting territories are situated. Different microhabitat uses by the sexes, perhaps due to constraints imposed by egg laying, nest guarding, or different energy requirements of males and females, probably occur in other intertidal and subtidal fishes and warrant further investigation.

## E. Spatial Patterns at the Individual Level

Some intertidal fish species can be said to show spatial organization down to the scale of individual tidepools or to groups of tidepools; that is, they show site fidelity and, in some cases, homing behavior to certain tidepools (e.g., Williams, 1957; Gibson, 1967; Green, 1971b, 1973; Khoo, 1974; Craik, 1981; Yoshiyama *et al.,* 1992; Gibson, Chapter 6, this volume). Such site-oriented behavior may be common in communities largely composed of resident species rather than transients or even seasonal occupants of the intertidal zone. In the central California–British Columbia region, where intertidal communities contain a large and stable resident component (Green, 1971a; Yoshiyama, 1981; Grossman, 1982), true homing behavior to specific tidepools is shown by a number of cottids (viz., *O. maculosus, O. snyderi, C. globiceps*). Some degree of site fidelity, that is, the tendency for individual fish to remain associated with specific sites for extended periods, occurs in several additional species, both cottids (*C. analis*) and non-cottids (e.g., *Gobiesox maeandricus*), and these species have the potential to exhibit homing behavior as well. Such localizing behaviors presumably have adaptive value, and their presence in fish communities undoubtedly reflects the nature of the habitat as well as the types of fishes present; for example, if different microhabitats differ significantly in the amount of food or protection they afford, the fish may be expected to concentrate at the better microhabitats, whereas in a more homogeneous environment there may be no advantage to fish to remain localized at any particular spot.

The complex nature of rocky shores means that numerous microhabitats are available for occupation and community structure of any one shore or tidepool may be governed by the range and type of microhabitats available. The presence of boulders, for example, seems to be important for clingfishes (Ibáñez *et al.,* 1989), weed for clinids (Marsh *et al.,* 1978), and holes in the rock for eels (Mahon and Mahon, 1994) and some blennies (Koppel, 1988; Kotrschal, 1988). Such cryptic species are usually residents; transients are more often species that occupy open water positions in pools (Potts, 1980; Mahon and Mahon, 1994) and are rarely found out of water, sheltering under stones. In extreme cases, some fish species have symbiotic relationships with invertebrates and consequently have very specific habitat requirements (Mahon and Mahon, 1994) (Figure 1).

## IV. Temporal Variation in Community Structure

In addition to variation in space, communities may also change markedly on time scales that vary from hours and days to years and longer.

### A. Variation in Community Structure over Tidal and Diel Cycles

The major change that takes place in community structure on a short time scale is caused by the influx of numerous individuals and species into the intertidal zone as the tide rises and by their movement offshore as it ebbs. The change in structure is most marked on mudflats (Ali and Hussain, 1990; Abou-Seedo, 1992), salt marshes (Cain and Dean, 1976; Rountree and Able, 1993; Cattrijsse *et al.*, 1994), and sandy beaches (Tyler, 1971; Gibson, 1973; Gibson *et al.,* 1996) because fish are absent from the intertidal zone at low tide so that the high-tide community is essentially the same as the subtidal community. On rocky shores the high-tide community consists of the residents augmented by the tidal visitors (e.g., Thomson and Lehner, 1976; Black and Miller, 1991). Such tidal changes in community structure may be modulated by the day/night cycle so that the high-tide community during the day may be different from that at night. In most cases, more species and individuals are found intertidally at night but such a conclusion is based on studies of fish communities on sandy beaches (Horn, 1980; Ross *et al.,* 1987; Gibson *et al.,* 1996) or mudflats (Wright, 1989). There seem to be no studies of the nocturnal high-tide community on rocky shores but a parallel can be seen in the crepuscular changeover of species seen on coral reefs (Helfman, 1993).

### B. Seasonal Variation in Community Structure

With one exception (Prochazka, 1996), marked seasonal changes in structure have been detected in all intertidal fish communities that have been studied for more than a few months. The nature and causes of these structural changes vary from location to location. In most temperate regions, the abundance of species and individuals is generally highest in late spring and summer due to the influx of juveniles of both resident and transient species following their breeding seasons. This influx of juveniles also has a profound effect on the age structure and, because not all species recruit at the same time, on the species composition of the community. The exact timing of these peaks in abundance varies considerably. In California (Grossman, 1986a; Moring, 1986) and New Zealand (Willis and Roberts, 1996), diversity and abundance is highest in spring whereas elsewhere it may occur in summer or autumn, as in the Gulf of California (summer–fall period; Thomson and Lehner, 1976), the Azores (third quarter of the year; Santos *et al.,* 1994), and Maine (June to September; Moring, 1990). On the southwest coast of Portugal, catches of most species peak in winter and early spring (Beja, 1995). The subsequent decrease in numbers and species after the peak may be the result of a variety of factors. Predation is probably a major cause but declining temperature is also a common correlate. The almost complete lack of seasonality in species composition on the west coast of South Africa is a result of the absence of transients in this area (Prochazka, 1996). Mortality caused by extreme low temperatures seems to be uncommon (Chang *et al.,* 1977; Lee, 1980) although it has been recorded in the Gulf of California (Thomson and Lehner, 1976) where the southerly elements of the fauna, that is, the least cold-adapted, were most affected. Other factors reputed to be responsible for seasonal changes in community structure are productivity in California

(Grossman, 1982) and wind effects in New Zealand (Willis and Roberts, 1996) and Taiwan (Chang *et al.,* 1977). Not all species in the community react to seasonality in environmental variables in the same way. Faunal affinity may play a large part as tropical species are more susceptible to low temperatures than temperate forms, to the extent that the latter may respond more consistently to seasonal temperature change and impart stability to communities that consist of both forms (Thomson and Lehner, 1976). In a general sense also, seasonal changes in community structure are likely to decrease with decreasing latitude (cf. for example, Chang *et al.,* 1973, and Moring, 1990). An interesting example of differential response to the same factor is provided by Willis and Roberts (1996) in North Island, New Zealand. Here the community is dominated by the Tripterygiidae and Gobiesocidae. Two species of triplefins are numerically dominant in the summer but are less common in the winter when dominance is assumed by two species of clingfishes (Figure 4). The change in dominance is caused not so much by an increase in clingfish numbers but rather by a decline in the abundance of the triplefins, possibly as the result of high mortality in winter storms.

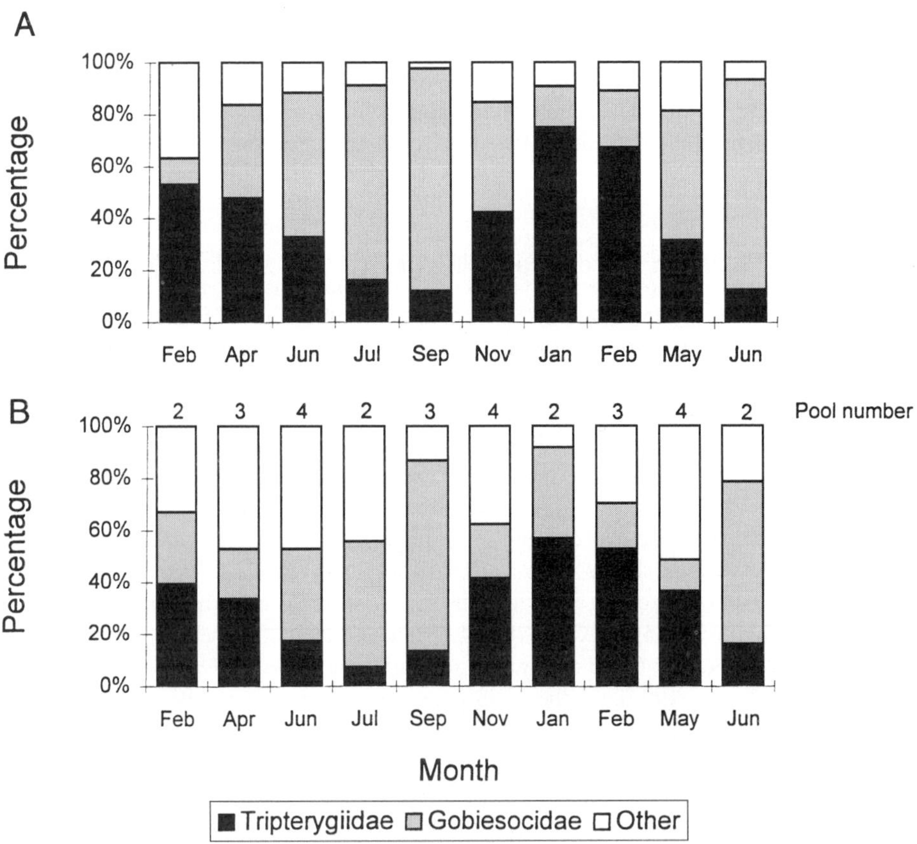


**Figure 4.** Seasonal changes in the composition of the fish community of New Zealand tidepools. (A) A single pool sampled bimonthly except in winter and summer, (B) three pools (Nos. 2, 3, 4 at top) sampled every 5 months. Reproduced from Willis and Roberts (1996) with kind permission from Kluwer Academic Publishers.

## C. Stability, Resilience, and Persistence of Community Structure

On a longer time scale of years and decades the question arises of whether communities remain the same from year to year, that is, whether community structure is persistent. Studies designed to answer this question clearly require considerable lengths of time to carry out and consequently are relatively rare. Nevertheless, sufficient information is available to arrive at some general conclusions. The most comprehensive data come from studies on the west coast of North America. The rocky intertidal fish community at Dillon Beach in central California has been described as deterministically structured as manifested by the constancy (persistence) over time (42 months) of the relative abundances of species (Grossman 1982, 1986b). Grossman (1986b) reported that the fish assemblage did not undergo large-scale changes in structure due to unpredictable environmental perturbations. The Dillon Beach fish assemblage could be typical, at least in its general features, of intertidal fish communities in the central California–British Columbia region. However, the extent to which it exemplifies other assemblages is as yet undetermined and will require long-term studies at other localities (Grossman, 1986b). The fish assemblage at another central California locality (Pescadero Point) 115 km south of Dillon Beach similarly did not show evidence of major stochastic changes in structure during 3 years of study (Yoshiyama, 1981, personal observation). Although Yoshiyama *et al.* (1986) noted some constancy within California rocky intertidal fish communities compared at times several years apart, they also reported a substantial decrease in abundance of *Oligocottus snyderi* (a primary resident species) at Pescadero Point sampled 7 years after Yoshiyama's (1981) initial survey and apparent changes in abundance of other species at that and other localities, including Dillon Beach. Indeed, it is expected that some degree of change in species abundances must occur from year to year, reflecting variable spawning or recruitment success as environmental conditions vary over time, as has been observed for intertidal fish assemblages monitored for extended periods (8+ years) in other geographical regions (i.e., northern Gulf of California, Thomson and Lehner, 1976; Britain, Jones and Clare, 1977; New England, Collette, 1986; Australia, Lardner *et al.*, 1993). Grossman's (1982, 1986b) data for Dillon Beach, for example, show that the top three species were not always the same ones from year to year, nor was the relative order of abundance constant. Nonetheless, the overall statistical consistency of that assemblage is striking. Similarly, in Collette's (1986) study of two New England tidepools over 19 years, two species (*Pholis gunnellus* and *Tautogolabrus adspersus*) were consistently the most numerous even though their abundances fluctuated considerably from year to year (Figure 5). Comparable results were obtained by Mahon and Mahon (1994) in Barbados where the rank order of species abundance and the relationships among species richness, abundance, and biomass with pool size did not differ among three sampling occasions spaced over 6 years. It is also evident from published studies and from subsequent observations, which in aggregate span 2 decades, that there is some general constancy of assemblage structure on the California–British Columbia outer coast north of San Francisco Bay where the predominant, or at least most conspicuous, tidepool fishes usually are *Oligocottus maculosus* and *O. snyderi,* while south of San Francisco Bay to Point Conception they are *Clinocottus analis* and *O. snyderi* (Green, 1971a; Grossman, 1982; Yoshiyama *et al.*, 1986). Furthermore, on the central California coast south of San Francisco Bay, the stichaeids *Xiphister atropurpureus, Cebidichthys violaceus,* and *Anoplarchus purpurescens* are consistently prominent faunal components in certain areas of boulder fields exposed by the tide (Barton, 1982; Yoshiyama *et al.*, 1986; R.M.Y. personal observation; M. Horn, personal communication). Relative

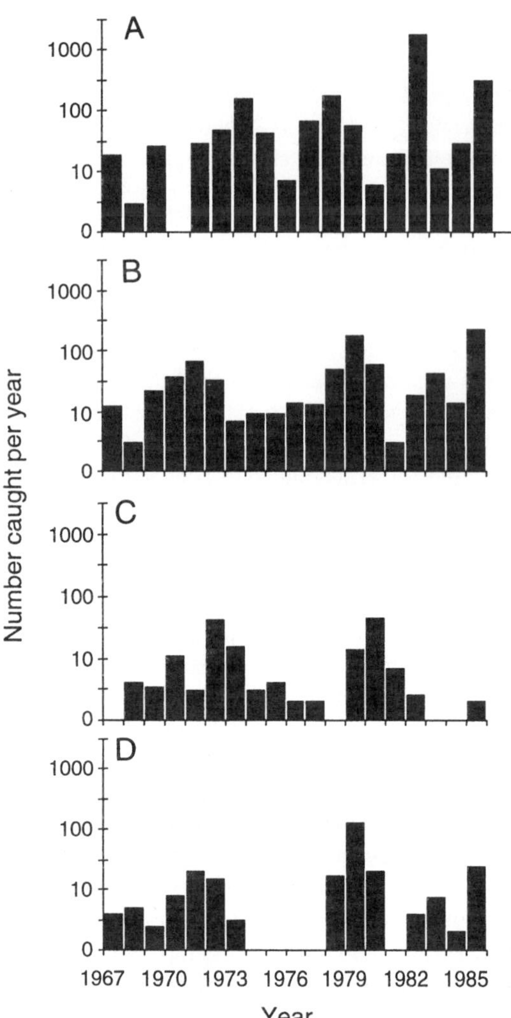


**Figure 5.** Annual variation in the numbers of four dominant species in two New England tidepools. One collection was made each year except 1969, 1982, and 1983 when two collections were made. The average of the two collections is plotted for these three years. Note logarithmic scale. (A) *Tautogolabrus adspersus,* (B) *Pholis gunnellus,* (C) *Gasterosteus aculeatus,* (D) *Myoxocephalus aenaeus.* From data in Collette (1986).

species abundances do fluctuate, but apparently not to the extent that resident species are very common in some years but rare in others. However, that perception must depend to some degree on the time span over which the fauna are observed. Studies in recent decades have clearly shown that *O. snyderi* is one of the most abundant and readily captured intertidal cottids of the central California–Oregon coast, yet that species was reported by Greeley (1899) over 90 years ago to be "Found in all kinds of pools from San Francisco to

Monterey Bay, but nowhere common." Thus, relative abundances of rocky intertidal fishes may not change appreciably over several years or even decades, but perhaps over the course of a century the numerical fluctuations are significant. Whether intertidal fish communities are deterministic in structure (sensu Grossman, 1982), as opposed to stochastically varying, therefore, may depend on the temporal scale of observation.

Related to the topic of constancy, or persistence, of community structure is that of stability and resilience, that is, the ability of an assemblage to maintain its structure despite environmental uncertainty and to recover from periodic disruptions of that structure. Studies in which tidepools are experimentally defaunated have direct bearing on this theme, as do observational surveys that monitor species assemblages over time and their responses to natural perturbations (e.g., Thomson and Lehner, 1976; Jones and Clare, 1977). For example, Grossman (1982) reported that the resident fish assemblage in an experimental area (30 × 17 m) at Dillon Beach recovered from repeated defaunations over a 29-month period, and hence was resilient to such perturbation. Shorter term defaunation experiments at other localities indicate that recolonization of depopulated tidepools can commence within one tidal cycle (1 day) and be well under way by 2 weeks (Collette, 1986; Mistry *et al.,* 1989; Lardner *et al.,* 1993), but full recovery of the fish assemblages takes appreciably longer (weeks to months; Williams, 1957; Cross, 1981; Matson *et al.,* 1986; Willis and Roberts, 1996) and the speed at which they recover may depend on the time of year (Beckley, 1985b; Mahon and Mahon, 1994; Willis and Roberts, 1996). The initial colonizers are usually the commonest species in the community. Polivka and Chotkowski (1998) observed that after one or two removals of fish from tidepools within a 90-day period, the species composition of recolonized pools remained dominated by the same species that were predominant in the initial sampling, although the presence and abundance of the less common species was unpredictable. Restoration of species diversity to original levels in the tidepools required 60–90 days but there was some indication of local depletion of potential colonists, which in two cottid species comprised primarily juveniles. Such perturbations could have different effects on different species, although the effects may be mostly temporary or limited in spatial extent. Nonetheless, certain intertidal fish species that are uncommon or highly patchy in distribution may be less able than others to recover from population reductions caused by extraneous perturbations, however infrequent those reductions may be. The scale at which recolonization is measured is also relevant because samples from small pools are likely to be more variable than those from larger ones.

It must be borne in mind that any assessment of whether a natural community is resilient to perturbations depends on many factors, such as the size of the experimental area (e.g., sizes of defaunated tidepools) in relation to the amount of adjoining habitat locally available, the nature and severity of the perturbation, and the proximity, direction and sizes of neighboring intertidal habitat patches that might act as a source of colonizers. Thus, comparisons between intertidal fish communities in different geographical regions, and even among localities within the same region, are meaningful only to the extent that the extraneous factors and attending circumstances are accounted for. Even at a single locality, a species assemblage may be viewed as resilient to perturbations only up to a certain magnitude; that is, a limited number of defaunated tidepools may be readily recolonized by fish from surrounding tidepools. That same assemblage may recover only slowly from a massive or persistent loss of fish from most of the available habitat, as might be caused by catastrophic oil spills or pollutant discharges [e.g., Alaskan intertidal fishes required at least 2 years to recover after the *Exxon Valdez* disaster (Barber *et al.,* 1995)], or by continuous heavy "harvesting" of the fishes and associated algae and invertebrate resources (Moring,

1983). As further noted by Moring (1976), the relatively restricted lateral movements of intertidal cottids (and probably many other resident intertidal fishes; Gibson, Chapter 6, this volume) may limit their ability to recover from local population depletion.

As a general summary, the available evidence indicates that the structure of intertidal fish assemblages changes regularly at a variety of time scales from hours to months in step with the rhythm of tidal, diel and seasonal cycles. In the longer term, assemblages are persistent and resilient to change such that general taxonomic structure is predictable from year to year.

## V. Interspecific Interactions

Interactions between species have been long known to affect significantly the structure of rocky intertidal invertebrate and algal assemblages of temperate regions (e.g., Connell, 1961, 1970; Dayton, 1971; Paine, 1980). Likewise, the interactions between fish species in tropical and subtropical shallow marine habitats are often pronounced and easily seen (e.g., coral reefs, Itzkowitz, 1974; Sale, 1978; mangrove mudflats, MacNae, 1968; Nursall, 1981; Clayton, 1993). Curiously, such species interactions, particularly interference competition as manifested by agonistic behaviors, seem to vary in occurrence in temperate rocky intertidal fish communities. These communities often contain high abundances of at least several species, as well as relatively high species diversity, and some degree of overt competitive interactions might reasonably be expected. The apparent rarity of behavioral interactions between species perhaps can be partly attributed to the difficulty of observing fish in rocky intertidal habitats, due to of the nature of the environment (e.g., high turbulence, limited times of accessibility) and of the fishes themselves (i.e., usually cryptic species that hide among algae and rocks but are highly motile and able to evade prolonged scrutiny). The most extensive observations on this subject have been made on sculpins on the Pacific coast of North America and we take these observations as a case study with which studies in other areas can be compared.

### A. Pacific North American Cottids

Numerous *in situ* observations of intertidal fishes at areas on Vancouver Island indicate the absence of agonistic interactions between the predominant cottid, *Oligocottus maculosus,* and the closely related *O. snyderi* and other species (Green, 1971a,c). Nakamura (1976) also reported the absence of aggressive behavior during laboratory and field studies of *O. maculosus* and *O. snyderi* from that same region. Cross (1981) observed no intra- or interspecific aggression or territorial behavior (during the nonreproductive season) in the cottids and pricklebacks of Washington inner coast areas, and he concluded that any competition that occurs among those species is exploitative (i.e., the mutual use of limited resources) rather than behavioral. Pfister's (1995) studies at Washington outer coast sites demonstrated a negative effect of *O. maculosus* on the growth rates of *Clinocottus globiceps,* and that the presence of *C. globiceps* appeared to influence the migration of *C. embryum* out of experimental tidepools. In concordance with previous studies, Pfister reported that those interspecific effects seemed to be mediated by resource competition rather than through agonistic interactions. Similarly, Grossman (1986a) suggested that the intertidal fish assemblage at Dillon Beach, California, probably was structured mostly by exploitation competition, given the apparent lack of "interference mechanisms."

However, aggressive tendencies of one cottid species, *Clinocottus analis,* in fact have been directly observed in aquarium-held fish, which, while subject to the constraints of observations in artificial settings, provide information on the potential for such interactions to occur in the natural environment. *C. analis* is highly aggressive toward *O. snyderi* and probably other cottid species, as well as to conspecifics (R.M.Y., personal observation; S. Norton, personal communication), and dominant individuals continually harass and will eventually kill subordinates. Although Richkus (1981) observed that *C. analis* held in an artificial pool showed a strong tendency to aggregate, interpretation of that result must take into account the unnatural setting, short observation periods, and limited acclimation time of the fish in that study; the observed aggregative tendencies of the fish may have been partly a fright or "anxiety" response due to the experimental conditions and disturbance. While *C. analis* undoubtedly show aggregative behavior in the natural environment under some circumstances (Richkus, 1978; R.M.Y., personal observation), this does not necessarily imply that interference mechanisms are absent as suggested by Grossman (1986a). Detailed observations in natural settings are needed to determine if interspecies aggression by *C. analis* is truly absent, rather than as yet undetected.

Within most of its range from Baja California to northern California (Miller and Lea, 1972), *Clinocottus analis* is the predominant cottid and probably the most abundant resident fish species in the middle and high intertidal pools, although it is also common in low intertidal pools and large individuals may occur in the shallow subtidal zone at some southern California localities (Williams, 1957; Yoshiyama, 1981, personal observation; Wells, 1986). The predominance of *C. analis* in areas south of San Francisco Bay may be the principal reason why the smaller *Oligocottus maculosus* is essentially absent in that region. *O. maculosus* occupies an intertidal niche in areas north of San Francisco Bay similar to that of *C. analis* to the south. *O. maculosus* can often occur in a broad vertical range of intertidal habitat but generally is most abundant in upper intertidal pools, at least from northern California to British Columbia (Nakamura, 1976; R.M.Y., personal observation). *O. maculosus* may be the single most common resident intertidal fish from northern California to localities in southeastern Alaska (Jordan and Gilbert, 1881; Moring, 1976). Yet, it is only rarely found south of San Francisco Bay (Yoshiyama, 1981; Yoshiyama *et al.*, 1986). There are no obvious physical factors that would seem to restrict the occurrence of *O. maculosus* southward along the central California coast, at least down to Point Conception and the Channel Islands off southern California. Aggressive exclusion by *C. analis* appears to be a highly likely causative factor and deserves further investigation.

Several field studies have been done on *C. analis* on the coast of southern California (e.g., Williams, 1957; Richkus, 1978; Wells, 1986), but none has considered the topic of its behavioral interactions with other species, probably because most other intertidal fishes in that region are far less numerous and *C. analis* is often the only fish species of consequence in the mid and upper intertidal pools (Wells, 1986; R.M.Y., personal observation). Nor have there been any substantial *in situ* behavioural observations conducted on *C. analis* north of Point Conception, demarcating central and southern California, where *C. analis* co-occurs with a diverse array of intertidal fishes. Experimental field studies involving the manipulation of species densities and microdistributional patterns would be very useful for clarifying interspecies dynamics, particularly in the geographical boundary area where the north–south switchover in numerical predominance of *O. maculosus* to *C. analis* occurs (i.e., within the coastal strip extending just north and south of San Francisco Bay). It is likely that mass experimental removals of single species (viz., *C. analis*) from selected localities for limited time periods and subsequent intensive monitoring of the fauna would be necessary to demonstrate unequivocally the negative influence of dominants such as *C. analis* on ecologically or behaviorally subordinate species. Such manipulations would require careful planning and execution to avoid excessive disruptions of the natural fauna. A start in this direction has been made with experiments in which removal of *O. maculosus* from tidepools increased the growth of *Clinocottus globiceps,* although the reverse was not true nor was recolonization of pools by either species affected by the presence of the other (Pfister, 1995).

## B. Other Species

If aggression by *C. analis* does exert a significant influence on the structure of intertidal fish assemblages in California, it would seem to be an exception among the Cottidae because in no other intertidal cottids have appreciable levels of interspecies agonism been reported, even for aquarium situations. The behavior of non-cottid rocky intertidal fish in the California region has been little studied. Laboratory experiments suggest aggressive interactions between some stichaeid species (*Anoplarchus, Cebidichthys, Xiphister* spp.; Jones, 1981), but for the most part, interspecific aggression among non-cottid fishes held in aquaria appears to be infrequent (R.M.Y., personal observation). Distributions of stichaeid and pholid fishes during low-tide periods, when they hide under aerially exposed rocks and algae, do not show any obvious patterns of spatial segregation indicative of aggressive displacement or territoriality and at least some of these species have been observed sheltering under the same individual rocks (e.g., *Xiphister mucosus* with *X. atropurpureus,* Barton, 1982; *Anoplarchus purpurescens* with *X. atropurpureus, Apodichthys* (=*Xererpes*) *fucorum,* and *Cebidichthys violaceus,* R.M.Y., personal observation; M. Horn, personal communication).

In blennies and other groups, however, in which agonistic behavior and differences in microhabitat preference are well developed, interspecific interactions do seem to be responsible for differences in distribution patterns, and hence community structure, at least in some locations. In the Mediterranean, for example, where numerous blenny species live in very shallow water, *Blennius canevae* and *B. incognitus* both live in holes but the former species is much more specific in its choice than the latter with the result that available refuges are partitioned between the species (Koppel, 1988). Comparable partitioning of available refuges has also been described for the New Zealand species *Acanthoclinus fuscus* (Plesiopidae) and *Forsterygion nigripenne robustum* (Tripterygiidae) (Mayr and Berger, 1992) and the southern Californian bennies *Hypsoblennius gilberti* and *H. jenkinsi* (Stephens *et al.,* 1970). In these three cases, the species with the narrowest habitat niche was the most aggressive. Furthermore, experimental studies have shown that in the two species of *Hypsoblennius* studied by Stephens *et al.* (1970) and in two sympatric *Gobiosoma* (Gobiidae) species (Hoese, 1966), the choice of habitat is altered by the presence of the other species. However, in a wider field study of the distribution and substratum preferences of several Mediterranean blenny species, although each species had a clear preference for a substratum type, interspecific competition for space was believed unlikely because the populations were evidently below the levels at which the habitat was likely to be saturated (Macpherson, 1994).

# VI. Intraspecific Interactions

Interactions between members of the same species, although unlikely to affect the species composition of the community, can have an effect on the size and age structure of single-species populations within the community. Here again few, if any, intraspecific interactions have been recorded within the Cottidae in contrast to the Blenniidae, Clinidae, and Gobiidae (reviewed in Gibson, 1969, 1982; Santos and Nash, 1996). In these three families aggression and territoriality is often limited to the reproductive season but evidence for its existence outside the breeding season suggests that it may be present at a low level year-round. Such behavior is usually considered responsible for distributing individuals throughout the available space and as a means of population regulation. Marsh *et al.* (1978), for example, have described aggressive interactions between adult and juvenile *Clinus superciliosus* and suggest that such behavior could result in smaller fish being forced into more open areas where they may be liable to increased predation. This aggressive behavior of adults toward juveniles may also play a role in maintaining the clearly defined vertical size partitioning of this species on the shore.

Other observations of agonistic behavior in the field are rare and its importance in nature is open to question. It may be present, but difficult to observe in cryptic species that are usually quiescent during low tide when most observations are made. At low tide many individuals of several species can often be found in the same small pool or under the same rock and these observations have been used to infer that the existence of agonistic behavior would not allow such situations to arise. Under these conditions, however, individuals are likely to be visually isolated and inactive. Agonistic behavior relies on visual stimuli and is dependent on encountering other individuals; consequently it is most likely to occur over the high-tide period when fish are most active. It has been argued for *Lipophrys* (*Blennius*) *pholis* that such behavior may fluctuate cyclically, driven in part by the endogenous tidally phased activity rhythm, resulting in the dispersion of individuals at high tide and allowing them to aggregate into suitable refuges at low tide (Gibson, 1967, 1968a). The existence of endogenous tidal rhythmicity in most resident intertidal families (Gibson, 1982) and the existence of well-developed agonistic behavior patterns in many species, albeit in the laboratory, suggests that such rhythmically fluctuating agonism may be widespread.

## VII. Predation

Predation, although a highly conspicuous and important factor in some fish communities, seems to occur relatively infrequently among the resident fish species of rocky intertidal assemblages. Of the resident cottids in central California–British Columbia, only *Artedius lateralis* (Yoshiyama, 1980, 1981) and juvenile *Scorpaenichthys marmoratus* (O'Connell, 1953), where present, may be described as significant piscivores. In central California, the primary fish prey of *A. lateralis* seem to be water column forms such as juvenile rockfish (*Sebastes* spp.) and probably surfperch (*Micrometrus* spp.), and also pricklebacks (e.g., *Cebidichthys violaceus*). Even so, fish compose at most a fraction of the diet of *A. lateralis* (e.g., up to 33% by volume and 7% by weight at two central California localities; Yoshiyama, 1980; Grossman, 1986a), and *A. lateralis* itself is a minor component of the fauna at most localities surveyed (Green, 1971a; Yoshiyama, 1981; Yoshiyama *et al.*, 1986). Similarly, fish on the Washington inner coast, primarily larval and juvenile cottids but including gunnels, make up only a minor part of the diet of *A. lateralis* (Cross, 1981). Juvenile *Scorpaenichthys marmoratus,* which are as large or larger than adults of other cottids, consume significant quantities of common tidepool fishes such as cottids, pricklebacks, and *Gibbonsia* spp. (e.g., ~53% by weight; O'Connell, 1953); several other resident intertidal species (i.e., *C. analis, O. maculosus, Sebastes rastrelliger*) also reportedly are piscivorous to some small degree (Mitchell, 1953; Nakamura, 1971; Grossman, 1986b; Wells, 1986). However, most predation upon these rocky intertidal fish assemblages probably is exerted by "external" predators, that is, large subtidal fishes (e.g., adult *Scorpaenichthys marmoratus, Enophrys bison, Ophiodon elongatus*), perhaps diving birds that forage in the intertidal zone during high tide (MacGinitie and MacGinitie, 1949; Robertson, 1974; Cross, 1981; Wells, 1986), and at some localities wading birds (great blue herons *Ardea herodias* and snowy egrets *Leucophoys thula*) (Cross, 1981; Yoshiyama, 1981; Pierce and Pierson, 1990), small mammalian predators (Pillsbury, 1957; Nakamura, 1976), and garter snakes *Thamnophis sirtalis* (Batts, 1961). The influence of such external predators upon the local abundance and microhabitat distribution of intertidal fishes is essentially unknown, but may be substantial. Terrestrial predators could deplete concentrations of stranded fishes in tidally exposed rock fields or significantly reduce fish numbers in easily accessible tidepools, whereas large subtidal fishes may cull the intertidal fish species to various degrees depending on the latter's microhabitat affinities. It is conceivable that incursions of subtidal predators during high tide may affect the distribution of intertidal species during both high and low tides. Limited experimental evidence indicates that the water-borne odor of the piscivorous *Hexagrammos decagrammus* induces behavioral changes in the tidepool sculpin *Oligocottus maculosus* in the laboratory and also may cause their abandonment of tidepools (Chotkowski, 1994). Yet, there also may be periods when predation by certain intertidal fish species upon others is significant, viz, when the previously "planktonic" postlarval young recruit en masse into the intertidal habitat. Such recruitment pulses occur, for example, during the late spring and early summer on the central California and southern Oregon coast for a number of cottid (e.g., *O. snyderi*) and stichaeid species (*Anoplarchus, Xiphister* spp.) (R.M.Y., unpublished data), and it would be surprising if the young recruits are not preyed upon by intertidal fishes that have relatively catholic diets (e.g., *C. analis, Gobiesox maeandricus, Hexagrammos* spp.) or are known piscivores (i.e., *A. lateralis, S. marmoratus*). Indeed, the somewhat atypical cottid *Ascelichthys rhodorus,* which is often abundant in the same boulder patch habitat favored by stichaeids (R.M.Y., personal observation; M. Chotkowski, personal communication), includes young-of-the-year stichaeids in its natural diet at least occasionally on the central Oregon coast (R.M.Y., unpublished data).

Elsewhere, several tidepool fish species reportedly feed on other fishes on the coast of Taiwan (Chang and Lee, 1969), but the identity of the prey was not recorded and in only two species did the diet contain more than 10% fish. In other regions also, fish seem to form a very minor proportion of the diet of species caught intertidally at low tide (Gibson, 1968b, 1972; Varas and Ojeda, 1990) so that predation by resident species seems an unlikely source of mortality, with the possible exception of *Paralabrax* in the Gulf of California (Thomson and Lehner, 1976). Consequently, as in California, most predation is probably caused by visiting predators such as larger fish, and in some locations by mammals (Kruuk *et al.,* 1988; Beja, 1995) and birds (Mahon and Mahon, 1994).

## VIII. Overview

Fish communities in intertidal environments consist essentially of two components, residents and visitors, and the basic composition of the community on any one shore is dependent on the nature of the substratum. Rocky shores, by virtue of the amount of cover they provide, have the richest resident communities. These resident communities are regularly supplemented by the influx of visitors at high tide. On shores where cover is limited or absent, resident species are rare and the intertidal zone is occupied by fishes only at high tide. Resident species are usually small, cryptic forms with short life spans, although the juveniles of many other species may also make use of intertidal areas as nursery grounds before moving into deeper water. The small size of the resident species can be considered to be one of the adaptations that enable them to live in shallow, turbulent water. Most intertidal fishes so far studied in any detail are carnivorous and feed mostly on crustaceans. Herbivores appear to be rare but such a conclusion may be biased by the lack of dietary studies on tropical species.

The structure of intertidal fish communities varies both spatially and temporally on a variety of scales. The scale of spatial variation ranges from vertical zonation patterns over a few meters on rocky shores, through differences between shores caused by variation in habitat structure, to latitudinal and global patterns of species distribution and community structure. It is remarkable, however, that resident intertidal fish communities, and particularly those on rocky shores, always consist of species belonging to a relatively few families. It is therefore possible to predict with some accuracy the probable composition of resident communities in any part of the world. Any description of community structure must also take into account changes that take place over time. Temporal change is particularly marked in intertidal communities of motile animals such as fishes that can respond to fluctuating environmental conditions by changing their location. As with spatial variation, temporal variation in community structure can take place on various time scales, the most marked of which are tidal and seasonal. To a large extent such temporal changes are regular and predictable, at least in outline. On a longer time scale, the little evidence that is currently available suggests that although relative and absolute abundances may change with variation in recruitment success, the species composition of communities changes little over many years at any one place. Perhaps this is to be expected in higher latitudes where the species pool is small, but the phenomenon has also been recorded at lower latitudes and so to this extent it may be regarded as generally applicable. Intertidal fish communities also seem to be stable; that is, they regain their original structure after major environmental perturbations such as storms or oil spills. Recolonization after such events takes place both by the movement of individuals into the affected area from elsewhere and by larval settlement. Replacement by the movement of individuals of resident species is likely to be slow because such species are generally poor swimmers and tend to be restricted in their movements, particularly in the older age groups. Consequently, most recolonization results from movement of small individuals and recruitment from the larval stages. The rate of recolonization will hence depend on the time of year that depopulation takes place because replacement will be faster at times when environmental conditions are favorable and/or larvae are settling. To some extent there is a conflict between larval dispersal on the one hand and the need to ensure that the larvae regain their position in the limited intertidal habitat prior to settlement on the other. Here again, evidence is limited but it does seem that the larval stages possess behavioral mechanisms that enable them to resist displacement out to sea and away from the adult habitat.

Other mechanisms involved in the control of community structure and population size also seem to be poorly known. The major structuring forces operative in many other communities, predation and competition, have been little studied but predation from within the resident intertidal fish community seems to be relatively unimportant. In contrast, predation pressure imposed by external predators such as visiting subtidal fishes, seabirds, wading birds, and terrestrial mammals may be considerable. The quantitative evaluation of such external predation on the abundance and distribution of intertidal fishes, during both high and low tides, would demonstrate the relative importance of predation as a selective (i.e., evolutionary) factor for intertidal fish assemblages, as well as elucidate the ecological linkages between the intertidal and adjoining ecosystems. In the absence of any evidence to the contrary, it may be inferred that resource partitioning among species limits intertidal fish community size and structure. The main limiting factors seem to be food and space, although the role of agonistic behavior and territoriality in controlling population density is ambiguous as it is more prevalent in some groups than others. The interplay among microhabitat choice, food specialization, potential competition between species, and predation pressure undoubtedly has influenced the evolution and population abundance of intertidal fishes. However, the mechanisms involved and their selective importance are only beginning to be discerned; we have yet to determine quantitatively how certain feeding specializations or behavioral attributes (e.g., homing behavior, air breathing) affect the fitness of individuals and populations and hence the composition and dynamics of the fish community and the ecosystem of which they are a part. Some future avenues of study might include: (1) molecular studies to re-evaluate the validity of earlier subspecies designations (e.g., within *Clinocottus analis* and *Anoplarchus purpurescens* on the California coast; Hubbs, 1926a,b); significant genetic differentiation of populations (subspecies) would suggest isolation between the intertidal fish communities at different localities that those populations compose. (2) Tagging and tracking studies to elucidate the nature and extent of movements of individual fish over the tidal cycle, both within the intertidal zone and across the intertidal–subtidal interface, and to discern possible seasonal shifts in activity and distribution. (3) *In situ* observations on behavioral interactions of individuals, both within and between species. On a final speculative note, molecular studies that allow age determinations of species, that is, by using "molecular clocks" (Hillis and Moritz, 1990; Avise, 1994), perhaps could provide rough estimates of the relative ages of intertidal fish assemblages and yield insights into the sequence by which species were "assembled" to produce the communities we see today. Clearly there is much still to be learned about intertidal fish communities.

## References


Abou-Seedo, F. S. (1992). Abundance of fish caught by stake traps (hadra) in the intertidal zone in Doha, Kuwait Bay. *J. Univ. Kuwait (Sci.)* **19,** 91–99.

Ali, T. S., and Hussain, N. A. (1990). Composition and seasonal fluctuation of intertidal fish assemblage in Kohr al-Zubair, Northwestern Arabian Gulf. *J. Appl. Ichthyol.* **6,** 24–36.

Armstrong, J., Staude, C. P., Thom, R. M., and Chew, K. K. (1976). Habitats and relative abundances of the intertidal macrofauna at five Puget Sound beaches in the Seattle area. *Syesis* **9,** 277–290.

Arruda, L. M. (1979). Specific composition and relative abundance of intertidal fish at two places on the Portuguese coast (Sesimbra and Magoito, 1977–78). *Arq. Mus. Bocage* **6,** 325–342.

Atkinson, C. E. (1939). Notes on the life history of the tidepool johnny (*Oligocottus maculosus*). *Copeia* **1939,** 23–30.

Avise, J. C. (1994). "Molecular Markers, Natural History and Evolution." Chapman & Hall, New York.

Barber, W. E., McDonald, L. M., Erickson, W. P., and Vallarino, M. (1995). Effect of the *Exxon Valdez* oil spill on intertidal fish: A field study. *Trans. Am. Fish. Soc.* **124,** 461–475.

Barton, M. G. (1982). Intertidal vertical distribution and diets of five species of central California stichaeoid fishes. *Calif. Fish Game* **68,** 174–182.

Batts, B. S. (1961). Intertidal fish as food of the common garter snake. *Copeia* **1961,** 350–351.

Beckley, L. E. (1985a). The fish community of East Cape tidal pools and an assessment of the nursery function of this habitat. *S. Afr. J. Zool.* **20,** 21–27.

Beckley, L. E. (1985b). Tide-pool fishes: Recolonization after experimental elimination. *J. Exp. Mar. Biol. Ecol.* **85,** 287–295.

Begon, M., Harper, J. L., and Townsend, C. R. (1990). "Ecology. Individuals, Populations and Communities." Blackwell Scientific, London.

Beja, P. R. (1995). Structure and seasonal fluctuations of rocky shore littoral fish assemblages in south-western Portugal: Implications for otter prey availability. *J. Mar. Biol. Assoc. U.K.* **75,** 833–847.

Bennett, B. A. (1987). The rock-pool fish community of Koppie Alleen and an assessment of the importance of Cape rock-pools as nurseries for juvenile fish. *S. Afr. J. Zool.* **22,** 25–32.

Bennett, B. A., and Griffiths, C. L. (1984). Factors affecting the distribution, abundance and diversity of rock pool fishes on the Cape Peninsula, South Africa. *S. Afr. J. Zool.* **19,** 97–104.

Bennett, B. A., Griffiths, C. L., and Penrith, M.-L. (1983). The diets of littoral fish from the Cape Peninsula. *S. Afr. J. Zool.* **18,** 343–352.

Black, R., and Miller, R. J. (1991). Use of the intertidal zone by fish in Nova Scotia. *Environ. Biol. Fish.* **31,** 109–121.

Breder, C. M., Jr. (1948). Observations on coloration in reference to behaviour in tide-pool and other marine shore fishes. *Bull. Am. Mus. Nat. Hist.* **92,** 285–311.

Briggs, J. C. (1974). "Marine Zoogeography." McGraw-Hill, New York.

Bussing, W. A. (1972). Recolonisation of a population of supratidal fishes at Eniwetok Atoll, Marshall Islands. *Atoll Res. Bull.* **154,** 1–7.

Cain, R. L., and Dean, J. M. (1976). Annual occurrence, abundance and diversity of fish in a South Carolina intertidal creek. *Mar. Biol.* **36,** 369–379.

Cattrijsse, A., Makwaia, E. D., Dankwa, H. R., Hamerlynck, O., and Hemminga, M. A. (1994). Nekton communities of an intertidal creek of a European estuarine brackish marsh. *Mar. Ecol. Prog. Ser.* **109,** 195–208.

Chadwick, E. M. P. (1976). A comparison of growth and abundance for tidal pool fishes in California and British Columbia. *J. Fish Biol.* **8,** 27–34.

Chang, K.-H., and Lee, S.-C. (1969). Stomach contents analysis of some intertidal fishes of Taiwan. *Bull. Inst. Zool. Acad. Sin.* **8,** 71–77.

Chang, K.-H., Lee, S.-C., Lee, J.-C., and Chen, C.-P. (1973). Ecological study on some intertidal fishes of Taiwan. *Bull. Inst. Zool. Acad. Sin.* **12,** 45–50.

Chang, K.-H., Lee, S.-C., and Wu, W.-L. (1977). Fishes of reef limestone platform at Maopitou, Taiwan: Diversity and abundance. *Bull. Inst. Zool. Acad. Sin.* **16,** 9–21.

Chotkowski, M. A. (1994). "The Behavioural Ecology and Population Dynamics of the Intertidal Fishes of the Northeast Pacific." PhD dissertation, University of California, Los Angeles.

Christensen, M. S., and Winterbottom, R. (1981). A correction factor for, and its application to, visual censuses of littoral fish. *S. Afr. J. Zool.* **16,** 73–79.

Clayton, D. A. (1993). Mudskippers. *Oceanogr. Mar. Biol. Annu. Rev.* **31,** 507–577.

Collette, B. B. (1986). Resilience of the fish assemblage in New England tidepools. *Fish. Bull.* **84** 200–204.

Connell, J. H. (1961). The influence of interspecific competition and other factors on the distribution of the barnacle *Chthamalus stellatus. Ecology* **42,** 710–723.

Connell, J. H. (1970). A predator-prey system in the marine intertidal region. I. *Balanus glandula* and several predatory species of *Thais. Ecol. Monogr.* **40,** 49–78.

Crabtree, R. E., and Dean, J. M. (1982). The structure of two South Carolina estuarine tide pool fish assemblages. *Estuaries* **5,** 2–9.

Craik, G. J. S. (1981). The effects of age and length on homing performance in the intertidal cottid, *Oligocottus maculosus* Girard. *Can. J. Zool.* **59,** 598–604.

Cross, J. N. (1981). "Structure of a Rocky Intertidal Fish Assemblage." PhD thesis, University of Washington, Seattle.

Davenport, J., and Rees, E. I. S. (1993). Observations on neuston and floating weed patches in the Irish Sea. *Estuar. Coast. Shelf Sci.* **36,** 395–411.

Dayton, P. K. (1971). Competition, disturbance, and community organization: The provision and subsequent utilization of space in a rocky intertidal community. *Ecol. Monogr.* **41,** 351–389.

DeMartini, E. E. (1969). A correlative study of the ecology and comparative feeding mechanism morphology of the Embiotocidae (surf-fishes) as evidence of the family's adaptive radiation into available ecological niches. *Wasmann J. Biol.* **27,** 177–247.

DeMartini, E. E. (1978). Spatial aspects of reproduction in buffalo sculpin, *Enophrys bison. Environ. Biol. Fish.* **3,** 331–336.

DeMartini, E. E., and Patten, B. J. (1979). Egg guarding and reproductive biology of the red Irish lord, *Hemilepidotus hemilepidotus* (Tilesius). *Syesis* **12,** 41–55.

Freeman M. C., Neally, M. C. N., and Grossman, G. D. (1985). Aspects of the life history of the fluffy sculpin, *Oligocottus snyderi*. *Fish. Bull.* **83,** 645–655.

Gibson, R. N. (1967). Studies on the movements of littoral fish. *J. Anim. Ecol.* **36,** 215–234.

Gibson, R. N. (1968a). The agonistic behaviour of juvenile *Blennius pholis* L. (Teleostei). *Behaviour* **30,** 192–217.

Gibson, R. N. (1968b). The food and feeding relationships of littoral fish in the Banyuls region. *Vie Mil. Ser. A* **19,** 447–456.

Gibson, R. N. (1969). The biology and behaviour of littoral fish. *Oceanogr. Mar. Biol. Annu. Rev.* **7,** 367–410.

Gibson, R. N. (1970). Observations on the biology of the giant goby *Gobius cobitis* Pallas. *J. Fish Biol.* **2,** 281–288.

Gibson, R. N. (1972). The vertical distribution and feeding relationships of intertidal fish on the Atlantic coast of France. *J. Anim. Ecol.* **41,** 189–207.

Gibson, R. N. (1973). The intertidal movements and distribution of young fish on a sandy beach with special reference to the plaice (*Pleuronectes platessa* L.). *J. Exp. Mar. Biol. Ecol.* **12,** 79–102.

Gibson, R. N. (1982). Recent studies on the biology of intertidal fishes. *Oceanogr. Mar. Biol. Annu. Rev.* **20,** 363–414.

Gibson, R. N. (1988). Patterns of movement in intertidal fishes. *In* "Behavioural Adaptations to Intertidal Life" (G. Chelazzi and M. Vannini, Eds.), pp. 55–63, Plenum Press, New York.

Gibson, R. N., Robb, L., Burrows, M. T., and Ansell, A. D. (1996). Tidal, diel and longer term changes in the distribution of fishes on a Scottish sandy beach. *Mar. Ecol. Prog. Ser.* **130,** 1–17.

Greely, A. W. (1899). Notes on the tide-pool fishes of California, with a description of four new species. *Bull. U.S. Fish. Comm.* **19,** 7–20.

Green, J. M. (1971a). Local distribution of *Oligocottus maculosus* Girard and other tidepool cottids of the west coast of Vancouver Island, British Columbia. *Can. J. Zool.* **49,** 1111–1128.

Green, J. M. (1971b). High tide movements and homing behaviour of the tidepool sculpin *Oligocottus maculosus*. *J. Fish. Res. Bd Can.* **28,** 383–389.

Green, J. M. (1971c). Field and laboratory activity patterns of the tidepool cottid *Oligocottus maculosus* Girard. *Can. J. Zool.* **49,** 255–264.

Green, J. M. (1973). Evidence for homing in the mosshead sculpin (*Clinocottus globiceps*). *J. Fish. Res. Bd. Can.* **30,** 129–130.

Grossman, G. D. (1979). Symbiotic burrow-occupying behavior in the bay goby, *Lepidogobius lepidus*. *Calif. Fish Game* **65,** 122–124.

Grossman, G. D. (1982). Dynamics and organization of a rocky intertidal fish assemblage: The persistence and resilience of taxocene structure. *Am. Nat.* **119,** 611–637.

Grossman, G. D. (1986a). Food resource partitioning in a rocky intertidal fish assemblage. *J. Zool. London (B)* **1,** 317–355.

Grossman, G. D. (1986b). Long term persistence in a rocky intertidal fish assemblage. *Environ. Biol. Fish.* **15,** 315–317.

Helfman, G. S. (1993). Fish behaviour by day, night and twilight. *In* "Behaviour of Teleost Fishes" (T. J. Pitcher, Ed.), 2nd ed., pp. 479–512. Chapman and Hall, London.

Herald, E. S. (1941). A systematic analysis of variation in the western American pipefish, *Syngnathus californiensis*. *Stanford Ichthyol. Bull.* **2,** 49–73.

Hillis, D. M., and Moritz, C. (Eds.) (1990). "Molecular Systematics." Sinauer Associates, Sunderland, MA.

Hoese, H. D. (1966). Habitat segregation in aquaria between two sympatric species of *Gobiosoma*. *Pub. Inst. Mar. Sci. Univ. Tex.* **11,** 7–11.

Horn, M. H. (1980). Diel and seasonal variation in abundance and diversity of shallow-water fish populations in Morro Bay, California. *Fish. Bull.* **78,** 759–770,

Horn, M. H. (1983). Optimal diets in complex environments: Feeding strategies of two herbivorous fishes from a temperate rocky intertidal zone. *Oecologia (Berlin)* **58,** 345–350.

Horn, M. H., and Allen, L. G. (1978). A distributional analysis of California coastal marine fishes. *J. Biogeog.* **5,** 23–42.

Horn, M. H., Murray, S. N., and Edwards, T. W. (1982). Dietary selectivity in the field and food preferences in the laboratory for two herbivorous fishes (*Cebidichthys violaceus* and *Xiphister mucosus*) from a temperate intertidal zone. *Mar. Biol.* **67,** 237–246.

Horn, M. H., Murray, S. N., and Seapy, R. R. (1983). Seasonal structure of a central California rocky intertidal community in relation to environmental variations. *Bull. S. Calif. Acad. Sci.* **82,** 79–94.

Horn, M. H., and Riegle, K. C. (1981). Evaporative water loss and intertidal vertical distribution in relation to body size and morphology of stichaeoid fishes from California. *J. Exp. Mar. Biol. Ecol.* **50,** 273–288.

Hubbs, C. L. (1921). The ecology and life-history of *Amphigonopterus aurora* and of other viviparous perches of California. *Biol. Bull.* **40,** 181–208.

Hubbs, C. L. (1926a). Notes on the blennioid fishes of western North America. *Pap. Mich. Acad. Sci. Arts Lett.* **7,** 351–394.

Hubbs, C. L. (1926b). A revision of the fishes of the subfamily Oligocottinae. *Occ. Pap. Mus. Zool. Univ. Mich. Ann Arbor* **171.**

Hubbs, C. L. (1948). Changes in the fish fauna of western North America correlated with changes in ocean temperature. *J. Mar. Res.* **7,** 459–482.

Ibáñez, M., Miguel, I., San Millán, D., and Ripa, I. (1989). Intertidal ichthyofauna of the Spanish Atlantic coast. *Sci. Mar.* **53,** 451–455.

Itzkowitz, M. (1974). A behavioural reconnaissance of some Jamaican reef fishes. *Zool. J. Linn. Soc.* **55,** 87–118.

Jones, D., and Clare, J. (1977). Annual and long-term fluctuations in the abundance of fish species inhabiting an intertidal mussel bed in Morecambe Bay, Lancashire. *Zool. J. Linn. Soc.* **60,** 117–172.

Jones, J. A. (1981). "Competition for Substrates in Laboratory Experiments between *Anoplarchus purpurescens* (Pisces, Sticheidae) and Three Related Species from the Central California Rocky Intertidal Zone." MA thesis. California State University, Fullerton.

Jordan, D. S., and Gilbert, C. H. (1881). Notes on fishes of the Pacific coast of the United States. *Proc. U.S. Nat. Mus.* **4,** 29–70.

Khoo, H. W. (1974). Sensory basis of homing in the intertidal fish *Oligocottus maculosus* Girard. *Can. J. Zool.* **52,** 1023–1029.

Kingsford, M. J. (1992). Drift algae and small fish in the coastal waters of northeastern New Zealand. *Mar. Ecol. Prog. Ser.* **80,** 41–55.

Koppel, V. H. (1988). Habitat selection and space partitioning among two Mediterranean blenniid species. *Pub. Stn. Zool. Napoli I. Mar. Ecol.* **9,** 329–346.

Kotrschal, K. (1988). Blennies and endolithic bivalves: Differential utilisation of shelter in Adriatic Blenniidae (Pisces: Teleostei). *Pub. Stn. Zool. Napoli I. Mar. Ecol.* **9,** 253–269.

Kruuk, H., Nolet, B., and French, D. (1988). Fluctuations in numbers and activity of inshore demersal fishes in Shetland. *J. Mar. Biol. Assoc. U.K.* **68,** 601–617.

Lardner, R., Ivantsoff, W., and Crowley, L. E. L. M. (1993). Recolonisation by fishes of a rocky intertidal pool following repeated defaunation. *Austr. Zool.* **29,** 85–92.

Lee, S.-L. (1980). Intertidal fishes of the rocky pools at Lanyu (Botel Tobago), Taiwan. *Bull. Inst. Zool. Acad. Sin.* **19,** 1–13.

MacGinitie, G. E., and MacGinitie, N. (1949). "Natural History of Marine Animals." McGraw-Hill, New York.

MacNae, W. (1968). A general account of the fauna and flora of mangrove swamps and forests in the Indo-West-Pacific region. *Adv. Mar. Biol.* **6,** 73–270.

Macpherson, E. (1994). Substrate utilisation in a Mediterranean littoral fish community. *Mar. Ecol. Prog. Ser.* **114,** 211–218.

Mahon, R., and Mahon, S. D. (1994). Structure and resilience of a tidepool fish assemblage at Barbados. *Environ. Biol. Fish.* **41,** 171–190.

Marliave, J. B. (1981). High intertidal spawning under rockweed, *Fucus distichus,* by the sharpnose sculpin, *Clinocottus acuticeps. Can. J. Zool.* **59,** 1122–1125.

Marliave, J. B. (1986). Lack of planktonic dispersal of rocky intertidal fish larvae. *Trans. Am. Fish. Soc.* **115,** 149–154.

Marliave, J. B., and DeMartini, E. D. (1977). Parental behavior of intertidal fishes of the stichaeid genus Xiphister. *Can. J. Zool.* **55,** 60–63.

Marsh, B., Crowe, T. M., and Siegfried, W. R. (1978). Species richness and abundance of clinid fish (Teleostei: Clinidae) in intertidal rock pools. *Zool. Afr.* **13,** 283–291.

Marshall, W. H., and Echeverria, T. W. (1992). Age, length, weight, reproductive cycle and fecundity of the monkeyface prickleback (*Cebidichthys violaceus*). *Calif. Fish Game* **78,** 57–64.

Matson, R. H., Crabtree, C. B., and Haglund T. R. (1986). Ichthyofaunal composition and recolonization in a central California tidepool. *Calif. Fish Game* **72,** 227–231.

Matyushin, V. M. (1982). To the intertidal fish fauna of the Eastern Kamchatka. *Biol. Moria* **2,** 60–63.

Mayr, M., and Berger, A. (1992). Territoriality and microhabitat selection in two intertidal New Zealand fish. *J. Fish Biol.* **40,** 243–256.

Meffe, G. K., and Carroll, C. R. (1994). "Principles of Conservation Biology." Sinauer Associates, Sunderland, MA.

Mgaya, Y. D. (1992). Density and production of *Clinocottus globiceps* and *Oligocottus maculosus* (Cottidae) in tidepools at Helby Island, British Columbia. *Mar. Ecol. Prog. Ser.* **85,** 219–225.

Miller, D. J., and Lea, R. N. (1972). "Guide to the Coastal Marine Fishes of California." *Calif. Fish Game Bull.* **157.**

Miller, K. A., and Marshall, W. H. (1987). Food habits of large monkeyface prickleback, *Cebidichthys violaceus. Calif. Fish Game* **73,** 37–44.

Mistry, S. D., Lizerbram, E. K., and Parton, E. R. (1989). Short-term ichthyofaunal recruitment in northern California tidepools. *Copeia* **1989,** 1081–1084.

Mitchell, D. F. (1953). An analysis of stomach contents of California tide pool fishes. *Am. Midl. Nat.* **49,** 862–871.

Montgomery, W. L. (1977). Diet and gut morphology in fishes, with special reference to the monkeyface prickleback, *Cebidichthys violaceus* (Stichaeidae: Blennioidei). *Copeia* **1977,** 178–182.

Moring, J. R. (1976). Estimates of population size for tidepool sculpins, *Oligocottus maculosus,* and other intertidal fishes, Trinidad Bay, Humboldt County, California. *Calif. Fish Game* **62,** 65–72.

Moring, J. R. (1979). Age structure of a tidepool sculpin, *Oligocottus maculosus,* population in northern California. *Calif. Fish Game* **65,** 111–113.

Moring, J. R. (1983). Human factors affecting rocky intertidal fishes. *In* "Proceedings of the Third Symposium on Coastal and Ocean Management, Coastal Zone '83" (O. T. Magoon and H. Converse Eds.), Vol. 2, pp. 1572–1585. American Society of Civil Engineers, New York.

Moring, J. R. (1986). Seasonal presence of tidepool fish species in a rocky intertidal zone of northern California, USA. *Hydrobiologia* **134,** 21–27.

Moring, J. R. (1990). Seasonal absence of fishes in tidepools of a boreal environment (Maine, USA). *Hydrobiologia* **194,** 163–168.

Muñoz, A. A., and Ojeda, F. P. (1997). Feeding guild structure of a rocky intertidal fish assemblage in central Chile. *Environ. Biol. Fish.* **49,** 471–479.

Nakamura, R. (1971). Food of two cohabiting tidepool Cottidae. *J. Fish. Res. Bd. Can.* **28,** 928–932.

Nakamura, R. (1976). Experimental assessment of factors influencing microhabitat selection by the two tidepools fishes *Oligocottus maculosus* and *O. snyderi. Mar. Biol.* **37,** 97–104.

Nursall, J. R. (1981). Behavior and habitat affecting the distribution of five species of sympatric mudskippers in Queensland. *Bull. Mar. Sci.* **31,** 730–735.

O'Connell, C. P. (1953). The life history of the cabezon *Scorpaenichthys marmoratus* (Ayres). *Calif. Fish Game Bull.* **93,** 1–76.

Paine, R. T. (1980). Food webs: Linkage, interaction strength and community infrastructure. *J. Anim. Ecol.* **49,** 667–685.

Pequeño, G., and Lamilla, J. (1995). Peces intramareales de la costa de Llanquihue (Chile): Composicion taxonomica, abundancia relativa y gradiente de distribucion longitudinal. *Rev. Biol. Mar. Valparaiso* **30,** 7–27.

Pequeño, G., Lamilla, J., Lloris, D., and Rucabando, J. (1995). Comparacion entre las ictiofaunas intramareales de los extremos austral y boreal de los canales Patagonicos. *Rev. Biol. Mar. Valparaiso* **30,** 155–177.

Pfister, C. A. (1995). Estimating competition coefficients from census data: A test with field manipulations of tidepool fishes. *Am. Nat.* **146,** 271–291.

Pierce, B. E., and Pierson, K. B. (1990). Growth and reproduction of the tidepool sculpin *Oligocottus maculosus. Japan. J. Ichthyol.* **36,** 410–417.

Pillsbury, R. W. (1957). Avoidance of poisonous eggs of the marine fish *Scorpaenichthys marmoratus* by predators. *Copeia* **1957,** 251–252.

Pinchuk, V. I. (1976a). Littoral ichthyofauna of the Kuril Islands. *Sov. J. Mar. Biol.* **2,** 107–112.

Pinchuk, V. I. (1976b). Ichthyofauna of the littoral of the Komandorskie Islands. *Sov. J. Mar. Biol.* **2,** 292–300.

Polivka, K. M., and Chotkowski, M. A. (1998). The effects of repeated tidepool defaunation on recolonization by Northeast Pacific intertidal fishes. *Copeia,* in press.

Potts, G. W. (1980). The littoral fishes of Little Cayman (West Indies). *Atoll Res. Bull.* **241,** 43–52.

Prochazka, K. (1994). The reproductive biology of intertidal klipfish (Perciformes: Clinidae) in South Africa. *S. Afr. J. Zool.* **29,** 244–251.

Prochazka, K. (1996). Seasonal patterns in a temperate intertidal fish community on the west coast of South Africa. *Environ. Biol. Fish.* **45,** 133–140.

Prochazka, K., and Griffiths, C. L. (1992). The intertidal fish fauna of the west coast of South Africa—Species, community and biogeographic patterns. *S. Afr. J. Zool.* **27,** 115–120.

Richkus, W. A. (1978). A quantitative study of intertidepool movement of the wooly sculpin *Clinocottus analis. Mar. Biol.* **49,** 277–284.

Richkus, W. A. (1981). Laboratory studies of intraspecific behavioral interactions and factors influencing tidepool selection of the wooly sculpin, *Clinocottus analis. Calif. Fish Game* **67,** 187–195.

Ricketts, E. F., Calvin, J., Hedgpeth, J. W., and Phillips, D. W. (1985). "Between Pacific Tides." 5th ed. Stanford University Press, Stanford.

Robertson, I. (1974). The food of nesting double-crested and pelagic cormorants at Mandarte Island, British Columbia, with notes on feeding ecology. *Condor* **76,** 346–348.

Ross, S. T., McMichael, R. H., Jr., and Ruple, D. L. (1987). Seasonal and diel variations in the standing crop of fishes and macroinvertebrates from a Gulf of Mexico surf zone. *Estuar. Coast. Shelf Sci.* **25,** 391–412.

Rountree, R. A., and Able, K. W. (1993). Diel variation in decapod crustacean and fish assemblages in New Jersey polyhaline marsh creeks. *Estuar. Coast. Shelf Sci.* **37,** 181–201.

Sale, P. F. (1978). Reef fishes and other vertebrates: A comparison of social structures. *In* "Contrasts in Behavior. Adaptations in the Aquatic and Terrestrial Environments" (E. S. Reese and F. J. Lighter Eds.), pp. 313–346. Wiley, New York.

Santos, R. S., and Nash, R. D. M. (1996). Seasonal variations of injuries suffered by individuals of the Azorean rock-pool blenny (*Parablennius sanguinolentus parvicornis*). *Copeia* **1996,** 216–219.

Santos, R. S., Nash, R. D. M., and Hawkins, S. J. (1994). Fish assemblages on intertidal shores of the island of Faial, Azores. *Arquipelago, Bull. Univ. Azores, Life Mar. Sci.* **12A,** 87–100.

Smale, M. J., and Buxton, C. D. (1989). The subtidal gully fish community of the eastern Cape and the role of this habitat as a nursery area. *S. Afr. J. Zool.* **24,** 58–67.

Stephens, J. S., Johnson, R. K., Key, G. S., and McCosker, J. E. (1970). The comparative ecology of three sympatric species of California blennies of the genus *Hypsoblennius* Gill (Teleostomi, Blenniidae). *Ecol. Monogr.* **40,** 213–233.

Stephenson, T. A., and Stephenson, A. (1972). "Life Between Tidemarks on Rocky Shores." Freeman, San Francisco.

Stepien, C. A. (1990). Population structure, diets and biogeographic relationships of a rocky intertidal fish assemblage in central Chile: High levels of herbivory in a temperate system. *Bull. Mar. Sci.* **47,** 598–612.

Stepien, C. A., Phillips, H., Adler, J. A., and Mangold, P. J. (1991). Biogeographic relationships of a rocky intertidal fish assemblage in an area of cold water upwelling off Baja California, Mexico. *Pacific Sci.* **1991,** 63–71.

Stepien, C. A., and Rosenblatt, R. H. (1991). Patterns of gene flow and genetic divergence in the northeastern Pacific Clinidae (Teleostei: Blennioidei), based on allozyme and morphological data. *Copeia* **1991,** 873–896.

Thomson, D. S., and Lehner C. E. (1976). Resilience of a rocky intertidal fish community in a physically unstable environment. *J. Exp. Mar. Biol. Ecol.* **22,** 1–29.

Tyler, A. V. (1971). Surges of winter flounder, *Pseudopleuronectes americanus,* into the intertidal zone. *J. Fish. Res. Bd. Can.* **28,** 1727–1732.

van der Veer, H. W., and Bergman, M. J. N. (1986). Development of tidally related behaviour of a newly settled 0-group plaice (*Pleuronectes platessa*) population in the western Wadden sea. *Mar. Ecol. Prog. Ser.* **31,** 121–129.

Varas, E., and Ojeda, F. P. (1990). Intertidal fish assemblages of the central Chilean coast: Diversity, abundance and trophic patterns. *Rev. Biol. Mar.* **25,** 59–70.

Veith, W. J., and Cornish, D. A. (1986). Ovarian adaptations in the viviparous teleosts *Clinus superciliosus* and *C. dorsalis* (Perciformes: Clinidae). *S. Afr. J. Zool.* 21, 343–347.

Vincent, A., Ahnesjö, I. and Berglund, A. (1994). Operational sex ratios and behavioural sex differences in a pipefish population. *Behav. Ecol. Sociobiol.* **34,** 435–442

Waples, R. S. (1987). A multispecies approach to the analysis of gene flow in marine shore fishes. *Evolution* **43,** 385–400.

Weaver, P. L. (1970). Species diversity and ecology of tidepool fishes in three Pacific coastal areas of Costa Rica. *Rev. Biol. Trop.* **17,** 165–185.

Wells, A. W. (1986). Aspects of ecology and life history of the woolly sculpin, *Clinocottus analis,* from southern California. *Calif. Fish Game* **72,** 213–226.

Wheeler, A. (1970). Notes on a collection of shore fishes from Guernsey, Channel Islands. *J. Fish Biol.* **2,** 323–328.

Williams, G. C. (1954). Differential vertical distribution of the sexes in *Gibbonsia elegans* with remarks on two nominal subspecies of this fish. *Copeia* **1954,** 267–273.

Williams, G. C. (1957). Homing behavior of California rocky shore fishes. *Univ. Calif. Pub. Zool.* **59,** 249–284.

Willis, T. J., and Roberts, C. D. (1996). Recolonisation and recruitment of fishes to intertidal rockpools at Wellington, New Zealand. *Environ. Biol. Fish.* **47,** 329–343.

Wright, J. M. (1989). Diel variation and seasonal consistency in the fish assemblage of the non-estuarine Sulaibikhat Bay, Kuwait. *Mar. Biol.* **102,** 135–142.

Yoshiyama, R. M. (1980). Food habits of three species of rocky intertidal sculpins (Cottidae) in central California. *Copeia* **1980,** 515–525.

Yoshiyama, R. M. (1981). Distribution and abundance patterns of rocky intertidal fishes in central California. *Envir. Biol. Fish.* **6,** 315–332.

Yoshiyama, R. M., Gaylord, K. B., Philippart, M. T., Moore, T. R., Jordan, J. R., Coon, C. C., Schalk, L. L., Valpey, C. J., and Tosques, I. (1992). Homing behavior and site fidelity in intertidal sculpins (Pisces: Cottidae). *J. Exp. Mar. Biol. Ecol.* **160,** 115–130.

Yoshiyama, R. M., and Sassaman, C. (1987). Geographical patterns of allozymic variation in three species of intertidal sculpins. *Environ. Biol. Fish.* **20,** 203–218.

Yoshiyama, R. M., Sassaman, C., and Lea, R. N. (1986). Rocky intertidal fish communities of California: Temporal and spatial variation. *Environ. Biol. Fish.* **17,** 23–40.

Yoshiyama, R. M., Sassaman, C., and Lea, R. N. (1987). Species composition of rocky intertidal and subtidal fish assemblages in central and northern California, British Columbia-southeast Alaska. *Bull. S. Calif. Acad. Sci.* **86,** 136–144.

Zander, C. D. (1967). Beiträge zur Ökologie und Biologie litoralbewohnender Salariidae und Gobiidae (Pisces) aus dem Roten Meer. *"Meteor" Forschungsergeb. Reihe D* **2,** 69–84.

# 14

# Systematics of Intertidal Fishes

Michael A. Chotkowski

*Department of Biology, University of California, Los Angeles, California*

Donald G. Buth

*Department of Biology, University of California, Los Angeles, California*

Kim Prochazka

*Zoology Department, University of Cape Town, Rondebosch, South Africa*

## I. Introduction

The intertidal zone, specifically the rocky intertidal zone, would appear as "a few fine lines and would hardly be noticeable" on a map of the world (Brehaut, 1982). The patchiness of this habitat raises the question of relationships among its inhabitants. It is known that intertidal fishes are generally small in size and have a limited range of body plans (Horn, Chapter 16, this volume). These initial observations suggest that intertidal fishes may be members of only a few taxonomic groups. Horn (Chapter 16, this volume) compared the intertidal fish assemblages reported from six locations, five of which occur on different continental landmasses. These assemblages were "almost completely different" in regard to species present and "only slightly more similar" in regard to the genera that were represented. Only at the level of family were groups shared among these disparate locations.

Modern classifications of fishes (e.g., Nelson, 1994) are based on principles of phylogenetic systematics (Hennig, 1966; Wiley, 1981). "Relationships" in phylogenetic systematics are strictly genealogical, not based on measures of overall similarity. Phylogenetic classifications communicate evolutionary history.

We raise the following questions in regard to intertidal fishes: (1) Which fishes have been reported from the intertidal zone? (2) What are the phylogenetic relationships among the intertidal fishes; that is, what distinct lineages are represented? (3) What inferences can be made about the evolution of intertidal fishes based on their phylogenetic relationships? To address these questions, we have selected 44 published studies and examined data from three unpublished sources that have reported intertidal ichthyofaunal assemblages at 77 different locations distributed around the world.

*Intertidal Fishes: Life in Two Worlds*

## II. The Intertidal Ichthyofauna

### A. Limits to Our Survey

Lists of intertidal fishes were obtained from 47 sources including published reports, unpublished dissertations, and other unpublished data made available to us. These particular studies were chosen because they had goals or methods in common. We sought studies that surveyed entire intertidal ichthyofaunal assemblages rather than studies aimed at particular species. An exception included fishes explicitly identified as intertidal species in the checklist of Wheeler (1994). Collections employed chemical means, or removal of tidepool water, to obtain the fishes. Such collections were usually limited to low tidal conditions and over rocky substrate where discrete pools of water would remain.

Collections conducted under these conditions would be expected to be biased in favor of resident species; tidal visitors that move into the intertidal zone at high tide to feed would be expected to have left the area before collection commenced. However, because most of these sources intended to quantify the ichthyofauna, *all* species found were reported. Thus, trapped tidal visitors, transients, and "strays" were included in these lists of intertidal fishes. Our survey, therefore, is not limited to resident species.

### B. An Annotated List of Intertidal Fishes

A descriptive paragraph is provided for each family of fishes represented in the sample of collections analyzed in this chapter. Classification information other than ordinal membership is not provided; refer to Table 1 or Nelson (1994). In some cases it was clear to the authors that a given species merely strayed into the intertidal zone and was collected; this was often true where only one or a few observations were noted for a family. In many other cases, accounts in the literature and the number of observations indicate that a family has a real presence in the intertidal zone. The authors' perception of each family's importance determined the extent of the descriptions of each family that have been provided. At least the distribution and number of species has been provided. Much of the basic information in this list was obtained from Nelson (1994); in addition, information about lower depth limits for many families was obtained from Weitzman (1997). To highlight our reliance on Nelson (1994), specific citation to the work was inserted in some cases where there exists disagreement with Nelson or ambiguity in accounts by other authorities. Each descriptive paragraph is followed by a list of species that appeared in the sample of collections examined here. The parenthetical numbers provided after each taxon refer to the appendix, which reports the latitude, site, and authority for each collection. The reader should interpret these references as citations of the relevant authors' work.

#### Order Orectolobiformes

Family Ginglymostomatidae. Nurse sharks.

The three species of nurse sharks are widely distributed in tropical coastal waters of the Atlantic, Indian, and Pacific oceans.

Intertidal observation: *Ginglymostoma cirratum* (COS3).

#### Order Anguilliformes

Family Anguillidae. Freshwater eels.

Usually catadromous, freshwater eels are found worldwide in tropical and temperate seas except eastern Pacific and southern Atlantic oceans. Adults live in fresh or estuarine waters.

**Table 1.** Phyletic Diversity of Extant Intertidal Fishes

| Taxon | Total species | Strictly freshwater species | Species that use freshwater | Strictly marine species | Rocky intertidal species |
|---|---|---|---|---|---|
| Class Myxini | | | | | |
| Order Myxiniformes | 43 | 0 | 0 | 43 | 0 |
| Class Cephalaspidimorphi | | | | | |
| Order Petromyzontiformes | 41 | 32 | 9 | 0 | 0 |
| Class Chondrichthyes | | | | | |
| Subclass Holocephali | | | | | |
| Order Chimaeriformes | 31 | 0 | 0 | 31 | 0 |
| Subclass Elasmobranchii | | | | | |
| Order Heterodontiformes | 8 | 0 | 0 | 8 | 0 |
| Order Orectolobiformes | 31 | 0 | 0 | 31 | 1 |
| Order Carchariniformes | 208 | 1 | 7 | 200 | 0 |
| Order Lamniformes | 16 | 0 | 0 | 16 | 0 |
| Order Hexanchiformes | 5 | 0 | 0 | 5 | 0 |
| Order Squaliformes | 74 | 0 | 0 | 74 | 0 |
| Order Squatiniformes | 12 | 0 | 0 | 12 | 0 |
| Order Pristiophoriformes | 5 | 0 | 1 | 4 | 0 |
| Order Rajiformes | 456 | 24 | 4 | 428 | 0 |
| Class Sarcopterygii | | | | | |
| Subclass Coelacanthimorpha | | | | | |
| Order Coelacanthiformes | 1 | 0 | 0 | 1 | 0 |
| Subclass Porolepimorpha and Dipnoi | | | | | |
| Order Ceratodontiformes | 1 | 0 | 1 | 0 | 0 |
| Order Lepidosireniformes | 5 | 5 | 0 | 0 | 0 |
| Class Actinopterygii | | | | | |
| Subclass Chondrostei | | | | | |
| Order Polypteriformes | 10 | 10 | 0 | 0 | 0 |
| Order Acipenseriformes | 26 | 14 | 12 | 0 | 0 |
| Subclass Neopterygii | | | | | |
| Order Semionotiformes | 7 | 6 | 1 | 0 | 0 |
| Order Amiiformes | 1 | 1 | 0 | 0 | 0 |
| Division Teleostei | | | | | |
| Subdivision Osteoglossomorpha | | | | | |
| Order Osteoglossiformes | 217 | 217 | 0 | 0 | 0 |
| Subdivision Elopomorpha | | | | | |
| Order Elopiformes | 8 | 0 | 7 | 1 | 0 |
| Order Albuliformes | 29 | 0 | 0 | 29 | 0 |
| Order Anguilliformes | 738 | 6 | 20 | 712 | 36 |
| Order Saccopharyngiformes | 26 | 0 | 0 | 26 | 0 |
| Subdivision Clupeomorpha | | | | | |
| Order Clupeiformes | 357 | 72 | 8 | 277 | 5 |
| Subdivision Euteleostei | | | | | |
| Superorder Ostariophysi | | | | | |
| Order Gonorhynchiformes | 35 | 28 | 1 | 6 | 0 |
| Order Cypriniformes | 2662 | 2662 | 0 | 0 | 0 |
| Order Characiformes | 1343 | 1343 | 0 | 0 | 0 |
| Order Siluriformes | 2405 | 2280 | 7 | 118 | 3 |
| Order Gymnotiformes | 62 | 62 | 0 | 0 | 0 |
| Superorder Protacanthopterygii | | | | | |
| Order Esociformes | 10 | 10 | 0 | 0 | 0 |
| Order Osmeriformes | 236 | 42 | 29 | 165 | 1 |
| Order Salmoniformes | 66 | 45 | 21 | 0 | 0 |

*continues*

**Table 1.** *Continued*

| Taxon | Total species | Strictly freshwater species | Species that use freshwater | Strictly marine species | Rocky intertidal species |
|---|---|---|---|---|---|
| Superorder Stenopterygii | | | | | |
| Order Stomiiformes | 321 | 0 | 0 | 321 | 0 |
| Order Ateleopodiformes | 12 | 0 | 0 | 12 | 0 |
| Superorder Cyclosquamata | | | | | |
| Order Aulopiformes | 219 | 0 | 0 | 219 | 1 |
| Superorder Scopelomorpha | | | | | |
| Order Myctophiformes | 241 | 0 | 0 | 241 | 0 |
| Superorder Lampridiomorpha | | | | | |
| Order Lampridiformes | 19 | 0 | 0 | 19 | 0 |
| Superorder Polymixiomorpha | | | | | |
| Order Polymixiiformes | 5 | 0 | 0 | 5 | 0 |
| Superorder Paracanthopterygii | | | | | |
| Order Percopsiformes | 9 | 9 | 0 | 0 | 0 |
| Order Ophidiiformes | 355 | 5 | 1 | 349 | 7 |
| Order Gadiformes | 482 | 1 | 1 | 480 | 11 |
| Order Batrachoidiformes | 69 | 5 | 1 | 63 | 2 |
| Order Lophiiformes | 297 | 0 | 0 | 297 | 4 |
| Superorder Acanthopterygii | | | | | |
| Series Mugilomorpha | | | | | |
| Order Mugiliformes | 66 | 1 | 6 | 59 | 13 |
| Series Atherinomorpha | | | | | |
| Order Atheriniformes | 285 | 146 | 25 | 114 | 8 |
| Order Beloniformes | 191 | 51 | 5 | 135 | 2 |
| Order Cyprinodontiformes | 807 | 794 | 11 | 2 | 4 |
| Series Percomorpha | | | | | |
| Order Stephanoberyciformes | 86 | 0 | 0 | 86 | 0 |
| Order Beryciformes | 123 | 0 | 0 | 123 | 13 |
| Order Zeiformes | 39 | 0 | 0 | 39 | 1 |
| Order Gasterosteiformes | 257 | 19 | 22 | 216 | 18 |
| Order Synbranchiformes | 87 | 84 | 3 | 0 | 0 |
| Order Scorpaeniformes | 1271 | 52 | 10 | 1209 | 67 |
| Order Perciformes | 9293 | 1922 | 263 | 7108 | 479 |
| Order Pleuronectiformes | 570 | 4 | 16 | 550 | 12 |
| Order Tetraodontiformes | 339 | 12 | 8 | 309 | 14 |
| Totals | 24,618 | 9966 | 500 | 14,153 | 702 |

*Note.* Table adapted from Nelson (1994). Intertidal species summarizes fishes collected in the intertidal zone in the 86 datasets treated in this chapter.

*Anguilla rostrata* adults have been observed on the sea floor as deep as 2000 m. One genus, 15 species; 2 species have been observed in intertidal zone, almost certainly as transients, in the North Atlantic and occasionally the central Atlantic (Dooley *et al.,* 1985).

Intertidal observations: *Anguilla anguilla* (A4–A6), *A. rostrata* (A1, A2).

## Family Moringuidae. Spaghetti eels.

This small family of burrowing eels occupies nearshore mainly sandy bottom habitats in the tropical Indo-Pacific and western Atlantic oceans. *Moringua microchir* has been observed in freshwater (Dingerkus and Séret, 1992).

Intertidal observations: *Moringua abbreviata* (TAI1), *M. edwardsi* (A3).

### Family Muraenidae. Moray eels.

Moray eels are abundant and diverse in tropical and temperate nearshore habitats. Most species are marine, but some are found in freshwater or are euryhaline. Moray eels commonly occupy interstitial spaces in coral or stony reefs in shallow nearshore and probably intertidal habitats (pp. 104–206 in Böhlke, 1989).

Intertidal observations: *Echidna catenata* (A3), *E. nebulosa* (COS3, SAF16, SAF19, SEY1, TAI1), *E. polyzona* (SAF19, TAI1), *Enchelycore ramosa* (NZ1), *Enchelycore* spp. (A3), *Gymnothorax dovii* (COS1, COS3), *G. eurostus* (EAS1, SAF22), *G. fimbriatus* (TAI1), *G. flavimarginatus* (SAF18, SAF19, TAI1), *G. meleagris* (TAI1), *G. petelli* (TAI1), *G. pictus* (TAI1), *G. prasinus* (AUS1, NZ3), *G. thyrsoideus* (TAI1), *G. undulatus* (SAF12, SAF15, SAF17–SAF20, SAF22, SEY1, TAI1), *G. clepsydra* (COS3), *G.* spp. (A3), *Muraena helena* (A10), *M. lentiginosa* (COS1, COS3), *Siderea pictus* (SEY1), *Uropterygius micropterus* (TAI1), *U. necturus* (COS1).

### Family Ophichthidae. Snake eels and worm eels.

Snake and worm eels are diverse in the tropical and temperate oceans. Most species occur in coastal waters and some enter freshwater. Ophichthids dig burrows in soft substrata (Nelson, 1994).

Intertidal observations: *Ahlia egmontis* (A3), *Muraenichthys iredalei* (AUS1), *Myrichthys acuminatus* (A3), *M. maculosus* (TAI1), *Myrophis* spp. (A3), *Opistognathus maxillosus* (A3).

### Family Congridae. Conger eels.

Conger eels are diverse in tropical to temperate waters worldwide. They occupy a variety of habitats; most members of *Conger* live in sandy bottom nearshore habitats, but other genera are known from depths up to 1800 m (Weitzman, 1997).

Intertidal observations: *Conger cinereus* (EAS1, TAI1), *C. conger* (A5, A4), *C. oceanicus* (A1), *C. verreauxi* (NZ5, NZ10, NZ14), *C. wilsoni* (NZ2–NZ4, SAF11, SAF12, SAF13, SAF19).

## Order Clupeiformes

### Family Engraulidae. Anchovies.

Anchovies are locally abundant in nearshore and epipelagic marine and freshwater habitats in the Atlantic, Indian, and Pacific. About 17 of 139 species occupy freshwater. Most engraulids are pelagic planktivores, and it is likely that engraulids are random or transient visitors to the intertidal zone.

Intertidal observation: *Thryssa kammalensis* (TAI1).

### Family Clupeidae. Herrings.

Herrings, like anchovies, are locally abundant in nearshore and epipelagic marine and freshwater habitats worldwide. Herrings are more abundant in tropical waters. About 50 of 181 clupeid species live in freshwater (Nelson, 1994); many of the others are euryhaline and regularly enter brackish or freshwater. Some are anadromous. Most species are pelagic planktivores. Herrings are probably transient or random visitors to the intertidal zone.

Intertidal observations: *Clupea harengus* (A1, A4), *Herklotsichthys abbreviata* (AUS1), *H. quadrimaculata* (TAI1), *Spratelloides delicatulus* (TAI1), *Sprattus sprattus* (A4).

## Order Siluriformes

### Family Ariidae. Sea catfishes.

Ariid and plotosid catfishes are among the very few ostariophysans found in marine waters. The much larger Ariidae (approximately 120 species; Nelson, 1994) is a primarily tropical and subtropical family with a worldwide distribution in nearshore waters.

Intertidal observations: *Galeichthys ater* (10), *G. feliceps* (SAF9, SAF12–SAF14).

### Family Plotosidae. Eeltail catfishes.

Approximately half of plotosid catfish species occur in each of marine and freshwater habitats. The family is distributed in the Indian and western Pacific oceans.

Intertidal observations: *Plotosus nkunga* (SAF21, SAF22).

## Order Osmeriformes

### Family Osmeridae. Smelts.

This small family (13 species) has a holarctic distribution in freshwater and also occurs in the North Pacific, North Atlantic, and Arctic. Some smelts are anadromous. Smelts are generally pelagic fishes that are unlikely to be more than random visitors to the intertidal zone. *Mallotus villosus,* which did not appear on this list, spawns in the high intertidal zone (Templeman, 1948).

Intertidal observation: *Osmerus eperlanus* (A4).

## Order Aulopiformes

### Family Synodontidae. Lizardfishes.

Lizardfishes are widely distributed in shallow tropical to warm temperate marine waters in the Atlantic, Indian, and Pacific. There are about 55 species (Nelson, 1994). Although lizardfishes are benthic and often abundant near shore, they apparently rarely enter the intertidal zone.

Intertidal observation: *Saurida gracilis* (AUS1).

## Order Ophidiiformes

### Family Bythitidae. Viviparous brotulas.

The viviparous brotulas are a mainly marine family widely distributed through the Atlantic, Indian, and Pacific. About 5 of 86 species (Weitzman, 1997) occupy freshwater. In contrast to the other ophidiiform families, bythitids are mainly found in shallow, nearshore waters. Although shallow water bythitids are generally benthic and interstitial, several benthopelagic species occur in deeper waters (to at least 2600 m; Weitzman, 1997).

Intertidal observations: *Bidenichthys consobrinus* (NZ2, NZ3), *Brosmodorsalis persicinus* (NZ2–NZ4), *Dermatopsis macrodon* (AUS1, NZ2), *Dinematichthys iluocoeteoides* (SEY1, TAI1), *Ogilbia ventralis* (COS1, COS3), *O.* spp. (A3), *O.* spp. n. (MEX2).

## Order Gadiformes

### Family Moridae. Morid cods or moras.

The morids are a marine family of about 98 species with a worldwide distribution mainly in deep water. While morid species living in shallow water are benthic, deepwater species are typically pelagic or benthopelagic (Weitzman, 1997). *Lotella rhacinus* is a common member of the New Zealand rocky intertidal fauna.

Intertidal observations: *Lotella rhacinus* (NZ2, NZ4, NZ7, NZ8, NZ9, NZ10, NZ13, NZ16), *Pseudophycis bachus* (NZ8, NZ13), *P. breviuscula* (NZ2–NZ4).

### Family Phycidae. Phycid hakes.

Phycid hakes are known from marine waters of the Atlantic and also from off New Zealand and Japan. The family occupies a wide band of depths, from the intertidal zone to at least 2000 m (Weitzman, 1997). *Ciliata* and *Gaidropsarus* spp. (commonly called "rocklings") are locally abundant resident rocky intertidal fishes in the northeastern Atlantic and Mediterranean. See Kotrschal (Chapter 7, this volume) for a discussion of the unique anterior dorsal fin sensory systems possessed by *Ciliata* and *Gaidropsarus*.

Intertidal observations: *Ciliata mustela* (A4–A6, A11), *C. septentrionalis* (A4), *Gaidropsarus guttatus* (A9, A10), *G. mediterraneus* (A4, A6, A8, A11), *G. novaezelandiae* (NZ7–NZ9, NZ12, NZ14), *Urophycis tenuis* (A1).

### Family Gadidae. Cods.

Cods are widely distributed in the Northern Hemisphere, including about 29 species in the Arctic, Atlantic, and Pacific Oceans and one freshwater species with a holarctic distribution. Cods are most diverse in the Atlantic (Nelson, 1994). Cods are found in depths from the intertidal zone to about 1000 m (Weitzman, 1997). Juveniles of several cod species are seasonally abundant in nearshore waters, often entering the intertidal zone.

Intertidal observations: *Merlangius merlangus* (A4), *Microgadus tomcod* (A2), *Pollachius virens* (A1, A4), *Theragra chalcogramma* (BC1).

## Order Batrachoidiformes

### Family Batrachoididae. Toadfishes.

Toadfishes are widely distributed in the Atlantic, Indian, and Pacific oceans. Most of the 69 species are marine, but there are a few freshwater taxa. Toadfishes are benthic fishes that usually live on sediment bottoms in coastal waters; some aggregate seasonally in shallow subtidal waters to reproduce.

Intertidal observations: *Aphos porosus* (CHL1), *Batrichthys apiatus* (SAF7), *Porichthys notatus* (CA13).

## Order Lophiiformes

### Family Antennariidae. Frogfishes.

Antennariids are benthic anglerfishes distributed worldwide in tropical and subtropical seas (except the Mediterranean). Unlike most other lophiiform fishes, antennariids generally live in shallow coastal waters. There are 43 species (Nelson, 1994), of which 24 are assigned to

*Antennarius,* a genus characterized by deep, globose morphology and prominent illicium (Pietsch and Grobecker, 1987).

Intertidal observations: *Antennarius multiocellatus* (A3), *A. nummifer* (SAF20), *A. sanguineus* (COS1), *A. tridens* (NZ2, NZ4).

## Order Mugiliformes

### Family Mugilidae. Mullets.

Mullets are distributed worldwide in marine and brackish coastal waters; a few are freshwater. Most associate with nearshore habitats having sediment substrata. Mullets are fusiform, powerfully swimming fishes that may reach large sizes (up to 90 cm; Nelson, 1994) and are probably tidal transients in the intertidal zone. There are about 66 described mugilids at present.

Intertidal observations: *Aldrichetta forsteri* (NZ2–NZ5, NZ8, NZ9, NZ13), *Chaenomugil proboscidens* (COS1, COS3), *Chelon labrosus* (A4, A10), *Liza dumerili* (SAF13), *L. macrolepis* (TAI1), *L. richardsoni* (SAF10, SAF11, SAF13, SAF14), *L. saliens* (A9), *L. tricuspidens* (SAF13), *Mugil cephalus* (AUS1, COS1, COS3), *M. curema* (MEX2), *M. liza* (A3), *Myxus capensis* (SAF14), *M. elongatus* (AUS1).

## Order Atheriniformes

### Family Atherinidae. Silversides.

Atherinids are found in and near tropical to temperate seas worldwide. About one-third of the 165 described atherinid species are freshwater, and many of the marine species are euryhaline. Atherinids are fusiform, surface-oriented fishes that occupy nearshore or pelagic habitats; maximum size is 60 cm (Nelson, 1994). Schooling atherinids are often locally very abundant near shore. *Menidia menidia, Leuresthes tenuis,* and *L. sardina* (the last two not represented in this list) use the high sandy intertidal zone for spawning (Clark, 1938; Walker, 1952). Most marine atherinids are probably tidal transients in the intertidal zone.

Intertidal observations: *Allanetta harringtonensis* (A3), *A. woodwardi* (TAI1), *Atherina breviceps* (SAF10, SAF13), *A. presbyter* (A4), *Atherinops affinis* (CA12, CA13), *Atherion elymus* (TAI1), *Menidia menidia* (A1, A2).

### Family Notocheiridae.

Notocheirids are exclusively marine and occur off South Africa, India, Japan, Australia, Hawaii, and Chile. There are six species.

Intertidal observations: *I. natalensis* (SAF20, SAF21), *I. rhothophilus* (AUS1).

## Order Beloniformes

### Family Belonidae. Needlefishes.

Needlefishes have a worldwide distribution in epipelagic tropical and temperate waters. About 10 of 32 species are limited to freshwater (Nelson, 1994). Needlefishes are long, fusiform, surface-oriented epipelagic fishes; a few are probably tidal transient visitors to the intertidal zone.

Intertidal observations: *Belone persimilis* (TAI1), *Tylosurus incisus* (TAI1).

## Order Cyprinodontiformes

### Family Fundulidae. Topminnows and killifishes.

Fundulids are found in freshwater, brackish water, and marine habitats in and around North and Central America. *Fundulus heteroclitus* lives primarily in marine shore habitats, including estuaries (Nelson, 1994). Some fundulids living these habitats spawn high in the intertidal zone.

Intertidal observations: *Fundulus heteroclitus* (A1, A2).

### Family Anablepidae.

This small family is primarily found in freshwater and brackish water and only rarely in marine habitats. One of the three genera (*Anableps*) possesses elevated eyes with divided pupils, allowing the fish to see both in the water and above it while at the surface. *Oxyzygonectes* is a monotypic genus confined to the Pacific coast of Costa Rica.

Intertidal observation: *Oxyzygonectes dowi* (COS2).

### Family Poeciliidae.

This large family has a tropical distribution in freshwater and brackish water. Few of the 293 described species occur in marine habitats, although many are somewhat euryhaline.

Intertidal observations: *Poecilia sphenops* (COS2), *Poeciliopsis turrubarensis* (COS2).

## Order Beryciformes

### Family Trachichthyidae. Roughies or slimeheads.

Trachichthyids are confined to marine waters, but are widely distributed in the Atlantic, Indian, and Pacific oceans. Most of the approximately 33 species occur in deep water (up to 1500 m; Weitzman, 1997).

Intertidal observations: *Optivus elongatus* (NZ1, NZ2, NZ4), *Paratrachichthys trailli* (NZ10, NZ16).

### Family Holocentridae. Squirrelfishes.

Squirrelfishes are confined to the tropical marine waters of the Atlantic, Indian, and Pacific basins. Most of the 65 described species occur close to shore, hiding in crevices and under coral or rocky reefs. Adults of most species remain near shelter on the bottom.

Intertidal observations: *Adioryx bullisi* (A3), *A. ruber* (TAI1), *A. vexillarius* (A3), *Holocentrus ascensionis* (A3), *H. lacteoguttatus* (HAW1), *H. rufus* (A3), *H. spinifer* (AUS1), *H. suborbitalis* (COS1, COS3), *Neoniphon sammara* (TAI1), *Sargocentron punctatissimum* (SAF22), *S. wilhelmi* (EAS1).

## Order Zeiformes

### Family Zeidae. Dories.

The Zeidae is a small (13 species) family of marine fishes distributed throughout the Atlantic, Indian, and Pacific Oceans.

Intertidal observations: *Zeus faber* (NZ3, NZ4).

## Order Gasterosteiformes

### Family Gasterosteidae. Sticklebacks.

This small family of about 7 species is widely distributed in the Northern Hemisphere. Sticklebacks occur in shallow freshwater, brackish water, and marine habitats (Wooton, 1976). Some populations of *Gasterosteus aculeatus* are anadromous. *G. wheatlandi* is primarily a marine fish, occurring in coastal marine habitats and a few freshwater locations. *Spinachia spinachia* is exclusively a coastal marine fish.

Intertidal observations: *Apeltes quadracus* (A1), *Gasterosteus aculeatus* (A1, A2, A4), *G. wheatlandi* (A1), *Pungitius pungitius* (A1, A2), *Spinachia spinachia* (A4).

### Family Syngnathidae. Pipefishes and seahorses.

Pipefishes and seahorses occur in marine and brackish water in the Atlantic, Indian, and Pacific basins. Some species occur in freshwater. Although the greatest syngnathid diversities lie in the tropics, pipefishes range into cool waters as far as Alaska and Tierra del Fuego in the Americas (Nelson, 1994). Most of the 215 syngnathid species are confined to shallow waters, where they associate with substratum features such as macroalgae, hydrozoans, or corals. Although syngnathids are commonly collected in the intertidal zone, the nature and extent of their use of it is not known.

Intertidal observations: *Choeroichthys sculpus* (TAI1), *Festucalex gibbsi* (SEY1), *Hippocampus abdominalis* (NZ2, NZ3, NZ5, NZ7–NZ10, NZ13), *Leptonotus elevatus* (NZ10, NZ10, NZ13, NZ16, NZ17), *L. norae* (NZ4, NZ7), *Lissocampus filum* (NZ2–NZ10, NZ13, NZ16), *Nannocampus elegans* (SAF12, SAF20), *Nerophis lumbriciformis* (A4, A6, A11), *Stigmatopora macropterygia* (NZ7, NZ8, NZ13), *S. nigra* (NZ2, NZ10), *Syngnathus acus* (A4, SAF13), *S. dunkeri* (A3), *S. fuscus* (A2), *S. leptorhynchus* (CA13), *S. rostellatus* (A4), *S. typhle* (A4)

### Family Fistulariidae. Cornetfishes.

Cornetfishes are widely distributed in shallow tropical marine waters. There are four species.

Intertidal observation: *Fistularia petimba* (TAI1).

## Order Scorpaeniformes

### Family Scorpaenidae. Scorpionfishes.

The Scorpaenidae is a large family (388 spp.) with a worldwide distribution in tropical and temperate marine waters. Few scorpionfishes occur in freshwater. Scorpionfishes occur in all depths from shore to about 2200 m (Weitzman, 1997). The family is diverse, with considerable variation in morphology, behavior, and reproductive natural history. Various representatives of the family are partial intertidal residents (sensu Gibson, 1982), using the intertidal zone as a refugium from predation during juvenile development, while many others are tidal transients. There probably exist scorpionfishes that are full residents of the intertidal zone.

Intertidal observations: *Centropogon australis* (AUS1), *C. marmoratus* (AUS1), *Coccotropsis gymnoderma* (SAF13, SAF14), *Dendrochirus brachypterus* (AUS1), *D. zebra* (AUS1), *Helicolenus percoides* (NZ7, NZ8), *Parascorpaena mcadamsi* (SAF19), *Pterois miles* (SAF19), *P. radiata* (TAI1), *Scorpaena albobrunnea* (TAI1), *S. cardinalis* (AUS1), *S. mystes* (MEX2), *S. papillosus* (NZ2–NZ5, NZ7, NZ9, NZ14), *S. plumieri* (A3), *Scorpaenodes carribaeus* (A3), *S. guamensis* (TAI1), *S. xyris* (COS1), *Scorpaenopsis cirrhosa* (TAI1), *S. diapolus* (TAI1), *Sebastes atrovirens* (CA13), *S. carnatus* (CA13), *S. chrysomelas* (CA13), *S. dalli* (CA13), *S. flavidus* (CA3), *S. melanops* (CA1, CA3, CA5, CA7, CA13), *S. mystinus* (CA1, CA3, CA13), *S. pinniger* (CA13), *S. rastrelliger* (CA3–CA7, CA13), *Sebastes* sp. (CA8), *Synanceia verrucosa* (SEY1).

### Family Congiopodidae. Racehorses, pigfishes, or horsefishes.

Congiopodids occur in marine waters of the Southern Hemisphere. *Congiopodus leucopaecilis* is a New Zealand intertidal endemic with a southern distribution (Paulin and Roberts, 1993).

Intertidal observations: *Congiopodus leucopaecilus* (NZ10, NZ8, NZ14, NZ16).

### Family Platycephalidae. Flatheads.

Flatheads occur in marine and sometimes brackish water primarily in the Indo-Pacific region. There are about 60 species.

Intertidal observations: *Platycephalus fuscus* (AUS1).

### Family Hexagrammidae. Greenlings.

Greenlings are exclusively marine and the family is the most species-rich family of fishes endemic to the North Pacific. There are 11 species, most of them occurring in shallow, nearshore waters. Hexagrammids are common tidal transients along North Pacific coasts.

Intertidal observations: *Hexagrammos decagrammus* (CA1, CA3–CA9, CA13), *H. lagocephalus* (CA8, CA13), *H. superciliosus* (CA3, CA5, CA7, CA9), *Oxylebius pictus* (CA13).

### Family Cottidae. Sculpins.

The sculpins form a large (300 spp.) family of marine and freshwater fishes with a wide distribution in the Northern Hemisphere. Four deepwater marine species occur in the Southern Hemisphere (Nelson, 1990). Most of the sculpins are marine, and most of these occur in nearshore waters. The family probably first appeared in the North Pacific and only invaded the Arctic and Atlantic oceans during the "Great Transarctic Biotic Interchange" 3.5 My (Briggs, 1995). A few subtidal species have transitional ranges extending into both Pacific and Atlantic oceans. Freshwater sculpins are holarctic in distribution and are diverse in both nearctic and palearctic regions. Most are members of the species-rich and morphologically conservative genus *Cottus,* which also includes several species found in coastal estuaries and rarely in the rocky intertidal zone. Cottids dominate the rocky intertidal zone in the North Pacific. Four genera, *Artedius, Ascelichthys, Clinocottus,* and *Oligocottus,* contain resident rocky intertidal fishes. These and several other genera contain partial residents (sensu Gibson, 1982) of the rocky intertidal zone; in most such cases juveniles dwell in the intertidal zone.

Intertidal observations: *Artedius corallinus* (MEX1), *A. fenestralis* (BC1, BC8, CA1), *A. harringtoni* (BC1, BC4, BC8, CA3, CA5, CA13), *A. lateralis* (BC1, BC2, BC4–BC8,

CA1–CA13, MEX1, OR1, OR2, WA1), *A. notospilotus* (CA5, CA13), *Ascelichthys rhodorus* (BC4, BC5, BC7, BC8, CA3, CA4, CA7, OR1, WA1), *Clinocottus acuticeps* (BC8, CA1–CA3, CA5, CA7, OR1, WA1), *C. analis* (CA2–CA14, MEX1), *C. embryum* (BC3, BC4, BC6, BC7, BC8, CA3, WA1), *C. globiceps* (BC3, BC4, BC6–BC8, CA1–CA9, CA13, OR1, OR2, WA1), *C. recalvus* (CA3, CA8, CA9, CA11, CA13), *Enophrys bison* (BC1, BC2, BC7, BC8, CA3, CA5, CA8, OR1), *Hemilepidotus hemilepidotus* (BC1, BC2, BC4, BC5, BC7, BC8), *H. spinosus* (BC8, CA1, CA3, CA8, CA13), *Jordania zonope* (BC8, CA3), *Leptocottus armatus* (BC8), *Myoxocephalus aenus* (A1, A2), *M. octodecemspinosus* (A1), *M. scorpius* (A1, A4), *Oligocottus maculosus* (BC1–BC3, BC6–BC8, CA1–CA5, CA7–CA9, OR1, OR2, WA1), *O. rimensis* (BC4, BC8, CA3, CA8, CA10–CA13, OR2, WA1), *O. rubellio* (CA3, CA5, CA8, CA10, CA11, CA13, MEX1), *O. snyderi* (CA1–CA13, BC4–BC8, MEX1, OR1, OR2, WA1), *Orthonopias triacis* (CA13), *Scorpaenichthys marmoratus* (CA1–CA10, CA13, MEX1, OR1, OR2), *Synchirus gilli* (CA13), *Taurulus bubalis* (A4, A6), *T. lilljeborgi* (A4).

### Family Hemitripteridae.

All eight hemitripterid species are marine. Seven occur in the North Pacific, while one occurs in the northwestern Atlantic. *Blepsias cirrhosus* is a northeastern Pacific species that commonly occurs in the intertidal zone where extensive macroalgal cover exists.

Intertidal observations: *Blepsias cirrhosus* (BC2, BC5, BC8, CA13).

### Family Cyclopteridae. Lumpfishes.

Lumpfishes occur in the cold-temperate and cooler waters of the Northern Hemisphere. The family is exclusively marine. *Cyclopterus lumpus* spawns in rocky substrata in nearshore and sometimes intertidal waters; juveniles occur in the intertidal zone (Wheeler, 1994).

Intertidal observations: *Cyclopterus lumpus* (A1, A2, A4, A5).

### Family Liparididae (Liparidae). Snailfishes.

Snailfishes are exclusively marine and have a worldwide distribution in warm, cool, and cold water, although they are rare in the Indian Ocean. The family also has an exceptionally wide depth range, occurring from the intertidal zone to more than 7000 m depth (Nelson, 1994). About 195 snailfish species have been described. Most snailfishes are benthic. Most snailfishes possess a pelvic disc (similar to that seen in gobiesocids and gobiids) capable of attaching to surfaces by suction. Several members of the genus *Liparis* are probably resident intertidal fishes.

Intertidal observations: *Liparis atlanticus* (A1, A2), *L. callyodon* (BC6), *L. cyclopus* (BC2), *L. florae* (BC1, BC2, BC4–BC7, CA1, CA3, CA5, CA7, CA13, MEX1, WA1), *L. fucensis* (CA3), *L. liparis* (A5), *L. montagui* (A4–A6), *L. mucosus* (CA4, CA13), *L. rutteri* (CA3), *Polypera greeni* (BC2).

## Order Perciformes

### Family Chandidae. Asiatic glassfishes.

Chandids occur in marine, brackish, and freshwater in the Indo-West Pacific region. There are about 41 species.

Intertidal observation: *Velambassis jacksoniensis* (AUS1).

## Family Moronidae. Temperate basses.

Temperate basses occur in brackish, freshwater, and marine coastal areas of North America (Atlantic and Gulf of Mexico coasts), Europe and northern Africa. There are about six species in the family.

Intertidal observation: *Dicentrarchus labrax* (A4).

## Family Serranidae. Sea basses.

Sea basses occur mainly in marine waters in all tropical and temperate seas. A few are known from freshwater. Most representatives of this large (449 species; Nelson, 1994) family are piscivorous reef fishes; many live in shallow, nearshore waters.

Intertidal observations: *Acanthistius cinctus* (AUS1, NZ1), *A. ocellatus* (AUS1), *A. sebastoides* (SAF12, SAF13), *Cephalopholis argus* (TAI1), *Ellerkeldia huntii* (NZ5, NZ10, NZ16), *Epinephelus adscenionis* (A3), *E. andersoni* (SAF19), *E. caeruleopunctatus* (TAI1), *E. cyanopodus* (AUS1), *E. daemelii* (AUS1, NZ1, NZ2), *E. faveatus* (SAF19), *E. flavocaerulus* (SAF12), *E. fuscoguttatus* (SAF20), *E. gauza* (SAF11–SAF14), *E. hexagonatus* (TAI1), *E. labriformis* (COS1, COS3), *E. marginatus* (A10, SAF16–SAF19, SAF21), *E. rivulatus* (AUS1), *E. septemfasciatus* (AUS1), *E. spiniger* (SAF12), *E. tauvina* (AUS1, TAI1), *Grammistes sexlineatus* (SAF19, SAF22, SEY1, TAI1), *Paralabrax maculatofasciatus* (MEX2), *Rypticus saponaceus* (A3).

## Family Pseudochromidae. Dottybacks.

This family of about 98 species is confined to the tropical Indo-Pacific; most are small (<11 cm) benthic coral reef fishes.

Intertidal observations: *Halidesmus scapularis* (SAF9, SAF11, SAF13, SAF14), *Labracinus melanotaenia* (TAI1), *Pseudochromis tapeinosoma* (TAI1).

## Family Plesiopidae.

Plesiopids occur exclusively in marine habitats in the Indo-West Pacific. All are small (<20 cm) and occur in nearshore waters shallower than 70 m depth (Smith-Vaniz and Johnson, 1990; Nelson, 1994), with several *Acanthoclinus* species residing in the intertidal zone.

Intertidal observations: *Acanthoclinus fuscus* (NZ2–NZ6, NZ7–NZ9, NZ14, NZ13, NZ16), *A. littoreus* (NZ2–NZ5, NZ7–NZ9, NZ14, NZ16), *A. marilynae* (NZ2, NZ4, NZ14, NZ16), *A. rua* (NZ7, NZ10, NZ13, NZ14), *Belonepterygion fasciolatum* (TAI1), *Plesiops coeruleolineatus* (SEY1, TAI1), *P. melas* (TAI1), *P. multisquamatus* (SAF21), *P. nigricans* (TAI1), *Trachinops taeniatus* (AUS1).

## Family Opistognathidae. Jawfishes.

Opistognathids occur in marine habitats in parts of the warm latitudes of the Atlantic, Indian, and Pacific oceans. The family includes at least 60 species (Nelson, 1994).

Intertidal observation: *Opistognathus maxillosus* (A3).

## Family Apogonidae. Cardinalfishes.

Cardinalfishes occur in tropical waters of the Atlantic, Indian, and Pacific oceans. All the approximately 207 representatives are small (<20 cm, most less than 10 cm) and most live in shallow nearshore habitats (Nelson, 1994); however, one genus, the *Glossamia,* occurs in exclusively in freshwater in New Guinea (Allen, 1991).

Intertidal observations: *Apogon angustatus* (TAI1), *A. coccineus* (TAI1), *A. cookii* (SEY1), *A. maculatus* (A3), *A. nubilis* (TAI1), *A. retrosella* (MEX2), *A. robustus* (TAI1), *A. taeniophorus* (SAF20), *Fowleria isostigma* (TAI1).

## Family Sillaginidae. Sillagos (whitings, smelt-whitings).

Sillaginids occur in coastal marine, brackish, and rarely freshwater habitats in the Indo-West Pacific region. Juveniles of several species commonly enter estuaries (Nelson, 1994). The family contains about 31 species.

Intertidal observations: *Sillago ciliata* (AUS1), *S. maculata* (AUS1).

## Family Carangidae. Jacks and pompanos.

Carangids occur mainly in marine habitats in the Atlantic, Indian, and Pacific oceans. Many of the approximately 140 species occur in nearshore habitats.

Intertidal observations: *Caranx latus* (A3), *Trachinotus botla* (SAF20).

## Family Lutjanidae. Snappers and fusiliers.

The Lutjanidae is an amphitropical and subtropical family of mainly benthopelagic reef fishes. None occurs deeper than about 550 m (Nelson, 1994). A few taxa are known to occur only in freshwater or brackish water. Juveniles of several marine species of *Lutjanus* enter brackish and freshwater habitats during development.

Intertidal observations: *Lutjanus agrentiventris* (MEX2, COS2), *L. aratus* (COS3), *L. monostigma* (TAI1), *L. novemfasciatus* (COS2).

## Family Gerreidae. Mojarras.

Gerreids occur mainly in marine habitats in most warm seas. A few of the approximately 40 species are found in brackish or rarely freshwater habitats.

Intertidal observations: *Eucinostomus dowi* (MEX2), *Eucinostomus* sp.n. (MEX2).

## Family Haemulidae. Grunts.

Grunts occur in marine, brackish, and occasionally freshwater habitats in the Atlantic, Indian, and Pacific oceans. Many are euryhaline. Most of the approximately 150 described grunts live in shallow coastal waters.

Intertidal observations: *Anisotremus davidsoni* (MEX2), *Haemulon aurolineatum* (A3), *Pomadasys leuciscus* (COS3), *P. olivaceum* (SAF11, SAF12).

## Family Sparidae. Porgies.

Porgies are mainly confined to coastal marine habitats. About 5 of 100 species are known from brackish or freshwater. The family contributes several species to the South African rocky intertidal fauna.

Intertidal observations: *Diplodus cervinus* (SAF11, SAF12, SAF13, SAF14, SAF19), *D. sargus* (A10, SAF10–SAF14, SAF16, SAF18, SAF19), *Lithognathus lithognathus*

(SAF12), *L. mormyrus* (SAF11), *Rhabdosargus globiceps* (SAF14, SAF11), *R. holubi* (SAF11–SAF14), *R. sarba* (AUS1), *Sarpa salpa* (SAF10–SAF13).

### Family Nemipteridae. Threadfin breams.

The nemipterids are a marine family of about 62 species distributed in the tropical and subtropical marine waters of the Indo-West Pacific region.

Intertidal observation: *Scolopsis cancellatus* (TAI1).

### Family Sciaenidae. Drums or croakers.

Sciaenids occur in marine, brackish, and freshwater habitats in the Atlantic, Indian, and Pacific. Most of the 270 extant representatives occur in shallow waters along the continental margins. About 28 species are known from freshwater, and several marine species are known to enter estuaries (Nelson, 1994). Sciaenids are not often observed in the rocky intertidal zone, with most preferring sediment bottoms.

Intertidal observation: *Umbrina capensis* (SAF13).

### Family Mullidae. Goatfishes.

Almost all of the 55 goatfish species occur in marine waters of the Atlantic, Indian, and Pacific oceans; a few enter brackish water. Many mullids occur in nearshore waters.

Intertidal observations: *Parupeneus barberinus* (TAI1), *P. fraterculus* (TAI1), *P. signatus* (AUS1), *Pseudupeneus maculatus* (A3), *P. rubescens* (SAF12).

### Family Pempheridae. Sweepers.

Sweepers occur in marine and brackish waters of the western Atlantic, Indian, and Pacific oceans. The family contains about 25 species.

Intertidal observations: *Pempheris adspersus* (NZ2, NZ3), *P. compressa* (AUS1), *P. schwenkii* (SAF20).

### Family Monodactylidae. Moonfishes or fingerfishes.

Monodactylids occur in marine, brackish, and freshwater habitats in western Africa and the Indo-Pacific region. There are about five species in the family. The three species of *Monodactylus* often ascend rivers, and some populations may live in freshwater (Nelson, 1994).

Intertidal observation: *Monodactylus falciformis* (SAF11).

### Family Chaetodontidae. Butterflyfishes.

Butterflyfishes are an exclusively marine family of about 114 described species occurring in tropical to temperate waters of the Atlantic, Indian, and Pacific oceans. The family is most diverse in the tropics. Butterflyfishes usually have extremely compressed bodies. Most species are associated with coral reefs less than 20 m in depth, although a few are found as deep as 200 m.

Intertidal observations: *Chaetodon adieregastos* (TAI1), *C. auriga* (SAF13, SAF19, TAI1), *C. citrinellus* (AUS1, TAI1), *C. humeralis* (COS1), *C. lunula* (SAF17, SAF19,

SAF22, TAI1), *C. marleyi* (SAF13, SAF14), *C. striatus* (A3), *C. vagabundus* (AUS1, TAI1).

## Family Pomacanthidae. Angelfishes.

Angelfishes have a worldwide, exclusively marine distribution in tropical waters. The highest species richness is in the western Pacific. Most of the approximately 74 species occur in shallow water (<20 m).

Intertidal observations: *Pomacanthus paru* (A3), *P. semicirculatus* (TAI1), *P. zonipectus* (COS3).

## Family Kyphosidae. Sea chubs.

The sea chubs include about 42 species of fishes that occur in nearshore marine waters in the tropical and temperate Atlantic, Indian, and Pacific oceans. The status of the family has been shuffled several times, and in fact the present assemblage may not be monophyletic (see discussion in Nelson, 1994). Two subfamilies, the Girellinae and Kyphosinae, are usually herbivorous; the other three (Scorpidinae, Microcanthinae, and Parascorpidinae) are carnivorous. Juveniles of many kyphosid species are found in the intertidal zone.

Intertidal observations: *Atypichthys strigatus* (AUS1), *Girella cyanea* (AUS1), *G. elevata* (AUS1), *G. fimbriata* (NZ1), *G. laevifrons* (CHL1), *G. nigricans* (CA10, CA12–CA14, MEX1), *G. tricuspidata* (AUS1, NZ2–NZ4), *G. simplicidens* (MEX2), *Graus nigra* (CHL1), *Hermosilla azurea* (MEX2), *Kyphosus cinerascens* (SAF21), *K. sydneyanus* (AUS1), *Microcanthus strigatus* (AUS1), *Neoscorpis lithophilus* (SAF13), *Scorpis lineolatus* (AUS1).

## Family Kuhliidae. Flagtails.

Flagtails occupy marine, brackish, and freshwater habitats mainly in the Indo-Pacific region. Most of the eight species occupy marine and/or brackish water. Within their ranges, *Kuhlia mugil, K. taeniura,* and perhaps others are locally common in the intertidal zone.

Intertidal observations: *Kuhlia mugil* (SAF10, SAF11, SAF13, SAF18, SAF22), *K. taeniura* (AUS1, SAF12, SAF14).

## Family Cirrhitidae. Hawkfishes.

Hawkfishes are widely distributed in tropical marine waters of the Atlantic, Indian, and Pacific oceans. The highest species richness is in the Indo-Pacific region. The family contains about 32 species.

Intertidal observations: *Cirrhitus pinnulatus* (SAF20, SAF21), *C. rivulatus* (COS3), *Dactylophora nigricans* (AUS1).

## Family Chironemidae. Kelpfishes.

This small family (4 species) occurs in coastal marine waters of Australia and New Zealand. At least one species, *Chironemus marmoratus,* is frequently observed in the intertidal zone.

Intertidal observations: *Chironemus marmoratus* (AUS1, NZ2, NZ4, NZ5).

## Family Aplodactylidae. Marblefishes.

Marblefishes are confined to coastal marine waters of southern Australia, New Zealand, Peru, and Chile (Nelson, 1994). *Aplodactylus arctidens* is a common New Zealand rocky intertidal fish.

Intertidal observations: *Aplodactylus arctidens* (NZ2, NZ4, NZ5, NZ10, NZ12, NZ14), *Crinodus lophodon* (AUS1).

## Family Cheilodactylidae. Morwongs.

The Cheilodactylidae is an exclusively marine family that is widely distributed in parts of the Southern and Northern Hemispheres. There are about 18 species, the largest of which reaches about 1 m in length. Various species are commonly observed constituents in the rocky intertidal faunas of southern Africa and New Zealand.

Intertidal observations: *Cheilodactylus fasciatus* (SAF9–SAF11, SAF13, SAF14), *C. fuscus* (AUS1), *C. spectabilis* (NZ2, NZ3, NZ5), *C. variegatus* (CHL1), *Chirodactylus brachydactylus* (SAF10–SAF14), *Nemadactylus macropterus* (NZ4, NZ8, NZ12–NZ14).

## Family Latridae. Trumpeters.

Most trumpeters are known from coastal marine waters and occur in southern Australia, New Zealand, Chile, and in the southern Atlantic Ocean. There are about nine species (Gon and Heemstra, 1987) in three genera, two of which are commonly observed in the New Zealand rocky intertidal zone.

Intertidal observations: *Latridopsis ciliaris* (NZ10, NZ8, NZ9, NZ14), *Mendosoma lineatum* (NZ8, NZ9, NZ14, NZ16, NZ17).

## Family Embiotocidae. Surfperches.

Surfperches occur in the coastal North Pacific. All but 1 of 24 species are confined to marine waters. *Micrometrus aurora* is a resident intertidal fish; *M. minimus* also commonly appears in intertidal collections, although the species is most often observed in association with shallow subtidal macrophyte beds.

Intertidal observations: *Amphistichus argenteus* (CA13), *Brachyistius frenatus* (CA13), *Cymatogaster aggregata* (MEX1), *Embiotoca jacksoni* (CA13, MEX1), *E. lateralis* (CA3, CA5, CA13), *Hyperprosopon ellipticum* (CA5), *Micrometrus aurora* (CA8, CA9, CA12, CA13, MEX1), *M. minimus* (CA5, CA13, MEX1), *Rhacochilus vacca* (CA13).

## Family Pomacentridae. Damselfishes.

The damselfishes form a diverse family of shallow-water marine reef fishes. The family is known from all tropical seas, but is most diverse in the Indo-Pacific region. Most damselfishes are small (maximum size is 35 cm). Many are associated with coral reefs.

Intertidal observations: *Abudefduf biocellatus* (AUS1), *A. coelestinus* (TAI1), *A. immaculatus* (AUS1), *A. imparipennis* (HAW1), *A. notatus* (SAF15–SAF19, SAF21, SAF22, TAI1), *A. saxatilis* (COS1, COS3, SAF13), *A. septemfasciatus* (SAF18, TAI1), *A. sexfasciatus* (SAF20), *A. sindonis* (HAW1), *A. sordidus* (AUS1, HAW1, SAF13, SAF15, SAF17–SAF19, SEY1, TAI1), *A. troschelii* (MEX2), *A. vaigiensis* (AUS1, SAF12, SAF18, SAF19, SAF22, TAI1), *Abudefduf* spp. (A3), *Chrysiptera glauca* (SEY1), *C. unimaculata* (SAF19,

SAF22), *C. glaucus* (TAI1), *C. leucopomus* (TAI1), *C. uniocellatus* (TAI1), *Eupomacentrus acapulcoensis* (COS1, COS3), *E. dorsopunicans* (A3), *E. flavilatus* (COS1, COS3), *E. rectifraenum* (MEX2), *Microspathodon dorsalis* (COS3), *Nexilarius concolor* (COS1, COS3), *Parma alboscapularis* (NZ1–NZ5), *P. microlepis* (AUS1), *P. unifasciata* (AUS1), *Plectroglyphidodon leucozona* (AUS1, SAF15, SAF17–SAF22, TAI1), *Stegastes fasciolatus* (EAS1), *S. partitus* (A3).

### Family Labridae. Wrasses.

Wrasses form the second largest family of marine fishes and have a worldwide distribution in tropical to temperate waters. Most wrasses are associated with shallow coastal reefs. The family is morphologically and ecologically very diverse. Most wrasses are relatively small, but one reaches 2.3 m in length. Various labrid subfamilies and tribes have been recognized, but the monophyly of these groups is generally uncertain (see discussion in Nelson, 1994). Several wrasses are probably resident intertidal fishes. However, the large number of wrasse observations is in large part due to the very high diversity and abundance of this family in the nearshore waters of many regions.

Intertidal observations: *Achoerodus viridis* (AUS1), *Anampses caeruleopunctatus* (TAI1), *Centrolabrus trutta* (A9, A10), *C. exoletus* (A4), *Cheilinus trilobatus* (TAI1), *Crenilabrus melops* (A4, A6), *Ctenolabrus rupestris* (A4), *Halichoeres bivittatus* (A3), *H. centiquadrus* (TAI1), *H. dispilus* (COS1, MEX2), *H. leparensis* (TAI1), *H. maculipinna* (A3), *H. margaritaceus* (TAI1), *H. marginatus* (SEY1, TAI1), *H. miniatus* (TAI1), *H. nebulosa* (AUS1, TAI1), *H. pictus* (A3), *H. radiatus* (A3), *H. scapularis* (SAF22), *H. sellifer* (COS1), *H. semicinctus* (MEX2), *Hemigymnus melapterus* (TAI1), *Labroides dimidiatus* (TAI1), *Labrus bergylta* (A4), *Macropharyngodon meleagris* (TAI1), *Notolabrus celidotus* (NZ2–NZ6, NZ8–NZ10, NZ13, NZ14, NZ16), *N. fucicola* (NZ2, NZ4, NZ5, NZ8–NZ10, NZ14, NZ16), *N. gymnogenis* (AUS1), *Oxyjulis californica* (CA13), *Pictilabrus laticlavus* (AUS1), *Pseudojulis notospilus* (COS1, COS3), *Pseudolabrus guntheri* (AUS1), *P. miles* (NZ9, NZ13, NZ14, NZ16), *Stethojulis bandanensis* (TAI1), *S. interrupta* (SAF12), *S. strigiventer* (SAF19), *S. trilineata* (SAF12, TAI1), *Symphodus melops* (A11), *Tautogolabrus adspersus* (A1, A2), *Thalassoma amblycephalus* (TAI1), *T. bifasciatum* (A3), *T. hardwickei* (TAI1), *T. lucasanum* (COS3), *T. pavo* (A10), *T. purpureum* (SAF15–SAF21, SEY1, TAI1), *T. quinquevittata* (TAI1), *T. trilobatus* (SAF21, SAF17), *T. umbrostigma* (EAS1, HAW1).

### Family Odacidae.

This small family (12 species) is confined to coastal marine waters of Australia and New Zealand. Two of four species of *Odax,* including *O. pullus,* are endemic to New Zealand, while the other odacids occur in southern Australia.

Intertidal observations: *Odax pullus* (NZ2–NZ6, NZ9, NZ10, NZ13).

### Family Scaridae. Parrotfishes.

Parrotfishes form an exclusively marine family of about 83 species with a mainly tropical distribution. Because they are herbivorous, parrotfishes are confined to shallow waters and are generally associated with coral reefs.

Intertidal observations: *Scarus lepidus* (TAI1), *Sparisoma* spp. (A3).

## Family Zoarcidae. Eelpouts.

Eelpouts are usually benthic fishes with a worldwide distribution. There are about 220 species. Members of the family occupy habitats ranging from the shallow subtidal (usually in high latitudes) to abyssal depths.

Intertidal observation: *Zoarces viviparus* (A4).

## Family Stichaeidae. Pricklebacks.

The Stichaeidae is a strictly marine family of about 65 species, most of which occur in the North Pacific. A few are found in the North Atlantic, and most of these have transitional distributions including some part of the Arctic and Pacific oceans. Stichaeids occur at various depths from the intertidal zone to about 700 m (Weitzman, 1997). Four genera contain one or more resident intertidal fishes: *Anoplarchus, Cebidichthys, Ulvaria,* and *Xiphister.* Intertidal stichaeids are usually eel-like or taeniform in morphology and most often occupy interstices of rocky habitats or are associated with macroalgae. Some of the subtidal stichaeids have a less eel-like morphology and more distinct caudal fin. See Horn (Chapter 16, this volume) for a discussion of stichaeid morphology and its correlates. The monkeyface prickleback *Cebidichthys violaceus* may be the largest of all resident intertidal fishes, reaching at least 76 cm in length (Eschmeyer *et al.,* 1983).

Intertidal observations: *Anoplarchus insignis* (BC2), *A. purpurescens* (CA1, CA3–CA5, CA7–CA9, CA12, CA13, BC1–BC3, BC6, BC7, WA1), *Cebidichthys violaceus* (CA1, CA3–CA5, CA7–CA9, CA12, CA13), *Ernogrammus walkeri* (CA13), *Phytichthys chirus* (BC2, BC4–BC7, CA3, CA4, CA13, WA1), *Plagiogrammus hopkinsi* (CA13), *Ulvaria subbifurcata* (A2, A1), *Xiphister atropurpureus* (BC1, BC2, BC4–BC7, CA1, CA3, CA4, CA8, CA9, CA11–CA13, WA1), *X. mucosus* (CA3–CA5, CA8, CA11–CA13, WA1).

## Family Cryptacanthodidae. Wrymouths.

The four wrymouth species are all marine and occur in the northwest Atlantic and northern Pacific oceans.

Intertidal observation: *Cryptacanthodes maculatus* (A1).

## Family Pholididae (Pholidae). Gunnels.

Gunnels form an exclusively marine family of about 14 species widely distributed in the North Atlantic and North Pacific. Gunnels are morphologically similar to pricklebacks and often occur in the same habitats. Two genera, *Apodichthys* and *Pholis,* contain resident intertidal fishes.

Intertidal observations: *Apodichthys flavidus* (BC1, BC2, BC5, BC7, CA1–CA9, CA13, OR1, OR2, WA1), *A. fucorum* (CA1, CA3–CA5, CA7–CA9, CA12, CA13, MEX1), *Pholis gunnellus* (A1, A2, A4, A5), *P. laeta* (BC7, WA1), *P. ornata* (CA1), *P. schultzi* (CA13), *Ulvicola santaerosae* (CA13).

## Family Scytalinidae. Graveldiver.

This family is monotypic. *Scytalina cerdale* occurs in the intertidal zone and shallow subtidal waters of the subarctic and cold-temperate northeastern Pacific. It lives in interstitial spaces within rocky, sandy, and gravelly substrata (Eschmeyer *et al.,* 1983).

Intertidal observations: *Scytalina cerdale* (BC2, CA3, CA13).

## Family Bovichthyidae.

Bovichthyids occur in marine waters of southern Australia, New Zealand, and southern South America; freshwater representatives are known from southeastern Australia and Tasmania. There are about 11 species. *Bovichthys variegatus* is a common rocky intertidal fish with a southern distribution in New Zealand (Paulin and Roberts, 1993).

Intertidal observations: *Bovichthys chilensis* (CHL1), *B. variegatus* (NZ7–NZ9, NZ13, NZ14, NZ16, NZ17).

## Family Nototheniidae. Cod icefishes.

Nototheniids are diverse and important fishes in the shallow coastal waters of Antarctica, New Zealand, and southern South America. Nototheniids are morphologically similar to hexagrammids but occur at a wider range of depths, from the intertidal zone to about 1600 m (Weitzman, 1997). *Paranotothenia angusta* is a component of the southern New Zealand rocky intertidal fauna.

Intertidal observations: *Paranotothenia angustata* (NZ7, NZ13, NZ15–NZ17).

## Family Pinguipedidae. Sandperches.

Sandperches have a marine distribution along the Atlantic coasts of South America and Africa, in the Indo-Pacific region, and off Chile. There are about 50 species. *Parapercis colias* is a component of the New Zealand rocky intertidal fauna.

Intertidal observations: *Parapercis cephalopunctatus* (TAI1), *P. colias* (NZ5, NZ10, NZ8, NZ13, NZ16).

## Family Creedidae. Sandburrowers.

Sandburrowers form a small family (about 16 species) of benthic fishes occurring mainly in the Indo-West Pacific region. Sandburrowers are all small, with none exceeding 8 cm in length. Two taxa, *Limnichthys polyactis* and *Tewara cranwellae,* are New Zealand rocky intertidal endemics.

Intertidal observations: *Apodocreedia vanderhorsti* (SAF20, SAF22), *Limnichthys nitidus* (SAF20, SAF22), *L. polyactis* (NZ2, NZ5, NZ10, NZ8, NZ14, NZ15), *Tewara cranwellae* (NZ2–NZ4, NZ6, NZ8, NZ9, NZ16).

## Family Percophidae. Duckbills.

Percophids are found in marine waters of the Atlantic Ocean and the Indo-West Pacific and southeastern Pacific regions. The family has approximately 19 species.

Intertidal observations: *Hemerocoetes monoterygius* (NZ8, NZ14).

## Family Leptoscopidae. Southern sandfishes.

Leptoscopids have a marine and occasionally estuarine distribution in Australian and New Zealand waters. There are about four species.

Intertidal observation: *Crapatulus arenarius* (AUS1).

## Family Ammodytidae. Sand lances.

Sand lances occupy marine habitats including the Arctic, Atlantic, Indian, and Pacific oceans. There are about 18 species. Several species occur in inshore waters, where they are usually associated with sandy substrata.

Intertidal observations: *Ammodytes americanus* (A1), *A. tobianus* (A4), *Hyperoplus lanceolatus* (A4).

## Family Trachinidae. Weeverfishes.

Weeverfishes are most common in the Mediterranean Sea, but also occur in marine habitats in the eastern Atlantic Ocean and the Black Sea. There are four species. Weeverfishes commonly occur in nearshore waters, where they bury themselves in sandy substrata.

Intertidal observation: *Echiichthys vipera* (A4).

## Family Uranoscopidae. Stargazers.

Most stargazers are known from marine waters, but are also occasionally found in estuaries. They occur in the Atlantic, Indian, and Pacific oceans. *Genyagnus monopterygius* is an intertidal fish with a widespread distribution in New Zealand waters (Paulin and Roberts, 1993).

Intertidal observations: *Genyagnus monopterygius* (NZ3–NZ6, NZ10).

## Family Tripterygiidae. Triplefin blennies.

Triplefin blennies are widely distributed in the Atlantic, Indian, and Pacific oceans. Most occur in shallow tropical coastal waters. One of 115 species is occasionally observed in estuaries. Most species reach maximum lengths of less than 6 cm, although one reaches 25 cm. Triplefins are found in the intertidal faunas of many regions, but they appear to be most diverse in New Zealand.

Intertidal observations: *Axoclinus lucillae* (COS3), *Bellapiscis lesleyae* (NZ4–NZ6, NZ8–NZ10, NZ14, NZ16), *B. medius* (NZ2, NZ4–NZ10, NZ13, NZ14), *Blennodon dorsale* (NZ3, NZ4, NZ12), *Cryptichthys jojettae* (NZ4, NZ6, NZ10, NZ8, NZ14, NZ16), *Forsterygion lapillum* (NZ2–NZ6, NZ8–NZ10, NZ16), *F. malcolmi* (NZ14), *F. varium* (NZ2–NZ5, NZ9, NZ10, NZ13, NZ14, NZ16), *Gilloblennius abditus* (NZ2, NZ16), *G. tripennis* (NZ2, NZ4, NZ5, NZ10, NZ14, NZ16), *Helcogramma obtusirostre* (SAF16–SAF22), *Karalepis stewarti* (NZ2, NZ4, NZ5, NZ14), *Lepidoblennius haplodactylus* (AUS1), *Norfolkia striaticeps* (AUS1), *Notoclinops caerulepunctus* (NZ9), *N. segmentatus* (NZ2, NZ4, NZ5, NZ7, NZ9, NZ14, NZ16), *N. yaldwyni* (NZ14), *Notoclinus compressus* (NZ2, NZ4, NZ5, NZ7–NZ9, NZ16), *N. fenestratus* (NZ2, NZ4, NZ5, NZ7–NZ10, NZ14, NZ16), *Ruanoho decemdigitatus* (NZ2, NZ4, NZ5, NZ7, NZ9, NZ10, NZ13, NZ14, NZ16), *R. whero* (SAF17, NZ2, NZ4–NZ6, NZ13, NZ14, NZ16), *Tripterygion chilensis* (CHL1), *T. cunninghami* (CHL1), *T. fuscipectoris* (TAI1), *T. jenningsi* (NZ17), *T. nigripinnis* (NZ3, NZ5, NZ8, NZ10, NZ13), *T. robustum* (NZ2, NZ5–NZ10, NZ16), *Tripterygion* sp. (TAI1), *Vauclusella annulata* (AUS1), *V. rufopilea* (NZ1).

## Family Dactyloscopidae. Sand stargazers.

Sand stargazers are small mainly marine fishes that occur in warm-temperate and tropical Atlantic and Pacific waters along the coasts of the Americas. There are about 41 species.

Intertidal observation: *Dactyloscopus pectoralis* (MEX2).

## Family Labrisomidae. Labrisomids.

Most of the approximately 102 labrisomid species occur in the tropical waters of the Atlantic and Pacific oceans. The family is allied with the Clinidae, and there is considerable morphological similarity between labrisomids and the Clinini. The Labrisomidae may be paraphyletic (Stepien, 1992; Stepien *et al.,* 1993). Labrisomids are small, benthic fishes that often appear in intertidal collections. Two genera, *Mnierpes* and *Dialommus* (the last not appearing here), have amphibious representatives.

Intertidal observations: *Auchenionchus microcirrhis* (CHL1), *Exerpes asper* (MEX2), *Labrisomus bucciferus* (A3), *L. gobio* (A3), *L. guppyi* (A3), *L. nigricinctus* (A3), *L. nuchipinnis* (A3, SAF1), *L. xanti* (MEX2), *Malacoctenus aurolineatus* (A3), *M. costaricanus* (COS3), *M. ebisui* (COS3), *M. erdmani* (A3), *M. gigas* (MEX2), *M. gilli* (A3), *M. triangulatus* (A3), *M. zonifer* (COS1, COS3), *Mnierpes macrocephalus* (COS1, COS3), *Paraclinus beebi* (COS3), *P. cingulatus* (A3), *P. integripinnis* (CA14), *P. mexicanus* (COS3), *P. nigripinnis* (A3), *P. sini* (MEX2), *Starksia sluteri* (A3), *Starksia* sp. (A3), *Stathmonotus stahli* (A3).

## Family Clinidae. Clinids.

Clinids are a marine family of about 73 species that has an antitropical distribution in both hemispheres. Clinids are generally small; the largest, the shallow-subtidal *Heterostichus rostratus,* reaches 60 cm in length. Clinids form a prominent subset of resident intertidal fishes in several regions, but are most diverse in southern Africa.

Intertidal observations: *Blennioclinus brachycephalus* (SAF3, SAF6–SAF14, SAF18), *B. stella* (SAF10, SAF13, SAF14, SAF18), *Blennophis anguillaris* (SAF3–SAF7, SAF9), *Cirrhibarbis capensis* (SAF7–SAF11, SAF13, SAF14), *Climacoporus navalis* (SAF11, SAF13, SAF14), *Clinus acuminatus* (SAF3–SAF10), *C. agilis* (SAF3–SAF7, SAF11), *C. berrisfordi* (SAF7, SAF9, SAF10), *C. brevicristatus* (SAF6), *C. cottoides* (SAF5–SAF14), *C. heterodon* (SAF3–SAF7), *C. nematopterus* (10), *C. rotundifrons* (SAF5, SAF7), *C. superciliosus* (SAF2–SAF14), *C. taurus* (SAF4, SAF7, SAF9), *C. venustris* (SAF4, SAF6, SAF7), *Cristiceps aurantiacus* (AUS1, NZ2, NZ4, NZ5), *Ericentrus rubrus* (NZ2, NZ4–NZ6, NZ7–NZ9), *Fucomimus mus* (SAF13, SAF9), *Gibbonsia elegans* (CA13, CA14, MEX1), *G. metzi* (CA3–CA9, CA12, CA13, MEX1), *G. montereyensis* (CA3–CA5, CA7, CA8, CA12, CA13, MEX1), *Heteroclinus perspicillatus* (AUS1), *H. whiteleggii* (AUS1), *Heterostichus rostratus* (CA13, MEX1), *Muraenoclinus dorsalis* (SAF3–SAF10, SAF14), *Pavoclinus graminis* (SAF10, SAF11, SAF13, SAF18), *P. laurentii* (SAF13, SAF15, SAF17–SAF21), *P. pavo* (SAF3, SAF7, SAF9–SAF11, MEX1), *Petraites heptaeolus* (AUS1), *P. nasutus* (AUS1), *Springeratus xanthosoma* (TAI1), *Xenopoclinus leprosus* (SAF5, SAF7).

## Family Chaenopsidae. Pikeblennies, tubeblennies, or flagblennies.

Chaenopsids are small marine fishes (the largest reaches 15 cm) occurring in warm-temperate and tropical waters of North and South America.

Intertidal observation: *Emblemaria hypacanthus* (MEX2).

## Family Blenniidae. Combtooth blennies.

Combtooth blennies are found in marine, brackish, and occasionally freshwater habitats in the Atlantic, Indian, and Pacific oceans. Most of the 345 species occur in the tropical and subtropical latitudes. Combtooth blennies are among the world's dominant families of intertidal fishes. Several tribes of blenniids are recognized by Nelson (1994), based on a multitude of studies. Five of six tribes (only the monotypic Phenablenniini is excepted) have intertidal representation. Intertidal combtooth blennies occur over most intertidal substrata. Certain genera of salariin mudskippers (*Alticus, Andamia, Praealticus*) are often described as supratidal rather than intertidal fishes [see chapters in this volume by Kotrschal (Chapter 7), Martin and Bridges (Chapter 4), and Zander *et al.* (Chapter 3) for discussions of aspects of combtooth blenny biology].

Intertidal observations: *Andamia pacifica* (TAI1), *Antennablennius australis* (SAF20–SAF22), *A. bifilum* (SAF15–SAF20, SAF22), *Aspidontus taeniatus* (SAF14), *Blenniella periopthalmus* (SAF22), *Blennius gattorugine* (A11), *B. incognitus* (A11), *B. sanguinolentus* (A10, A11), *B. trigloides* (A11), *Cirripectes alboapicalis* (NZ1), *C. sebae* (TAI1), *Coryphoblennius galerita* (A4, A6–A11), *Entomacrodus caudofasciatus* (TAI1), *E. cymatobiotus* (NZ1), *E. decussatus* (TAI1), *E. marmoratus* (HAW1), *E. niuafoouensis* (NZ1), *E. striatus* (SAF20–SAF22, SEY1, TAI1), *Hirculops cornifer* (SAF19), *Hypsoblennius brevipinnis* (COS3), *H. exstochilus* (A3), *H. gentilis* (MEX2), *H. gilberti* (CA13, CA14), *H. sordidus* (CHL1), *Istiblennius bilitonensis* (TAI1), *I. dussumieri* (SAF15–SAF19, SAF22, SEY1), *I. edentulus* (AUS1, SAF12, SAF15, SAF18, SAF19, SAF22, SEY1, TAI1), *I. gibbifrons* (SAF19), *I. impudens* (SAF19), *I. lineatus* (TAI1), *I. periopthatlmus* (SAF19, TAI1, SEY1), *I. unicolor* (SAF15, SAF20, SAF22), *I. zebra* (HAW1), *Lipophrys* (=*Blennius*) *pholis* (A4–A11), *L. velifer* (SAF1), *Omobranchus banditus* (SAF12, SAF15–SAF19), *O. woodi* (SAF13), *Ophioblennius atlanticus* (A3, SAF1), *O. steindachneri* (COS1, COS3), *Parablennius cornutus* (SAF2, SAF7, SAF10–SAF15, SAF17, SAF19), *P. gattorugine* (A4, A7, A8), *P. intermedius* (AUS1), *P. laticlavius* (NZ2–NZ6), *P. pilicornis* (SAF1, SAF2, SAF11–SAF13, SAF17), *P. trigloides* (A7, A8), *Paralipophrys trigloides* (A9), *Parenchelyurus hepburni* (TAI1, SEY1), *Pereulixia kosiensis* (SAF20–SAF22), *Pictiblennius sanguinolentus* (A9), *Plagiotremus tapeinosoma* (NZ2), *Praealticus tanegasimae* (TAI1), *Salarias fasciatus* (TAI1), *Scartella emarginata* (SAF1, SAF2, SAF12, SAF13, SAF16, SAF17, SAF19, SAF22), *Scartichthys viridis* (CHL1).

### Family Gobiesocidae. Clingfishes.

Clingfishes form a mostly marine family of morphologically unusual fishes found primarily in the intertidal and shallow subtidal waters of the Atlantic, Indian, and Pacific oceans. The family has a geographically widespread presence in the intertidal zone, with highest species richness probably occurring in New Zealand. There are approximately 120 described clingfish species, most of which are less than 7 cm in length. However, two clingfishes, the South African *Chorisochismus dentex* and the Chilean *Sicyaces sanguineus,* reach about 30 cm and more than 2 kg in weight, making them among the largest resident (or partial resident) rocky intertidal fishes. Clingfishes have a large pelvic sucker and are dorsoventrally compressed, enabling them to withstand strong waves and tidal currents. Probably many clingfishes are able to emerge during low tides.

Intertidal observations: *Apletodon dentatus* (A11, A4), *A. microcephalus* (A6), *A. pellegrini* (SAF2, SAF11, SAF14), *Arcos rhodospilus* (COS3), *A. rubiginosus* (A3), *Chorisochismus dentex* (SAF2–SAF7, SAF9–SAF11, SAF13, SAF14), *Dellichthys morelandi* (NZ2–NZ6, NZ8–NZ10, NZ16), *Diplecogaster bimaculata* (A4), *Diplocrepis puniceus* (NZ4–NZ9, NZ11–NZ13, NZ16), *Eckloniaichthys scylliorhiniceps* (SAF3, SAF5–SAF7), *Gastrocyathus gracilis* (NZ2–NZ5, NZ8–NZ10, NZ12, NZ16), *Gastrocymba quadrira-*

*diata* (NZ17), *Gastroscyphus hectoris* (NZ2, NZ4, NZ8, NZ9, NZ12, NZ14, NZ16), *Gobiesox daedaleus* (COS1, COS3), *G. marmoratus* (CHL1), *G. meandricus* (BC2, CA1, CA3–CA5, CA8, CA12, CA13, WA1), *G. potamius* (COS2), *G. rhessodon* (CA13, CA14, MEX1), *Haplocylix littoreus* (NZ2, NZ3–NZ5, NZ8–NZ10, NZ13, NZ16), *Lepadogaster candollei* (A11, A4), *L. lepadogaster* (A4, A6, A7, A11), *Rimicola eigenmanni* (MEX1), *R. muscarum* (CA13), *Tomicodon petersi* (COS1, COS3), *Trachelochismus melobesia* (NZ2–NZ5, NZ8, NZ13, NZ16), *T. pinnulatus* (NZ2–NZ6, NZ8–NZ10, NZ13, NZ16).

## Family Callionymidae. Dragonets.

All but 2 of about 130 callionymid species occur in marine waters; 2 enter rivers. The family has a tropical distribution with maximum species richness in the Indo-West Pacific region.

Intertidal observations: *Callionymus lyra* (A5), *Draculo celetus* (CA13).

## Family Eleotridae. Sleepers.

Eleotrids occur in marine, brackish, and freshwater habitats in most tropical and subtropical regions. Most species occur in shallow coastal waters. There are about 150 species in the family. *Grahamichthys radiata* is a widely distributed New Zealand intertidal endemic.

Intertidal observations: *Eleotris picta* (COS2), *Gobiomorus maculatus* (COS2), *Grahamichthys radiata* (NZ4, NZ8–NZ10, NZ13, NZ14).

## Family Gobiidae. Gobies.

The Gobiidae is a large family of about 1875 species, most of which are marine. Indeed, the family contains more marine fishes than any other. Most of the freshwater gobiids occur on oceanic islands. The Gobiidae and the closely allied Eleotridae consist mainly of small, benthic fishes that occupy nearshore habitats in tropical and subtropical waters. Like the gobiesocids and liparids, most gobiids possess a pelvic disc capable of attaching to solid substrata by suction. Many gobies are euryhaline and eurythermal. Gobies are among the world's dominant families in number of intertidal representatives; the family also contains many amphibious fishes. Three genera of rockskippers, *Boleophthalmus, Periophthalmus,* and *Periophthalmodon,* have representatives capable of spending large proportions of their time in terrestrial habitats, where they are able to move about, forage, and maintain physiological homeostasis very effectively [see Martin and Bridges (Chapter 4, this volume) for a discussion of air-breathing by intertidal fishes; Kotrschal (Chapter 7, this volume) for a discussion of the sensory systems of intertidal and supratidal fishes; and Zander *et al.* (Chapter 3, this volume) for within-intertidal zone distributional correlates]. Although many intertidal gobies have a terete morphology (see Horn, Chapter 16, this volume) with large fan-like fins and well-developed, elevated eyes, there are also a number of eel-like genera that in some cases are blind or even lack eyes (e.g., *Typhlogobius*). These differences in morphology presumably have ecological and within-intertidal zone distributional correlates.

Intertidal observations: *Acentrogobius ornatus* (TAI1), *Arenigobius bifrenatus* (AUS1), *Aruma histrio* (MEX2), *Asterropteryx semipunctatus* (TAI1, SEY1), *Awaous transandeanus* (COS2), *Barbulifer antennatus* (A3), *Bathygobius albopunctatus* (SAF18), *B. andrei* (COS2), *B. cocosensis* (SAF18), *B. cotticeps* (SAF20), *B. curacao* (A3), *B. cyclopterus*

(SAF18), *B. fuscus* (AUS1, HAW1, SEY1, TAI1), *B. krefftii* (AUS1), *B. ramosus* (COS1, COS3), *Caffrogobius caffer* (SAF6, SAF9, SAF11–SAF14), C. *natalensis* (SAF15–SAF17), *C. nudiceps* (SAF3), *C. saldanha* (SAF3, SAF10, SAF12–SAF14), *Callogobius sclateri* (TAI1), *Coryogalops oculata* (SEY1), *C. william* (SAF13, SAF18), *Coryphopterus nicholsi* (BC2), *Eviota abax* (TAI1), *E. prasina* (SEY1), *Evorthodus minutus* (COS2), *Favonigobius melanobranchus* (SAF20), *F. lateralis* (AUS1), *Fusigobius neophytus* (TAI1), *Ginsburgellus novemlineatus* (A3), *Gnatholepis knighti* (TAI1), *Gobionellus microdon* (COS2), *G. sagittula* (COS2), *Gobiopsis atrata* (NZ2, NZ4–NZ6, NZ8, NZ13, NZ14, NZ16), *Gobiosoma chiquita* (MEX2), *G. hildbrandi* (A3), *Gobius casamancus* (SAF1), *G. cobitus* (A4, A6, A7), *G. couchi* (A4), *G. niger* (A4), *G. paganellus* (A4–A6, A9, A10), *Gobius* sp. (A11), *Gobiusculus flavescens* (A4), *Gobulus hancocki* (COS1), *Gymneleotris seminudus* (COS3), *Heterelectris zonata* (SAF18, SAF19, SAF20), *Istigobius ornatus* (SEY1), *Kelloggella oligolepis* (HAW1), *Lythrypnus* sp. (A3), *Pomatoschistus microps* (A4, A5), *P. minutus* (A4), *P. pictus* (A4, A5), *Priolepis cincta* (SAF19–SAF22), *P. cinctus* (SAF12), *Psammogobius knysnaensis* (SAF7, SAF10), *Riukiuia* sp. (TAI1), *Typhlogobius californiensis* (CA13), *Zonogobius eugenius* (TAI1), *Z. semidoliatus* (TAI1).

### Family Microdesmidae. Wormfishes and dartfishes.

Microdesmids are eel-like small fishes, none greater than 30 cm in length, that occupy a variety of nearshore tropical habitats including coral reefs and sediment bottom shores. Representatives occur at depths from the intertidal zone to about 40 m (Nelson, 1994).

Intertidal observations: *Cerdale floridana* (A3), *Clarkichthys bilineata* (COS3), *Parioglossus dotui* (TAI1).

### Family Scatophagidae. Scats.

This small family (four species) is known mainly from marine and brackish waters of the Indo-West Pacific region. *Scatophagus tetracanthus* can reproduce in freshwater (Nelson, 1994).

Intertidal observation: *Scatophagus argus* (AUS1).

### Family Siganidae. Rabbitfishes.

The Siganidae is a mainly marine family distributed in the tropical Indo-West Pacific and eastern Mediterranean regions. There are about 27 species.

Intertidal observation: *Siganus spinus* (TAI1).

### Family Acanthuridae. Surgeonfishes.

Surgeonfishes occur in tropical and subtropical waters worldwide, but are absent from the Mediterranean Sea. Most surgeonfishes are shallow-water, reef-associated fishes. There are approximately 72 species. *Acanthurus triostegus* is one of the few species names reported from the intertidal zone at geographically widely separated sites.

Intertidal observations: *Acanthurus bahianus* (A3), *A. grammoptilus* (AUS1), *A. leucopareius* (EAS1), *A. lineatus* (TAI1), *A. nigrofuscus* (TAI1), *A. sandvicensis* (HAW1), *A. triostegus* (AUS1, SAF14, SAF15, SAF17–SAF19, SAF22, TAI1), *A. xanthopterus* (TAI1), *Ctenochaetus striatus* (TAI1), *Prionurus microlepidotus* (AUS1).

## Order Pleuronectiformes

### Family Bothidae. Lefteye flounders.

Lefteye flounders occur exclusively in marine waters of the Atlantic, Indian, and Pacific Oceans, in depths from shallow subtidal to the continental slopes. There are about 115 species. Juveniles of some lefteye flounders occasionally enter the intertidal zone.

Intertidal observations: *Bothus mancus* (TAI1), *Engyprosopon grandisquamma* (AUS1).

### Family Scophthalmidae.

Scophthalmids occur in the northern Atlantic and also the Mediterranean and Black Seas. The family is exclusively marine and includes about 18 species. One species, *Zeugopterus punctatus,* differs from most flatfishes in preferring rocky substrata (Wheeler, 1994) and occasionally may be found in the rocky intertidal zone of the English Channel and Atlantic coasts of England. Scophthalmids occur from the shallow subtidal to about 840 m depth (Weitzman, 1997). As is the case with other pleuronectiform fishes, only juveniles of most scophthalmid species are likely to be observed in the intertidal zone.

Intertidal observations: *Scophthalmus maximus* (A4), *S. rhombus* (A4), *Zeugopterus punctatus* (A4).

### Family Paralichthyidae.

Most of the approximately 85 paralichthyids are marine; a few enter rivers. The family is distributed in the Atlantic, Indian, and Pacific oceans. Juvenile *Citharichthys stigmaeus* are occasionally observed in the rocky intertidal zone in the northeastern Pacific, usually in surge channels having patches of sandy substrata (M.A.C., personal observation).

Intertidal observations: *Citharichthys gilberti* (COS2), *C. stigmaeus* (CA1, CA5, CA13).

### Family Pleuronectidae. Righteye flounders.

Pleuronectids have a marine distribution in the Arctic, Atlantic, Indian, and Pacific oceans. A few species enter brackish or freshwater habitats. There are about 93 species. Juveniles occasionally enter the intertidal zone, usually in areas with sandy substrata.

Intertidal observations: *Pleuronectes flesus* (A4), *P. platessa* (A4), *Limanda limanda* (A4), *Pseudopleuronectes americanus* (A1).

### Family Soleidae. Soles.

Soles occur in tropical to temperate latitudes, especially in the eastern Atlantic and western and southern Pacific. There are about 89 species, many of which occur in coastal waters.

Intertidal observations: *Heteromycteris capensis* (SAF12), *Pardachirus hedleyi* (AUS1), *Synaptura marginata* (SAF12), *S. nigra* (AUS1).

### Family Cynoglossidae. Tonguefishes.

Tonguefishes have a tropical and subtropical distribution. There are about 110 species.

Intertidal observation: *Paraplagusia unicolor* (AUS1).

## Order Tetraodontiformes

### Family Balistidae. Triggerfishes.

Triggerfishes occur in the Atlantic, Indian, and Pacific oceans, generally in shallow waters and often in association with coral reefs. There are about 40 species.

Intertidal observation: *Rhinecanthus verrucosus* (TAI1).

### Family Monacanthidae. Filefishes.

There are about 95 species of filefishes known from the Atlantic, Indian, and Pacific oceans. Filefishes are exclusively marine. Most filefishes occur in nearshore reef habitats. The family achieves its greatest diversity in Australia. One taxon, *Parika scaber,* routinely appears in New Zealand intertidal collections.

Intertidal observations: *Monacanthus chinensis* (AUS1), *Parika scaber* (NZ2–NZ5, NZ8, NZ10, NZ16).

### Family Ostraciidae. Boxfishes, cowfishes, and trunkfishes.

Ostraciids occur in tropical waters of the Atlantic, Indian, and Pacific oceans. The family is exclusively marine and contains about 33 species.

Intertidal observation: *Lactoria diaphana* (AUS1).

### Family Tetraodontidae. Puffers.

Puffers are widely distributed in tropical and subtropical waters of the Atlantic, Indian, and Pacific oceans. Most species are marine, but several are euryhaline or occur in freshwater. Most of the approximately 121 species are small, but one reaches 90 cm (Nelson, 1994). Puffers are often associated with coral reefs, but some species occur in sediment bottom habitats and there are a number of partial resident or transient intertidal representatives. *Takifugu niphobles,* which does not appear on this list, spawns high in the intertidal zone (Yamahira, 1996).

Intertidal observations: *Amblyrhynchotes honckenii* (SAF12–SAF14), *Arothron inconditus* (SAF13, SAF14), *Canthigaster benetti* (TAI1), *C. rostrata* (A3), *Sphoeroides spengleri* (A3), *Tetractenos glaber* (AUS1), *T. hamiltoni* (AUS1), *Tetraodon hispidus* (TAI1), *Torquigener pleurogramma* (AUS1).

### Family Diodontidae. Porcupinefishes.

Diodontids are exclusively marine and are distributed in the Atlantic, Indian, and Pacific oceans. There are about 19 species. Adult diodontids live in shallow subtidal waters, while juveniles are pelagic.

Intertidal observation: *Diodon holacanthus* (TAI1).

## III. Taxonomic Summary of Intertidal Fishes

We have summarized the taxonomic distribution of reported rocky intertidal fishes at the levels of class and order (Table 1) following Nelson (1994). Because many rocky intertidal areas of the world have yet to be surveyed and our literature search was selective, the counts of intertidal species are undoubtedly underestimates. Nevertheless some general observations can be made. For example, we note that, with a single elasmobranch exception, the Classes Myxini, Cephalaspidomorphi, Chondrichthyes, and Sarcopterygii do not presently contribute to the rocky intertidal ichthyofauna. Within the Class Actinopterygii, strictly freshwater species are (obviously) not reported from the intertidal zone, even as strays; diadromous species are negligible components of the intertidal ichthyofauna.

Is the rocky intertidal fish assemblage simply a random draw from the remaining strictly marine groups of actinopterygiians? Statistically speaking, the answer is "no." After consolidating orders (see Table 1) along phylogenetic lines to assure adequate expected numbers of species in each cell, a chi-square test revealed that the distribution of intertidal fishes differs from the overall distribution of fish species among the actinopterygian orders ($X^2 = 75.8$, $df = 9$, $P < 0.001$). Intertidal fish species are concentrated within the two most derived superorders of the Euteleostei, the Paracanthopterygii and the Acanthopterygii. Within the latter group, the percomorph orders Scorpaeniformes and Perciformes contain the most intertidal species.

To examine to what degree phylogenetic constraint may have played a role in this taxonomic asymmetry, we have partitioned the distribution of rocky intertidal fishes by family in the Scorpaeniformes and Perciformes (Table 2). Eight of 25 scorpaeniform families contain species that inhabit the intertidal zone. The distribution of these eight families shows no phylogenetic pattern within the order; they range from the ancestral Scorpaenidae to the highly derived Cyclopteridae and Liparidae. Within any of these eight scorpaeniform families, no more than 27% of marine species are rocky intertidal species (mean = 9.6%).

**Table 2.** Summary of Extant Scorpaeniform and Perciform Marine Species (from Nelson, 1994) and the Subset Reported in Studies Cited Here (See Annotated List in Text)

| Order | Family | Marine species | Rocky intertidal species |
|---|---|---|---|
| Scorpaeniformes | Scorpaenidae | 388 | 25 |
| | Congiopodidae | 9 | 1 |
| | Platycephalidae | 60 | 1 |
| | Hexagrammidae | 11 | 3 |
| | Cottidae | 266 | 26 |
| | Hemitripteridae | 8 | 1 |
| | Cyclopteridae | 28 | 1 |
| | Liparidae | 195 | 9 |
| Perciformes | Chandidae | 20 | 1 |
| | Moronidae | 4 | 1 |
| | Serranidae | 449 | 24 |
| | Pseudochromidae | 98 | 3 |
| | Plesiopidae | 26 | 10 |
| | Opistognathidae[a] | 60 | 1 |
| | Apogonidae | 207 | 9 |
| | Sillaginidae | 31 | 2 |
| | Carangidae | 140 | 2 |
| | Lutjanidae | 125 | 4 |
| | Gerreidae | 40 | 2 |
| | Haemulidae | 150 | 4 |

*continues*

**Table 2.** *Continued*

| Order | Family | Marine species | Rocky intertidal species |
|---|---|---|---|
| Perciformes *continued* | Sparidae | 100 | 10 |
| | Nemipteridae | 62 | 1 |
| | Sciaenidae | 242 | 1 |
| | Mullidae | 55 | 5 |
| | Pempheridae | 25 | 3 |
| | Monodactylidae | 5 | 1 |
| | Chaetodontidae | 114 | 8 |
| | Pomacanthidae | 74 | 3 |
| | Kyphosidae | 42 | 15 |
| | Kuhliidae | 7 | 2 |
| | Cirrhitidae | 32 | 3 |
| | Chironemidae | 4 | 1 |
| | Aplodactylidae | 5 | 2 |
| | Cheilodactylidae | 18 | 6 |
| | Latridae | 9 | 2 |
| | Embiotocidae | 23 | 6 |
| | Pomacentridae[a] | 315 | 30 |
| | Labridae[a] | 500 | 44 |
| | Odacidae | 12 | 1 |
| | Scaridae | 83 | 2 |
| | Zoarcidae | 220 | 1 |
| | Stichaeidae | 65 | 7 |
| | Cryptacanthodidae | 4 | 1 |
| | Pholidae | 14 | 5 |
| | Scytalinidae | 1 | 1 |
| | Bovichthyidae | 11 | 2 |
| | Nototheniidae | 50 | 1 |
| | Pinguipedidae | 50 | 2 |
| | Creedidae | 16 | 4 |
| | Percophidae | 40 | 1 |
| | Leptoscopidae | 4 | 1 |
| | Ammodytidae | 18 | 1 |
| | Trachinidae | 4 | 1 |
| | Uranoscopidae | 50 | 1 |
| | Tripterygidae | 115 | 30 |
| | Dactyloscopidae | 41 | 1 |
| | Labrisomidae | 102 | 26 |
| | Clinidae | 73 | 33 |
| | Chaenopsidae | 56 | 1 |
| | Blenniidae | 345 | 55 |
| | Gobiesocidae[b] | 120 | 24 |
| | Callionymidae | 130 | 2 |
| | Eleotridae[b] | 150 | 3 |
| | Gobiidae | 1875 | 54 |
| | Microdesmidae | 60 | 3 |
| | Scatophagidae | 4 | 1 |
| | Siganidae | 27 | 1 |
| | Acanthuridae | 72 | 10 |
| Totals | | 6770 | 480 |

*Note.* Families having no rocky intertidal representatives in these studies are not listed.
[a]Number of species given by Nelson (1994) explicitly described as an underestimate.
[b]Includes some freshwater species (Nelson, 1994)

Rocky intertidal perciform species exhibit a phylogenetic distributional pattern similar to that of their scorpaeniform counterparts. Fifty-nine of 148 perciform families contain intertidal species. The distribution of the 59 families shows little phylogenetic pattern within the order; they range from the more ancestral families (e.g., Chandidae) through the more derived groups (e.g., Gobiidae and Acanthuridae); however, there are no intertidal representatives among the 20 families that are considered to be the most derived within the order.

Despite the wide phylogenetic distribution of the perciform families that contain intertidal species, there are two subgroups in which the numbers of intertidal species are noteworthy: the suborder Labroidei has intertidal species in all of its marine families (that is, five of six families) that sum to nearly 12% of the intertidal species in our annotated list. Similarly, all six families of the marine suborder Blennioidei contain some intertidal species that sum to nearly 21% of those in our list. In addition to the monotypic Scytalinidae, several perciform families contain a relatively high proportion of intertidal species including cheilodactylids (33%), kyphosids (35%), pholids (35%), aplodactylids (40%), and clinids (45%); no other polytypic perciform family has more than 30% of its marine species represented in the intertidal (mean = 14.3%).

Thus, the distribution of intertidal fish species in the phylogenetic classification of Nelson (1994) is the result of little, if any, phylogenetic constraint. Orders, families, and most genera are not made up exclusively of intertidal species; if such exclusive distributional patterns existed, they would support hypotheses of phyletic radiation (i.e., constraint) within the intertidal zone. Although most intertidal species are placed in only two percomorph orders, they are widely distributed at the familial level and usually have nonintertidal species among their closest relatives.

## IV. Phylogenetic Relationships among Intertidal Fishes

In general, rocky intertidal fish species are acanthopterygians. In this study, 11 of 13 acanthopterygian orders are represented by at least one species that occurs in the intertidal zone (Table 1). However, the wide taxonomic distribution of intertidal species among the Euteleostei shows that at least some ichthyofaunal components of the intertidal zone may be quite distantly related to the others.

Those orders that contain the most intertidal species (Scorpaeniformes and Perciformes) have no polytypic families endemic to the intertidal zone (Table 2). The nearest approach to an intertidal family is the Gobiesocidae, which is almost entirely distributed in the intertidal zone and shallow subtidal waters. However, most species occupy transitional habitats including portions of both the intertidal and shallow subtidal zones, and some genera (for example, *Rimicola* in the northeastern Pacific) are more commonly observed in association with nearshore macroalgae than in the intertidal zone.

## V. Conclusions

The intertidal ichthyofaunal assemblage has a heterogeneous origin. However, the percomorph fishes, among the most advanced of the Euteleostei, contribute the greatest proportion of species to this assemblage. We agree with Horn (Chapter 16, this volume) that the three body plans that predominate among intertidal fishes are largely the result of exaptations of the nonintertidal relatives of this assemblage. This interpretation supports a scenario of repeated invasions of the intertidal and subsequent speciation, with a greater number of these invasions coming from among the exapted percomorph groups. During the Tertiary, the percomorphs have been among the most successful marine groups, and it should be no surprise that they have been able to extend their success by including the intertidal zone among the habitats into which they have radiated. Yet, intertidal fish species are always the minority in polytypic orders or families, possibly because of the limited availability of intertidal habitat. We believe the most parsimonious explanation is that the intertidal fauna has accrued through repeated invasions of the intertidal zone rather than the intertidal zone functioning as the source of origin for recent species found in subtidal waters. Invasion hypotheses can be tested when complete species-level phylogenies are available for the scorpaeniform and perciform orders.

**Appendix.** Key to Rocky Intertidal Fish Studies Used in This Chapter (Latitudes are approximate)

| Identifier | Latitude | No. Species | Location | Reference |
|---|---|---|---|---|
| A1 | 45° N | 22 | Schoodic Peninsula, Maine | Moring, 1990 |
| A2 | 42° N | 13 | Broad Sound, New England | Collette, 1988 |
| A3 | 13° N | 62 | Martins Bay, Barbados | Mahon and Mahon, 1994 |
| A4 | 50–58° N | 54 | British Isles intertidal fishes checklist | Wheeler, 1994 |
| A5 | 54° N | 12 | Morecambe Bay, west coast of England | Jones and Clare, 1977 |
| A6 | 50° N | 13 | Roscoff, France | Gibson, 1972 |
| A7 | 39° N | 6 | Sesimbra, Portugal | Arruda, 1979a |
| A8 | 39° N | 5 | Magoito, Portugal | Arruda, 1979a |
| A9 | 38° N | 8 | Azores | Arruda, 1979b |
| A10 | 38° N | 11 | Azores | Patzner *et al.*, 1992 |
| A11 | 43° N | 14 | Mompas, Spain | Ibanez *et al.*, 1986 |
| SAF1 | 15° S | 6 | Namibe (Mocamedes), Angola | Kensley and Penrith, 1973 |
| SAF2 | 18° S | 6 | Rocky Point, northern Namibia | Penrith and Kensley, 1970b |
| SAF3 | 27° S | 13 | Luderitz, Namibia | Penrith and Kensley, 1970a |
| SAF4 | 31° S | 9 | Groenrivier, west coast of South Africa | Prochazka and Griffiths, 1992 |
| SAF5 | 33° S | 12 | Cape St Martin, west coast of South Africa | Prochazka and Griffiths, 1992 |
| SAF6 | 34° S | 13 | Cape Town, South Africa | Bennett and Griffiths, 1984 |
| SAF7 | 34° S | 20 | Cape Town, South Africa | Prochazka, 1996 |
| SAF8 | 34° S | 6 | False Bay, South Africa | Marsh *et al.*, 1978 |
| SAF9 | 34° S | 17 | False Bay, South Africa | Bennett and Griffiths, 1984 |
| SAF10 | 34° S | 21 | Koppie Alleen, south coast of South Africa | Bennett, 1987 |
| SAF11 | 34° S | 31 | Tsitsikamma, South Africa | Burger, 1990 |
| SAF12 | 34° S | 35 | Port Alfred, South Africa | Christensen and Winterbottom, 1981 |
| SAF13 | 34° S | 44 | Port Elizabeth, South Africa | Beckley, 1985a |
| SAF14 | 34° S | 31 | Port Elizabeth, South Africa | Beckley, 1985b |
| SAF15 | 30° S | 14 | Pennington, Natal, South Africa | Buxton *et al.*, unpublished |
| SAF16 | 30° S | 11 | Kelso, Natal, South Africa | Buxton *et al.*, unpublished |

*continues*

**Appendix.** *Continued*

| Identifier | Latitude | No. Species | Location | Reference |
|---|---|---|---|---|
| SAF17 | 30° S | 19 | Park Rynie, Natal, South Africa | Buxton *et al.*, unpublished |
| SAF18 | 30° S | 26 | Clansthal, Natal, South Africa | Buxton *et al.*, unpublished |
| SAF19 | 30° S | 37 | Durban, Natal, South Africa | Beckley, unpublished |
| SAF20 | 27° S | 25 | Mbibi, Natal, South Africa | Buxton *et al.*, unpublished |
| SAF21 | 27° S | 16 | Black Rock, Natal, South Africa | Buxton *et al.*, unpublished |
| SAF22 | 27° S | 26 | Bhanga Nek, Natal, South Africa | Buxton *et al.*, unpublished |
| SEY1 | 5° S | 23 | Seychelles | Winterbottom, unpublished |
| TAI1 | 23° N | 122 | Sanhsientai, Taiwan | Lee, 1980 |
| EAS1 | 27° S | 6 | Easter Island | Duhart and Ojeda, 1994 |
| HAW1 | 20° N | 10 | Hawaii | Gosline, 1965 |
| BC1 | 55° N | 11 | Gnarled Islands, British Columbia | Peden and Wilson, 1976 |
| BC2 | 55° N | 16 | Arniston Point, Dundas Island, British Columbia | Peden and Wilson, 1976 |
| BC3 | 54° N | 4 | Welcome Island, British Columbia | Peden and Wilson, 1976 |
| BC4 | 54° N | 11 | Welcome Island, British Columbia | Peden and Wilson, 1976 |
| BC5 | 54° N | 10 | Welcome Harbour, British Columbia | Peden and Wilson, 1976 |
| BC6 | 54° N | 10 | Welcome Harbour, British Columbia | Peden and Wilson, 1976 |
| BC7 | 54° N | 14 | White Rocks, British Columbia | Peden and Wilson, 1976 |
| BC8 | 49° N | 16 | Vancouver Island, British Columbia | Green, 1971 |
| WA1 | 48° N | 16 | Washington | Cross, 1982 |
| OR1 | 43° N | 9 | Cape Arago, Oregon | Yoshiyama *et al.*, 1986 |
| OR2 | 42° N | 7 | Brookings, Oregon | Yoshiyama *et al.*, 1986 |
| CA1 | 41° N | 20 | Trinidad Bay, California | Moring, 1986 |
| CA2 | 40° N | 8 | Cape Mendocino, California | Yoshiyama *et al.*, 1986 |
| CA3 | 39° N | 37 | Point Arena, California | Yoshiyama *et al.*, 1986 |
| CA4 | 38° N | 20 | Bodega Head, California | Polivka and Chotkowski, 1998 |
| CA5 | 38° N | 38 | Dillon Beach, California | Grossman, 1982 |
| CA6 | 38° N | 9 | Dillon Beach, California | Grossman, 1986 |
| CA7 | 38° N | 19 | Dillon Beach, California | Yoshiyama *et al.*, 1986 |
| CA8 | 37° N | 24 | San Mateo, California | Yoshiyama, 1981 |
| CA9 | 37° N | 18 | Pescadero Point, California | Yoshiyama *et al.*, 1986 |
| CA10 | 36° N | 7 | Piedras Blancas, California | Yoshiyama *et al.*, 1986 |
| CA11 | 36° N | 10 | Soberanes Point, California | Yoshiyama *et al.*, 1986 |
| CA12 | 35° N | 15 | San Simeon, California | Matson *et al.*, 1986 |
| CA13 | 35° N | 61 | San Simeon, California | Chotkowski and Buth, in review |
| CA14 | 34° N | 6 | Southern California | Cross, 1982 |
| MEX1 | 32° N | 19 | Punta Clara, Baja California, Mexico | Stepien *et al.*, 1991 |
| MEX2 | 31° N | 24 | Puerto Penasco, Gulf of California | Thomson and Lehner, 1976 |
| COS1 | 10° N | 25 | Playas del Coco, Costa Rica | Weaver, 1970 |
| COS2 | 10° N | 14 | Rincon de Osa, Costa Rica | Weaver, 1970 |
| COS3 | 10° N | 36 | Tamarindo, Costa Rica | Weaver, 1970 |

*continues*

**Appendix.** *Continued*

| Identifier | Latitude | No. Species | Location | Reference |
|---|---|---|---|---|
| CHL1 | 33° S | 11 | Quintay, Chile | Varas and Ojeda, 1990 |
| NZ1 | 30° S | 10 | Kermadec Islands | Paulin and Roberts, 1993 |
| NZ2 | 35° S | 52 | Northland, New Zealand | Paulin and Roberts, 1993 |
| NZ3 | 36° S | 31 | Auckland, New Zealand | Paulin and Roberts, 1993 |
| NZ4 | 37° S | 50 | Bay of Plenty, New Zealand | Paulin and Roberts, 1993 |
| NZ5 | 39° S | 42 | Gisborne-Hawke Bay, New Zealand | Paulin and Roberts, 1993 |
| NZ6 | 39° S | 18 | Taranaki, New Zealand | Paulin and Roberts, 1993 |
| NZ7 | 40° S | 18 | Chatham Islands | Paulin and Roberts, 1993 |
| NZ8 | 41° S | 42 | Wairarapa, New Zealand | Paulin and Roberts, 1993 |
| NZ9 | 41° S | 36 | Wellington, New Zealand | Paulin and Roberts, 1993 |
| NZ10 | 41° S | 33 | Marlborough, New Zealand | Paulin and Roberts, 1993 |
| NZ11 | 42° S | 1 | Westland, New Zealand | Paulin and Roberts, 1993 |
| NZ12 | 44° S | 8 | Canterbury, New Zealand | Paulin and Roberts, 1993 |
| NZ13 | 45° S | 29 | Fiordland, New Zealand | Paulin and Roberts, 1993 |
| NZ14 | 46° S | 31 | Otago, New Zealand | Paulin and Roberts, 1993 |
| NZ15 | 47° S | 2 | Southland, New Zealand | Paulin and Roberts, 1993 |
| NZ16 | 47° S | 38 | Stewart and Snares Islands, New Zealand | Paulin and Roberts, 1993 |
| NZ17 | 50° S | 6 | Auckland Islands, New Zealand | Paulin and Roberts, 1993 |
| AUS1 | 32° S | 83 | Seal Rocks, New South Wales, Australia | Lardner *et al.*, 1993 |

## Acknowledgments

We thank L. E. Beckley, P. Buxton, and R. Winterbottom for permission to use their unpublished data. M. H. Horn, K. L. M. Martin, S. Sumida, and M. Busby provided useful advice on improving the manuscript. M.A.C. thanks James Clegg, Director, and the staff of the UC Davis Bodega Marine Laboratory for use of space and facilities.

## References

Allen, G. R. (1991). "Field Guide to the Freshwater Fishes of New Guinea." Christensen Research Institute, Madang, Papua New Guinea.

Arruda, L. M. (1979a). Specific composition and relative abundance of intertidal fish at two places on the Portuguese coast (Sesimbra and Magoito, 1977–78). *Arquiv. Mus. Bocage 2ª Ser.* **6,** 325–342.

Arruda, L. M. (1979b). On the study of a sample of fish captured in the tidal range at Azores. *Bolm Soc. Port. Ciênc. Nat.* **19,** 5–36.

Beckley, L. E. (1985a). The fish community of East Cape tidal pools and an assessment of the nursery function of this habitat. *S. Afr. J. Zool.* **20,** 21–27.

Beckley, L. E. (1985b). Tide-pool fishes: Recolonisation after experimental elimination. *J. Exp. Mar. Biol. Ecol.* **85,** 287–295.

Bennett, B. A. (1987). The rock-pool fish community of Koppie Alleen and an assessment of the importance of Cape rock-pools as nurseries for juvenile fish. *S. Afr. J. Zool.* **22**(1), 25–32.

Bennett, B. A., and Griffiths, C. L. (1984). Factors affecting the distribution, abundance and diversity of rock-pool fishes on the Cape Peninsula, South Africa. *S. Afr. J. Zool.* **19,** 97–104.

Böhlke, E. B. (Ed.). (1989). "Fishes of the Western North Atlantic," Part 9, Vol. 1, "Orders Anguilliformes and Saccopharyngiformes," pp. 1–655; Vol. 2, "Leptocephali," pp. 657–1055. Sears Foundation for Marine Research, Memoir (Yale University), New Haven.

Brehaut, R. (1982). "Ecology of Rocky Shores." Studies in Biology #139. Edward Arnold, London.

Briggs, J. C. (1995). "Global Biogeography." Elsevier, Amsterdam.

Burger, L. F. (1990). "The Distributional Patterns and Community Structure of the Tsitsikamma Rocky Littoral Ichthyofauna." Unpublished MSc thesis, Rhodes University, Grahamstown, South Africa.

Chotkowski, M. A., and Buth, D. G. (in review). Fishes of the San Simeon Reef, 1948–1997.

Christensen, M. S., and Winterbottom, R. (1981). A correction factor for, and its application to, visual census of littoral fish. *S. Afr. J. Zool.* **16,** 73–79.

Clark, F. N. (1938). Grunion in southern California. *Calif. Fish Game* **24,** 49–54.

Collette, B. B. (1988). Resilience of the fish assemblage in New England tidepools. *U.S. Fish. Bull.* **84,** 200–204.

Cross, J. N. (1982). Resource partitioning in three rocky intertidal fish assemblages. *In* "Gutshop '81: Fish Food Habit Studies. Proc. 3rd Pacific Workshop" (G. M. Caillett and C. A. Simenstad, Eds.), pp. 142–150. Washington Sea Grant Publication, University of Washington, Seattle.

Dingerkus, G., and Séret, B. (1992). First record of *Moringua microchir* for New Caledonia and from freshwater (Teleostei: Anguilliformes: Moringuidae). *Cybium* **16,** 175–176.

Dooley, J. K., Van Tassell, J., and Brito, A. (1985). An annotated checklist of the shorefishes of the Canary Islands. *Am. Mus. Novit.* **2824,** 1–49.

Duhart, M., and Ojeda, F. P. (1994). Ichthyological characterisation of intertidal pools, and trophic analysis of herbivore subtidal fishes of Easter Island. *Med. Amb.* **12**(1), 32–40.

Eschmeyer, W. N., Herald, E. S., and Hammann, H. (1983). "Pacific Coast Fishes." Houghton Mifflin, Boston.

Gibson, R. N. (1972). The vertical distribution and feeding relationships of intertidal fish on the Atlantic coast of France. *J. Anim. Ecol.* **41,** 189–207.

Gibson, R. N. (1982). Recent studies on the biology of intertidal fishes. *Oceanogr. Mar. Biol. Annu. Rev.* **20,** 363–414.

Gon, O., and Heemstra, P. C. (1987). *Mendosoma lineatum* Guichenot 1848, first record in the Atlantic Ocean, with a re-evaluation of the taxonomic status of other species of the genus *Mendosoma* (Pisces, Latridae). *Cybium* **11,** 183–193.

Gosline, W. A. (1965). Vertical zonation of inshore fishes in the upper water layers of the Hawaiian Islands. *Ecology* **46**(6), 823–831.

Green, J. M. (1971). Local distribution of *Oligocottus maculosus* Girard and other tidepool cottids of the west coast of Vancouver Island, British Columbia. *Can. J. Zool.* **49,** 1111–1128.

Grossman, G. D. (1982). Dynamics and organisation of a rocky intertidal fish assemblage: The persistence and resilience of taxocene structure. *Am. Nat.* **119**(5), 611–637.

Grossman, G. D. (1986). Long term persistence in a rocky intertidal fish assemblage. *Environ. Biol. Fish.* **15**(4), 315–317.

Hennig, W. (1966). "Phylogenetic Systematics." University of Illinois Press, Urbana, IL.

Ibanez, M., Miguel, I., and Eizmendi, A. (1986). Ichthyofauna in tide-pools. I. Methodology and preliminary results. *Lurralde* **9,** 159–164.

Jones, D., and Clare, J. (1977). Annual and long-term fluctuations in the abundance of fish species inhabiting an intertidal mussel bed in Morecambe Bay, Lancashire. *Zool. J. Linn. Soc.* **60,** 117–172.

Kensley, B. F., and Penrith, M-L. (1973). The constitution of the intertidal fauna of rocky shores of Moçamedes, southern Angola. *Cimbebasia* (A) **2**(9), 113–123.

Lardner, R., Ivanstoff, W., and Crowley, L. E. L. M. (1993). Recolonisation by fishes of a rocky intertidal pool following repeated defaunation. *Austr. Zool.* **29**(1–2), 85–92.

Lee, S-C. (1980). Intertidal fishes of a rocky pool of the Sanhsientai, eastern Taiwan. *Bull. Inst. Zool. Acad. Sin.* **19**(1), 19–26.

Mahon, R., and Mahon, S. D. (1994). Structure and resilience of a tidepool fish assemblage at Barbados. *Environ. Biol. Fish.* **41,** 171–190.

Marsh, B., Crowe, T. M., and Siegfried, W. R. (1978). Species richness and abundance of clinid fish (Teleostei: Clinidae) in intertidal rock pools. *Zool. Afr.* **13**(2), 283–291.

Matson, R. H., Crabtree, C. B., and Haglund, T. R. (1986). Ichthyofaunal composition and recolonisation in a central Californian tidepool. *Calif. Fish Game* **72**(4), 227–231.

Moring, J. R. (1986). Seasonal presence of tidepool fish species in a rocky intertidal zone of northern California, USA. *Hydrobiologia* **134,** 21–27.

Moring, J. R. (1990). Seasonal absence of fishes in tidepools of a boreal environment (Maine, USA). *Hydrobiologia* **194,** 163–168.

Nelson, J. S. (1990). Redescription of *Antipodocottus elegans* (Scorpaeniformes: Cottidae) from Australia, with comments on the genus. *Copeia* **1990,** 840–846.

Nelson, J. S. (1994). "Fishes of the World." Wiley, New York.

Patzner, R. A., Santos, R. S., Ré, P., and Nash, R. D. M. (1992). Littoral fishes of the Azores: An annotated checklist of fishes observed during the Expedition Azores 1989. *Arquipélago Life Earth Sci.* **10,** 101–111.

Paulin, C., and Roberts, C. (1993). Biogeography of New Zealand rockpool fishes. *In* "Proceedings of the Second International Temperate Reef Symposium, 7–10 January 1992, Auckland, New Zealand" (C. N. Battershill *et al.*, Eds.). National Institute of Water and Atmospheric Research Ltd.

Peden, A. E., and Wilson, D. E. (1976). Distribution of intertidal and subtidal fishes of northern British Columbia and southeastern Alaska. *Syesis* **9,** 221–248.

Penrith, M-L., and Kensley, B. F. (1970a). The constitution of the intertidal fauna of rocky shores of South West Africa. Part I. Lüderitzbucht. *Cimbebasia* (A) **1**(9), 191–237.

Penrith, M-L., and Kensley, B. (1970b). The constitution of the fauna of rocky intertidal shores of South West Africa. Part II. Rocky Point. *Cimbebasia* (A) **1**(10), 243–265.

Pietsch, T. W., and Grobecker, D. B. (1987). "Frogfishes of the World: Systematics, Zoogeography, and Behavioral Ecology." Stanford University Press, Palo Alto.

Polivka, K. M., and Chotkowski, M. A. (1998). The effects of repeated tidepol defaunation on recolonization by northeast Pacific intertidal fishes. *Copeia* **2,** 456–462.

Prochazka, K. (1996). Seasonal patterns in a temperate intertidal fish community on the west coast of South Africa. *Environ. Biol. Fish.* **45,** 133–140.

Prochazka, K., and Griffiths, C. L. (1992). The intertidal fish fauna of the west coast of South Africa—Species, community and biogeographic patterns. *S. Afr. J. Zool.* **27**(3), 115–120.

Smith-Vaniz, W. F., and Johnson, G. D. (1990). Two new species of Acanthoclininae (Pisces: Plesiopidae) with a synopsis and phylogeny of the subfamily. *Proc. Acad. Nat. Sci. Phila.* **142,** 211–260.

Stepien, C. A. (1992). Evolution and biogeography of the Clinidae (Teleostei: Blennioidei). *Copeia* **1992**(2), 375–392.

Stepien, C. A., Phillips, H., Adler, J. A., and Mangold, P. J. (1991). Biogeographic relationships of a rocky intertidal fish assemblage in an area of cold water upwelling off Baja California, Mexico. *Pacific Sci.* **45**(1), 63–71.

Stepien, C. A., Dixon, M. T., and Hillis, D. M. (1993). Evolutionary relationships of the blennioid fish families Clinidae, Labrisomidae and Chaenopsidae: Congruency between DNA sequence and allozyme data. *Bull. Mar. Sci.* **52,** 496–515.

Templeman, W. (1948). The life history of the capelin (*Mallotus villosus*) in Newfoundland waters. *Bull. Newfoundland Govt. Lab. St. John's* **17,** 1–155.

Thomson, D. A., and Lehner, C. E. (1976). Resilience of a rocky intertidal fish community in a physically unstable environment. *J. Exp. Mar. Biol. Ecol.* **22,** 1–29.

Varas, E., and Ojeda, F. P. (1990). Intertidal fish assemblages of the central Chilean coast: Diversity, abundance and trophic patters. *Rev. Biol. Mar.* **25**(2), 59–70.

Walker, B. W. (1952). A guide to the grunion. *Calif. Fish Game* **38,** 409–420.

Weaver, P. L. (1970). Species diversity and ecology of tidepool fishes in three Pacific coastal areas of Costa Rica. *Rev. Biol. Trop.* **17**(2), 165–185.

Weitzman, S. H. (1997). Systematics of deep-sea fishes. *In* "Deep-Sea Fishes" (D. J. Randall and A. P. Farrell, Eds.), pp. 43–77. Academic Press, San Diego.

Wheeler, A. (1994). Field key to the shore fishes of the British Isles. *Field Stud.* **8,** 481–521.

Wiley, E. O. (1981). "Phylogenetics: The Theory and Practice of Phylogenetic Systematics." Wiley, New York.

Wooton, R. J. (1976). "The Biology of the Sticklebacks." Academic Press, San Diego.

Yamahira, K. (1996). The role of intertidal egg deposition on survival of the puffer, *Takifugu niphobles* (Jordan and Snyder) embryos. *J. Exp. Mar. Biol. Ecol.* **198,** 291–306.

Yoshiyama, R. M. (1981). Distribution and abundance patterns of rocky intertidal fishes in central California. *Environ. Biol. Fish.* **6**(3/4), 315–332.

Yoshiyama, R. M., Sassaman, C., and Lea, R. N. (1986). Rocky intertidal fish communities of California: Temporal and spatial variation. *Environ. Biol. Fish.* **17**(1), 23–40.

# 15

# Biogeography of Rocky Intertidal Fishes

Kim Prochazka

*Zoology Department, University of Cape Town, Rondebosch, South Africa*

Michael A. Chotkowski

*Department of Biology, University of California, Los Angeles, California*

Donald G. Buth

*Department of Biology, University of California, Los Angeles, California*

## I. Introduction

Modern biogeographers strive not only to find patterns of distribution of organisms, but also to elucidate the factors responsible for the establishment and maintenance of these patterns. However, the search for process is a secondary step to the establishment of pattern (Myers and Giller, 1988). Because the patterns of intertidal fishes have not been described before, the purpose of this chapter is to provide an account of the overarching distributional patterns of intertidal fishes. The coverage has been limited to rocky intertidal habitats because most of the available data are concentrated there.

Only data from primary sources have been used. These include data collected using some kind of ichthyocide or anesthetic, or by emptying pools of their water to collect the fish. Other methods, such as catching fish by hand, dip-netting only visible, moving fishes, often fail to capture or record a large proportion of species. Data from publications such as field guides cannot be used for this type of analysis because distribution maps or descriptions are often colored by the author's own biogeographical interpretation and bias, and data for individual species frequently are given without citation. Also, field guide accounts often state species range limits for some species that are based on isolated observations of individual fishes taken far outside their usual range.

The major problems encountered in analyses presented here are biases arising from uneven sampling of intertidal fishes on a geographical basis, and differences in the intensity of sampling among localities. Comparison of faunal lists from different locations and by different researchers is further compromised by the diversity of sampling methods employed. Because resident rocky intertidal fishes are generally cryptic (in the temperate


*Intertidal Fishes: Life in Two Worlds*

latitudes, at least) and negatively buoyant, they are not easily sampled (see sampling discussion in Gibson, Chapter 2, this volume), and the techniques that have been used to capture these fishes vary greatly in their efficacy under different conditions.

Although a large literature describing the nearshore fishes of most of the world's coasts exists, extensive areas still have not been rigorously sampled specifically for intertidal fishes, and we consequently know little about them. Much of the western Pacific, parts of the Indian Ocean, the arctic regions, and most maritime island regions fall into this category. Some regions, including a large proportion of the tropics and subtropics, have very small tidal excursions, which might discourage field workers from separately sampling the intertidal zone.

Intertidal fishes can be classified according to Gibson (1982) as fully or partially resident or transient species. While it would be valuable to compare the biogeography of true and partial residents and transient species, information needed to classify most fish species found in the intertidal zone is unavailable. It is particularly difficult to distinguish between true and partial resident species, and, although patterns of residency exist among genera and families for which we have data, we cannot consistently infer the residency category for fishes reported in collections taken at many remote sites. For this reason, analyses presented here include all species captured in the intertidal zone.

## II. Scope of Study and Methodology

Species lists were acquired from 47 sources, including published papers, theses, and unpublished data. Besides their use in this chapter, the collections data were used to develop a systematic list of intertidal fishes (Chotkowski *et al.,* Chapter 14, this volume). Only data from collections that satisfied three criteria were used. First, the sampling locality had to be unambiguously limited to the intertidal zone. A great many ichthyofaunal surveys were not used because both intertidal and shallow subtidal habitats were sampled, or the described methods left us in doubt whether all fishes came from between the tidemarks. Second, only collections intended to sample the fauna exhaustively were used; data collected in the course of taxon-specific research were not included. Third, only samples collected using dip nets, ichthyocide, anesthetics, bailing, or combinations of these methods were used.

Eighty-five discrete collections taken at latitudes ranging from 55° N to 47° S were used (Table 1). Almost all collections were obtained in temperate latitudes (Figure 1). Approximately 75% of the collections were taken in three regions (southern Africa, New Zealand, and the west coast of the United States). Because some studies provided quantitative accounts of species taken while others provided only presence/absence data, we were unable to develop quantitative descriptions of all faunas. Instead, we limited our enquiry to questions that could be answered using presence/absence data. We asked three questions. First, in terms of numbers of species, what families of fishes dominate the rocky intertidal zone in various parts of the world? Second, are there distinct intertidal ichthyofaunas associated with contiguous sequences of biogeographical provinces, just as there are for other intertidal organisms? Third, is there a relationship between the species richness of an intertidal fauna and its latitude?

To determine which families of fishes predominate in various parts of the world, we tabulated the numbers of fishes representing the 22 most abundant families (representing, on average, about 75% of fish species collected) within each of 16 regions approximately corresponding to accepted biogeographical units (Table 2). Because density or relative

**Table 1.** Collection Records Examined in This Study

| Region | DESIG | Site within region | Latitude | Species | Reference |
|---|---|---|---|---|---|
| South Africa–Indian Ocean (SAFR-I) | SAF12 | Port Alfred, South Africa[a] | 34° S | 35 | Christensen and Winterbottom, 1981 |
| | SAF13 | Port Elizabeth, South Africa | 34° S | 44 | Beckley, 1985a |
| | SAF14 | Port Elizabeth, South Africa | 34° S | 31 | Beckley, 1985b |
| | SAF15 | Pennington, Natal, South Africa | 30° S | 14 | Buxton *et al.*, unpublished |
| | SAF16 | Kelso, Natal, South Africa | 30° S | 11 | Buxton *et al.*, unpublished |
| | SAF17 | Park Rynie, Natal, South Africa | 30° S | 19 | Buxton *et al.*, unpublished |
| | SAF18 | Clansthal, Natal, South Africa | 30° S | 26 | Buxton *et al.*, unpublished |
| | SAF19 | Durban, Natal, South Africa | 30° S | 37 | Beckley, unpublished |
| | SAF20 | Mbibi, Natal, South Africa | 27° S | 25 | Buxton *et al.*, unpublished |
| | SAF21 | Black Rock, Natal, South Africa | 27° S | 16 | Buxton *et al.*, unpublished |
| | SAF22 | Bhanga Nek, Natal, South Africa | 27° S | 26 | Buxton *et al.*, unpublished |
| Seychelles Islands (SEY) | SEY1 | | 5° S | 23 | Winterbottom, unpublished |
| Taiwan (TAI) | TAI1 | | 23° N | 122 | Lee, 1980 |
| Easter Island (EAS) | EAS1 | | 27° S | 6 | Duhart and Ojeda, 1994 |
| Hawaiian Islands (HAW) | HAW1 | | 20° N | 10 | Gosline, 1965 |
| New South Wales, Australia (AUS) | AUS1 | | 32° S | 83 | Lardner *et al.*, 1993 |
| New Zealand, Kermadec Islands (NZ-K) | NZ1 | | 30° S | 10 | Paulin and Roberts, 1993 |
| New Zealand, Main (NZ-M) | NZ2 | Northland, New Zealand | 35° S | 52 | Paulin and Roberts, 1993 |
| | NZ3 | Auckland, New Zealand | 36° S | 31 | Paulin and Roberts, 1993 |
| | NZ4 | Bay of Plenty, New Zealand | 37° S | 50 | Paulin and Roberts, 1993 |
| | NZ5 | Gisborne-Hawke Bay, New Zealand | 39° S | 42 | Paulin and Roberts, 1993 |
| | NZ6 | Taranaki, New Zealand | 39° S | 18 | Paulin and Roberts, 1993 |
| | NZ7 | Chatham Islands | 40° S | 18 | Paulin and Roberts, 1993 |
| | NZ8 | Wairarapa, New Zealand | 41° S | 42 | Paulin and Roberts, 1993 |
| | NZ9 | Wellington, New Zealand | 41° S | 36 | Paulin and Roberts, 1993 |
| | NZ10 | Marlborough, New Zealand | 41° S | 33 | Paulin and Roberts, 1993 |
| | NZ11 | Westland, New Zealand | 42° S | 1 | Paulin and Roberts, 1993 |
| | NZ12 | Canterbury, New Zealand | 44° S | 8 | Paulin and Roberts, 1993 |
| | NZ13 | Fiordland, New Zealand | 45° S | 29 | Paulin and Roberts, 1993 |
| | NZ14 | Otago, New Zealand | 46° S | 31 | Paulin and Roberts, 1993 |
| | NZ15 | Southland, New Zealand | 47° S | 2 | Paulin and Roberts, 1993 |

| | | | | | |
|---|---|---|---|---|---|
| New Zealand, Subantarctic Islands (NZ-SA) | NZ16 | Stewart and Snares Islands | 47° S | 38 | Paulin and Roberts, 1993 |
| | NZ17 | Auckland Islands, New Zealand | 50° S | 6 | Paulin and Roberts, 1993 |
| Cold-Temperate Northeastern Pacific CTNEP | BC1 | Gnarled Islands, British Columbia | 55° N | 11 | Peden and Wilson, 1976 |
| | BC2 | Arniston Point, Dundas Island, British Columbia | 55° N | 16 | Peden and Wilson, 1976 |
| | BC3 | Welcome Island, British Columbia | 54° N | 4 | Peden and Wilson, 1976 |
| | BC4 | Welcome Island, British Columbia | 54° N | 11 | Peden and Wilson, 1976 |
| | BC5 | Welcome Harbour, British Columbia | 54° N | 10 | Peden and Wilson, 1976 |
| | BC6 | Welcome Harbour, British Columbia | 54° N | 10 | Peden and Wilson, 1976 |
| | BC7 | White Rocks, British Columbia | 54° N | 14 | Peden and Wilson, 1976 |
| | BC8 | Vancouver Island, British Columbia | 49° N | 16 | Green, 1971 |
| | WA1 | Washington | 48° N | 16 | Cross, 1982 |
| | OR1 | Cape Arago, Oregon | 43° N | 9 | Yoshiyama *et al.*, 1986 |
| | OR2 | Brookings, Oregon | 42° N | 7 | Yoshiyama *et al.*, 1986 |
| | CA1 | Trinidad Bay, California | 41° N | 20 | Moring, 1986 |
| | CA2 | Cape Mendocino, California | 40° N | 8 | Yoshiyama *et al.*, 1986 |
| | CA3 | Point Arena, California | 39° N | 37 | Yoshiyama *et al.*, 1986 |
| | CA4 | Bodega Head, California | 38° N | 20 | Polivka and Chotkowski, 1998 |
| | CA5 | Dillon Beach, California | 38° N | 38 | Grossman, 1982 |
| | CA6 | Dillon Beach, California | 38° N | 9 | Grossman, 1986 |
| | CA7 | Dillon Beach, California | 38° N | 19 | Yoshiyama *et al.*, 1986 |
| | CA8 | San Mateo, California | 37° N | 24 | Yoshiyama, 1981 |
| | CA9 | Prescadero Point, California | 37° N | 18 | Yoshiyama *et al.*, 1986 |
| | CA10 | Piedras Blancas, California | 36° N | 7 | Yoshiyama *et al.*, 1986 |
| | CA11 | Soberanes Point, California | 36° N | 10 | Yoshiyama *et al.*, 1986 |
| | CA12 | San Simeon, California | 35° N | 15 | Matson *et al.*, 1986 |
| | CA13 | San Simeon, California | 35° N | 61 | Chotkowski and Buth, in review |
| Warm-Temperate and Subtropical Northeastern Pacific (WTSTNEP) | CA14 | Southern California | 34° N | 6 | Cross, 1982 |
| | MEX1 | Punta Clara, Baja California, Mexico | 32° N | 19 | Stepien *et al.*, 1991 |
| | MEX2 | Puerto Penasco, Gulf of California | 31° N | 24 | Thomson and Lehner, 1976 |
| | COS1 | Playas del Coco, Costa Rica | 10° N | 25 | Weaver, 1970 |
| | COS2 | Rincon de Osa, Costa Rica | 10° N | 14 | Weaver, 1970 |
| | COS3 | Tamarindo, Costa Rica | 10° N | 36 | Weaver, 1970 |

*continues*

**Table 1.** *Continued*

| Region | DESIG | Site within region | Latitude | Species | Reference |
|---|---|---|---|---|---|
| Temperate Southeastern Pacific (TESEP) | CHL1 | Quintay, Chile | 33° S | 11 | Varas and Ojeda, 1990 |
| Temperate Northwestern Atlantic (TENWA) | A1 | Schoodic Peninsula, Maine | 45° N | 22 | Moring, 1990 |
| | A2 | Broad Sound, New England | 42° N | 13 | Collette, 1988 |
| Temperate Northeastern Atlantic (TENEA) | A5 | Morecambe Bay, west coast of England | 54° N | 12 | Jones and Clare, 1977 |
| | A6 | Roscoff, France | 50° N | 13 | Gibson, 1972 |
| | A7 | Sesimbra, Portugal | 39° N | 6 | Arruda, 1979a |
| | A8 | Magoito, Portugal | 39° N | 5 | Arruda, 1979a |
| | A9 | Azores Islands | 38° N | 8 | Arruda, 1979b |
| | A10 | Azores Islands | 38° N | 11 | Patzner *et al.*, 1992 |
| | A11 | Mompas, Spain | 43° N | 14 | Ibanez *et al.*, 1986 |
| Tropical Northwestern Atlantic (TRNWA) | A3 | Martins Bay, Barbados | 13° N | 62 | Mahon and Mahon, 1994 |
| Southern Africa–Atlantic (SAFR-A) | SAF1 | Namibe (Mocamedes), Angola | 15° S | 6 | Kensley and Penrith, 1973 |
| | SAF2 | Rocky Point, northern Namibia | 18° S | 6 | Penrith and Kensley, 1970b |
| | SAF3 | Luderitz, Namibia | 27° S | 13 | Penrith and Kensley, 1970a |
| | SAF4 | Groenrivier, west coast of South Africa | 31° S | 9 | Prochazka and Griffiths, 1992 |
| | SAF5 | Cape St. Martin, west coast of South Africa | 33° S | 12 | Prochazka and Griffiths, 1992 |
| | SAF6 | Cape Town, South Africa | 34° S | 13 | Bennett and Griffiths, 1984 |
| | SAF7 | Cape Town, South Africa | 34° S | 20 | Prochazka, 1996 |
| | SAF8 | False Bay, South Africa | 34° S | 6 | Marsh *et al.*, 1978 |
| | SAF9 | False Bay, South Africa | 34° S | 17 | Bennett and Griffiths, 1984 |
| | SAF10 | Koppie Alleen, south coast of South Africa | 34° S | 21 | Bennett, 1987 |
| | SAF11 | Tsitsikamma, South Africa | 34° S | 31 | Burger, 1990 |
| | SAF12 | Port Alfred, South Africa[a] | 34° S | 35 | Christensen and Winterbottom, 1981 |

*Note.* DESIG, site designation. Site designations are the same as those used by Chotkowski *et al.*, Chapter 14, this volume. Species, total number of species recorded at site in study.

[a]Lies in both the Southern Africa–Atlantic (SAFR-A) region and the Southern Africa–Indian Ocean (SAFR-I) regions.

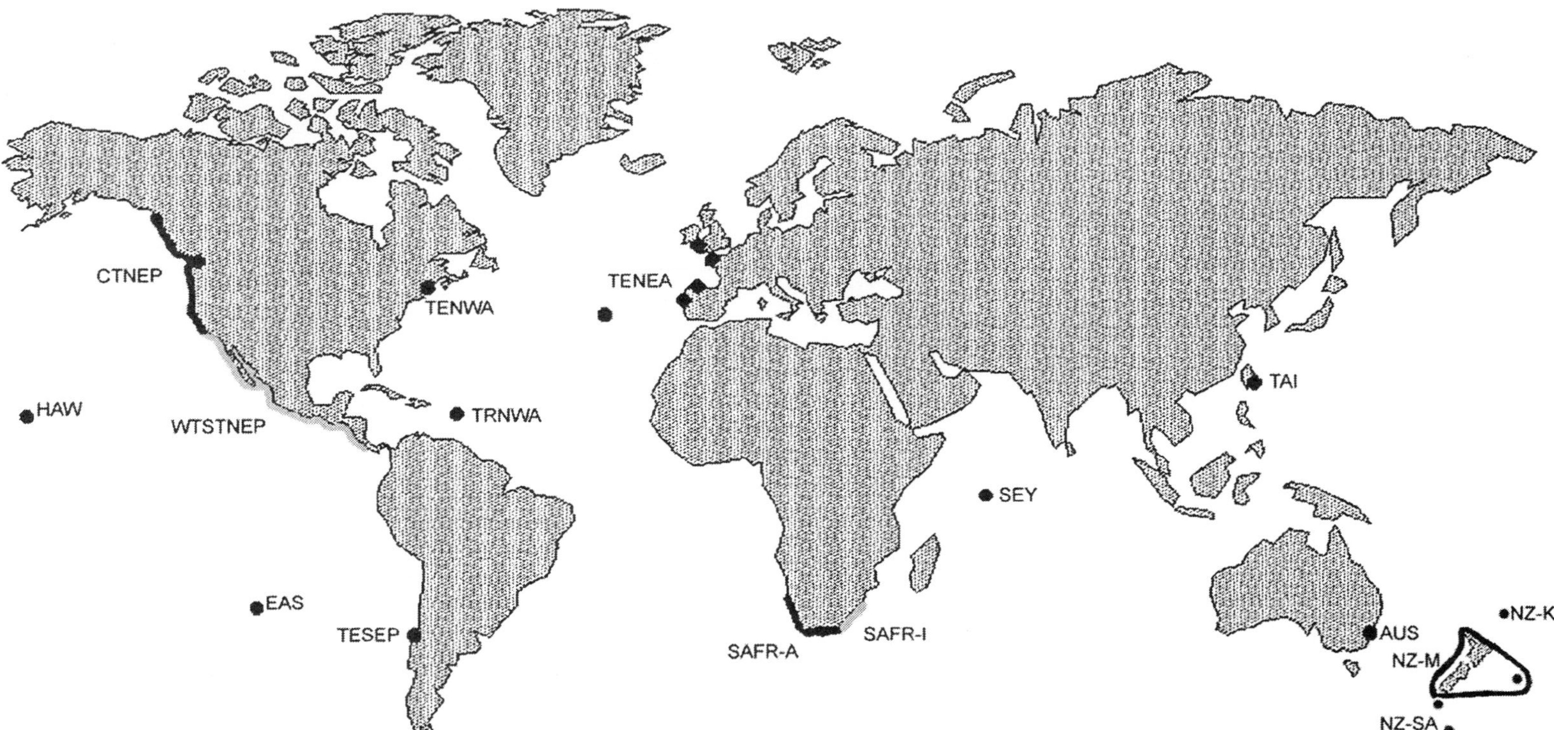


**Figure 1.** Map of the world showing locations of intertidal samples examined in this chapter. Three regions (southern Africa, New Zealand, and the northeastern Pacific) represented by 15 or more collections are indicated by solid bars rather than filled circles. The southern African region is described in Atlantic and Indian Ocean subregions, which are indicated by black and gray bars, respectively. The northeastern Pacific is described in cold-temperate and warm-temperate/subtropical subregions, which also are denoted by black and gray, respectively.

**Table 2.** Numbers of Fish Species Collected in the Rocky Intertidal Zone in 16 Regions

| | Indo-West Pacific region | | | | | | | | | Eastern Pacific region | | | Atlantic Ocean region | | | |
|---|---|---|---|---|---|---|---|---|---|---|---|---|---|---|---|---|
| | SAFR-I | SEY | TAI | EAS | HAW | AUS | NZ-K | NZ-M | NZ-SA | CTNEP | WTSTNEP | TESEP | TENWA | TENEA | TRNWA | SAFR-A |
| No. of Sites: | 11 | 1 | 1 | 1 | 1 | 1 | 1 | 14 | 2 | 24 | 6 | 1 | 2 | 7 | 1 | 11 |
| Muraenidae | 5 | 3 | 10 | 1 | 0 | 1 | 1 | 1 | 0 | 0 | 5 | 0 | 0 | 1 | 3 | 0 |
| Mugilidae | 4 | 0 | 1 | 0 | 0 | 2 | 0 | 1 | 0 | 0 | 3 | 0 | 0 | 2 | 1 | 1 |
| Holocentridae | 1 | 0 | 2 | 1 | 1 | 1 | 0 | 0 | 0 | 0 | 1 | 0 | 0 | 0 | 4 | 0 |
| Syngnathidae | 2 | 1 | 1 | 0 | 0 | 0 | 0 | 6 | 2 | 1 | 0 | 0 | 1 | 1 | 1 | 0 |
| Scorpaenidae | 3 | 1 | 5 | 0 | 0 | 5 | 0 | 2 | 0 | 20 | 2 | 0 | 0 | 0 | 2 | 0 |
| Cottidae | 0 | 0 | 0 | 0 | 0 | 0 | 0 | 0 | 0 | 22 | 6 | 0 | 3 | 1 | 0 | 0 |
| Liparidae | 0 | 0 | 0 | 0 | 0 | 0 | 0 | 0 | 0 | 7 | 1 | 0 | 1 | 2 | 0 | 0 |
| Serranidae | 9 | 1 | 5 | 0 | 0 | 7 | 2 | 2 | 1 | 0 | 2 | 0 | 0 | 1 | 2 | 1 |
| Plesiopidae | 1 | 1 | 4 | 0 | 0 | 1 | 0 | 4 | 3 | 0 | 0 | 0 | 0 | 0 | 0 | 0 |
| Apogonidae | 1 | 1 | 5 | 0 | 0 | 0 | 0 | 0 | 0 | 0 | 1 | 0 | 0 | 0 | 1 | 0 |
| Sparidae | 7 | 0 | 0 | 0 | 0 | 1 | 0 | 0 | 0 | 0 | 0 | 0 | 0 | 1 | 0 | 7 |
| Kyphosidae | 2 | 0 | 0 | 0 | 0 | 7 | 1 | 1 | 0 | 1 | 3 | 2 | 0 | 0 | 0 | 0 |
| Pomacentridae | 8 | 2 | 9 | 1 | 3 | 7 | 1 | 1 | 0 | 0 | 7 | 0 | 0 | 0 | 3 | 0 |
| Labridae | 6 | 2 | 17 | 1 | 1 | 5 | 0 | 3 | 3 | 1 | 5 | 0 | 1 | 4 | 5 | 0 |
| Tripterygiidae | 2 | 0 | 2 | 0 | 0 | 3 | 1 | 19 | 13 | 0 | 1 | 2 | 0 | 0 | 0 | 0 |
| Labrisomidae | 0 | 0 | 0 | 0 | 0 | 0 | 0 | 0 | 0 | 0 | 11 | 1 | 0 | 0 | 14 | 1 |
| Clinidae | 10 | 0 | 1 | 0 | 0 | 5 | 0 | 2 | 0 | 4 | 4 | 0 | 0 | 0 | 0 | 21 |
| Blenniidae | 18 | 5 | 12 | 0 | 2 | 2 | 3 | 2 | 0 | 1 | 4 | 2 | 0 | 11 | 2 | 5 |
| Gobiesocidae | 2 | 0 | 0 | 0 | 0 | 0 | 0 | 7 | 8 | 3 | 6 | 1 | 0 | 4 | 1 | 3 |
| Gobiidae | 12 | 5 | 10 | 0 | 2 | 4 | 0 | 1 | 1 | 2 | 10 | 0 | 0 | 5 | 5 | 5 |
| Acanthuridae | 1 | 0 | 5 | 1 | 1 | 3 | 0 | 0 | 0 | 0 | 0 | 0 | 0 | 0 | 1 | 0 |
| Tetraodontidae | 2 | 0 | 2 | 0 | 0 | 3 | 0 | 0 | 0 | 0 | 0 | 0 | 0 | 0 | 2 | 0 |
| Other Families | 25 | 1 | 31 | 1 | 0 | 26 | 1 | 30 | 9 | 33 | 28 | 3 | 18 | 8 | 15 | 11 |
| Species total | 121 | 23 | 122 | 6 | 10 | 83 | 10 | 82 | 40 | 95 | 100 | 11 | 24 | 41 | 62 | 55 |

*Note.* Only the 22 families with greatest overall representation are listed. "Other Families" category provides number of species not in families named in table. Data summarized from 44 published studies and 3 unpublished datasets listed in Table 1. SAFR-I, Southern Africa–Indian Ocean; SEY, Seychelles Islands; TAI, Taiwan; EAS, Easter Island; HAW, Hawaiian Islands; AUS, New South Wales, Australia; NZ-K, New Zealand, Kermadec Islands; NZ-M, New Zealand, Main (North and South Islands, Chatham Islands); NZ-SA, New Zealand, Subantarctic Islands; CTNEP, Cold-Temperate Northeastern Pacific; WTSTNEP, Warm-Temperate and Subtropical Northeastern Pacific; TESEP, Temperate Southeastern Pacific; TENWA, Temperate Northwestern Atlantic; TENEA, Temperate Northeastern Atlantic; TRNWA, Tropical Northwestern Atlantic; SAFR-A, Southern Africa–Atlantic.

abundance data were often not available, we were restricted to comparing numbers of species among families and sites. The reader should not infer that a heavily represented family is also a numerically abundant family.

To measure congruence between patterns of ichthyofaunal composition and biogeographical regions, we selected a regional example, southern Africa, for analysis. We subjected datasets recording presence/absence of intertidal fishes at 22 regional sites to analyses that revealed the similarity relationships among the sites. A clustering method that computed similarity based on all data was used in conjunction with a stronger Wagner procedure that computed similarity based only on species occurring at two or more sites.

## III. Overview of Faunas

In our sample of studies, the world's rocky intertidal ichthyofauna is dominated by Blenniidae (55 species), Gobiidae (54 species), and Labridae (44 species), with nine other families represented by 22 or more species (Figure 2). However, different regional faunas are dominated by different groups of fishes (Table 2). In general, subtropical and tropical intertidal collections yielded more representatives of typically warm-water families, such as Labridae, Pomacentridae, and Blenniidae, whereas the temperate collections were regionally distinctive. For instance, South Africa features a large array of Clinidae, while New Zealand has a rich fauna of Tripterygiidae. The northeastern Pacific is dominated by the primarily cold-temperate and northern Pacific Cottidae and Stichaeidae.

In the following general descriptions of regional rocky intertidal fish faunas, we usually rely on Ekman (1953) and Briggs (1995) as authorities for names of biogeographical regions. Because of the patchy distribution of samples, we have consolidated adjacent biogeographical regions in some cases. In general, we have not distinguished resident from transient species (sensu Gibson, 1982).

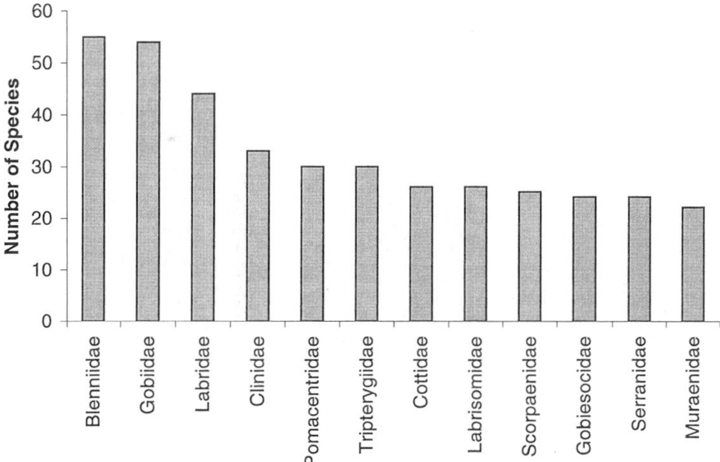


**Figure 2.** Families with the largest numbers of different species appearing in the 47 datasets examined in this chapter. See Table 1 for a summary of the collections.

## A. Indo-West Pacific

The Indo-West Pacific region is exceedingly large, including the entire Indian Ocean from the east coast of southern Africa to Taiwan and the Hawaiian Islands in the northeast, approximately to Sydney on the east coast of Australia, and eastward to Easter Island (Briggs, 1995). Numbers of species reported at a site range from a low of 1 species at Westland, New Zealand (NZ11), to a high of 122 species in Taiwan (TAI1), and averaged about 29 species per site over 33 sites in the area as a whole (Table 1). In numbers of species present, Indo-West Pacific faunas are generally dominated by blenniids, gobiids, labrids, pomacentrids, serranids, and muraenids, although the temperate subregions are individually distinct (Table 2). Because the region is very large, we present brief accounts of four subregions.

### 1. Southwestern Indian Ocean

The biogeography of rocky shore seaweeds and invertebrates in southern Africa is relatively well studied (Stephenson, 1939, 1944, 1948; Bolton, 1986; Emanuel *et al.*, 1992); these studies, especially Emanuel *et al.* (1992), strongly support the division of southern Africa into five biogeographical provinces. Two of these lie in Indian Ocean waters and are described below. The biogeography of coastal fishes in southern Africa has been less thoroughly studied, although the distributional patterns of rocky intertidal fishes have been documented by Prochazka (1996) in a thesis chapter dealing with rocky intertidal fishes.

The sequence of samples examined here spans part or all of two southeastern African marine biogeographical provinces. East Coast Warm-Temperate Province stretches from Port Alfred to the area around, and including, Durban, South Africa (sites SAF12–SAF19 in Table 1). Species richness is similar to that of the Cool-Temperate South Coast Province (described in Section III.C.4), with records of 59 species representing 20 families. The fauna contains a large proportion of partial residents and transient species. The Subtropical East Coast Province extends from north of Durban at least as far as the border between South Africa and Moçambique (sites SAF20–SAF22 in Table 1). This province is characterized by having a large proportion of tropical species, which are brought southward in the Moçambique Current.

A total of 121 species representing 35 families occurred at 16 sites (SAF12–SAF22 in Table 1) spanning all or part of two biogeographical provinces in southeastern Africa, including the East Coast Warm-Temperate Province and the Subtropical East Coast Province. The fauna is generally dominated by blenniids (18 species), gobiids (12 species), and clinids (10 species) (Table 2). Several generally tropical families, including the serranids (9 species), pomacentrids (8 species), sparids (7 species), labrids (6 species), and muraenids (5 species), are represented by five or more species, making the southeastern African provinces for which we have data more similar to the Central Indo-West Pacific than to the temperate southwestern regions of Africa discussed in Section III.C.4. The southwestern African fauna is richest in clinids, with smaller numbers of blenniids, gobiids, and gobiesocids (Table 2).

The Seychelles fauna is dominated by blennids, gobiids, and muraenids, which account for 13 of 23 reported species (SEY1, Table 2). Two pomacentrids and two labrids are also present. This oceanic fauna differs from the Subtropical East Coast African fauna in lacking mugilids, sparids, and kyphosids, and, with 23 reported species, is less species-rich than 7 of 11 southeastern African sites.

## 2. Central Pacific

In this chapter, the Central Pacific region is represented by the Hawaiian Islands (HAW1, Table 1) and Easter Island (EAS1, Table 1). There are few landmasses in this region, and those that are present are relatively small, of volcanic origin, and widely dispersed. The fauna is not as diverse as that of the eastern Indo-West Pacific. Only 10 species, including 3 pomacentrids, 2 blenniids, and 2 gobiids, were recorded from Hawaii (Table 2), while the highly depauperate Easter Island fauna has a single species each of Muraenidae, Congridae, Holocentridae, Pomacentridae, Labridae, and Acanthuridae. The Hawaiian Islands and Easter Island intertidal faunas share the labrid *Thalassoma umbrostigma.*

## 3. Warm-Temperate Southeastern Australia

The single Australian study examined here (AUS1, Table 1), from warm-temperate waters at Seal Rocks, New South Wales, reveals a diverse fauna without any overwhelmingly dominant families. The tropical families Serranidae and Pomacentridae and the tropical-temperate family Kyphosidae are each represented by seven species (Table 2). The widespread scorpaenids and tropical labrids are both represented by five species. Interestingly, only three tripterygiids were recorded, and no gobiesocids were reported, although these families are diverse and widely distributed across the Tasman Sea in New Zealand.

## 4. New Zealand

New Zealand has a rich rockpool (= rocky intertidal) fish fauna, with 94 species having been recorded around its coastline (Paulin and Roberts, 1993). The overall biogeography of the fauna has been relatively well described. Paulin and Roberts (1993) recognize New Zealand rocky intertidal fishes as either widespread, northern, or southern in distribution. Widespread fishes are those that occur at many locations between the northern tip of the North Island and Stewart Island (sites NZ2–NZ15 in Table 1; Figure 3), whereas northern fishes are generally limited to regions north of Cook Strait (the passage separating the North and South Islands) and centered between the northern tip and easternmost point of the North Island or occur at the Kermadec Islands (site NZ1 in Table 1). Southern fishes occur generally south of Cook Strait and usually are present at the Snares Islands and some subantarctic islands (sites NZ16 and NZ17 in Table 1). Most New Zealand rocky intertidal fishes are widespread (59.6%), whereas 29.8% have northern distributions and 10.6% have southern distributions. The families with the most representatives include the Tripterygiidae (21 species), Gobiesocidae (9 species), Blenniidae (5 species), Syngnathidae (4 species), and Plesiopidae (4 species) (Table 2). The northern distributional category also includes representatives of a number of generally tropical families (Table 2), whereas the southern category includes a species of Nototheniidae, a family with an austral, largely Antarctic distribution. Faunas in the more southerly latitudes were generally somewhat depauperate compared to more northerly latitudes in the New Zealand region.

Paulin and Roberts (1993) attributed the substantial faunal differences between the North and South Islands and some of the offshore island groups to regional hydrography. The Kermadec Islands are located 800 km north of New Zealand and lie in a warmer water mass strongly influenced by the East Australian current and its associated ichthyoplankton. The subantarctic groups lie 300–900 km south of the New Zealand mainland and, like the Kermadecs, are hydrographically isolated from it. However, whereas the Kermadecs are

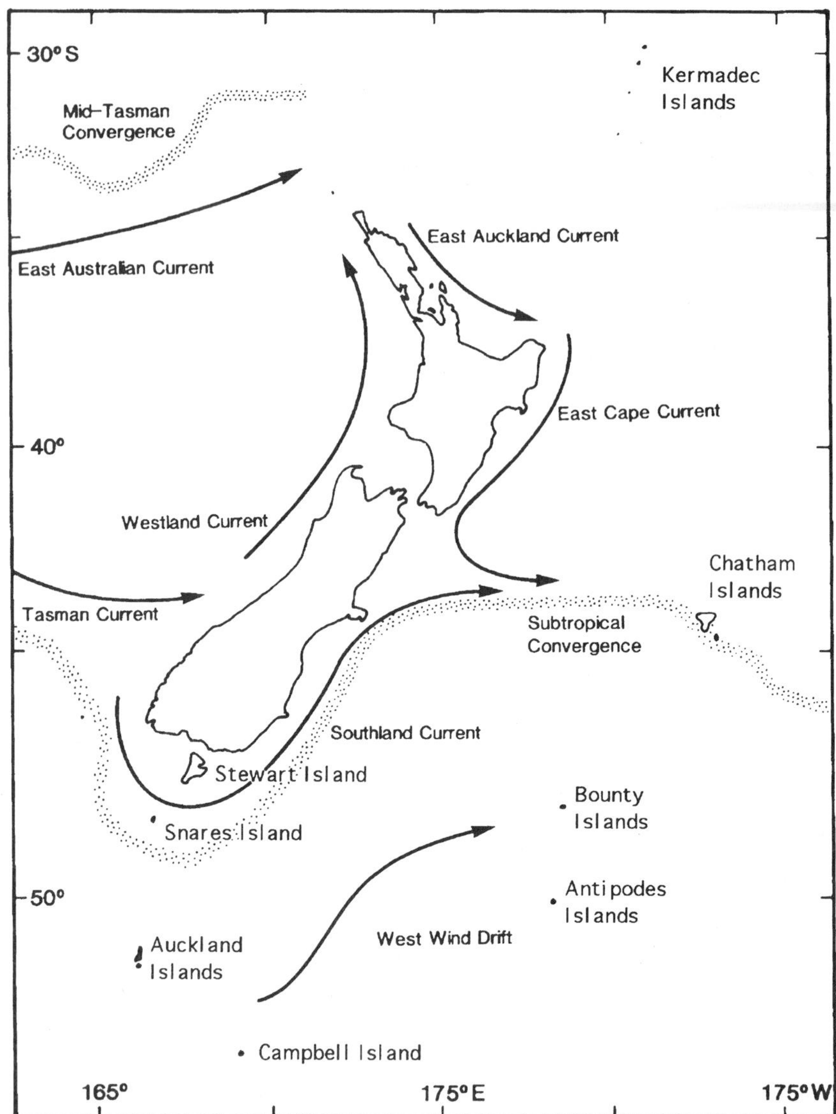


**Figure 3.** Map of New Zealand showing pertinent hydrographic features and offshore islands. Modified from Paulin and Roberts (1992), with permission of C. D. Paulin.

bathed by the East Australian Current, the subantarctic groups lie in the cool West Wind Drift. Both the Kermadecs and the subantarctic islands have depauperate faunas with several rockpool fish species endemic to them. In contrast, the Chatham Islands lie 860 km east of the New Zealand mainland, but due to hydrographic conditions are not as isolated from the mainland as the Kermadecs or the subantarctic islands. Consequently, the Chatham Islands have a more species-rich rockpool fauna with a strong affinity to the New Zealand mainland. No fish species are known to be endemic to the Chatham Islands.

A striking feature of the New Zealand fish fauna is the difference in proportion of endemic species between rockpool and nonrockpool fishes. Of 1008 fish species that occur in

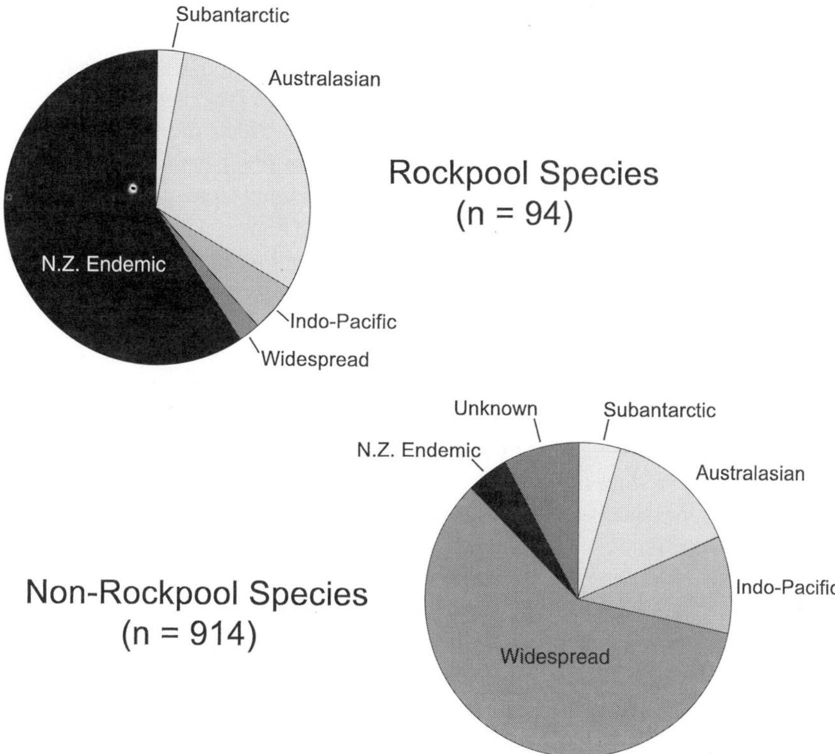


**Figure 4.** Pie diagrams showing the affinities of New Zealand fish species from rockpool and non-rockpool habitats. Modified from Paulin and Roberts (1993), with permission of C. D. Paulin.

the New Zealand Exclusive Economic Zone (Paulin *et al.,* 1989), which includes the North and South Islands and several archipelagoes in the surrounding seas, 94 (9.3% of the total fish fauna) appear in the intertidal zone (Paulin and Roberts, 1993). Overall, 58 species (61.7%) of rockpool fishes are endemic to New Zealand (Figure 4). By contrast, the non-rockpool fauna, which consists of 914 species, features 52 New Zealand endemics (5.7% of subtidal fishes), a percentage almost an order of magnitude lower than that for rockpool fishes. Of 15 intertidal species known from the isolated Kermadec Islands (800 km north of New Zealand), two are Kermadec endemics. The relatively very high levels of endemism among New Zealand rockpool fishes presumably reflects differences between typical intertidal fishes and their nonintertidal counterparts. Intertidal fishes are generally small, negatively buoyant, territorial, and short-lived, traits that reduce vagility and increase the likelihood of speciation (see Rosenblatt, 1963).

## B. Eastern Pacific

This region consists of the west coasts of the Americas. Although we have termed this region the Eastern Pacific, three subregions for which we have some data are sufficiently different in their intertidal fish faunas to warrant brief individual discussion. The Eastern Pacific region as a whole has a far lower diversity than the Indo-West Pacific, with an average of 16 species per site (all in CTNEP, WTSTNEP, and TESEP in Table 1).

## 1. Cold-Temperate Northeastern Pacific (CTNEP)

Briggs (1995) recognizes a broad Eastern Pacific Boreal Region, extending from Point Conception north through Alaska and then southwest through the Aleutian Islands. Briggs subdivided this cold-temperate region into two provinces, with the more northern Aleutian Province extending south to approximately the Queen Charlotte Islands (British Columbia, Canada) and the Oregonian Province continuing south to Point Conception.

Intertidal fish diversity is relatively low throughout this region, with an average reported species richness of 15 species per site in the Oregonian Province (sites BC1–CA13 in Table 1). The fauna is dominated by members of the families Cottidae (23 species) and Stichaeidae (5 species), which commonly represent 60–100% (by number) of fishes collected in the region, with several scorpaenids (15 species), liparids (6 species), embiotocids (6 species), pholids (4 species), clinids (4 species), and a gobiesocid also commonly occurring (Table 2).

A combination of a coastal bight and offshore islands south of Point Conception (southern California, USA) reduce the influence of the cool California Current on coastal waters in that region. There usually exists a substantial gradient of nearshore sea surface temperature across the Point. Consequently, Point Conception separates the warm- and cold-temperate regions of the northeast Pacific coast (Hayden *et al.,* 1984). The coastal and especially the bay-estuarine fish faunas of the Oregonian Province north of the Point are both distinct from their counterparts in the San Diegan Province, but the Point seems to be more of a boundary for southern species than for northern ones (Horn and Allen, 1976, 1978).

## 2. Warm-Temperate and Subtropical Northeastern Pacific (WTSTNEP)

These two regions extend from Point Conception south to Ecuador (Briggs, 1995), including the Galapagos Islands. Species richness is high, with an average of 25 species recorded per site (sites CA14–COS3 in Table 1). The faunas of this region are quite different from those in the cold-temperate northeastern Pacific, lacking cottids and stichaeids altogether. However, at the familial level they are similar to faunas of the Indo-West Pacific region because they are composed predominantly of pomacentrids, labrids, labrisomids, blenniids, and gobiids (Table 2), which have heavy representation in the Indo-West Pacific. We noted no intertidal fish species shared by the WTSTNEP and Indo-West Pacific regions.

## 3. Temperate-Southeastern Pacific (TESEP)

The western coast of South America is strongly affected by the cold Peru Current, which in conjunction with prevalent wind-driven upwelling keeps inshore sea surface temperatures low in the region. We found few unambiguously intertidal fish collections reported from this region. The fauna is relatively depauperate (Table 2). Varas and Ojeda (1990) recorded 11 species, including 2 kyphosids, 2 tripterygiids, 2 blenniids, 1 labrisomid, and 1 gobiesocid. The TESEP shares four families (Blenniidae, Gobiesocidae, Labridae and Tripterygiidae) and no species with the WTSTNEP.

Ocean currents and the large interval of open sea between the Central Pacific islands and the South American landmass form a strong barrier to dispersal of shallow-water fishes that do not have extensive pelagic larval phases (Briggs, 1974; Leis, 1984). Five families not represented in the Central Pacific occurred in the TESEP, whereas the former had 22 families not found in the TESEP. Nineteen species representing 5 families were collected from

Easter Island (Duhart and Ojeda, 1994), making the fauna of the TESEP relatively poorer in terms of families than the Central Pacific. Only the seven most numerically abundant species were presented by Duhart and Ojeda (1994), making comparison between Easter Island (EAS1) and the TESEP difficult (Table 2). Nevertheless, the faunas of these two areas appear very different, with the Central Pacific being dominated numerically by pomacentrids (80% of all individuals captured) and the TESEP by kyphosids (mean = 53%; Varas and Ojeda, 1990; Muñoz and Ojeda, 1997). Only the labrids (which constituted an average of 12% of the fauna of the Southeastern Pacific; Varas and Ojeda, 1990; Muñoz and Ojeda, 1997) were also included in the numerically dominant species at Easter Island (Central Pacific), where they made up only 4% of the individuals present. The TESEP shared only one family, the Blenniidae, with the Hawaiian Islands (Central Pacific) (Gosline, 1965). No species were shared between the Central Pacific and TESEP. As expected, there appears to be little affinity between the TESEP and the Central Pacific intertidal ichthyofaunas, even at the familial level.

## C. Atlantic Ocean

In the Atlantic Ocean as a whole, species diversity is almost as high as in the Indo-West Pacific (24 species per site, Table 2). However, this figure is elevated by the 62 species in Barbados, the only site in the Tropical Western Atlantic for which data are available (Table 2). With Barbados excluded, diversity becomes lower (11 species per site) than in either the Indo-West Pacific or the Eastern Pacific (Table 2).

### 1. Tropical Northwestern Atlantic (TRNWA)

Sixty-two species of intertidal fishes were collected from two rockpools in Barbados (Mahon and Mahon, 1994). Major elements in the fauna include Labridae (10 species), Blenniidae (8 species), Gobiidae (5 species), Holocentridae (4 species), Pomacentridae (3 species), Muraenidae (3 species), and Ophichthidae (3 species) (Table 2). Although the Acanthuridae are only represented by one species, this species is numerically the second most dominant. Because samples exist only from Barbados, the extent of this region cannot be defined.

The Isthmus of Panama presents a virtually insurmountable barrier to the dispersal of marine organisms between the eastern Pacific and western Atlantic (Briggs, 1995), and, in fact, the Tropical Western Atlantic intertidal ichthyofauna shows almost no affinity to the WTSTNEP (Table 2). Weaver (1970) recorded 16 species of intertidal fishes from Costa Rica. The Tropical Western Atlantic fauna is dominated numerically by, in descending order, Labridae, Acanthuridae, and Blenniidae (Mahon and Mahon, 1994), whereas that of the Central Eastern Pacific is dominated by members of the Labrisomidae, Gobiidae, Eleotridae, and Pomacentridae (Weaver, 1970). No species are shared between the two regions.

### 2. Northeastern Atlantic (TENEA)

This region extends from the Azores west to southern England in the north and Spain in the south. The fauna of this region is relatively depauperate, with an average species richness of only 10 species per site (Table 2). Major elements of the fauna include 6 species of blenniids, 4 species each of gobiids and gobiesocids, and 3 species each of gadids and labrids (Gibson, 1972; Jones and Clare, 1977; Arruda, 1979a,b; Ibanez *et al.*, 1986; Patzner *et al.*, 1992). Relationships of this region to the other Atlantic regions are discussed below.

## 3. Northwestern Atlantic (TENWA)

The Northwestern Atlantic includes at least the area between Cape Cod and northern Maine on the east coast of the United States. Species richness in the Northwest Atlantic is moderate, with an average of 18 species per site, and a total of 17 families being represented (Tables 1, 2). Major families include the Gasterosteidae (4 species), Gadidae (3 species), and Cottidae (3 species) (Collette, 1988; Moring, 1990).

The Northwestern and Northeastern Atlantic share eight families. With the exception of the Gadidae, which is represented by three species in each subregion, the dominant families (in terms of species richness) differ between the two regions. Within the eight families that are shared, two species, *Cyclopterus lumpus* and *Pholis gunnellus,* are common to both subregions.

## 4. Temperate South Atlantic (SAFR-A)

Two of five southern African biogeographical provinces (Stephenson, 1939, 1944, 1948; Bolton, 1986; Emanuel *et al.,* 1992) are discussed in Section III.A.1. The remaining three provinces are generally temperate and Atlantic in their influences. Overall, 55 species occurred at 12 sites (SAF1–SAF12 in Table 1) in the temperate Atlantic region, including 21 clinids, 7 sparids, 5 blenniids, 5 gobiids, 3 gobiesocids, and 14 other species of various families.

The Angolan Province (sites SAF1 and SAF2, Table 1) extends from northern Namibia northward. A paucity of northerly data makes the northern boundary of this province indeterminate. The fauna of this province is depauperate, with only six species recorded at each of the two sites. The fauna is dominated by blenniids (five species) and gobiesocids (two species), with one representative each of the Gobiidae, Labrisomidae, and Clinidae (Penrith and Kensley, 1970b; Kensley and Penrith, 1973). Personal observations (K.P.) indicate that juvenile serranids, especially *Epinephalus marginata,* are abundant in intertidal pools in southern Angola, although this species was not recorded by Kensley and Penrith (1973).

The Cold-Temperate West Coast, or Namaqua Province, extends from at least as far north as Lüderitz in Namibia south to the Cape of Good Hope and the adjacent False Bay (sites SAF3–SAF9, Table 1). There exists a large gap between Lüderitz and the Angolan Province boundary where the majority of this coastline is sandy, with very few rocky outcrops where rocky intertidal fishes might occur. Furthermore, this gap has not been closely studied because large portions of it are within high-security diamond mining areas.

The intertidal fish fauna of the Namaqua province is richer in species than that of the Angolan Province, 26 species having been recorded in rock pools in this province (Penrith and Kensley, 1970a; Marsh *et al.,* 1978; Bennett and Griffiths, 1984; Prochazka and Griffiths, 1992; Prochazka, 1996). The fauna is overwhelmingly dominated by members of the Clinidae, in terms of both species richness (17 species) and abundance. Also present are Gobiidae (4 species), Blenniidae (3 species), and four other families represented by one species. There are few partial residents, and intertidal fish samples taken at low tide within this biogeographic province are almost devoid of transient species.

The Cool-Temperate South Coast Province extends from the southern end of the Namaqua Province to Port Alfred (South Africa) in the east (SAF10–SAF12, Table 1). The fauna is much more diverse than that of either of the two west coast provinces, with 61 species from 24 families having been recorded (Christensen and Winterbottom, 1981; Beckley, 1985a,b; Bennett, 1987). The Clinidae remain the most numerically abundant and species-rich family present (13 species), especially in more westerly areas, and true residents constitute the bulk of the assemblages (Bennett, 1987). Toward the east, species richness increases, and Clinidae constitute an increasingly small proportion of both the species richness and abundance. The Sparidae (7 species), Gobiidae (5 species), and Blenniidae (3 species) are more diverse and numerous than in the west, with more abundant partial residents and transient species (Christensen and Winterbottom, 1981; Beckley, 1985a,b).

## IV. Similarity Relationships among Regional Intertidal Ichthyofaunas

To quantify differences in assemblage composition among geographically diverse ichthyofaunas, we used presence/absence data to cluster collection localities. Our major objective was to determine whether intertidal fish assemblage composition varies in a clear and consistent fashion across biogeographical boundaries, as it often does for other groups of organisms. Given that few intertidal fish species have wide distributions (e.g., Chotkowski *et al.,* Chapter 14, this volume; Horn, Chapter 16, this volume), a worldwide analysis of biogeographic relationships that employed presence–absence of species would not be informative. Either higher taxonomic categories (for instance, families) must be used as variables, or the geographic scope of any investigation must be limited to manageably small regions. New Zealand, the northeastern Pacific, and southern Africa are three areas of the world that have received the most attention in terms of surveys of intertidal fishes. Biogeographic relationships among populations within New Zealand and surrounding islands have been extensively studied (Paulin and Roberts, 1993). The northeastern Pacific is a tempting subject because more samples are available from it than from any other single region. However, the distribution of collections in that region is not uniform, and there are few strong biogeographical boundaries. Point Conception is one; see Horn and Allen (1978) for discussion of biogeographical subregions of the southern Oregonian Province. Southern Africa, by contrast, provides a sequence of distinctive biogeographical regions defined by nonfish biotas (e.g., Emanuel *et al.,* 1992), and sufficient fish sampling has been done to measure the congruence of intertidal fish assemblage patterns to them. Therefore, we have restricted our primary focus to patterns of distribution in the southern African fauna.

We used two different approaches to estimate biogeographic relationships among 22 locations reported from the southern African region. In an initial assessment of overall similarity, a matrix containing scores of presence (1) or absence (0) of particular species was prepared from the lists from 22 southern African locations. A similarity coefficient based on Euclidean distance was calculated for all pairwise comparisons of locations. These coefficients were clustered using Minitab software (Minitab, Inc., 1994) by an agglomerative hierarchical clustering of variables routine using an average linkage algorithm. Biogeographic relationships based on overall similarity are depicted in Figure 5. Similarity among geographic samples ranged from about 65 to 90%. Most, but not all, geographic neighbors clustered together.

A second analysis of relationships employed a maximum parsimony approach. Effective data reduction was accomplished by removing species from the aforementioned matrix that

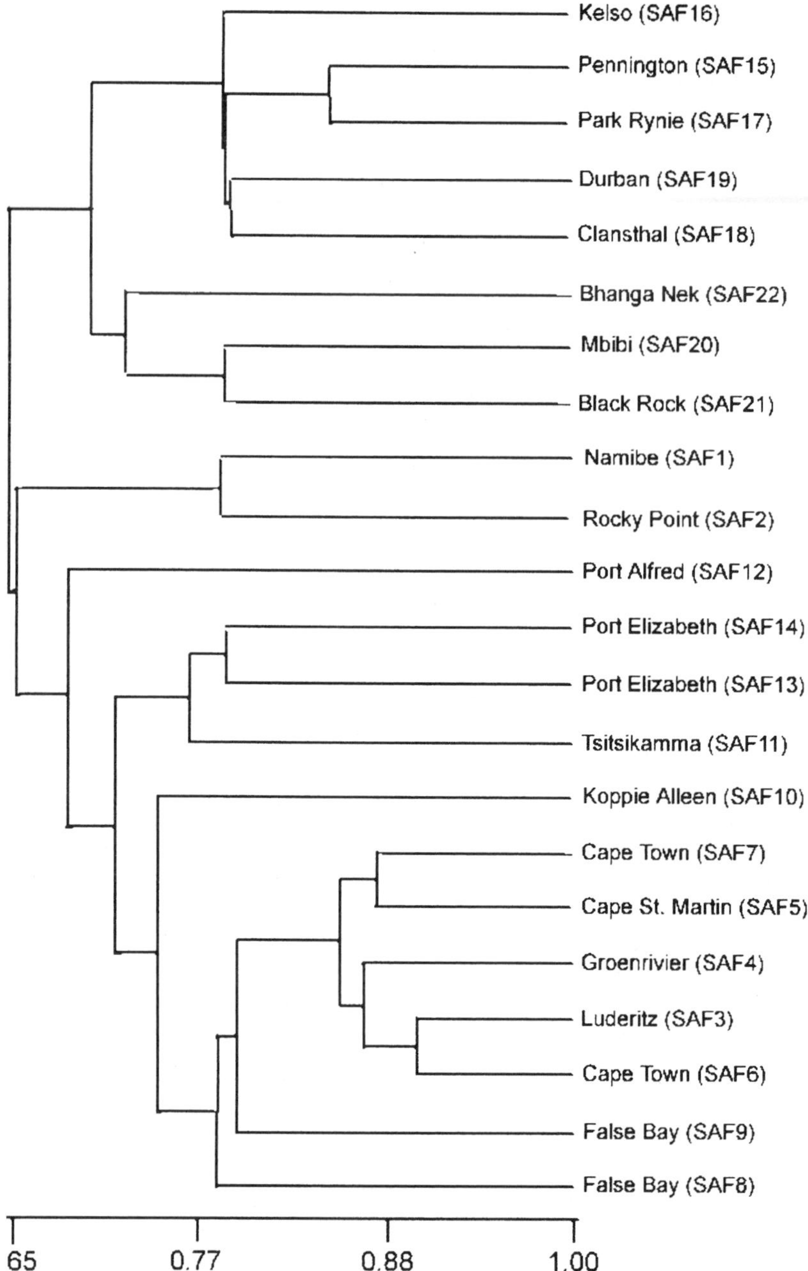


**Figure 5.** Overall similarity relationships among southern African intertidal fish assemblages reported from 22 locations. Geographic units are identified by name and site identifier. See Table 1 for citation of studies referenced.

were reported from only one location. This procedure is analogous to removing uninformative autapomorphic characters in a phylogenetic analysis. The reduced data matrix was subjected to a parsimony analysis using the "Phylogenetic Analysis Using Parsimony" (PAUP) program of Swofford (1985). Analytical options used included global branch swapping, multiple parsimony (to find all equally parsimonious trees), and Farris optimization (Farris, 1970). Ten minimum length unrooted Wagner trees were found; each was 178 steps in length with a consistency index (CI) of 0.427. A strict consensus tree constructed from these 10 topologies showed a polytomy involving only four locations (sites SAF3, SAF4, SAF5, and SAF6) that linked within a single cluster. One of the 10 equally parsimonious topologies is shown in Figure 6.

With only three exceptions, discussed below, the 22 locations can be arranged in four geographically consistent groups along the southern African coast (Figure 7). These four groups correspond to the Subtropical East Coast Province, the East Coast Warm-Temperate Province, the Cool-Temperate South Coast Province, and the Namaqua Province of Ekman (1953). Two exceptions to this group membership involve geographically distal locations from the Angolan Province: Namibe (site SAF1) and Rocky Point (site SAF2). It is interesting to note that these two geographic nearest neighbors were clustered together in terms of overall similarity and together clustered apart from the other, more distant, locations (Figure 5), yet this expected relationship was not recovered in the parsimony analysis (Figure 6). In the latter, the affinities of the Angolan Province locations were between the East Coast Provinces and the South Coast Province. The remaining exception to geographically consistent group membership involved the two samples from False Bay, one of which (site SAF9) had affinities with locations to the north (Namaqua Province) whereas the other (site SAF8) had affinities with locations to the east (Cool-Temperate South Coast Province). The geographic significance of such a small-scale, or inconsistent, partition cannot be assessed properly until additional studies are conducted.

Together, the two analyses provide an interesting picture of geographic relationships among southern African locations. The use of species-level data for comparison was successful in recovering relationships on this geographic scale. The fact that the parsimony analysis worked as well as it did is consistent with distribution patterns that were influenced more by vicariant events than by widespread dispersal, although the data do not support such a conclusion by themselves.

To test the feasibility of using higher taxonomic categories in a broader geographic analysis, a matrix was prepared scoring families of intertidal fishes as present or absent at nine widely separated locations including: Easter Island (EAS1, 27° S), Taiwan (TAI1, 23° N), Barbados (A3, 13° N), California (CA12, 35° N), New Zealand (NZ8, 41° S), Britain (A5, 54° N), Namibia (SAF3, 27° S), Chile (CHL1, 33° S), and Seychelles (SEY1, 5° S) (Table 1). Effective data reduction was accomplished by removing from the matrix families that were represented at only one of the locations. A parsimony analysis was conducted using the PAUP program (Swofford, 1985). With only nine locations, an exhaustive search of all possible tree topologies was performed. Fourteen minimum length unrooted Wagner trees were generated using this procedure; each was 57 steps in length with a CI of 0.579. Unfortunately, a strict consensus tree of these 14 trees provided little information. The Taiwan, Seychelles, and Barbados locations formed a group in all 14 trees; however, there is no resolution among the remaining six sites. The only resolution gained via a majority-rule consensus tree is the pairing of California and Chile.

Despite the higher CI compared to the southern Africa example, the low resolution of our second example suggests little promise in an approach that uses a higher taxonomic category of intertidal fishes to treat a very broad geographic area. It is interesting to note, however, that the three geographically disparate locations that did form a group in the parsimony analysis (Taiwan, Barbados, and Seychelles) are all tropical locations. This result was certainly due to the presence at those locations of circumtropical families

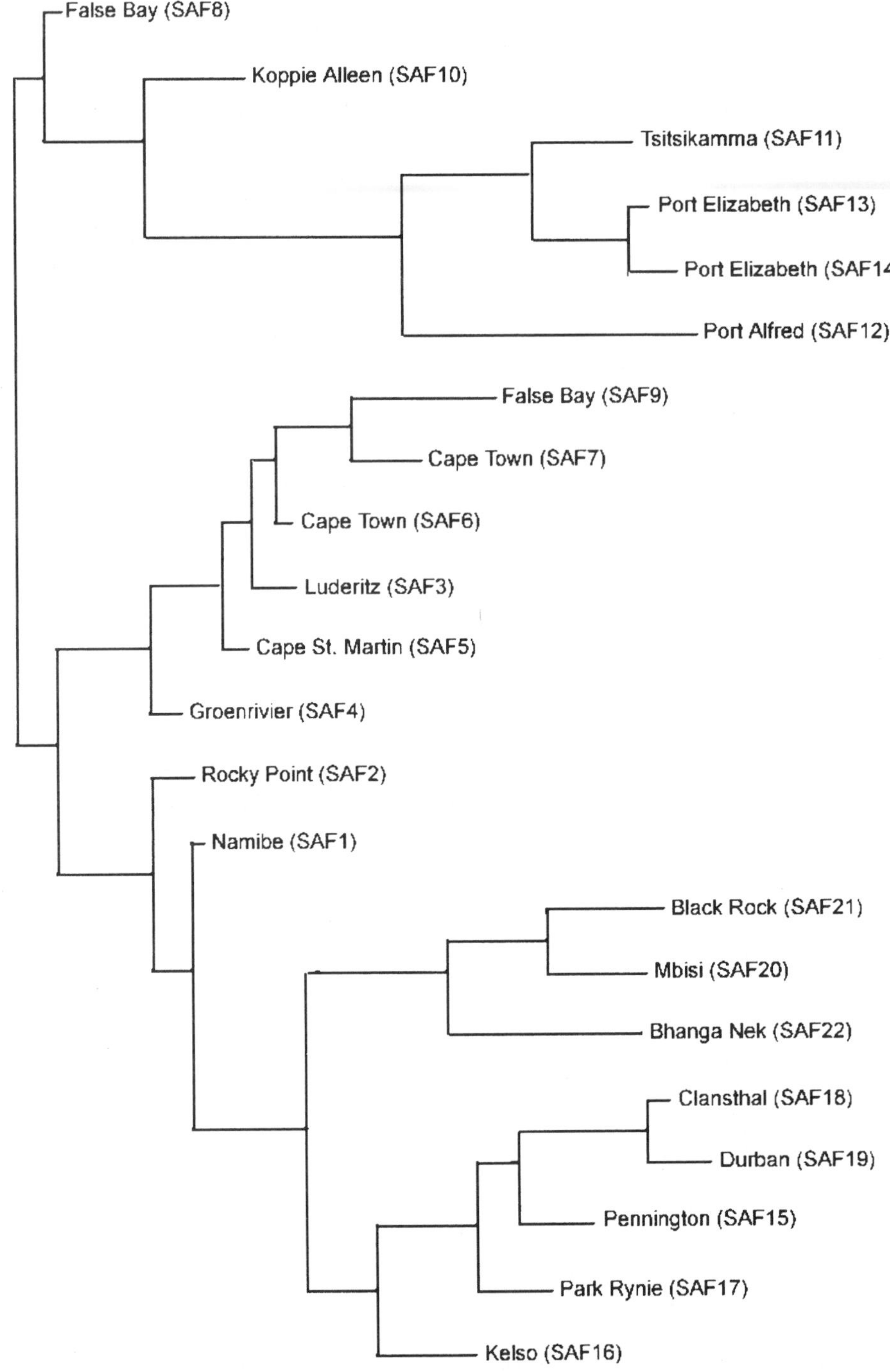


**Figure 6.** One of 10 equally parsimonious unrooted Wagner trees that depict relationships among southern African intertidal fish assemblages reported from the same 22 locations represented in Figure 5. This unrooted tree is drawn with a root at the midpoint of the longest branch length (patristic distance) in the network. Branch lengths reflect the amount of species presence/absence differences. Geographic units are identified by both name and site identifier.

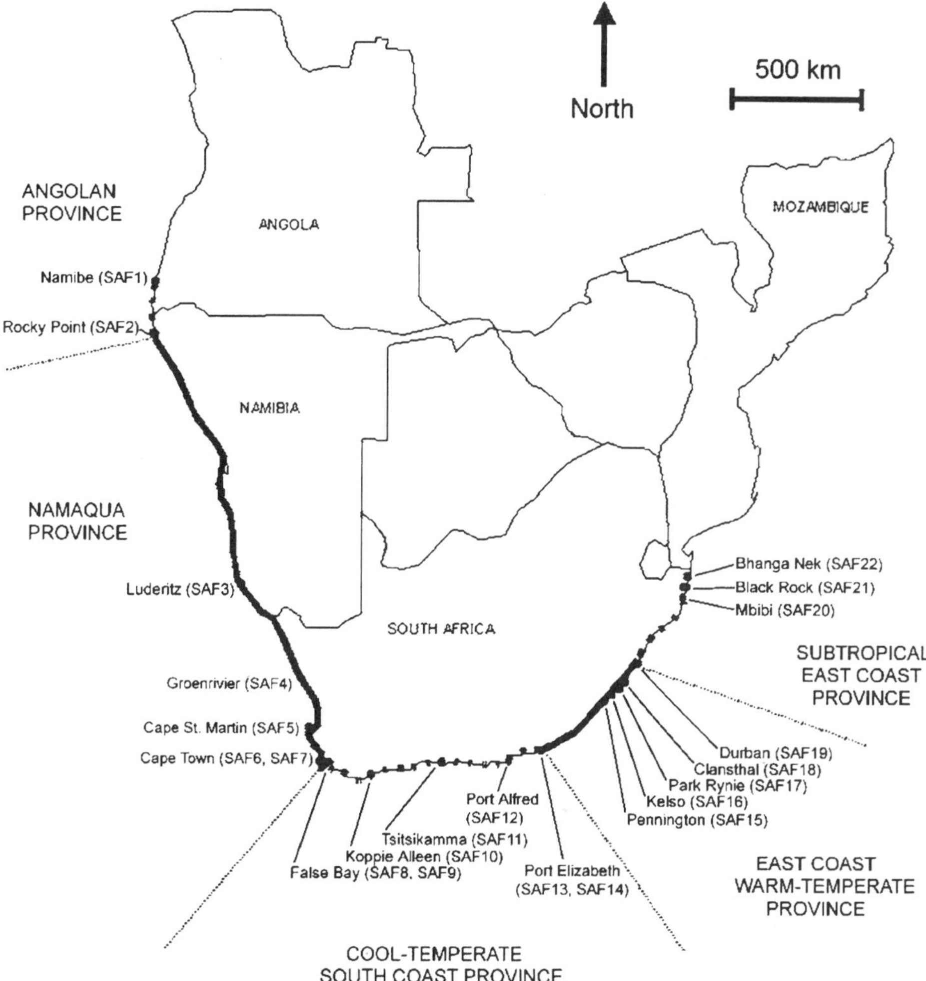


**Figure 7.** Distribution of intertidal study sites represented in Figures 5 and 6 showing their relationships in the southern African region. Geographic units are identified by both the name of the location and the site identifier. Biogeographical provinces correspond to groupings in both analyses, so additional group identifiers have been omitted.

(Muraenidae, Labridae, and to a lesser extent Holocentridae) that would be likely to distinguish tropical samples in any analysis.

## V. Latitudinal Patterns of Species Richness

The coral reefs typical of many tropical regions are more spatially heterogeneous than rock substrata in the temperate latitudes (Ebeling and Hixon, 1991). Because fish abundance and diversity may be proportionate to bottom relief (Ebeling *et al.,* 1980), additional spatial heterogeneity and other features provided by coral reefs may support more species-rich intertidal faunas. Are there more intertidal fishes in the tropics than in the temperate latitudes? Although we have inadequate data to propose an answer (see below), we confidently

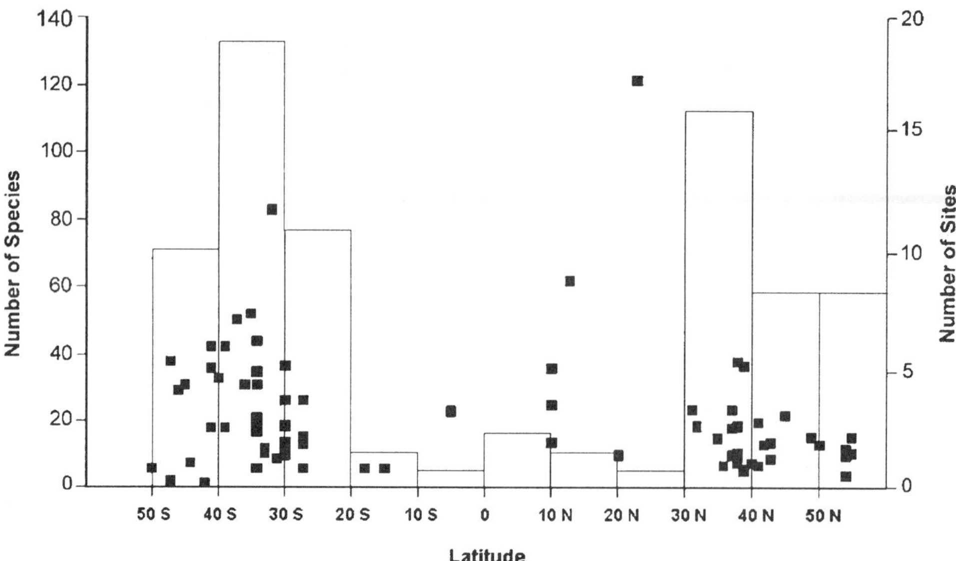


**Figure 8.** Plot of species richness of intertidal fish collections vs. latitude of collecting site. $N = 85$. The superimposed histogram represents the number of collection sites in each 10° band of latitude.

predict that more species are present in the tropical intertidal zone, if only because there are more nearshore fishes in the tropics than elsewhere, especially over coral reefs (Mead, 1970; Briggs, 1974; Ehrlich, 1975). However, this is not a particularly satisfying answer if we really want to know whether there are more *resident* intertidal fishes. Unfortunately, because authors generally do not identify resident fishes at each site, we necessarily weakened the question: Do fish collections from intertidal sites in the tropics contain more species than those from the temperate latitudes?

A plot of collection species richness versus latitude (Figure 8) reveals no clear trend, but does highlight an unfortunate lack of tropical data. Moreover, whereas two of the low-latitude sites, the Hawaiian Islands (20° N) and the Seychelles (5° S), are isolated archipelagoes far from sources of potential colonists, most of the higher latitude sites were on continental margins or in the relatively species-rich New Zealand region. The absence of adequate data from the tropical latitudes obviated any decisive statistical analysis bearing on questions involving the tropics. However, a regression analysis pooling observations from the Northern and Southern Hemispheres revealed that collection species richness (log-transformed to normalize the residuals) does decline significantly with increasing latitude (ln (No. species) = 3.39 − 0.017 × Latitude (°), adjusted $R^2 = 4.4\%$, $Pr < 0.05$). However, the conclusion that the slope is less than zero and the estimate of its modulus are strongly influenced by only four low-latitude observations: three in Costa Rica (COS1–COS3, about 10° N) and one at the Seychelles Islands (SEY1, about 5° S). More definitive analyses will have to await future work.

## VI. Conclusions

We have seen that there are considerable differences in species richness at the family level in intertidal ichthyofaunas in various parts of the world. Distribution patterns are primarily delineated by the major ocean masses and refined by local currents. The similarity relationships among the southern African assemblages examined in Section IV are closely congruent to groupings predicted by regional biogeography of shore seaweeds and invertebrates. The close match of assemblage groupings to biogeographical regions indicates that intertidal fish assemblage composition strongly depends on local conditions, as is generally the case for shore biotas in southern Africa.

We have also revisited work by Paulin and Roberts (1993) revealing much higher levels of endemism among New Zealand intertidal fishes than among subtidal fishes from that region. It appears that intertidal ichthyofaunas are generally more regionally distinct than their subtidal counterparts; in particular, we note that there are very few widely distributed intertidal fish species (Chotkowski *et al.,* Chapter 14, this volume). Most resident species are small, short-lived, and closely tied to shore habitats, so dispersal by adult migration across the open sea is unlikely. Furthermore, most intertidal fish deposit benthic eggs and do not have an extended planktonic larval phase in the pelagic zone (DeMartini, Chapter 8, this volume), so dispersal by early life history stages is also much less likely than among subtidal fishes.

Additional intertidal survey work will make it possible to move beyond a descriptive to an analytical approach to intertidal fish biogeography. Furthermore, we hope that this broad overview will stimulate detailed studies of the regional biogeography of these organisms, coupled with careful examinations of phyletic factors contributing to distributional patterns. In the interim, deficiencies in our knowledge of rocky intertidal fishes and their distributions remain clearly apparent. The most obvious of these is unequal distribution of published sampling studies. By far the greatest effort has been directed at temperate systems; New Zealand, southern Africa, and the northeastern Pacific have been most extensively investigated. By contrast, the specifically *intertidal* ichthyofauna of most tropical regions is not well known, preventing for now many interesting temperate/tropical comparisons.

## Acknowledgments

We thank Lynnath Beckley, Colin Buxton, and Rick Winterbottom for allowing us to use their unpublished data. Michael Horn, Karen Martin, and Chris Paulin provided many useful suggestions for improving the manuscript.

## References

Arruda, L. M. (1979a). On the study of a sample of fish captured in the tidal range at Azores. *Bolm Soc. Port. Ciênc. Nat.* **19,** 5–36.

Arruda, L. M. (1979b). Specific composition and relative abundance of intertidal fish at two places on the Portuguese coast (Sesimbra and Magoito, 1977–78). *Arquiv. Mus. Bocage 2ª Ser* **6,** 325–342.

Beckley, L. E. (1985a). The fish community of East Cape tidal pools and an assessment of the nursery function of this habitat. *S. Afr. J. Zool.* **20,** 21–27.

Beckley, L. E. (1985b). Tide-pool fishes: Recolonisation after experimental elimination. *J. Exp. Mar. Biol. Ecol.* **85,** 287–295.

Bennett, B. A. (1987). The rock-pool fish community of Koppie Alleen and an assessment of the importance of Cape rock-pools as nurseries for juvenile fish. *S. Afr. J. Zool.* **22**(1), 25–32.

Bennett, B. A., and Griffiths, C. L. (1984). Factors affecting the distribution, abundance and diversity of rock-pool fishes on the Cape Peninsula, South Africa. *S. Afr. J. Zool.* **19,** 97–104.

Bolton, J. J. (1986). Marine phytogeography of the Benguela upwelling region on the west coast of southern Africa: A temperature dependent approach. *Bot. Mar.* **29**(3), 251–256.

Briggs, J. C. (1974). "Marine Zoogeography." McGraw-Hill, New York.

Briggs, J. C. (1995). "Global Biogeography." Elsevier, Amsterdam.

Burger, L. F. (1990). "The Distributional Patterns and Community Structure of the Tsitsikamma Rocky Littoral Ichthyofauna." Unpublished MSc thesis, Rhodes University, Grahamstown, South Africa.

Christensen, M. S., and Winterbottom, R. (1981). A correction factor for, and its application to, visual census of littoral fish. *S. Afr. J. Zool.* **16,** 73–79.

Collette, B. B. (1988). Resilience of the fish assemblage in New England tidepools. *U.S. Fish. Bull.* **84,** 200–204.

Duhart, M., and Ojeda, F. P. (1994). Ichthyological characterisation of intertidal pools, and trophic analysis of herbivore subtidal fishes of Easter Island. *Med. Amb.* **12**(1), 32–40.

Ebeling, A. W., and Hixon, M. A. (1991). Tropical and temperate reef fishes: Comparison of community structures. *In* "The Ecology of Fishes on Coral Reefs" (P. F. Sale, Ed.), pp. 509–563. Academic Press, San Diego.

Ebeling, A. W., Larson, R. J., and Alevizon, W. S. (1980). Habitat groups and island-mainland distribution of kelp-bed fishes off Santa Barbara, California. *In* "The California Islands: Proceedings of a Multidisciplinary Symposium" (D. M. Power, Ed.), pp. 403–431. Santa Barbara Museum of Natural History, Santa Barbara, CA.

Ehrlich, P. R. (1975). The population biology of coral reef fishes. *Annu. Rev. Ecol. Syst.* **6,** 211–247.

Ekman, S. (1953). "Zoogeography of the Sea." Sidgwick and Jackson, London.

Emanuel, B. P., Bustamante, R. H., Branch, G. M., Eekhout, S., and Odendaal, F. J. (1992). A zoogeographic and functional approach to the selection of marine reserves on the west coast of South Africa. *S. Afr. J. Mar. Sci.* **12,** 341–354.

Farris, J. S. (1970). Methods for computing Wagner trees. *Syst. Zool.* **19,** 83–92.

Gibson, R. N. (1972). The vertical distribution and feeding relationships of intertidal fish on the Atlantic coast of France. *J. Anim. Ecol.* **41,** 189–207.

Gibson, R. N. (1982). Recent studies on the biology of intertidal fishes. *Oceanogr. Mar. Biol. Annu. Rev.* **20,** 363–414.

Gosline, W. A. (1965). Vertical zonation of inshore fishes in the upper water layers of the Hawaiian Islands. *Ecology* **46**(6), 823–831.

Hayden, B. P., Ray, G. C., and Dolan, R. (1984). Classification of coastal and marine environments. *Environ. Conserv.* **2,** 199–207.

Horn, M. H., and Allen, L. G. (1976). Numbers of species and faunal resemblance of marine fishes in California bays and estuaries. *Bull. S. Calif. Acad. Sci.* **75,** 159–170.

Horn, M. H., and Allen, L. G. (1978). A distributional analysis of California coastal marine fishes. *J. Biogeogr.* **5,** 23–42.

Ibanez, M., Miguel, I., and Eizmendi, A. (1986). Ichthyofauna in tide-pools. I. Methodology and preliminary results. *Lurralde* **9,** 159–164.

Jones, D., and Clare, J. (1977). Annual and long-term fluctuations in the abundance of fish species inhabiting an intertidal mussel bed in Morecambe Bay, Lancashire. *Zool. J. Linn. Soc.* **60,** 117–172.

Kensley, B. F., and Penrith, M-L. (1973). The constitution of the intertidal fauna of rocky shores of Moçamedes, southern Angola. *Cimbebasia* (A) **2**(9), 113–123.

Lardner, R., Ivanstoff, W., and Crowley, L. E. L. M. (1993). Recolonisation by fishes of a rocky intertidal pool following repeated defaunation. *Austr. Zool.* **29**(1–2), 85–92.

Lee, S-C. (1980). Intertidal fishes of a rocky pool of the Sanhsientai, eastern Taiwan. *Bull. Inst. Zool., Acad. Sin.* **19**(1), 19–26.

Leis, J. M. (1984). Larval fish dispersal and the East Pacific Barrier. *Oceanogr. Trop.* **19,** 181–192.

Mahon, R., and Mahon, S. D. (1994). Structure and resilience of a tidepool fish assemblage at Barbados. *Environ. Biol. Fish.* **41,** 171–190.

Marsh, B., Crowe, T. M., and Siegfried, W. W. (1978). Species richness and abundance of clinid fish (Teleostei: Clinidae) in intertidal rock pools. *Zool. Afr.* **13**(2), 283–291.

Mead, G. W. (1970). A history of South Pacific Fishes. *In* "Scientific Exploration of the South Pacific" (W. S. Wooster, Ed.), pp. 236–251. National Academy of Science, Washington, D.C.

Minitab Release 10.2. (1994). Minitab, Inc., State College, PA.

Moring, J. R. (1990). Seasonal absence of fishes in tidepools of a boreal environment (Maine, USA). *Hydrobiologia* **194,** 163–168.

Muñoz, A. A., and Ojeda, F. P. (1997). Feeding guild structure of a rocky intertidalfish assemblage in central Chile. *Environ. Biol. Fish.* **49**(4), 471–479.

Myers, A. A., and Giller, P. S. (1988). "Analytical Biogeography." Chapman and Hall, London.

Patzner, R. A., Santos, R. S., Ré, P., and Nash, R. D. M. (1992). Littoral fishes of the Azores: An annotated checklist of fishes observed during the "Expedition Azores 1989." *Arquipélago. Life Earth Sci.* **10,** 101–111.

Paulin, C. D., and Roberts, C. D. (1992). "The Rockpool Fishes of New Zealand, Te ika aaria o Aotearoa." Museum of New Zealand, Te Papa Tongarewa, Wellington, NZ.

Paulin, C., and Roberts, C. D. (1993). Biogeography of New Zealand rockpool fishes. *In* "Proceedings of the Second International Temperate Reef Symposium, 7–10 January 1992, Auckland, New Zealand" (C. N. Battershill *et al.*, Eds.). National Institute of Water and Atmospheric Research Ltd.

Paulin, C. D., Stewart, A. L., and Roberts, C. D. (1989). "New Zealand Fish, a Complete Guide." *Nat. Mus. N.Z. Misc. Ser.* **19.**

Penrith, M-L., and Kensley, B. F. (1970a). The constitution of the intertidal fauna of rocky shores of South West Africa. Part I. Lüderitzbucht. *Cimbebasia* (A) **1**(9), 191–237.

Penrith, M-L., and Kensley, B. (1970b). The constitution of the fauna of rocky intertidal shores of South West Africa. Part II. Rocky Point. *Cimbebasia* (A) **1**(10), 243–265.

Prochazka, K. (1996). Seasonal patterns in a temperate intertidal fish community on the west coast of South Africa. *Environ. Biol. Fish.* **45,** 133–140.

Prochazka, K., and Griffiths, C. L. (1992). The intertidal fish fauna of the west coast of South Africa—Species, community and biogeographic patterns. *S. Afr. J. Zool.* **27**(3), 115–120.

Rosenblatt, R. H. (1963). Some aspects of speciation in marine shore fishes. *In* "Speciation in the Sea" (J. P. Harding and N. Tebble, Eds.), pp. 171–180. Systematics Association Publication No. 5.

Stephenson, T. A. (1939). The constitution of the intertidal fauna and flora of South Africa, Part 1. *J. Linn. Soc. (Zool.)* **40**(1), 487–536.

Stephenson, T. A. (1944). The constitution of the intertidal fauna and flora of South Africa, Part 2. *Ann. Nat. Mus.* **10**(3), 261–358.

Stephenson, T. A. (1948). The constitution of the intertidal fauna and flora of South Africa, Part 3. *Ann. Nat. Mus.* **11**(2), 207–324.

Swofford, D. L. (1985). "Phylogenetic Analysis Using Parsimony, Version 2.4."

Thomson, D. A., and Lehner, C. E. (1976). Resilience of a rocky intertidal fish community in a physically unstable environment. *J. Exp. Mar. Biol. Ecol.* **22,** 1–29.

Varas, E., and Ojeda, F. P. (1990). Intertidal fish assemblages of the central Chilean coast: Diversity, abundance and trophic patters. *Rev. Biol. Mar.* **25**(2), 59–70.

Weaver, P. L. (1970). Species diversity and ecology of tidepool fishes in three Pacific coastal areas of Costa Rica. *Rev. Biol. Trop.* **17**(2), 165–185.

# 16

# Convergent Evolution and Community Convergence: Research Potential Using Intertidal Fishes

Michael H. Horn

*Department of Biological Science, California State University, Fullerton, California*

## I. Introduction

The purpose of this chapter is to demonstrate that intertidal fish faunas offer compelling and unexploited opportunities for the study of convergent evolution and community convergence. The only way that such studies can be properly conducted, however, is through phylogenetic analyses based on well-resolved phylogenies of the intertidal species and their closest relatives from other habitats. Phylogenies that would discern sister relationships between intertidal fish species and their relatives in nonintertidal habitats are largely lacking. In this chapter structural and functional traits that could serve as the basis for an analysis of convergent evolution are identified and elements that should be part of an appropriate research design are emphasized. The overarching importance of pursuing such a research program is that rigorous demonstration of convergent evolution has the power to address how fishes may or may not evolve, through adaptation, to increase their fitness in the intertidal environment.

Rocky shores supporting intertidal fish communities are commonly separated by stretches of sandy beaches, by latitudinal temperature gradients, or by ocean basins. Varying degrees of genetic isolation have been found within and among fish species including intertidal forms along coastlines and between mainland and island populations (e.g., Haldorson, 1980; Waples and Rosenblatt, 1987; Stepien and Rosenblatt, 1991; Yoshiyama and Sassaman, 1993), indicating different amounts of gene flow between populations. Intertidal fish communities are therefore isolated to varying degrees from one other (see Gibson and Yoshiyama, Chapter 13, and Chotkowski *et al.,* Chapter 14, this volume) and undergo development and evolution in some degree of isolation depending on the amount of gene flow among them. Intertidal fish faunas tend to be dominated by relatively few families, but



*Intertidal Fishes: Life in Two Worlds*

these families are represented in different geographic regions by different genera and other subfamilial groups. Resident intertidal fishes are generally recognized as small and expressing relatively few body plans. These morphological restrictions apparently imposed on them by the rocky intertidal habitat coupled with the low degree of relatedness of taxa from different sites may provide the conditions for convergent evolution to occur among members of intertidal fish faunas.

In the broad sense, convergent evolution refers to the acquisition of similar traits in unrelated organisms as driven by the selection pressures of similar environments separated in space or time. In the short term of convergence, ecologists have focused on the development of community structure under similar but spatially separated habitats and, in a deterministic framework, have proposed that such communities will converge to a common structure. This emphasis has been criticized because of challenges to the equilibrium paradigm and because communities themselves do not evolve but rather change as the products of ongoing evolutionary processes within constituent members of the community. The role of history has become recognized as a strong factor affecting community structure, leading to an interest in studying "replicate" rather than merely similar communities in different parts of the world (Samuels and Drake, 1997).

In the longer view of convergence linked inextricably to this role of history, evolutionary biologists have invoked a comparative approach using phylogenetic analysis of the different lineages that have evolved in similar but spatially separate environments to test for convergent evolution or convergent adaptation (Brooks and McLennan, 1991; Pagel, 1994; Larson and Losos, 1996). The outcomes of the interaction between phylogenetic and environmental information provide an adaptational framework for addressing convergent evolution (Brooks and McLennan, 1991). Similar, homologous, traits in either similar or dissimilar environments do not support the hypothesis of convergence because of phylogenetic constraints, nor do the appearance of similar but homoplastic traits in dissimilar environments because the convergence is not a result of environmental pressures. However, similar, homoplastic, traits occurring in similar environments do provide support for convergent adaptation. Environmentally driven change can also be hypothesized in cases where nonsimilar traits in a homologous transformation series occur in different environments; this pattern represents divergent adaptation.

The purpose of this chapter, again, is to present information pertinent to further research on convergent evolution in rocky intertidal fishes. The treatment is necessarily introductory and largely descriptive because too few sufficiently resolved phylogenies for fish groups, including intertidal fish species, are available to make rigorous tests of a convergent evolution hypothesis. Attention is directed toward both convergent adaptation in the long-term, evolutionary sense and community convergence in perhaps the shorter term, ecological sense. Section II focuses on taxonomic and body morphology comparisons of six intertidal fish communities widely separated from one another in the temperate zone of the Northern and Southern hemispheres (Figure 1). Section III concentrates on the prospects for convergent adaptation in intertidal fishes based on these body morphology (structural) traits in the six communities and on three functional traits recognized as widely occurring among resident species: desiccation tolerance, air-breathing, and parental care. Section IV discusses some of the ecological attributes of the six fish communities within the context of the community convergence concept. The intent in each section of the chapter is to stimulate further, quantitative, studies of convergence among intertidal fishes and to emphasize the importance of convergence in claiming that certain traits are adaptations to the intertidal environment.

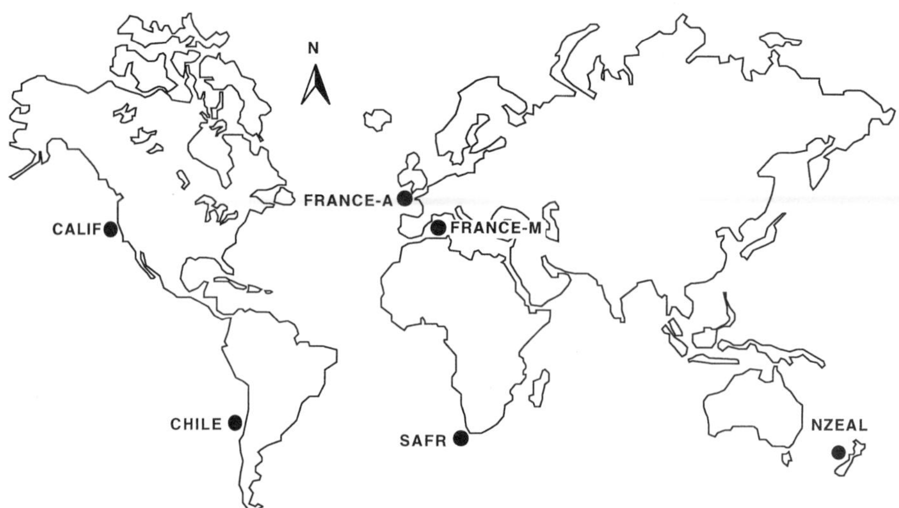


**Figure 1.** Map showing the location of six studies of temperate-zone intertidal fish communities (see Table 1). CALIF, California (Yoshiyama, 1981); CHILE, Chile (Muñoz and Ojeda, 1997); FRANCE-A, France, Atlantic (Gibson, 1972); FRANCE-M, France, Mediterranean (Gibson, 1968); NZEAL, New Zealand (Willis and Roberts, 1996); SAFR, South Africa (Prochazka, 1996).

# II. Description of Six Intertidal Fish Communities

## A. Taxonomic Composition

The fishes listed in Table 1 are the species encountered at the six widely separated study sites (Figure 1). The lists are the results of roughly similar collection methods (i.e., rotenone or quinaldine; see Gibson, Chapter 2, this volume) although the time frame and capture efforts vary considerably among the studies. As such, they represent, within limits, the main groups of intertidal fishes occupying temperate-zone habitats. Yet, because the six communities overlap little in taxonomic composition, they provide logical material for an examination of convergent adaptation and community convergence. Fourteen of the 22 families belong to the Perciformes, the largest teleostean order, and 5 of these perciform families (Clinidae, Tripterygiidae, Blenniidae, Gobiesocidae, and Gobiidae) plus the Cottidae (Scorpaeniformes) account for 69% of the species and 84% of the individuals captured at the six sites combined (Table 2). Members of the Clinidae alone comprise 18% of the species and 38% of the individuals collected, primarily because of their extraordinary species richness at the South African location. The Gobiesocidae is the only one of these families represented at all six sites, and 15% of the species captured at these sites are members of this family.

The six communities are almost completely different in taxonomic composition at the species level, only slightly more similar at the generic level, and considerably more so at the family level. Analysis of similarity based on presence/absence data using the Jaccard Index (Krebs, 1989) shows that the mean similarity value is 0.01 (±0.05 SD) at the species level with only two sites (FRANCE-A and FRANCE-M) having any species (5) in common; 0.03 (±0.06) at the generic level with the FRANCE-A and FRANCE-M sites having

**Table 1.** Species Lists, Size Ranges, and Abundance Rankings of Fishes Occupying the Rocky Intertidal Habitat at Six Temperate-Zone Sites

| Family<br>*Species* | Size range<br>(mm SL) | Abundance rank |
|---|---|---|
| 1. California (CALIF), 37° N (Yoshiyama, 1981) | | |
| Scorpaenidae | | |
| *Sebastes* sp.* | | 8 |
| Hexagrammidae | | |
| *Hexagrammos decagrammus** | 530 | 15 |
| *Hexagrammos lagocephalus* | 610 | 20 |
| Cottidae | | |
| *Artedius lateralis**[FND] | 30–127 | 5 |
| *Clinocottus analis* | 30–>60 | 2 |
| *Clinocottus globiceps* | 30–120 | 9 |
| *Clinocottus recalvus* | 130 | 19 |
| *Hemilepidotus spinosus* | 290 | 21 |
| *Oligocottus maculosus* | 30–90 | 14 |
| *Oligocottus rimensis* | 64 | 22 |
| *Oligocottus rubellio* | 100 | 23 |
| *Oligocottus snyderi**[TLP] | 30–>60 | 1 |
| *Scorpaenichthys marmoratus* | 30–223 | 10 |
| Embiotocidae | | |
| *Micrometrus aurora** | 180 | 17/18 |
| Stichaeidae | | |
| *Anoplarchus purpurescens* | 30–>90 | 3 |
| *Cebidichthys violaceus** | 30–>90 | 4 |
| *Xiphister atropurpureus* | 30–>180 | 6 |
| *Xiphister mucosus* | 580 | 16 |
| Pholidae | | |
| *Apodichthys flavidus* | 30–>180 | 12/13 |
| *Apodichthys (= Xererpes) fucorum**[,a] | 30–180 | 7 |
| Clinidae | | |
| *Gibbonsia metzi** | 30–>180 | 11 |
| *Gibbonsia montereyensis* | 30–180 | 12/13 |
| Gobiesocidae | | |
| *Gobiesox maeandricus** | 160 | 17/18 |
| 2. Chile (CHILE), 32–33° S (Muñoz and Ojeda, 1997) | | |
| Girellidae | | |
| *Girella laevifrons** | 90 (39–153) | 3 |
| *Graus nigra* | 94 (24–203) | 4 |
| Bovichthyidae | | |
| *Bovichtus chilensis** | 92 (45–188) | 7 |
| Tripterygiidae | | |
| *Tripterygion chilensis** | 54 (25–78) | 2 |
| *Tripterygion cunninghami* | 48 (33–59) | 8 |
| Labrisomidae | | |
| *Auchenionchus microcirrhus** | 96 (32–222) | 5 |
| *Auchenionchus variolosus* | 97 (57–180) | 9 |
| *Calliclinus geniguttatus* | 64 (28–122) | 6 |

*continues*

**Table 1.** *Continued*

| Family<br>*Species* | Size range<br>(mm SL) | Abundance rank | |
|---|---|---|---|
| Clinidae | | | |
| *Myxodes viridis** | 79 (25–135) | 10 | |
| Blenniidae | | | |
| *Hypsoblennius sordidus* | 68 (33–116) | 11/12 | |
| *Scartichthys viridis** | 147 (40–302) | 1 | |
| Gobiesocidae | | | |
| *Gobiesox marmoratus* | 61 (45–89) | 13 | |
| *Sicyases sanguineus** | 62 (41–84) | 11/12 | |
| 3. France, Atlantic (FRANCE-A), 49° N (Gibson, 1972) | | | |
| | | Sheltered | Exposed |
| Anguillidae | | | |
| *Anguilla anguilla** | 112–228 | 8 | |
| Phycidae[b] | | | |
| *Ciliata mustela** | 25–107 | 5/6 | |
| *Gaidropsarus mediterraneus* | 21–127 | | 5 |
| Syngnathidae | | | |
| *Nerophis lumbriciformis** | 65–137 | 4 | 4 |
| Cottidae | | | |
| *Taurulus bubalis** | 22–100 | 5/6 | 9 |
| Liparidae | | | |
| *Liparis montagui** | 12–38 | | 10 |
| Labridae | | | |
| *Symphodus (= Crenilabrus) melops**[c] | 55–105 | 9 | 7 |
| Blenniidae | | | |
| *Lipophrys (= Blennius) pholis**[d] | 30–150 | 2 | 1 |
| *Coryphoblennius galerita* | 32–60 | 7 | 8 |
| Gobiesocidae | | | |
| *Apletodon dentatus dentatus (= A. microcephalus)*[d] | 25–35 | | 11 |
| *Lepadogaster lepadogaster** | 36–73 | | 6 |
| Gobiidae | | | |
| *Gobius cobitis* | 31–192 | 3 | |
| *Gobius paganellus** | 32–104 | 1 | 2 |
| 4. France, Mediterranean (FRANCE-M), 43° N (Gibson, 1968) | | | |
| Phycidae[b] | | | |
| *Gaidropsarus mediterraneus** | 58–196 | 7/8 | |
| Sparidae | | | |
| *Diplodus (= Charax) puntazzo**[c] | 38 | 18/19 | |
| Labridae | | | |
| *Symphodus (= Crenilabrus) cinereus*[c] | 85 | 18/19 | |
| *Symphodus roissali (= Crenilabrus quinquemaculatus)**[c] | 68, 73 | 16/17 | |
| Tripterygiidae | | | |
| *Tripterygion tripteronotus** | 36–56 | 10 | |
| Clinidae | | | |
| *Clinitrachus (= Cristiceps) argentatus**[d] | 49–69 | 9 | |

**Table 1.** *Continued*

| Family<br>*Species* | Size range<br>(mm SL) | Abundance rank |
|---|---|---|
| Blenniidae | | |
| *Aidablennius sphynx*[d] | 35–76 | 7/8 |
| *Coryphoblennius galerita*[d] | 41–57 | 13/14/15 |
| *Parablennius gattorugine*[d] | 152, 182 | 16/17 |
| *Parablennius sanguinolentus**[,d] | 51–149 | 5 |
| *Parablennius zvonimiri*[d] | 31–61 | 11/12 |
| *Salaria pavo*[e] | 32–118 | 2/3 |
| Gobiesocidae | | |
| *Gouania wildenowi* | 38–52 | 13/14/15 |
| *Lepadogaster candollei** | 32–63 | 4 |
| *Lepadogaster lepadogaster* | 29–61 | 6 |
| Gobiidae | | |
| *Chromogobius quadrivittatus* | 33–45 | 13/14/15 |
| *Gobius bucchichii* | 35–99 | 2/3 |
| *Gobius cobitis** | 38–236 | 1 |
| *Gobius paganellus* | 61–110 | 11/12 |
| 5. New Zealand (NZEAL), 41° S (Willis and Roberts, 1996) | | |
| Phycidae[b] | | |
| *Gaidropsarus novaezelandiae** | 250 | 23 |
| Syngnathidae | | |
| *Lissocampus filum** | 115 | 20 |
| Plesiopidae | | |
| *Acanthoclinus fuscus** | 300 | 5 |
| *Acanthoclinus littoreus* | 155 | 9 |
| Labridae | | |
| *Notolabrus celidotus** | 250 | 7 |
| *Notolabrus fucicola* | 500 | 14 |
| Bovichthyidae | | |
| *Bovichtus variegatus** | 250 | 17 |
| Tripterygiidae | | |
| *Bellapiscis lesleyae* | 65 | 8 |
| *Bellapiscis medius** | 75 | 2 |
| *Forsterygion lapillum* | 50–110 | 3 |
| *Forsterygion varium* | 130 | 13 |
| *Gilloblennius tripennis* | 110 | 19 |
| *Grahamina capito* | 100? | 10 |
| *Notoclinops segmentatus* | 45 | 18 |
| *Notoclinops yaldwyni* | 55 | 25 |
| *Notoclinus compressus* | 90 | 22 |
| *Ruanoho decemdigitatus* | 105 | 16 |
| Clinidae | | |
| *Ericentrus rubrus** | 105 | 6 |
| Gobiesocidae | | |
| *Dellichthys morelandi* | 70 | 12 |
| *Diplocrepis puniceus* | 125 | 15 |
| *Gastrocyathus gracilis* | 45 | 24 |

*continues*

**Table 1.** *Continued*

| Family<br>*Species* | Size range<br>(mm SL) | Abundance rank |
|---|---|---|
| *Gastroscyphus hectoris* | 64 | 4 |
| *Haplocylix littoreus* | 150 | 21 |
| *Trachelochismus melobesia* | 50 | 11 |
| *Trachelochismus pinnulatus** | 100 | 1 |
| Gobiidae | | |
| *Gobiopsis atrata** | 85 | 26 |
| 6. South Africa (SAFR), 34° S (Prochazka, 1996) | | |
| Batrachoididae | | |
| *Batrichthys apiatus** | 100 | 19/20 |
| Clinidae | | |
| *Blennioclinus brachycephalus* | 150 | 11 |
| *Blennophis anguillaris* | 300 | 9/10 |
| *Cirrhibarbis capensis* | 350 | 12/13 |
| *Clinus acuminatus* | 130 | 5 |
| *Clinus agilis** | 100 | 2 |
| *Clinus berrisfordi* | 120 | 12/13 |
| *Clinus cottoides* | 150 | 4 |
| *Clinus heterodon* | 130 | 7 |
| *Clinus rotundifrons* | 100 | 17 |
| *Clinus superciliosus* | 300 | 1 |
| *Clinus taurus* | 230 | 18 |
| *Clinus venustris* | 120 | 8 |
| *Muraenoclinus dorsalis* | 100 | 3 |
| *Pavoclinus pavo* | 150 | 15 |
| *Xenopoclinus leprosus* | 80 | 17 |
| Blenniidae | | |
| *Parablennius cornutus** | 150 | 9/10 |
| Gobiesocidae | | |
| *Chorisochismus dentex** | 300 | 6 |
| *Eckloniaichthys scylliorhiniceps* | 35 | 14 |
| Gobiidae | | |
| *Psammogobius knysnaensis** | 70 | 19/20 |

*Note.* Species marked with an asterisk (*) are used to represent the family and its body plan in Figure 2.
[a]Taxonomy updated from Robins *et al.* (1991).
[b]Taxonomy updated from Nelson (1994).
[c]Taxonomy updated from Whitehead *et al.* (1986a).
[c]Taxonomy updated from Whitehead *et al.* (1986b).
[d]Taxonomy updated from Almada and Santos (1995).

the most genera in common; and 0.28 (±0.13) at the family level with, interestingly, the FRANCE-M and NZEAL sites having the highest similarity value (Table 3).

## B. Classification of Body Plans

Small body size characterizes the fishes of the six communities (Table 1), a feature shown in other chapters in this volume to be prevalent among intertidal fishes. For all six sites together, more than two-thirds (mean 69.7%, ±17.2 SD) of the fish species are less than

**Table 2.** Numbers and Proportions (%) of Species and Individuals Represented by Six Teleostean Families at Six Temperate-Zone Rocky Intertidal Sites

| Family | CALIF | CHILE | FRANCE-A | FRANCE-M | NZEAL | SAFR | Totals (%/%) |
|---|---|---|---|---|---|---|---|
| | Site: Species/individuals | | | | | | |
| Cottidae | 10/2584 | | 1/12 | | | | 11/2596 (10%/19%) |
| Clinidae | 2/79 | 1/24 | | 1/10 | 1/143 | 15/4965 | 20/5221 (18%/38%) |
| Tripterygiidae | | 2/135 | | 1/7 | 10/1374 | | 13/1516 (11%/11%) |
| Blenniidae | | | 2/148 | 6/97 | | 1/19 | 9/264 (8%/2%) |
| Gobiesocidae | 1/16 | 2/30 | 2/15 | 3/60 | 7/1138 | 2/423 | 17/1682 (15%/12%) |
| Gobiidae | | | 2/184 | 4/182 | 1/1 | 1/1 | 8/368 (7%/3%) |
| Totals (%/%) | 13/2679 (54/68) | 5/189 (79/96) | 7/359 (95/99) | 15/356 (57/72) | 19/2656 (38/29) | 19/5408 (73/83) | 78/11647 (69%/84%) |

*Note.* Percentages in the last column are the proportions that each family contributed to the total number of species and individuals at all six sites combined. Percentages in the bottom row are the proportions that all six families combined contributed to the total number of species and individuals at each site.

**Table 3.** Degree of Similarity at the Species (S), Genus (G), and Family (F) Levels of the Fishes Composing the Rocky Intertidal Communities at Six Temperate-Zone Sites

| | | CALIF | CHILE | FRANCE-A | FRANCE-M | NZEAL | SAFR |
|---|---|---|---|---|---|---|---|
| CALIF | S | — | | | | | |
| | G | — | | | | | |
| | F | — | | | | | |
| CHILE | S | 0.00 | — | | | | |
| | G | 0.04 | — | | | | |
| | F | 0.15 | — | | | | |
| FRANCE-A | S | 0.00 | 0.00 | — | | | |
| | G | 0.00 | 0.00 | — | | | |
| | F | 0.13 | 0.14 | — | | | |
| FRANCE-M | S | 0.00 | 0.00 | 0.19 | — | | |
| | G | 0.00 | 0.04 | 0.25 | — | | |
| | F | 0.14 | 0.36 | 0.42 | — | | |
| NZEAL | S | 0.00 | 0.00 | 0.00 | 0.00 | — | |
| | G | 0.00 | 0.03 | 0.03 | 0.03 | — | |
| | F | 0.13 | 0.33 | 0.38 | 0.55 | — | |
| SAFR | S | 0.00 | 0.00 | 0.00 | 0.00 | 0.00 | — |
| | G | 0.00 | 0.00 | 0.00 | 0.04 | 0.00 | — |
| | F | 0.18 | 0.30 | 0.27 | 0.44 | 0.27 | — |

*Note.* Based on the Jaccard Index of Similarity: $C_j = C/(A + B - C)$, where $A$ = number of species at locality A, $B$ = number of species at locality B, and $C$ = number of species common to A and B.

150 mm in length and 40.5% (±14.9) are less than 100 mm. The five most abundant species at all sites combined are even smaller with 87.2% (±10.0) less than 150 mm and 52.8% (±25.1) less than 100 mm in length.

Resident intertidal fishes generally seem to represent only a few patterns of body shape and paired fin arrangement, suggesting morphological convergence in response to the selection pressures of the environment. To address this observation further, I classified the species occurring in the six intertidal communities according to body plan based on the general premise that fish body shapes fall into one of three broad categories: (1) dorsoventrally flattened, (2) roughly cylindrical and tapered (terete), or (3) laterally compressed. The body may be shallow or deep and short or elongate, the latter especially if the fish is laterally compressed or terete in cross section.

This approach resulted in the designation of the following eight body plans to encompass the 109 fish species at the six intertidal sites (Figure 2):

1. Elongate body encased in bony rings (EBR). This body plan is represented only by species of the family Syngnathidae (pipefishes) and only at the FRANCE-A and NZEAL sites. Pipefishes were not among the most common intertidal fishes in these communities (Table 1).

2. Elongate, eel-shaped body (EEL). This body plan is represented by one species of the family Anguillidae (freshwater eels) at the FRANCE-A site only and by the Pholidae (gunnels) and Stichaeidae (pricklebacks) only at the CALIF site, where members of these two families were among the 10 most abundant species (Table 1).

3. Dorsoventrally flattened body without a pelvic suction disc (FND). This body plan is represented by a single species of the Batrachoididae (toadfishes) at the SAFR site and of the Cottidae (sculpins) at the CALIF and FRANCE-A sites. The sculpin *Artedius lateralis* was among the 10 most abundant species at the CALIF site (Table 1).

4. Dorsoventrally flattened body with a pelvic suction disc (FSD). This body plan is represented by one to several species of the Gobiesocidae (clingfishes) at all six sites and by one species of Liparidae (snailfishes) at the FRANCE-A site only. Clingfishes were among the most abundant species at the FRANCE-M, NZEAL, and SAFR sites (Table 1).

5. Terete (approximately cylindrical and tapered) body with fused ventral (pelvic) fins (TFV). This body plan is represented by one to several species of the family Gobiidae (gobies) at the FRANCE-A, FRANCE-M, NZEAL, and SAFR sites. Gobies were among the most abundant species at the FRANCE-A and FRANCE-M sites (Table 1).

6. Terete body with large, low-set pectoral fins (TLP). This body plan is represented by the Scorpaenidae (rockfishes), Hexagrammidae (greenlings), Cottidae (sculpins), Bovichthyidae (thornfishes), Tripterygiidae (triplefin blennies), and Blenniidae (combtooth blennies). The TLP body plan is exemplified in all six intertidal communities with combtooth blennies represented at four sites, triplefins at three sites, and thornfishes at two sites. Sculpins were among the most abundant species at the CALIF site, as were combtooth blennies at the CHILE, FRANCE-A, and FRANCE-M sites and triplefins at the CHILE and NZEAL sites (Table 1).

7. Laterally compressed, shallow body (LCS). This body plan is represented by the Phycidae (phycid hakes), Plesiopidae (roundheads including spiny basslets), Clinidae (clinids), and Labrisomidae (labrisomids). The LCS body plan is exemplified in all six intertidal communities with clinids represented at all sites except FRANCE-A, phycid hakes at three sites, and labrisomids and roundheads at one site each. Clinids were among the most abundant species at the NZEAL and SAFR sites, especially the latter where these fishes were

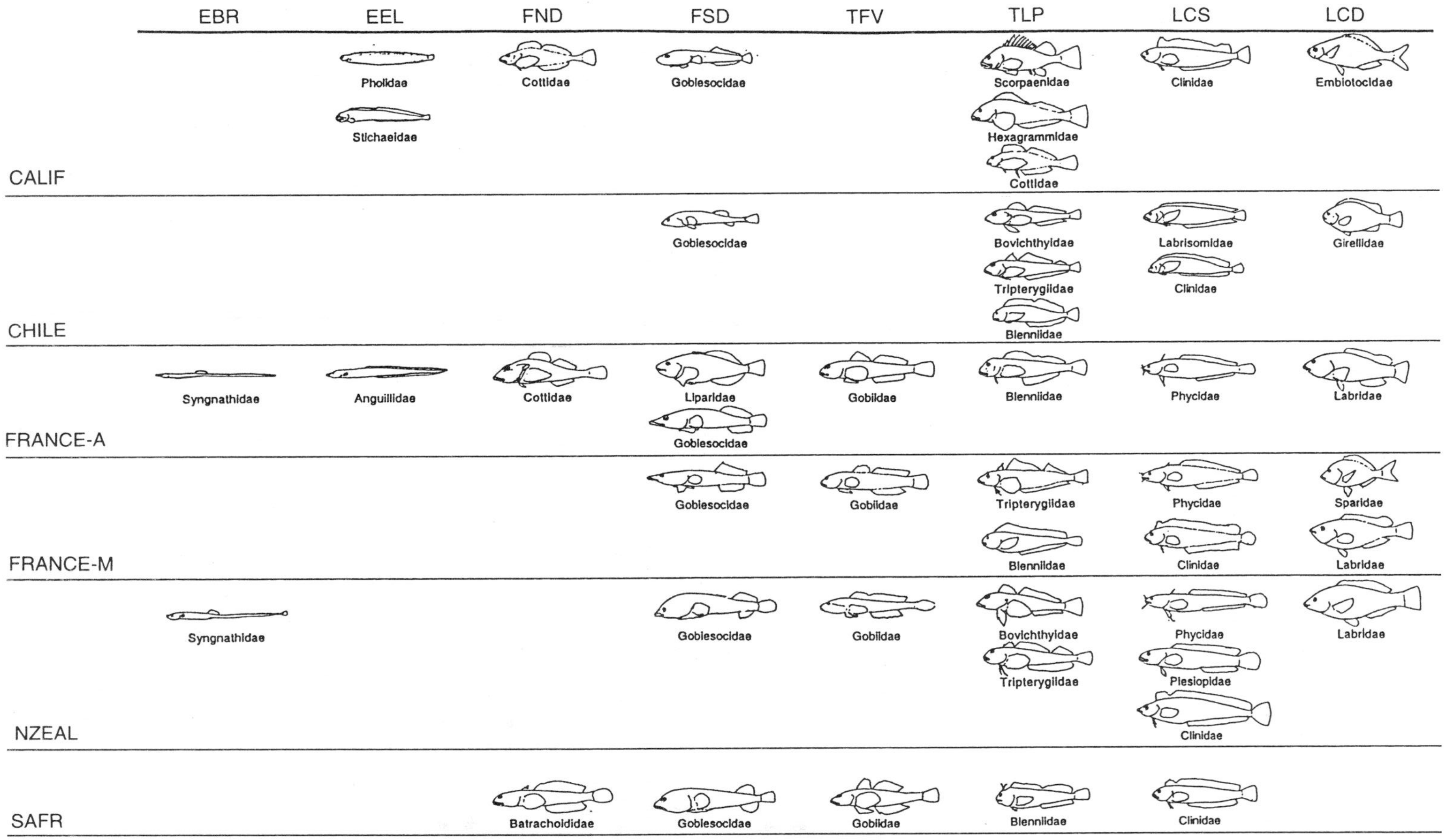

| | EBR | EEL | FND | FSD | TFV | TLP | LCS | LCD |
|---|---|---|---|---|---|---|---|---|
| CALIF | | Pholidae<br>Stichaeidae | Cottidae | Gobiesocidae | | Scorpaenidae<br>Hexagrammidae<br>Cottidae | Clinidae | Embiotocidae |
| CHILE | | | | Gobiesocidae | | Bovichthyidae<br>Tripterygiidae<br>Blenniidae | Labrisomidae<br>Clinidae | Girellidae |
| FRANCE-A | Syngnathidae | Anguillidae | Cottidae | Liparidae<br>Gobiesocidae | Gobiidae | Blenniidae | Phycidae | Labridae |
| FRANCE-M | | | | Gobiesocidae | Gobiidae | Tripterygiidae<br>Blenniidae | Phycidae<br>Clinidae | Sparidae<br>Labridae |
| NZEAL | Syngnathidae | | | Gobiesocidae | Gobiidae | Bovichthyidae<br>Tripterygiidae | Phycidae<br>Plesiopidae<br>Clinidae | Labridae |
| SAFR | | | Batrachoididae | Gobiesocidae | Gobiidae | Blenniidae | Clinidae | |

**Figure 2.** Outline drawings of fishes representing the designated body plans in each family of fishes from six temperate-zone intertidal communities (see Table 1). Body plan abbreviations: EBR, elongate with body rings; EEL, elongate, eel-like; FND, flattened without pelvic suction disc; FSD, flattened with pelvic suction disc; TFV, terete (cylindrical and tapered) with fused ventral (pelvic) fins; TLP, terete with large, low-set pectoral fins; LCS, laterally compressed, shallow-bodied; LCD, laterally compressed, deep-bodied. Species representing each family and body plan at each site are marked with an asterisk (*) in Table 1.

represented by 15 species including the five most abundant taxa in the community; labrisomids were among the most common species at the CHILE site (Table 1).

8. Laterally compressed, deep body (LCD). This body plan is represented by the Sparidae (porgies), Girellidae (nibblers), Embiotocidae (surfperches) and Labridae (wrasses) and is exemplified in all communities except SAFR with wrasses occurring at the FRANCE-A, FRANCE-M, and NZEAL sites and the other three families at one site each (Table 1). Nibblers were among the most abundant species at the CHILE site although, for the most part, fishes in this category are transient species in the intertidal zone.

## III. Convergent Adaptation

### A. Evidence from Structural Traits

The eight body plan designations, although subjective, suggest convergent adaptation. Just three body plans, TLP, LCS, and FSD, account for almost 75% of all the species in the six communities. For example, many species of Cottidae and most if not all species of Tripterygiidae possess the TLP body plan and are similar in size and appearance (Figure 2), yet they are members of different percomorph orders (Scorpaeniformes and Perciformes) and have different centers of distribution (see Table 1). Both sculpins and triplefin blennies use their large, low-set pectoral fins not only to prop up the body above the substratum but to grasp the bottom. Similarly, the pelvic fin is modified to form a suction disc in the Liparidae and Gobiesocidae, which are also members of the Scorpaeniformes and Perciformes, respectively, and represent the FSD body plan (Figure 2). The relatively flattened body and modified pelvic fins allow these fishes to adhere closely and tightly to rock surfaces. Many Gobiidae, including intertidal species, also possess a fused pelvic fin but because of their terete body shape and because the pelvic fin is less of a suction disc, they are classified as having the TFV body plan (Figure 2). No other family at the six sites possesses this body plan. The EEL body plan, represented at only two of the six sites, is perhaps best expressed by members of the Pholidae and Stichaeidae, which occur only in the Northern Hemisphere and mainly in the North Pacific (Nelson, 1994). The EBR body plan is also represented in only two of the six communities and is exemplified solely by the Syngnathidae, a family with its greatest species richness in warm-temperate to tropical waters (Nelson, 1994). Little convergence thus seems to have occurred on the EEL or EBR body plan by members of other families, at least at these six sites.

A problem that arises in this analysis is whether the traits mentioned are actual adaptations for intertidal life. Horn and Gibson (1988) point out that many intertidal fishes display features (e.g., small size, slender or flattened shape) that are characteristic of species living on the bottom in shallow, turbulent waters whether intertidal or subtidal or both. As stated by Almada and Santos (1995), if the presence of a character or suite of characters evolved in each lineage after colonization of rocky intertidal habitats, then intertidal species would be expected to differ from their subtidal sister taxa in these traits. The fishes in the different lineages would have independently evolved adaptations to the intertidal zone and thus show convergent adaptation as described by Brooks and McLennan (1991). In contrast, if the same characters are present in both intertidal and subtidal species of a given family that has some intertidal members, this pattern can be construed as evidence that the characters evolved before colonization of the intertidal habitat (assuming, of course, that intertidal species evolved from subtidal ancestors). These characters can be termed exaptations

(Gould and Vrba, 1982), traits that originated for some other selective or historical reason, unrelated to current function. Such traits have been coopted for uses unrelated to their origins (Larson and Losos, 1996).

In terms of body plans, intertidal fishes appear, as expected, to portray a combination of exaptational and adaptational characters. In the wide-ranging clinids, with the LCS body plan, the shallow, laterally compressed body and cryptic coloration appear to be adaptations for life among seaweeds (see Springer, 1995), but these traits exist in species from both subtidal and intertidal habitats. Several families, including scorpaenids, cottids, and blenniids, exhibit the TLP body plan, a terete body with large, low-set pectoral fins. These fins allow these mostly benthic fishes to prop up their bodies above the substratum, but, again, these characters are found in both subtidal and intertidal species. In both of these examples, the traits can be viewed as exaptations. Clingfishes, with their flattened body and ventral suction disc (FSD body plan), appear to be adapted for adhering to the substratum in shallow-water habitats with strong surge and wave action, whether in subtidal or intertidal habitats. In the xiphisterine stichaeids, the EEL body plan appears to be mainly characteristic of species that occupy the intertidal zone, whereas the more ancestral alectriine clade with approximately a TLP or an LCS body plan is largely confined to subtidal habitats (see Stoddard, 1985). The EEL body plan in the xiphisterine clade may thus represent an adaptation for dwelling in crevice or under-rock microhabitats (see Section III.B.1). In all of these examples, the research design to test for exaptation versus adaptation calls for a sufficiently resolved phylogeny so that the morphological traits can be compared by ecomorphological analysis (Norton *et al.,* 1995) and by performance tests (e.g., adherence strength in clingfishes) on intertidal species and their nearest subtidal sister taxa.

Studies showing that intertidal assemblages are largely distinct in species composition from subtidal assemblages (e.g., Yoshiyama *et al.,* 1987; Almada and Santos, 1995) support the scenario that adaptations to intertidal life are to be found among the fishes living there, especially the resident species. Brooks and McLennan (1991), however, point out that ecological and behavioral diversification often seems to lag behind morphological diversification and speciation. As a result, adaptation can be seen as more of a conservative, complementary, influence on evolution rather than the cause of diversification.

## B. Evidence from Functional Traits

What functional characters are found among intertidal species that hold promise for supporting the convergent evolution hypothesis? Traditionally, certain traits and capabilities of intertidal fishes have been labeled as adaptations without rigorously comparing these features to those found in subtidal relatives. Most of this work has focused loosely on within-group comparisons rather than on taxa based on robust phylogenetic trees, and little has been accomplished in testing for convergent adaptation between taxa belonging to different monophyletic groups. I discuss in turn below three types of functional traits—desiccation tolerance, air-breathing, and parental care—that are generally considered to be adaptations to the demands of intertidal existence. Particularly relevant published works are emphasized in each case.

### 1. Desiccation Tolerance

Dehydrating conditions are generally thought be one of the principal factors limiting the distribution of intertidal plants and animals. The vertical distribution of many intertidal fish species appears to be controlled largely by tolerance to desiccation (Bridges, 1993; Zander *et al.,* Chapter 3, this volume). Both gobiesocids (Eger, 1971) and stichaeoids (Horn and Riegle, 1981) inhabiting rocky shores can withstand high evaporative water loss and survive long periods in air. In both groups, vertical distribution is positively correlated with ability to survive in air. For example, the stichaeid *Cebidichthys violaceus* has a small surface to weight ratio, high water content, and long tolerance to emersion; accordingly, this fish has the highest, most exposed vertical distribution among the five species studied by Horn and Riegle (1981). Whether the characters associated with desiccation tolerance have converged in the distantly related gobiesocids and stichaeoids remains unclear because the appropriate experiments on survival and water loss capacities have not been carried out on the nearest subtidal sister species to determine the degree of contrast in traits of fishes from the two habitats.

The Xiphisterini, the clade with more derived characters within the subfamily Xiphisterinae (Stoddard, 1985), contains three of the four stichaeid species studied by Horn and Riegle (*Xiphister atropurpureus* and *X. mucosus* along with *C. violaceus*) and consists mainly of intertidal species with eel-shaped (EEL) bodies (Figure 2). Invasion of the intertidal zone by ancestral species in this clade may have led to the evolution of desiccation tolerance among its members. It is possible, however, that differences found by Horn and Riegle (1981) in survival and water loss capacities among some of these species may not be related to vertical distribution and exposure but to other selective forces of the intertidal environment such as turbulence, food supply, or predation.

The position of *Anoplarchus purpurescens,* the fourth stichaeid in Horn and Riegle's (1981) study, complicates the picture because it is a member of the more ancestral xiphisterine clade (tribe Alectriini) and is somewhat less eel-like (approaching the LCS body plan), yet its desiccation tolerance matches that of *C. violaceus.* The possibility exists that convergent evolution has occurred between members of the two tribes. On the other hand, it is perhaps as likely that characters conferring desiccation tolerance evolved in ancestors of the entire subfamily and are thus exaptational in the Alectriini and Xiphisterini. Further research on stichaeid interrelationships and on performance among intertidal species and their sister taxa in subtidal habitats is required to resolve these questions.

## 2. Air-Breathing

Air-breathing is recognized as one of several adaptive responses evolved by fishes living in habitats where oxygen levels may be severely depleted (Graham, 1997). Although marine air-breathing fishes generally lack the air-breathing organs found in their freshwater counterparts (Graham, 1976), many intertidal fishes of temperate as well as tropical shores can breathe air and share certain behavioral and morphological features (see Martin and Bridges, Chapter 4, this volume). Intertidal fishes are amphibious air breathers in that gas exchange occurs during exposure to air during receding tides or terrestrial sojourns (Graham, 1997; Martin and Bridges, Chapter 4, this volume). Five of the six main families of temperate-zone intertidal fishes (Cottidae, Blenniidae, Tripterygiidae, Gobiesocidae, and Gobiidae) contain air-breathing representatives; only the Clinidae apparently lack such species (Graham, 1997; see Martin and Bridges, Chapter 4, this volume).

Amphibious members of these families probably have converged in evolving an air-breathing specialization in response to some condition(s) of the intertidal environment. Martin (1996) proposes that air-breathing ability may have arisen independently in many species as an adaptive response to the periodic hypoxia that occurs in tidepools during nocturnal low tides or, alternatively, that air-breathing arose as a consequence of low metabolic rate and sedentary habit. In the latter case, air-breathing is an exaptation, an epiphenomenon of low activity and low energy demand, and thus not unique to intertidal species. Martin (1996) examined these alternatives by comparing the air-breathing ability of five species of cottids representing an ecological gradient from the highest intertidal zone to deep subtidal waters. She found that the intertidal cottids (*Ascelichthys rhodorus* and *Oligocottus maculosus*) have specific behavioral and physiological abilities (e.g., high respiratory gas exchange in air, lack of reliance on anaerobiosis) not present in the shallow subtidal and deeper water cottids (*Jordania zonope, Icelinus borealis,* and *Chitonotus pugetensis*).

Martin's work thus supports the adaptation hypothesis, but it is limited by lack of a resolved phylogeny that includes the species studied and shows the relationships between the intertidal and the subtidal taxa. In this case, performance as related to the intertidal habitat has been measured but cannot be placed in the context of character evolution. Research that links physiology and phylogeny has yet to be accomplished for any clades that include intertidal air-breathing species; thus, a rigorous test of the convergent evolution of air-breathing capability rests on future studies.

## 3. Parental Care

The reproductive behavior and life history of many different phyletic lines of rocky intertidal fishes fall within a narrow range of patterns (Gibson, 1969, 1982; see DeMartini, Chapter 8, Coleman, Chapter 9, and Pfister, Chapter 10, this volume). This limited variety of patterns has prompted Almada and Santos (1995) to evaluate the likelihood of convergent evolution of parental care among blennioid fishes in the intertidal habitat. They asked whether the behavioral patterns arose in different groups by convergent evolution in the process of colonizing the intertidal zone or whether the patterns were expressed as common features or preconditions (exaptations) that made these groups successful in colonizing the intertidal habitat. Almada and Santos concentrated on northeastern Atlantic blennioids in which intertidal species constitute about half of the total number of blennioid species in the region and assumed on the basis of parsimony that intertidal species evolved from nonintertidal ancestors.

These authors argue that if parental care evolved in each lineage *after* the colonization of rocky intertidal habitats then only intertidal species would show parental care; that is, there would be a contrast in behavior between intertidal and subtidal species and thus a true adaptation for intertidal life. On the other hand, if parental care is present in both intertidal and subtidal members of the most proximal outgroups with intertidal species, this lack of contrast between fishes of the two habitats would support the hypothesis that parental care evolved *before* colonization of the intertidal zone and thus is not an adaptation to the intertidal environment. Almada and Santos' (1995) behavioral survey led them to conclude that parental care in rocky intertidal species is better viewed as an exaptation because in most cases the trait is also present in subtidal species related to each resident intertidal species. (The authors were not explicit as to whether any of the "related" subtidal species are sister taxa to the intertidal species.) They suggested, however, that male courtship displays have been modified in response to the turbulence and strong wave action in the rocky intertidal zone to minimize swimming time and loss of contact with the substratum. These displays, then, may represent intertidal adaptations, but Almada and Santos emphasize that detailed cross phylogenetic and ecological studies are required for confirmation.

Gould, S. J., and Vrba, E. S. (1982). Exaptation—Missing term in the science of form. *Paleobiology* **8,** 4–15.

Graham, J. B. (1976). Respiratory adaptations of marine air-breathing fishes. *In* "Respiration of Amphibious Vertebrates" (G. M. Hughes, Ed.), pp. 165–187. Academic Press, London.

Graham, J. B. (1997). "Air-Breathing Fishes: Evolution, Diversity, and Adaptation." Academic Press, San Diego.

Gu, B., Schelske, C. L., and Hoyer, M. V. (1996). Stable isotopes of carbon and nitrogen as indicators of diet and trophic structure of the fish community in a shallow hypereutrophic lake. *J. Fish Biol.* **49,** 1233–1243.

Haldorson, L. (1980). Genetic isolation of Channel Islands fish populations: Evidence from two embiotocid species. *In* "The California Islands: Proceedings of a Multidisciplinary Symposium" (D. M. Power, Ed.), pp. 433–442. Santa Barbara Natural History Museum, Santa Barbara, CA.

Horn, M. H., and Gibson, R. N. (1988). Intertidal fishes. *Sci. Am.* **256,** 64–70.

Horn, M. H., and Riegle, K. C. (1981). Evaporative water loss and intertidal vertical distribution in relation to body size and morphology of stichaeoid fishes from California. *J. Exp. Mar. Biol. Ecol.* **50,** 273–288.

Krebs, C. J. (1989). "Ecological Methodology." Harper & Row, New York.

Larson, A., and Losos, J. B. (1996). Phylogenetic systematics of adaptation. *In* "Adaptation" (M. R. Rose and G. V. Lauder, Eds.), pp. 187–220. Academic Press, San Diego.

Martin, K. L. M. (1996). An ecological gradient in air-breathing ability among marine cottid fishes. *Physiol. Zool.* **69,** 1096–1113.

Muñoz, A. A., and Ojeda, F. P. (1997). Feeding guild structure of a rocky intertidal fish assemblage in central Chile. *Environ. Biol. Fish.* **49,** 471–479.

Nelson, J. S. (1994). "Fishes of the World," 3rd ed. Wiley, New York.

Norton, S. F., Luczkovich, J. J., and Motta, P. J. (1995). The role of ecomorphological studies in the comparative biology of fishes. *Environ. Biol. Fish.* **44,** 287–304.

Pagel, M. D. (1994). The adaptationist wager. *In* "Phylogenetics and Ecology" (P. Eggleton and R. I. Vane-Wright, Eds.), pp. 29–52. Academic Press, London.

Prochazka, K. (1996). Seasonal patterns in a temperate intertidal fish community on the west coast of South Africa. *Environ. Biol. Fish.* **45,** 133–140.

Robins, C. R., Bailey, R. M., Bond, C. E., Brooker, J. R., Lachner, E. A., Lea, R. N., and Scott, W. B. (1991). "Common and Scientific Names of Fishes from the United States and Canada," 5th ed. American Fisheries Society Spec. Pub. 20, Bethesda, MD.

Samuels, C. L., and Drake, J. A. (1997). Divergent perspectives on community convergence. *Trends Ecol. Evol.* **12,** 427–432.

Springer, V. G. (1995). Blennies. *In* "Encyclopedia of Fishes" (J. R. Paxton and W. N. Eschmeyer, Eds.), pp. 216–219. Academic Press, San Diego.

Stepien, C. A., and Rosenblatt, R. H. (1991). Patterns of gene flow and genetic divergence in the northeastern Pacific Clinidae (Teleostei: Blennioidei), based on allozyme and morphological data. *Copeia* **1991,** 873–896.

Stoddard, K. M. (1985). "A Phylogenetic Analysis of Some Prickleback Fishes (Teleostei, Stichaeidae, Xiphisterinae) from the North Pacific Ocean, with a Discussion of their Biogeography." MA thesis, California State University, Fullerton.

Vander Zanden, M. J., and Rasmussen, J. B. (1996). A trophic position model of pelagic food webs: Impact on contaminant bioaccumulation in lake trout. *Ecol. Monogr.* **66,** 451–477.

Waples, R. S., and Rosenblatt, R. H. (1987). Patterns of larval drift in southern California marine shore fishes inferred from allozyme data. *Fish. Bull.* **85,** 1–11.

Whitehead, P. J. P., Bauchot, M.-L., Hureau, J.-C., Nielsen, J., and Tortenese, E. (Eds.). (1986a). "Fishes of the North-eastern Atlantic and the Mediterranean," Vol. II. UNESCO, Paris.

Whitehead, P. J. P., Bauchot, M.-L., Hureau, J.-C., Nielsen, J., and Tortenese, E. (Eds.). (1986b). "Fishes of the North-eastern Atlantic and the Mediterranean," Vol. III. UNESCO, Paris.

Willis, T. J., and Roberts, C. D. (1996). Recolonisation and recruitment of fishes to intertidal rockpools at Wellington, New Zealand. *Environ. Biol. Fish.* **47,** 329–343.

Yoshiyama, R. M. (1981). Distribution and abundance patterns of rocky intertidal fishes in central California. *Environ. Biol. Fish.* **6,** 315–332.

Yoshiyama, R. M., and Sassaman, C. (1993). Levels of genetic variability in sculpins (Cottidae: Teleostei) of the North American Pacific coast and an assessment of potential correlates. *Biol. J. Linn. Soc.* **50,** 275–294.

Yoshiyama, R. M., Sassaman, C., and Lea, R. N. (1987). Species composition of rocky intertidal and subtidal fish assemblages in central and northern California, British Columbia-southeast Alaska. *Bull. S. Calif. Acad. Sci.* **86,** 136–144.

# 17

# The Fossil Record of the Intertidal Zone

Hans-Peter Schultze

*Institut für Paläontologie, Museum für Naturkunde, Berlin, Germany*

## I. Introduction

Depending on their exposure, coastal areas can have very different potentials to be documented in the fossil record. The areas that are least likely to be documented are rocky shores (Johnson, 1992; Johnson and McKerrow, 1995), because they are areas of erosion and abrasion and not of deposition. As a rule, these areas are characterized by invertebrates adapted to life as borers, encrusters, and clingers. Fishes have left no fossil record from rocky shore environments.

In contrast to rocky shores, coastal areas with a soft bottom are well documented in the fossil record. These deposits of the intertidal zone are characterized by inorganic sediment structures (tidal laminations and mud cracks), organic sediment structures (stromatolites, and tracks, and traces), mixed assemblages, and juvenile forms, and rarely by morphological or behavioral adaptations of vertebrates. Sedimentation rates are high in these areas; thus fossils, including fishes, are well preserved. I concentrate here on such deposits, first giving criteria for recognition and then discussing fossil examples through geologic time.

## II. Criteria for Recognizing Fossil Shores

### 1. Sediment Structures

The changing tide is reflected in rhythmic deposition of laminae. Ideally, one lamina is deposited by ebb and one by flood, separated by a slack lamina. Tidal rhythmites are characterized by major and minor laminations. They show more than one periodicity (Archer and Clark, 1992), in contrast to other laminated sediments (such as varves, which are characterized by seasonal variations in sedimentation rate and sediment composition occurring in lakes) and turbidites (which are characterized by graded bedding and occur in deeper marine or lacustrine waters). Minor laminations reflect daily or semidaily periodicities (Figure 1), whereas major laminations reflect neap–spring cycles (= fortnightly tidal

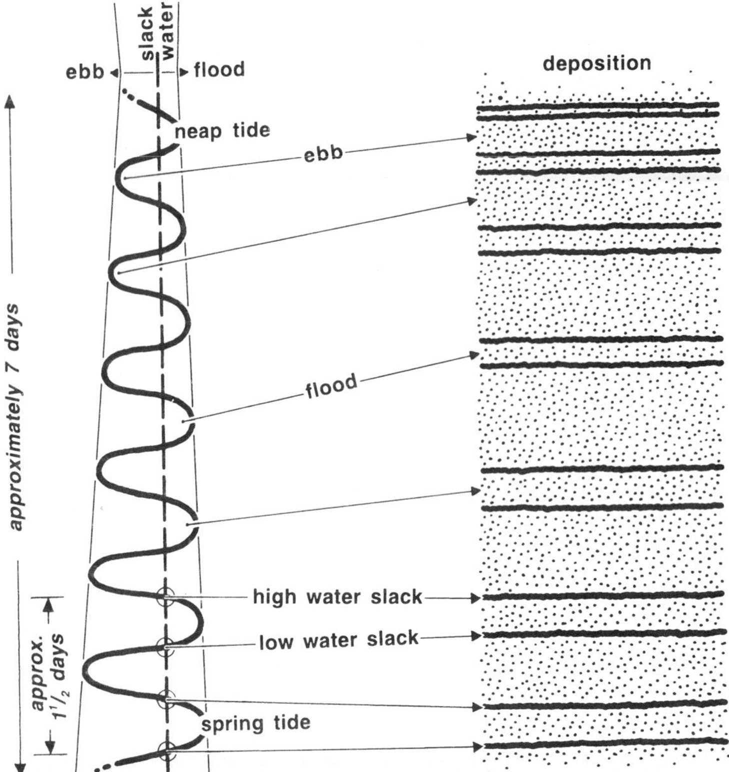


**Figure 1.** Schematic representation of a diurnal (one tide per day) half neap–spring–neap cycle and its appearance in the sediment (deposition). The dark bands (couplets) are tidal slack deposits, whereas the light bands represent sediments of ebb and flood, respectively. The daily periodicity (ebb–flood cycle) is overlain by 7-day periodicity (half neap–spring–neap cycle). Modified from Figs. 4 and 5 of Kuecher *et al.* (1990).

cycles). Not every tide deposits a lamina, so the number of laminations within each neap–spring cycle is not always complete (14 for daily or 28 for semidaily periods).

The fossil record contains many examples of tidal rhythmites, some of them extensively researched (Visser, 1980; Williams, 1989; Kuecher *et al.,* 1990; Archer, 1991, 1994; Feldman *et al.,* 1993; Chan *et al.,* 1994). Fossil fishes are often found in tidal sediments, but usually it is difficult to prove that they died and were buried in their habitat: They could have come from further offshore or from rivers.

Subaerial exposure of forming sediments causes shrinkage cracks (desiccation polygons) and so-called mudcracks, which form when silty to muddy sediments dry. Similar structures can form under water (syneresis cracks) and are difficult to distinguish from mudcracks in the fossil record. Nevertheless mudcracks are commonly used as indicators of subaerial exposure of forming sediments. They can be identified as desiccation cracks in combination with other features such as tracks.

## 2. Stromatolites

Stromatolites are laminated sediments that range from mats to discrete columnar mounds and are formed by cyanobacteria or algae, which bind detritus or precipitate calcium carbonate as a coating. They occur from subtidal marine through freshwater environments (Walter, 1976; Awramik, 1984; Riding, 1991). Stromatolites of the intertidal zone show laminations that are the result of the same forces that produce laminations in intertidal sediments. The same counting of daily increments as with tidal rhythmites is possible, with the same caution about missing laminations (Pannella, 1972, 1975). Stromatolites are known from the Precambrian, where they were very common, to the present. Their relative rarity in the Phanerozoic is believed to be a result of grazing by invertebrates or of substrate competition (Pratt, 1982). Stromatolites form today only where grazing invertebrates are absent, or where they can grow faster than they can be consumed as food.

For one occurrence of Recent stromatolites (Shark Bay, Australia), their association with fishes has been investigated (Lenanton, 1977). Certain fishes closely associate with the stromatolite environment, whereas others are more frequently found outside the stromatolite reef.

Two fossil stromatolite occurrences with fishes are recorded from the Lower Permian of New Mexico, U.S.A. (Toomey and Cys, 1977), and from the Upper Pennsylvanian of Kansas, U.S.A. (Sawin *et al.,* 1985; Chorn and Schultze, 1990). In New Mexico fish scales and teeth occur in a red shale surrounding the stromatolitic mounds, whereas in Kansas fish and amphibian bones and teeth are embedded both within and between the mounds (Figure 2).

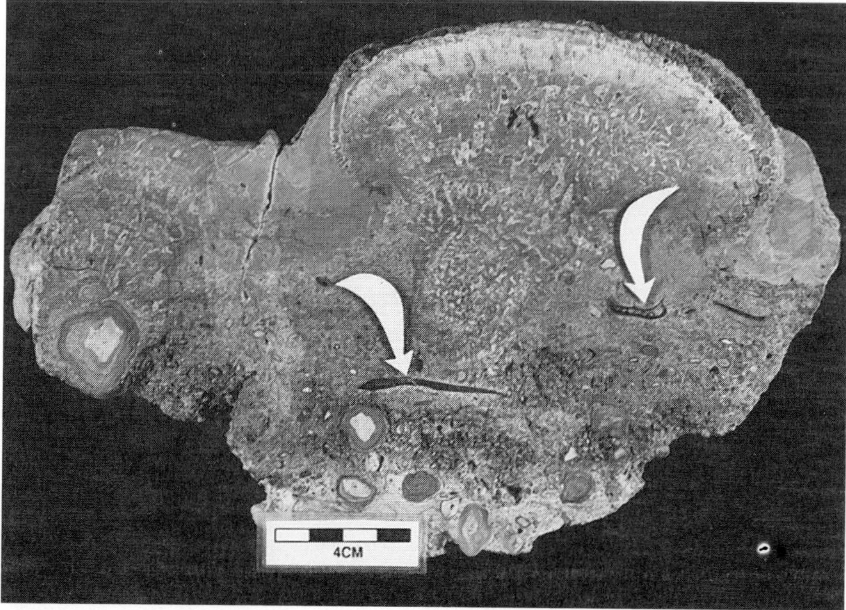


**Figure 2.** Fossil bones embedded in stromatolites of the Bern Limestone Formation, Upper Pennsylvanian, from Robinson, Brown County, Kansas, U.S.A. Arrows point to cross sections of lungfish (*Sagenodus copeanus*) bones. Photograph provided by Dr. J. Chorn, Lawrence, Kansas.

## 3. Tracks and Traces

Trackways, trails of terrestrial animals, indicate very shallow water or subaerial exposed sediments. A combination with fish trails, trace fossils, or sediment structures indicates that the trackways were formed under water. The tracks in Figure 3 show that an amphibian traversed a tidal area shortly before the water returned; while the area was submerged, a fish left its trail to the right of the amphibian trackway.

Trace fossils are the record of animal activities in or on the sediment (burrows, borings, trails). The distribution of trace fossils is a reliable indicator of water depth (Seilacher, 1967). A small-sized diverse trace fossil assemblage is characteristic of intertidal sediments (Hakes, 1985; and others in Curran, 1985).

**Figure 3.** Cooccurrence of trackways of walking (a) and swimming (b) amphibians with trail of a swimming fish (c) on one bedding surface demonstrates changing dry and water-covered phases of coastal zone. Stephanian, Upper Carboniferous; Emma Quarry, Puertollano, province Ciudad Real, southern Spain. Scale bar equals 5 cm. Courtesy Dr. R. Soler-Gijón.

## 4. Mixed Assemblages and Juvenile Forms

Many marine fishes reproduce in nearshore waters (Gunter, 1961; Christensen, 1978; Major, 1978; Toole, 1980); some pelagic fishes deposit their eggs near the coastline. In some cases, the larvae move even closer to the coastline after hatching and return to deeper water after attaining larger size. A good fossil example is the Upper Pennsylvanian deposits of Mazon Creek, Illinois, which contain mostly juvenile forms (Figure 4) in a tidal to deltaic depositional environment (Schultze, 1980, 1985b).

Mixed fish assemblages can result when freshwater fishes transported from rivers and lakes become mixed with marine fishes in an intertidal environment. Fishes are also often found with plants, insects, and terrestrial vertebrates (Schultze, 1996a). The fishes may or may not be autochthonous, whereas the terrestrial forms are brought in by wind or rivers. Mixed faunas have misled some researchers. Fishes in these mixed faunas were thought to be washed in from freshwater (see below), although their distribution (Schultze, 1985a) and growth series in the same deposits indicate that they lived, as well as died, in a marine coastal environment (Schultze *et al.*, 1994).

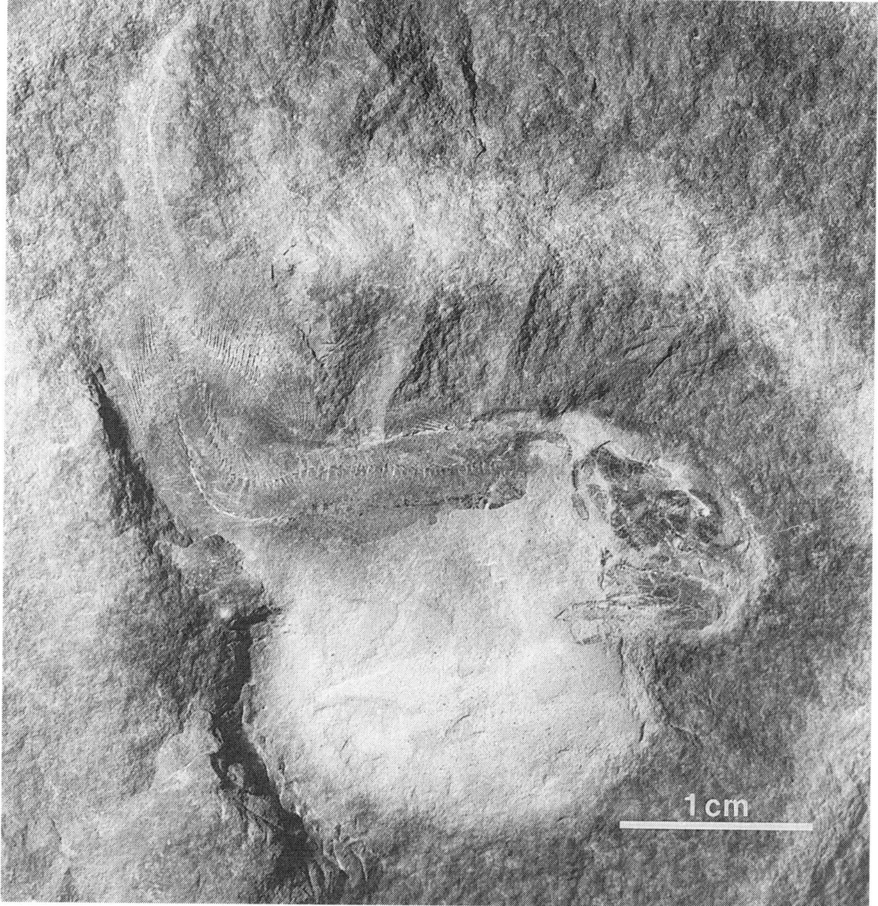


**Figure 4.** Juvenile actinistian (*Rhabdoderma exiguum*), with yolk sac, of the Upper Pennsylvanian coastal to deltaic deposits of Mazon Creek, Illinois, U.S.A.

## 5. Histological and Morphological Features

Cyclic growth increments in scales, otoliths, and bones of fishes have been used to determine their age. With a few exceptions (Pannella, 1971, 1974, 1980; Ross, 1983; Campana, 1984; Brown and Gruber, 1988), sequences of growth lines have not been correlated with tidal sequences. Pannella (1971, 1974, 1980) described two daily growth lines of 0.5–4 $\mu$m width in otoliths, which are grouped in fortnight groups of two 7–8 first-order bands of different composition and thickness. Two fortnight groups of different widths represent a monthly deposit. Campana (1984) interpreted the lunar pattern as a consequence of temperature change during a biweekly tidal cycle. The periodicities were not observed in otoliths of freshwater fishes (Campana, 1984). In a fossil example, Solder-Gijón (1994) demonstrated that the tidal sequence is reflected in the growth lines of shark spines from Upper Pennsylvanian stromatolites of Kansas. Figure 5 shows another example, a cranial roofing bone of a dipnoan from the same deposits. A lighter growth zone with two dark lines and a darker zone with two dark lines represent a time span of approximately 3 months. Four cycles of 3 months equal one annual cycle of bone deposition. The first 3-month period represents the time of most rapid growth either following a spawning phase, in similar fashion to some Recent osteichthyans (Clugston and Cooper, 1960), or because of favorable climatic conditions such as less dry seasonal conditions in the tropics (Meunier *et al.,* 1979; Meunier and Pascal, 1980; Lecomte *et al.,* 1986). The sequence of growth lines shows three periodicities superimposed over each other, a feature indicative of tidal periodicity in interaction with climatic fluctuations.

It is unfortunate that osteological features of extant intertidal fishes have not been better studied, so that they could be applied to fossils. One osteological feature that has been well

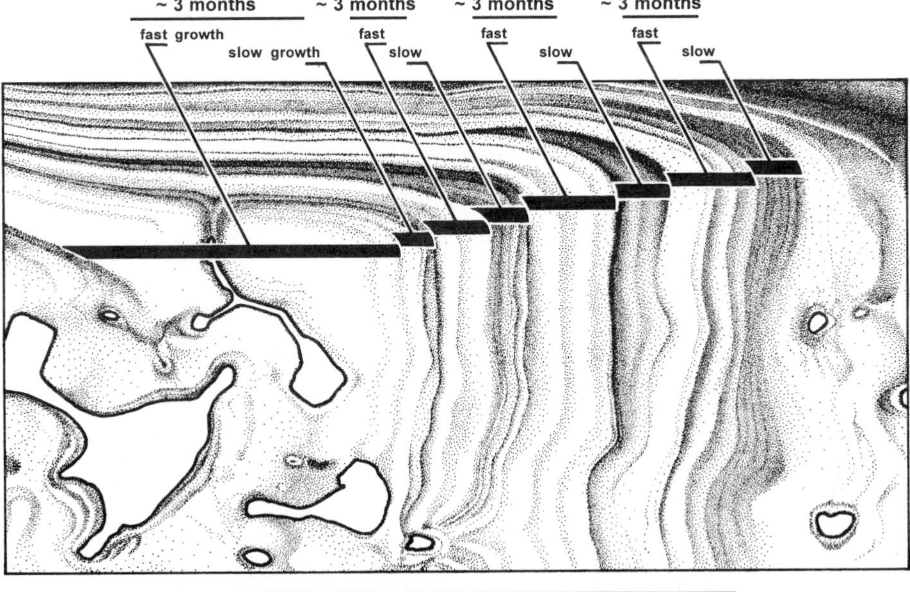


**Figure 5.** Cross section through the lateral part (dorsal surface at top) of a B-bone of *Sagenodus copeanus* of the Bern Limestone Formation, Upper Pennsylvanian, from Robinson, Brown County, Kansas, U.S.A. One year of growth, represented by four cycles of approximately 3 months.

described is found in the mudskippers. These are extant teleostean fishes adapted to life in the intertidal to supratidal environments of the African and Southeast Asian coasts. The eyes of adult mudskippers (*Periophthalmus*) lie extremely high on the skull, so that the fishes have stereoscopic vision in addition to being adapted to vision on land (Harms, 1929). To support the high position of the eyes, a median bony support ("eye brow") has developed on the skull by modification of the parietal bones (usually called "frontals") (Figure 6). A similar medial bony support (but formed by true frontals) for the eyes can be

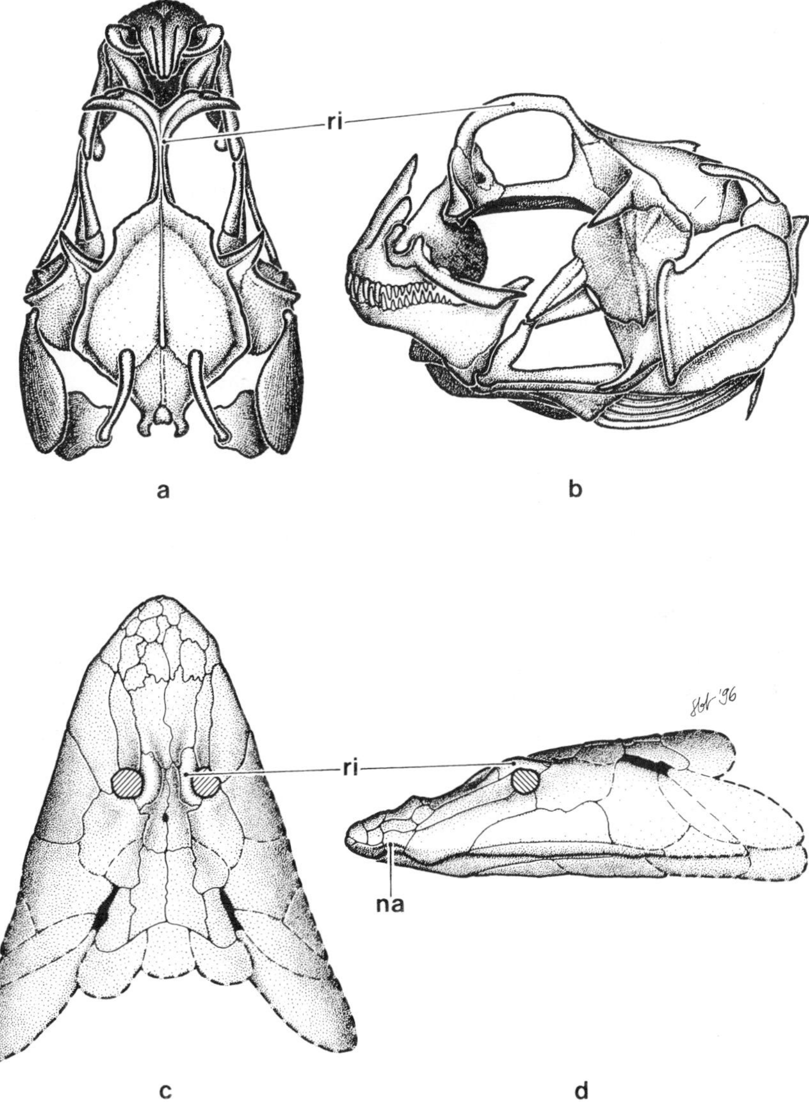


**Figure 6.** Lateral (b, d) and dorsal (a, c) views of skull of the extant mudskipper *Periophthalmus* (a, b) and the fossil elpistostegid *Panderichthys* (c, d). Abbreviations: na, external nasal opening; ri, rim ("eye brow") medial to the orbits. *Periophthalmus* after Gregory (1933).

found in fossil rhipidistians near the transition to tetrapods (elpistostegids; Schultze, 1996b, 1997). It is characteristic also for early tetrapods, which were amphibious. This adaptation to an amphibious lifestyle should not be confused with adaptations for floating at the water surface (e.g., crocodiles). In elpistostegids and early tetrapods the nasal opening was low in the snout to be close to the wet surface of the tidal flats. In crocodiles the nasal openings are situated as high as the eyes to reach above the water to breathe while the eyes observe the water surface.

Another morphological feature that may indicate an intertidal mode of life is a flattened ventral surface to the body, as in extant rays and flatfishes, which is an adaptation to lying on a soft bottom (Koepcke, 1973). Among fossils it can be found in osteostracan (cephalaspids) and galeaspid agnathans and in some placoderms (e.g., antiarchs).

## III. Freshwater–Marine Controversy

The lack of unique morphological features permitting us to recognize intertidal fishes combined with the lack of "typical" marine invertebrates at many fossil-fish localities has misled many authors into interpreting these localities as freshwater deposits. The cooccurrence of fishes, arthropods, and plants, but without calcereous marine invertebrates, at these localities is the result of chemical differentiation. Phosphatic and organic materials are preserved, whereas calcitic material has been dissolved. Negative arguments (the lack of . . .) have been used for postulating freshwater paleoenvironments, especially in the Paleozoic (Schultze, 1996a).

That the origin of vertebrates took place in freshwater is a concept based mainly on negative arguments (Romer and Grove, 1935; Romer, 1955; and other papers). Gross (1950), Denison (1956), and Robertson (1957) demonstrated that early vertebrates are more common and more diverse in marine deposits. These discussions centered around paleoenvironmental interpretation of the earliest vertebrate localities, where, because of chemical differentiation, the only invertebrates are the inarticulate brachiopod *Lingula* and arthropods. *Lingula* is a marine form that reaches into the brackish region. Gross (1950) used comparison of fossil assemblages, one of the strongest arguments besides detailed taphonomic investigations of localities, to determine the paleoenvironment (Schultze, 1996a).

This discussion is closely connected with the interpretation of the Devonian Old Red Sandstone, once commonly regarded to be continental freshwater deposits (Friend and Moody-Stuart, 1970, 1972; Allen and Dineley, 1976). However, more recent investigations (Goujet, 1984; Blieck, 1985) have demonstrated that at least some of the deposits are coastal marine. The composition of the whole fauna (Mark-Kurik, 1991; Schultze and Cloutier, 1996) shows similarities to faunas of undisputed marine paleoenvironments. The $^{87}Sr/^{86}Sr$ isotopic ratio of bones seems to be a useful tool in determining the paleoenvironment. Seawater has a very constant lower $^{87}Sr/^{86}Sr$ ratio (today: 0.7092) than freshwater, which has a variable, but with few exceptions higher, ratio (today: 0.710–0.730 average) (see Schmitz *et al.,* 1991, for procedure). The ratio is changing over geologic time and can be deduced from sediments. Fishes take up and concentrate strontium in their bones from the surrounding water and their food. The ratio of $^{87}Sr/^{86}Sr$ in bones and teeth corresponds closely to that of the surrounding water. By comparing the $^{87}Sr/^{86}Sr$ ratio of the water (deduced from the sediment) with that of fish bones, a reliable indication is given for the paleoenvironment of the fish. The fishes of the Old Red Sandstone gave mixed results, some indicating freshwater and others marine environments (Schmitz *et al.,* 1991).

Based on biased interpretation of deposits as freshwater, groups such as the xenacanth sharks have been used as an indicator of freshwater (Masson and Rust, 1984) even though their teeth are common in unequivocally marine sediments (Schultze, 1985a). The same is true of the heterostracan agnathans, which were demonstrated by Blieck (1985) to be indicators of a coastal marine environment. In any case, one should be careful in using fishes as indicators of a particular paleoenvironment. Fishes are mobile and some enter rivers, and thus they may indicate marine connectedness (confluence of fresh and marine waters) rather than a marine depositional environment. Fishes may be diadromous. Invertebrates, being less mobile, are better indicators of their general environment than fishes. Therefore, the paleoenvironment of the localities discussed below is based primarily on geological features such as tidal rhythmites and stromatolites and the occurrence of marine invertebrates, and only secondarily on the composition of the fish assemblage.

## IV. Fossil Examples

### 1. Ordovician (510–439 mill. years)

The earliest undoubted fossil vertebrates are of Ordovician age. They are known from Australia (475 mill. years), from Bolivia, and from many localities in North America (455 mill. years). The first Ordovician vertebrates were found at the end of the last century in the Harding Sandstone (Figure 7) of Colorado (Walcott, 1892). Lack of typical marine fossils in the Harding Sandstone caused Romer and Grove (1935) and others to postulate that these agnathans lived in a freshwater environment. However, the wide distribution and lithological features of the Harding Sandstone were later used to reinterpret it as a coastal marine or intertidal depositional environment (benthic assemblage 1 of Boucot and Janis, 1983). Even Graffin (1992), who interpreted a section of the Harding Sandstone as fluvial, also found agnathan remains in the intertidal facies. His conglomeratic facies may simply be a local coastal or estuarine deposit. The best evidence of a shallow-marine depositional environment is the repeated occurrence of early agnathans in marine limestones that contain terrigenous derived quartz sand (Darby, 1982). Darby (1982) argued for a habitat of the early vertebrates "in a nearshore zone on a sandy bottom" as filter feeders. A very nearshore, intertidal marine environment is generally accepted for the Ordovician vertebrates of the three fossil-producing areas in Bolivia, Australia, and North America (Elliott *et al.,* 1991).

### 2. Silurian (439–409 mill. years)

The surface of the Baltic region [Estonia (Figure 7), Latvia, and Lithuania] is formed of Paleozoic rocks (Ordovician, Silurian, and Devonian), in places not covered by Quaternary sediments. The subsurface also is well known from many boreholes. Based on this information, a detailed paleoecological interpretation of the Silurian sediments has been made (Märss and Ejnasto, 1978; Blieck, 1985). The deposition extends from lagoons to deeper water basins (Figure 8). Märss and Ejnasto (1978) divided the paleoenvironments into five zones (repeated in Blieck, 1985; Märss, 1986). Thelodonts, osteostracans, and anaspids occur predominantly in the lagoonal paleoenvironment together with eurypterids and inarticulate brachiopods, whereas heterostracans (and thelodonts in the latest Silurian, the Downtonian, or Pridoli) prefer the coastal area along with a variety of marine invertebrates.

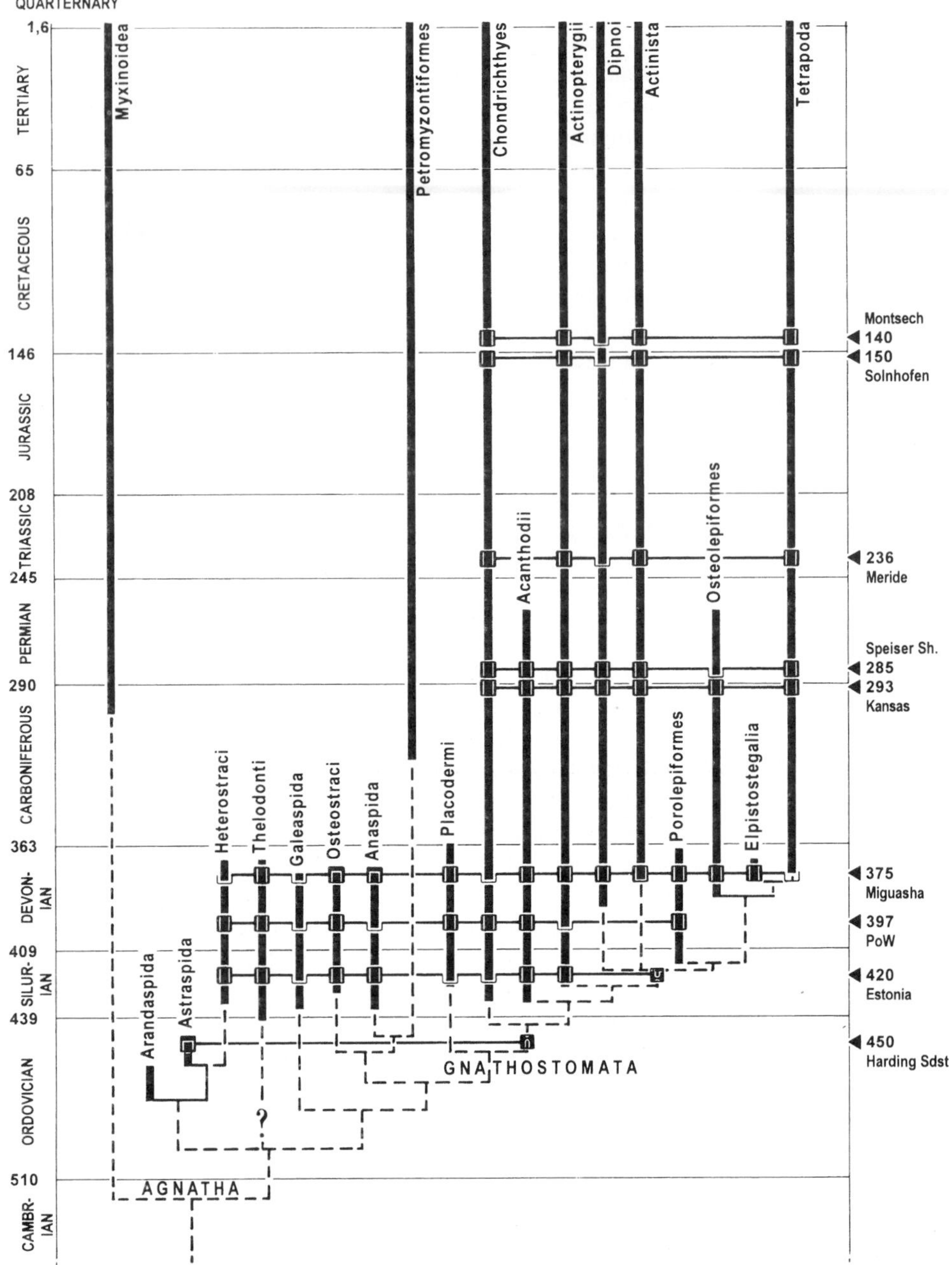


**Figure 7.** Distribution of intertidal localities throughout geological time (ages in millions of years; after Harland *et al.*, 1990) with represented vertebrate groups. Periods (Cambrian to Quarternary) on the left; localities mentioned in the text on the right.

The coastal zone has the highest diversity of vertebrates, with all major groups then living represented. Osteichthyans first emerge in this zone in the Ludlovian (Late Silurian). Acanthodians, the best swimmers of the time, occur mostly on the open platform; they can also be found in the coastal area and on the slopes of the basins (where thelodonts and hetero-

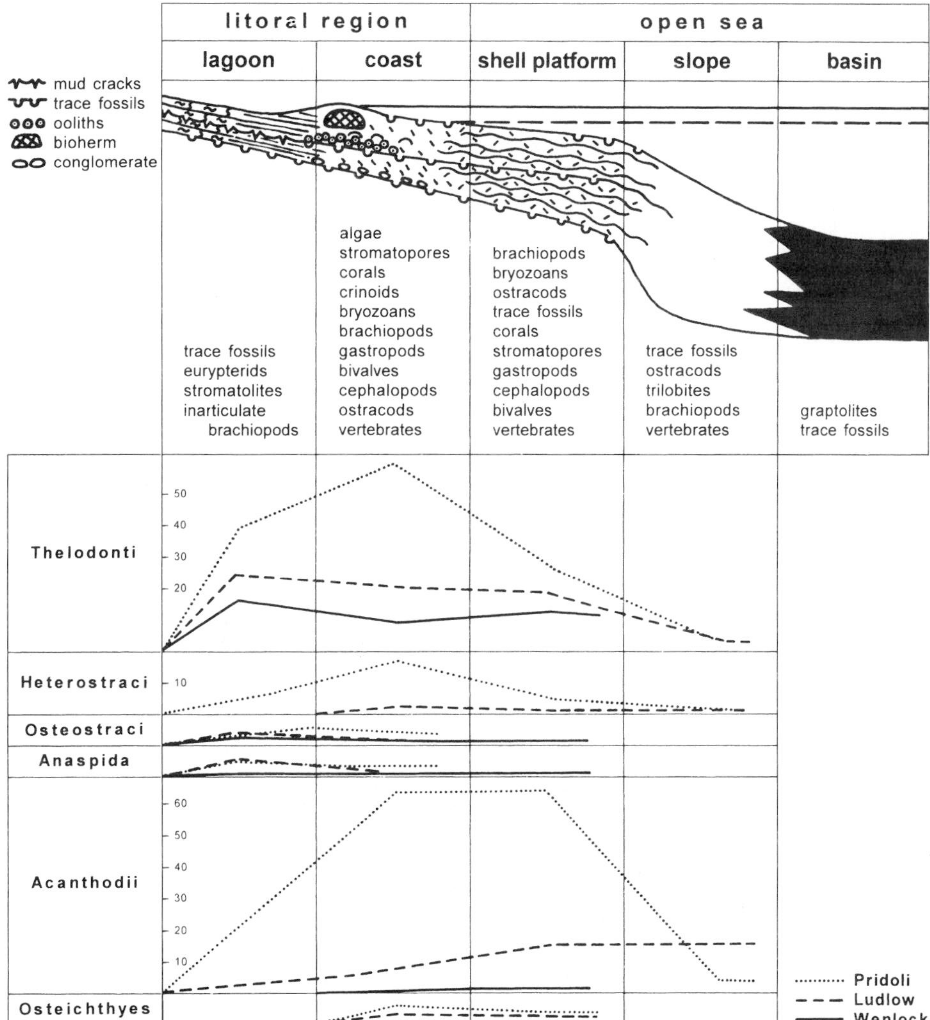


**Figure 8.** Distribution of vertebrates in Silurian marine sediments of Estonia. (Top) Geological section from coastal to basin regions; (bottom) distribution of vertebrates in the different regions from Wenlock to Pridoli (430–409 mill. years ago). Modified from Märss and Ejnasto (1978).

stracans also occur). Thelodonts occupy different water zones (Turner, 1992). *Turinia* may have been a bottom dweller, whereas other thelodonts (especially the recently described, laterally compressed forms with deep, symmetrical tails; Wilson and Caldwell, 1993) occupied the water column closer to the surface.

## 3. Devonian (409–363 mill. years)

Shallow-marine sediments rich in fossil fishes surround the northern margin of the Canadian Shield. Langenstrassen and Schultze (1996) demonstrated, for different Lower Devonian areas of the Canadian Arctic, a sequence from an open-marine environment, with a rich invertebrate and vertebrate fauna, to a near-shore environment (red sediments), with a restricted vertebrate and invertebrate fauna.

South of Canyon Fjord on Ellesmere Island, sedimentary structures (criterion 1) and trace fossils (criterion 3) indicate a very shallow-marine to supratidal depositional environment. Intercalated limestones show a diverse marine fauna with pteraspid and thelodont agnathans, placoderms, elasmobranchs, acanthodians, and porolepiforms. In contrast, the fauna is restricted in the red sandy littoral to deltaic sediments. Invertebrates are represented here by *Teichichnus* (a trace fossil), lingulid brachiopods, and a few ostracods. Cephalaspid agnathans, placoderms, acanthodians, and porolepiforms did reach this littoral intertidal to supratidal environment, but are less diverse.

At Drake Bay in the northwestern part of Prince-of-Wales Island (PoW in Figure 7) are richly fossiliferous marine Lower Devonian sediments, which change to red sandy and even conglomeratic toward the Boothia Uplift at the eastern side of the island. The rich fish fauna of the western region includes pteraspid agnathans, placoderms, acanthodians, and porolepiforms (*Powichthys*), which reach along Baring Channel into the red sediments. The placoderms, porolepiforms, and one heterostracan, *Ctenaspis,* reach farther east than the other fishes. *Ctenaspis* is considered to have been adapted to the sandy bottom (Dineley, 1976).

The Upper Devonian Escuminac Formation at Miguasha (Figure 7) at the southern border of the Gaspé Peninsula, eastern Quebec, is tectonically isolated and has long been interpreted as a freshwater deposit (Dineley and Williams, 1968). It contains one of the most diverse and abundant Paleozoic fish faunas (Schultze and Cloutier, 1996) including *Elpistostege,* a rhipidistian closely related to tetrapods (Schultze, 1996b). Faunal comparison (Schultze, 1972; Schultze and Cloutier, 1996), chemical analysis of the sediments (Vézina, 1991; Chidiac, 1996) and fish bones (Schmitz *et al.,* 1991), trace fossils (Maples, 1996), and acritarchs (Cloutier *et al.,* 1996) all indicate a coastal marine to estuarine environment. *Elpistostege* is one of the few forms that show morphological features indicative of intertidal or supratidal adaptation (criterion 5; Figure 6; Schultze, 1997). Its presence supports the intertidal interpretation for the depositional environment of the Escuminac Formation. Cephalaspid agnathans with a flattened ventral surface represent the last fossil record of the group, and the antiarch, *Bothriolepis canadensis,* with a flattened ventral surface (criterion 5) is the most common fish at the locality.

## 4. Carboniferous (363–290 mill. years)

During the Pennsylvanian (Late Carboniferous), a shallow sea covered the North American Midcontinent. The deposits on the margins of this sea in many cases clearly show tidal influence (Kuecher *et al.,* 1990; Archer and Clark, 1992; Feldman *et al.,* 1993; Archer, 1994).

The Mazon Creek area in north-central Illinois is famous for its plant and fish fossils. Kuecher *et al.* (1990) described tidal deposits (criterion 1) from there. Johnson and Richardson (1966) drew the coastline between the presumed freshwater Braidwood fauna and the marine Essex fauna. The Essex fauna is rich in marine invertebrates (scyphozoans, inarticulate brachiopods, bivalves, cephalopods, annelids, echinoderms, etc.), whereas arthropods are common in the Braidwood fauna (Baird *et al.,* 1985; Maples and Schultze, 1989). A comparable division can be observed in the vertebrates: agnathans and holocephalans occur exclusively in the Essex fauna and most tetrapods exclusively in the Braidwood fauna. However, most fishes (elasmobranchs, acanthodians, actinopterygians, actinistians, rhipidistians, and dipnoans) can be found in both environments. The dominance of juvenile forms is remarkable (criterion 4; Figure 4; Schultze, 1980, 1985b; Schultze and Bardack, 1987) in both faunas.

A mixture of marine and terrestrial forms (criterion 4) occurs at three upper Pennsylvanian localities in Kansas (Figure 7) (Schultze, 1996a): Hamilton (Cunningham *et al.,* 1994), Robinson (Sawin *et al.,* 1985; Chorn and Schultze, 1990), and Garnett (Reisz *et al.,* 1982). The sediments of Hamilton show tidal cyclicity (criterion 1; Cunningham *et al.,* 1994); the vertebrates are preserved in and around stromatolites (criterion 2) at Robinson; and Garnett is well known for its trackways (criterion 3; Reisz, 1990). Hamilton and Garnett are geographically limited deposits laid down in narrow estuarines. Both have a large terrestrial faunal component. In addition Hamilton has a rich fish fauna with xenacanth and hybodont elasmobranchs, juvenile to adult acanthodians, one rhipidistian, palaeonisciforms, and dipnoans. Except for three shark teeth, only actinistians are represented at Garnett. The fish fauna of Hamilton is also represented at Robinson, a very shallow area with stromatolites in the Midcontinent sea of the Late Carboniferous. Acanthodians are the most common fishes at Hamilton, whereas dipnoans, *Sagenodus,* and *Gnathorhiza* are preeminent at Robinson.

## 5. Permian (290–245 mill. years)

Cuffey (1967) constructed in detail the paleoenvironment of a sedimentary sequence (Wreford Megacyclothem) in time (in the Early Permian) and geographic extent (a narrow north–south stretch between Nebraska and Oklahoma). The lower part of the Wreford Megacyclothem is represented by the Speiser Shale (Figure 7). Schultze (1985a) presented the distribution of fishes in the tidal to supratidal sediments of the Wreford Megacyclothem in Kansas. Paleonisciform and deep-bodied platysomid actinopterygians, acanthodians, and cladodont, xenacanthid, and bradyodont elasmobranchs occur there together with bryozoans and brachiopods in intertidal deposits and with the lungfish *Gnathorhiza* at the intertidal–supratidal transition (Figure 9 of Schultze, 1985a). While most fishes are represented by isolated remains, complete specimens of the lungfish *Gnathorhiza* and a few tetrapods (*Diplocaulus, Brachydectes* [= *Lysorophus*], *Acroplous, Trimerorhachis*) are found in burrows. The burrows are not far from the coastline in either direction. The burrows may have been a protective adaptation for surviving low tide, as they are for many Recent intertidal animals. The close proximity of the coastline and large numbers of *Gnathorhiza* bones in tempestites (= storm deposits) at one locality (Bushong, Kansas) coupled with echinoderm remains and common *Gnathorhiza* bones in the marine stromatolites of Robinson, Kansas (see above), and other marine sediments indicate that we have here an intertidal adaptation of a lungfish. *Gnathorhiza* burrows in this environment do not represent estivation burrows in a freshwater environment as seen at present with *Protopterus* in Africa, which occurs far away from the sea.

Except for the lungfish *Gnathorhiza,* none of the Permian fishes show adaptation to the intertidal zone. Xenacanthid elasmobranchs, acanthodians, and palaeonisciform actinopterygians may have been able to enter freshwaters, nevertheless, they occur more often in coastal marine to intertidal deposits. They are only erroneously considered freshwater indicators.

## 6. Triassic (245–208 mill. years)

Many rich vertebrate localities are known in the Triassic of southern Switzerland (Zorn, 1971; Furrer, 1995; Meride in Figure 7) and northern Italy (Tintori, 1996). They yield a rich fish (elasmobranchs, actinopterygians, and actinistians) and aquatic amniote (sauropterygians, placodonts, and ichthyosaurs) fauna in addition to a few terrestrial to semiterrestrial amniotes (*Ticinosuchus, Tanystropheus*). These are deposits of coastal basins with intertidal sediments (criterion 1) and mud-cracked stromatolites (criterion 2). None of the fishes show any intertidal adaptation; only their occurrence in the intertidal sediments and cooccurrence with terrestrial amniotes indicate their ability to live in the intertidal zone. The actinopterygian *Prohalecites* is known from both adult and juvenile specimens (criterion 4).

## 7. Jurassic (208–146 mill. years)

The fish fauna of Solnhofen (Figure 7), southern Germany, is famous for its excellent preservation. Over a period of about 10 mill. years, finely laminated sediments (criterion 1) were deposited in more-or-less restricted embayments of the Late Jurassic sea. Tracks and traces (criterion 3), frequent occurrence of juvenile specimens (criterion 4), and the occurrence of tetrapods and insects are indicators of deposition close to the coastline (Walther, 1904; Barthel *et al.,* 1990; Viohl, 1996). Sharks, palaeonisciform to teleostean actinopterygians, and actinistians are represented. Sharks, fast-swimming pachycormiform actinopterygians, and actinistians are invaders of the lagoonal system. They and the deep-bodied pycnodontiform actinopterygians (inhabitants of nearby sponge reefs) are represented by adults, whereas juveniles of most teleost groups are common (criterion 4).

## 8. Cretaceous (146–65 mill. years)

Since the 1960s two Lower Cretaceous localities in Spain [Montsech (Figure 7) and Las Hoyas] have produced a fish fauna similar on higher taxic levels to the Late Jurassic fish fauna of Solnhofen. Yet many authors (see Martínez-Delclós, 1995) insist that this is a freshwater fauna, using arguments based on terrestrial forms (insects, tetrapods, plants) found at the sites. The occurrence of agglutinating foraminifera (Schairer and Janicke, 1970: *Glomospirella, Lituotuba, Haplophragmoides, Miliammina,* typical for coastal areas) and acritarchs (Barale, 1991) in Montsech and the $^{87}$Sr/$^{86}$Sr ratio in fish bones of Las Hoyas indicate a marine environment. A contrary view was advanced by Talbot *et al.* (1996) who interpreted the marine $^{87}$Sr/$^{86}$Sr values as altered by marine Jurassic rocks. The ostracode *Cypridea,* which occurs in Montsech as adults, is found from freshwater to marine environments. The occurrence of only adult *Cypridea* indicates that they were washed in (Schairer and Janicke, 1970). $^{87}$Sr/$^{86}$Sr values at Las Hoyas were found to be 0.707644 (Talbot *et al.,* 1996), which according to Schmitz *et al.* (1991) falls within the range expected for a marine environment; values higher than 0.71 indicate freshwater. The paleogeographic situation at both localities (Barale *et al.,* 1984; Fregenal-Martínez and Meléndez, 1996) supports a coastal, intertidal, and possibly lagoonal paleoenvironment despite the interpretations of the cited authors. Like the fishes at Solnhofen, juveniles are common for teleosts, whereas other fishes are mostly represented by adults.

## 9. Tertiary (65–1.6 mill. years)

Because of their recency, comparison of Tertiary with extant forms and thus unequivocal determinations of the paleoenvironment of Tertiary localities can be done. Gaudant (1979) gave a bathymetric distribution for fishes of the Miocene, where 80% of the teleostean suborders and 87% of shark families are identical with extant suborders and families. Gaudant (1989) gave examples of littoral fishes (*Spartelloides, Gobius, Trachurus,* and *Arnoglossus*) from Upper Miocene localities near Murcia, Spain.

## V. The Fossil Record of Intertidal Fishes

Localities with intertidal vertebrates occur more frequently in the Paleozoic, especially Lower Paleozoic, rocks. The first vertebrates appeared in the intertidal zone, and only later conquered the open-marine environment.

The first agnathans, arandaspids and astraspids, were intertidal vertebrates (Elliott *et al.*, 1991). According to Blieck (1985), heterostracans are indicators of coastal marine environments; they lived in lagoonal and littoral areas and some were able to reach the open shelf. The osteostracans and galeaspidomorphs, with their flattened ventral surfaces, were bottom dwellers. They were the most coastal-adapted agnathans and may have entered estuaries and rivers. However, they may not have been confined to the bottom as much (Afanassieva, 1992) as is postulated by most researchers. Thelodonts and anaspids had laterally compressed and relatively deep bodies, so one has to assume good swimming abilities. Nevertheless their remains are found mostly in lagoonal and coastal sediments. Some thelodonts (*Turinia*) have a dorsoventrally compressed body form indicating adaptation to bottom dwelling.

Silurian placoderms have recently been discovered in shallow-marine deposits of China, but otherwise the group is restricted to the Devonian. Early Devonian placoderms often occur in red sediments and have been previously considered freshwater forms. However, Goujet (1984) has shown that these forms occur in shallow-marine deposits. Also occurring in shallow-marine sediments are the arthrodires in the Lower Devonian deposits of the Canadian Arctic (see above). Within the placoderms, the antiarchs, with their flat ventral body surfaces, are the most coastal (bottom) adapted forms; they constitute the majority of fish fossils in the Late Devonian Escuminac Formation of Miguasha, Canada (see above).

The Early Silurian mongolepids, close relatives of chondrichthyans, occur in shallow-marine (lagoonal to deltaic) deposits (Karatajute-Talimaa and Predtechenskyi, 1995). Early chondrichthyans, known only by scales from Silurian and Lower Devonian deposits, are found in shallow-marine deposits, for example, in the Lower Devonian of the Canadian Arctic (see above). They do not seem to reach the intertidal to supratidal region. Some chondrichthyans, for example, xenacanths, are common in shallow-marine to freshwater deposits in the Carboniferous and Permian, but they cannot be used as freshwater indicators (Schultze, 1996a).

Acanthodians are a teleostome group known from the Early Silurian through the Late Permian. They occur in marine to freshwater environments and are found in intertidal areas such as those of the Upper Devonian Escuminac Formation (see above), and in the Carboniferous near-shore deposits of Kansas (see above).

The first osteichthyans appear in the Late Silurian. The earliest actinopterygians are represented by scales from Silurian and Lower Devonian marine near-shore deposits. The Middle to Late Devonian *Cheirolepis,* often cited as a freshwater inhabitant, is represented by scales in marine sediments and as complete specimens in the intertidal deposits of Miguasha, Canada (see above). Actinopterygians had already spread into open-marine environments by the Devonian; they also may have entered freshwaters by that time. Today they are widespread in all aquatic environments. It is not possible to assign environmental preferences to higher categories of actinopterygians; environmental specializations can be given only for species and perhaps some genera of actinopterygians. As for the lungfish, *Diabolepis,* a Chinese form close to dipnoans, occurs in shallow-marine deposits, whereas most Early Devonian dipnoans are found in open-marine deposits. *Uranolophus,* which occurs in channel deposits, is an exception. Devonian lungfish occur in open-marine to coastal deposits (see Miguasha above). This is also true for Carboniferous and Permian dipnoans (see Kansas above). It seems possible to use them as indicators of coastal marine deposits. Freshwater dipnoans are common in Mesozoic and Cenozoic deposits. The first actinistians occur in open-marine and coastal (Miguasha above) deposits and occasionally are found in intertidal coastal sediments (see Garnett, Kansas, above). Early rhipidistians occur in Lower Devonian coastal deposits such as those in the Canadian Arctic (rhipidistians). Many Devonian forms were considered freshwater (see above), but today are regarded as coastal marine (e.g., *Eusthenopteron* in the Escuminac Formation). They also occurred as intertidal fishes in Carboniferous sediments of Kansas.

The closest relatives of tetrapods, the Elpistostegalia, are fishes of the coastal zone. *Elpistostege* (Schultze, 1996b) occurs in the intertidal sediments of the Escuminac Formation in Miguasha, Quebec, Canada (Schultze and Cloutier, 1996). Complete specimens of a second genus, *Panderichthys,* have been found in sediments interpreted as submarine delta slopes in the clay pit of Lode, Latvia (Kurss, 1992). The Elpistostegalia and early tetrapods show an intertidal and supratidal adaptation in the high position of the orbits, comparable to the extant mudskippers (Figure 6) (Schultze, 1997). The earliest tetrapods possessed functional gills (Coates and Clack, 1991); they were not terrestrial animals but rather inhabitants of the intertidal zone transitional to dry land.

The intertidal zone is the place of origin of most fish groups. From there they spread seaward onto the shelf platform and into the open sea and landward to deltaic (and into the rivers) and supratidal areas. The tetrapods entered the terrestrial realm through the intertidal and supratidal zones.

## Acknowledgments

Ms. E. Siebert, Museum für Naturkunde, Berlin, prepared Figures 1, 3, 5–8. Dr. John Chorn, University of Kansas, Lawrence, Kansas, supplied the photograph for Figure 2, and Dr. Rodrigo Soler-Gijón, Universidad Complutense, Madrid, Spain, kindly gave permission to use his unpublished work for Figures 3 and 5. Prof. Dr. G. Arratia, Dr. A. Boucot, Dr. J. Chorn, Dr. C. G. Maples, and Dr. Rodrigo Soler-Gijón read the manuscript and suggested improvements; Drs. J. Zidek and M. Chotkowski corrected the English. Prof. Dr. H. Wilkens, Zoologisches Institut und Museum, Universität Hamburg, and Dr. W. W. Gettmann, Director, Löbbecke Museum & Aquazoo, Düsseldorf, supplied *Periophthalmus* specimens. I am grateful to these persons for their cooperation, which made the compilation of this chapter possible.

## References

Afanassieva, O. (1992). Some pecularities of osteostracan ecology. *In* "Fossil Fishes as Living Animals" (E. Mark-Kurik, Ed.), pp. 61–65. Academia 1, Academy of Sciences of Estonia, Tallinn.

Allen, J. R. L., and Dineley, D. L. (1976). The succession of the Lower Old Red Sandstone (Siluro-Devonian) along the Ross-Tewkesbury Spur Motorway (M.50), Hereford & Worcester. *Geol. J. Liverpool* **11,** 1–14.

Archer, A. W. (1991). Modeling of tidal rhythmites using modern tidal periodicities and implications for short-term sedimentation rates. *In* "Sedimentary Modeling: Computer Simulations for Improved Parameter Definition" (E. K. Franseen, W. L. Watney, C. G. St. C. Kendall, and W. C. Ross, Eds.), pp. 185–194. Kansas Geological Survey Bulletin 233, Lawrence, KS.

Archer, A. W. (1994). Extraction of sedimentological information via computer-based image analyses of gray shales in Carboniferous coal-bearing sections of Indiana and Kansas, USA. *Math. Geol.* **26,** 47–65.

Archer, A. W., and Clark, G. R., II (1992). Depositional environment of the *Dunbarella* beds: An exercise in paleoecology and sediment cyclicity. *In* "Geology and Paleontology of the Kinney Brick Quarry" (J. Zidek, Ed.), pp. 27–36. Late Resources, Bulletin 138. Socorro, New Mexico.

Awramik, S. M. (1984). Ancient stromatolites and microbial mats. *In* "Microbial Mats: Stromatolites." (Y. Cohen, R. W. Castenholz, and H. O. Halvorson, Eds.), MBL Lectures in Biology Vol. 3, pp. 1–22. Liss, New York.

Baird, G. C., Shabica, C. W., Anderson, J. L., and Richardson, E. S., Jr. (1985). Biota of a Pennsylvanian muddy coast: Habitats within the Mazonian delta complex, Northeast Illinois. *J. Paleontol.* **59,** 253–281.

Barale, G. (1991). La palinologia de les calcàries litogràfiques del Montsech. *In* "Les calcàries litogràfiques del Cretaci inferior del Montsec. Deu anys de campagyes paleontològiques" (X. Martínez-Deldòs, Ed.), pp. 71–72. Institut d'Estudis Ilerdencs, Lleida. [Pp. 49–50 in English section]

Barale, G., Blanc-Louvel, C., Buffetaut, E., Courtinat, B., Peybernes, B., Via Boada, L., and Wenz, S. (1984). Les gisements de calcaires lithographiques du Crétacé inférieur du Montsech (province de Lérida, Espagne). Considérations paléoécologiques. *Géob. Mém. Spécial* **8,** 275–283.

Barthel, K. W., Swinburne, N. H. M., and Conway Morris, S. (1990). "Solnhofen. A Study in Mesozoic Palaeontology." Cambridge University Press, Cambridge/New York.

Blieck, A. (1985). Paléoenvironnements des Hétérostracés, Vertébrés agnathes ordoviciens à dévoniens. *Bull. Mus. natl. Hist. nat. Paris* **7,** 143–155.

Boucot, A. J., and Janis, C. (1983). Environment of the early Paleozoic vertebrates. *Palaeogeogr. Palaeoclimatol. Palaeoecol.* **41,** 251–287.

Brown, C. A., and Gruber, S. H. (1988). Age assessment of the Lemon Shark, *Negaprion brevirostris,* using tetracycline validated vertebral centra. *Copeia* **1988,** 747–753.

Campana, S. E. (1984). Lunar cycles of otolith growth in the juvenile starry flounder *Platichthys stellatus. Mar. Biol.* **80,** 239–246.

Chan, M. A., Kvale, E. P., Archer, A. W., and Sonett, C. P. (1994). Oldest direct evidence of lunar-solar tidal forcing encoded in sedimentary rhythmites, Proterozoic Big Cottonwood Formation, central Utah. *Geology* **22,** 791–794.

Chidiac, Y. (1996). Paleoenvironmental interpretation of the Escuminac Formation based on geochemical evidence. *In* "Devonian Fishes and Plants of Miguasha, Quebec, Canada" (H.-P. Schultze and R. Cloutier, Eds.), pp. 47–53. Verlag Dr. Friedrich Pfeil, München.

Chorn, J., and Schultze, H.-P. (1990). The Robinson locality: Fauna and depositional environment. *In* "1990 Society of Vertebrate Paleontology: Upper Paleozoic of Eastern Kansas Excursion Guidebook" (C. R. Cunningham and C. G. Maples, Eds.), pp. 17–24. Kansas Geological Survey Open-file Report 90-24, Lawrence, KS.

Christensen, M. S. (1978). Trophic relationships in juveniles of three species of sparid fishes in the South African marine littoral. *U.S. Natl. Mar. Fish. Serv. Fish. Bull.* **76,** 389–401.

Cloutier, R., Loboziak, S., Candilier, A.-M., and Blieck, A. (1996). Biostratigraphy of the Upper Devonian Escuminac Formation, eastern Québec, Canada: A comparative study based on miospores and fishes. *Rev. Palaeobot. Palynol.* **93,** 191–215.

Clugston, J. P., and Cooper, E. L. (1960). Growth of the common eastern madtom, *Noturus insignis* in Central Pennsylvania. *Copeia* **1960,** 9–16.

Coates, M. I., and Clack, J. A. (1991). Fish-like gills and breathing in the earliest known tetrapod. *Nature* **352,** 234–236.

Cuffey, R. J. (1967). Bryzoan *Tabulipora carbonaria* in Wreford Megacyclothem (Lower Permian) of Kansas. *Univ. Kansas Paleontol. Contrib.* **43,** 1–96.

Cunningham, C. R., Feldman, H. R., Franseen, E. K., Gastaldo, R. A., Mapes, G., Maples, C. G., and Schultze, H.-P. (1994). The Upper Carboniferous (Stephanian) Hamilton Fossil-Lagerstätte (Kansas, U.S.A.): A valley-fill, tidally influenced deposit. *Lethaia* **26,** 225–236.

Curran, H. A. (Ed.) (1985). "Biogenic Structures: Their Use in Interpreting Depositional Environments." Society of Economic Paleontologists and Mineralogists, Special Publication No. 35, Tulsa, Oklahoma.

Darby, D. G. (1982). The early vertebrate *Astraspis,* habitat based on a lithologic association. *J. Paleontol.* **86,** 1187–1196.

Denison, R. H. (1956). A review of the habitat of the earliest vertebrates. *Fieldiana: Geol.* **11,** 359–457.

Dineley, D. L. (1976). New species of *Ctenaspis* (Ostracodermi) from the Devonian of Arctic Canada. *In* "Athlon-Essays on Palaeontology in Honour of Loris Shano Russell" (C. S. Churcher, Ed.), pp. 26–44. Royal Ontario Museum, Toronto.

Dineley, D. L., and Williams, B. P. J. (1968). Sedimentation and paleoecology of the Devonian Escuminac Formation and related strata, Escuminac Bay, Quebec. *Geol. Soc. Am. Spec. Pap.* **106,** 241–264.

Elliott, D. K., Blieck, A. R. M., and Gagnier, P.-Y. (1991). Ordovician vertebrates. *In* "Advances in Ordovician Geology" (C. R. Barnes and S. H. Williams, Eds.), pp. 93–106. Geological Survey of Canada, Paper 90-9.

Feldman, H. R., Archer, A. W., Kvale, E. P., Cunningham, C. R., Maples, C. G., and West, R. R. (1993). A tidal model of Carboniferous Konservat-Lagerstätten Formation. *Palaios* **8,** 485–498.

Fregenal Martínez, M. A., and Meléndez, N. (1996). I. Geology: Stratigraphy, basin evolution and geochemistry. I.1. Geological setting. *In* "Las Hoyas. A Lacustrine Konservat-Lagerstätte Cuenca, Spain. II. International

Symposium on Lithographic Limestones. Field Trip Guide Book" (N. Meléndez, Ed.), pp. 1–10. Modesto Escudero, Madrid.

Friend, P. F., and Moody-Stuart, M. (1970). Carbonate deposition on the river floodplains of the Wood Bay Formation (Devonian) of Spitsbergen. *Geol. Mag.* **107,** 181–195.

Friend, P. F., and Moody-Stuart, M. (1972). Sedimentation of the Wood Bay Formation (Devonian) of Spitsbergen: Regional analysis of a late orogenic basin. *Norsk Polarinst. Skrifter* **157,** 1–77.

Furrer, H. (1995). The Kalkschieferzone (Upper Meride Limestone; Ladinian) near Meride (Canton Ticino, Southern Switzerland) and the evolution of a Middle Triassic intraplatform basin. *Eclogae Geol. Helv.* **88,** 827–852.

Gaudant, J. (1979). Principes et methodes d'une paleoichthyologie bathymetrique. *Palaeogeogr. Palaeoclimatol. Palaeoecol.* **28,** 263–278.

Gaudant, J. (1989). Poissons téléostéens, bathymétrie et paléogéographie du Messinien d'Espagne méridionale. *Bull. Soc. Géol. France* **5,** 1161–1167.

Goujet, D. (1984). "Les Poissons Placodermes du Spitsberg. Arthrodires Dolichothoraci de la Formation de Wood Bay (Dévonien Inferieur)." Cahiers de Paléontologie. Centre National de la Recherche Scientifique, Paris.

Graffin, G. (1992). A new locality of fossiliferous Harding Sandstone: Evidence for freshwater Ordovician vertebrates. *J. Vert. Paleontol.* **12,** 1–10.

Gregory, W. K. (1933). Fish skulls. A study of the evolution of natural mechanisms. *Trans. Am. Philos. Soc.* **23,** I–VII + 75–481.

Gross, W. (1950). Die paläontologische und stratigraphische Bedeutung der Wirbeltierfaunen des Old Reds und der marinen altpaläozoischen Schichten. *Abhandl. Deutsch. Akad. Wissensch. Berlin, Math.-Naturwissensch. Klasse* **1949,** 1–130.

Gunter, G. (1961). Some relations of estuarine organisms to salinity. *Limnol. Oceanogr.* **6,** 182–190.

Hakes, W. G. (1985). Trace fossils from brackish-marine shales, Upper Pennsylvanian of Kansas, U.S.A. *In* "Biogenic Structures: Their Use in Interpreting Depositional Environments" (H. A. Curran, Ed.), pp. 21–35. Society of Economic Paleontologists and Mineralogists, Special Publication No. 35.

Harland, W. B., Armstrong, R. L., Cox, A. V., Craig, L. E., Smith, A. G., and Smith, D. G. (1990). "A Geologic Time Scale 1989." Cambridge University Press, Cambridge/New York.

Harms, J. W. (1929). Die Realisation von Genen und die consecutive Adaptation. I. Phasen in der Differenzierung der Anlagenkomplexe und die Frage der Landtierwerdung. *Zeitsch. Wiss. Zool.* **133,** 211–397.

Johnson, M. E. (1992). Studies on ancient rocky shores: A brief history and annotated bibliography. *J. Coast. Res.* **8,** 797–812.

Johnson, M. E., and McKerrow, W. S. (1995). The Sutton Stone: An Early Jurassic rocky shore deposit in South Wales. *Palaeontology* **38,** 529–541.

Johnson, R. G., and Richardson, E. S., Jr. (1966). A remarkable Pennsylvanian fauna from the Mazon Creek area, Illinois. *J. Geol.* **74,** 626–631.

Karatajute-Talimaa, V., and Predtechenskyi, N. (1995). The distribution of the vertebrates in the Late Ordovician and Early Silurian palaeobasins of the Siberian Platform. *Bull. Mus. natl. Hist. nat., 4. Sér., Sect. C: Sci. Terre* **27,** 39–55.

Koepcke, H.-W. (1973). "Die Lebensformen (Grundlagen zu einer universell gültigen biologischen Theorie)." Goecke & Evers, Krefeld, Germany.

Kuecher, G. J., Woodland, B. G., and Broadhurst, F. M. (1990). Evidence of deposition from individual tides and of tidal cycles from the Francis Creek Shale (host rock to the Mazon Creek Biota), Westphalian D (Pennsylvanian), northeastern Illinois. *Sediment. Geol.* **68,** 211–221.

Kurss, V. (1992). Depositional environment and burial conditions of fish remains in Baltic Middle Devonian. *In* "Fossil Fishes as Living Animals" (E. Mark-Kurik, Ed.), pp. 251–260. Academia 1, Academy of Sciences of Estonia, Tallinn.

Langenstrassen, F., and Schultze, H.-P. (1996). Unterdevonische Fischfunde aus Sedimenten des Flachmeerbereiches der kanadischen Arktis. *Neues J. Geol. Paläontol. Abhandl.* **201,** 33–93.

Lecomte, F., Meunier, F. J., and Rojas-Beltran, R. (1986). Données préliminaires sur la croissance de deux téléostéens de Guyane, *Arius proops* (Ariidae, Siluriformes) et *Leporinus friderici* (Anostomidae, Characoidei). *Cybium* **10,** 121–134.

Lenanton, R. C. J. (1977). Fishes from the hypersaline waters of the stromatolite zone of Shark Bay, Western Australia. *Copeia* **1977,** 389–390.

Major, P. F. (1978). Aspects of estuarine intertidal ecology of juvenile striped mullet, *Mugil cephalus,* in Hawaii. *U.S. Natl. Mar. Fish. Serv. Fish. Bull.* **76,** 299–314.

Maples, C. G. (1996). Paleoenvironmental significance of trace fossils in the Escuminac Formation. *In* "Devonian

Fishes and Plants of Miguasha, Quebec, Canada" (H.-P. Schultze and R. Cloutier, Eds.), pp. 114–119. Verlag Dr. Friedrich Pfeil, München.

Maples, C. G., and Schultze, H.-P. (1989). Preliminary comparison of the Pennsylvanian assemblage of Hamilton, Kansas, with marine and nonmarine contemporaneous assemblages. *In* "Regional Geology and Paleontology of Upper Paleozoic Hamilton Quarry Area in Southeastern Kansas" (G. Mapes and R. Mapes, Eds.), pp. 253–273. Kansas Geological Survey, Guidebook Series 6, Lawrence, Kansas. [For 1988]

Mark-Kurik, E. (1991). On the environment of Devonian fishes. *Proc. Estonian Acad. Sci. Geol.* **40,** 122–125.

Märss, T. (1986). ["Silurian Vertebrates of Estonia and West Latvia."] Academy of Sciences of the Estonian SSR Institute of Geology, Valgus, Tallinn. [In Russian with extensive English summary]

Märss, T., and Ejnasto, R. (1978). [Distribution of vertebrates in deposits of various facies in the North Baltic Silurian]. *Eesti NSV Teaduste Akad. Toimetised 27, Keemia, Geol.* **1,** 16–22. [In Russian]

Martínez-Delclós, X. (compl.) (1995). "Montsec and Mont-ral-Alcover, two Konservat-Lagerstätten. Catalonia, Spain. II. International Symposium on Lithographic Limestones. Pre-Symposium Field Trip." Institut d'Estudis Illerdencs, Lleida, Spain.

Masson, A. G., and Rust, B. R. (1984). Freshwater shark teeth as paleoenvironmental indicators in the Upper Pennsylvanian Morien Group of the Sydney Basin, Nova Scotia. *Can. J. Earth Sci.* **21,** 1151–1155.

Meunier, F. J., and Pascal, M. (1980). Quelques données comparatives sur la croissance périodique des tissus squelettiques chez les ostéichthyens. *Bull. Soc. Zool. France* **105,** 337–342.

Meunier, F. J., Pascal, M., and Loubens, G. (1979). Comparaison de méthodes squelettochronologiques et considerations fonctionelles sur le tissu osseux acellulaires d'un ostéichthyen du lagon néo-calédonien, *Lethrinus nebulosus* (Forskål, 1775). *Aquaculture* **17,** 137–157.

Pannella, G. (1971). Fish otoliths: Daily growth layers and periodical patterns. *Science* **173,** 1124–1127.

Pannella, G. (1972). Paleontological evidence on the earth's rotational history since early Precambrian. *Astrophys. Space Sci.* **16,** 212–237.

Pannella, G. (1974). Otolith growth pattern: An aid in age determination in temperate and tropical fishes. *In* "The Ageing of Fish" (T. B. Bagenal, Ed.), pp. 28–39. Unwin Brothers, Old Woking, Surrey, England.

Panella, G. (1975). Palaeontological clocks and the history of the earth's rotation. *In* "Growth Rhythms and the History of the Earth Rotation" (G. D. Rosenberg and S. K. Runcorn, Eds.), pp. 253–284. Wiley, London/New York.

Pannella, G. (1980). Growth patterns in fish sagittae. *In* "Skeletal Growth of Aquatic Organisms" (D. C. Rhoads and R. A. Lutz, Eds.), pp. 519–560. Plenum Press, New York.

Pratt, B. R. (1982). Stromatolite decline—A reconsideration. *Geology* **10,** 512–515.

Reisz, R. R. (1990). Geology and paleontology of the Garnett quarry. *In* "1990 Society of Vertebrate Paleontology: Upper Paleozoic of Eastern Kansas Excursion Guidebook" (C. R. Cunningham and C. G. Maples, Eds.), pp. 43–48. Kansas Geological Survey Open-file Report 90-24, Lawrence, KS.

Reisz, R. R., Heaton, M. J., and Pynn, B. R. (1982). Vertebrate fauna of Late Pennsylvanian Rock Lake Shale near Garnett, Kansas: Pelycosauria. *J. Paleontol.* **56,** 741–750.

Riding, R. (Ed.) (1991). "Calcareous Algae and Stromatolites." Springer Verlag, Berlin.

Robertson, J. D. (1957). The habitat of the early vertebrates. *Biol. Rev.* **32,** 156–187.

Romer, A. S. (1955). Fish origins—Fresh or salt water? *Pap. Mar. Biol. Oceanogr. Deep Sea Res.* **3** (Suppl.), 261–280.

Romer, A. S., and Grove, B. H. (1935). Environment of the early vertebrates. *Am. Midl. Nat.* **16,** 805–856.

Ross, R. M. (1983). Annual, semilunar, and diel reproductive rhythms in the Hawaiian labrid *Thalassoma duperrey. Mar. Biol.* **72,** 311–318.

Sawin, R. S., West, R. R., and Twiss, P. C. (1985). A stromatolite biostrome in the Upper Carboniferous of Northeast Kansas. *Compte Rendu, Neuv. Congr. Int. Stratigr. Géol. Carbon.* **5,** 361–372.

Schairer, G., and Janicke, V. (1970). Sedimentologisch-paläontologische Untersuchungen an den Plattenkalken der Sierra de Montsech (Prov. Lérida, NE-Spanien). *Neues J. Geol. Paläontol. Abhandl.* **135,** 171–189.

Schmitz, B., Åberg, G., Werdelin, L., Forey, P., and Bendix-Almgren, S. E. (1991). $^{87}Sr/^{86}Sr$, Na F, Sr, and La in skeletal fish debris as a measure of the paleosalinity of fossil-fish habitats. *Geol. Soc. Am. Bull.* **103,** 786–794.

Schultze, H.-P. (1972). New fossils from the lower Upper Devonian of Miguasha. *In* "Vertebrate Paleontology of Eastern Canada" (R. L. Carroll, E. S. Belt, D. L. Dineley, D. Baird, and D. C. McGregor, Eds.), p. 94. International Geological Congress 24th Session, Montreal 1972, Field Excursion A 59.

Schultze, H.-P. (1980). Eier legende und lebend gebärende Quastenflosser. *Nat. Mus.* **110,** 101–108.

Schultze, H.-P. (1985a). Marine to onshore vertebrates in the Lower Permian of Kansas and their paleoenvironmental implications. *Univ. Kansas Paleontol. Contrib.* **113,** 1–18.

Schultze, H.-P. (1985b). Reproduction and spawning sites of *Rhadoderma* (Pisces, Osteichthyes, Actinistia) in Pennsylvanian deposits of Illinois, USA. *Compte Rendu, Neuv. Congr. Int. Stratigr. Géol. Carbon.* **5,** 326–330.

Schultze, H.-P. (1996a). Terrestrial biota in coastal marine deposits: Fossil-Lagerstätten in the Pennsylvanian of Kansas, USA. *Palaeogeogr. Palaeoclimatol. Palaeoecol.* **119,** 255–273.

Schultze, H.-P. (1996b). The elpistostegid fish *Elpistostege,* the closest the Miguasha fauna comes to a tetrapod. *In* "Devonian Fishes and Plants of Miguasha, Quebec, Canada" (H.-P. Schultze and R. Cloutier, Eds.), pp. 316–327. Verlag Dr. Friedrich Pfeil, München.

Schultze, H.-P. (1997). Umweltbedingungen beim Übergang von Fisch zu Tetrapode. *Sitzungsber. Gesell. Naturforsch. Freunde Berlin* **36,** 59–77.

Schultze, H.-P., and Bardack, D. (1987). Diversity and size changes in palaeonisciform fishes (Actinopterygii, Pisces) from the Pennsylvanian of the Mazon Creek fauna, Illinois, USA. *J. Vert. Paleontol.* **7,** 1–23.

Schultze, H.-P., and Cloutier, R. (1996). Comparison of the Escuminac Formation ichthyofauna with other late Givetian/early Frasnian ichthyofaunas. *In* "Devonian Fishes and Plants of Miguasha, Quebec, Canada" (H.-P. Schultze and R. Cloutier, Eds.), pp. 348–368. Verlag Dr. Friedrich Pfeil, München.

Schultze, H.-P., Maples, C. G., and Cunningham, C. R. (1994). The Hamilton-Konservatlagerstätte: Stephanian terrestrial biota in a marginal-marine setting. *Trans. R. Soc. Edinburgh* **84,** 443–451.

Seilacher, A. (1967). Bathymetry of trace fossils. *Mar. Geol.* **5,** 413–428.

Soler-Gijón, R. (1994). Tidal evidence in the occipital spine of the xenacanth sharks from the Stephanian of Kansas. 64. Jahrestagung der Paläontologischen Gesellschaft, Vortrags-und Posterkurzfassungen: 60. Ungarische Geologische Gesellschaft, Budapest.

Talbot, M. R., Meléndez, N., and Fregenal-Martínez, M. A. (1996). The waters of the Las Hoyas lake: Sources and limnological characteristics. "Las Hoyas. A Lacustrine Konservat-Lagerstätte Cuenca, Spain. II. International Symposium on Lithographic Limestones. Field Trip Guide Book" (N. Meléndez, Ed.), pp. 11–16. Modesto Escudero, Madrid.

Tintori, A. (1996). The field excursion in Northern Italy. *In* "Mesozoic Fishes—Systematics and Paleoecology" (G. Arratia and G. Viohl, Eds.), pp. 567–575. Verlag Dr. Friedrich Pfeil, München.

Toole, C. L. (1980). Intertidal recruitment and feeding in relation to optimal utilization of nursery areas by juvenile English sole (*Parophrys vetulus:* Pleuronectidae). *Environ. Biol. Fish.* **5,** 383–390.

Toomey, D. F., and Cys, J. M. (1977). Spirorbid/algal stromatolites, a probable marginal marine occurrence from the Lower Permian of New Mexico, USA. *Neu. J. Geol. Paläontol. Monatshefte* **1977,** 311–342.

Turner, S. (1992). Thelodont lifestyles. *In* "Fossil Fishes as Living Animals" (E. Mark-Kurik, Ed.), pp. 21–40. Academia 1, Academy of Sciences of Estonia, Tallinn.

Vézina, D. (1991). Nouvelles observations sur l'environment sédimentaire de la Formation d'Escuminac (Dévonien supérieur, Frasnien), Québec, Canada. *Can. J. Earth Sci.* **28,** 225–230.

Viohl, G. (1996). The paleoenvironment of the Late Jurassic fishes from the southern Franconian Alb (Bavaria, Germany). *In* "Mesozoic Fishes—Systematics and Paleoecology" (G. Arratia and G. Viohl, Eds.), pp. 513–528. Verlag Dr. Friedrich Pfeil, München.

Visser, M. J. (1980). Neap-spring cycles reflected in Holocene subtidal large-scale bedform deposits: A preliminary note. *Geology* **8,** 543–546.

Walcott, C. D. (1892). Notes on the discovery of a vertebrate fauna in Silurian (Ordovician) strata. *Geol. Soc. Am. Bull.* **3,** 153–172.

Walter, M. R. (Ed.) (1976). "Stromatolites—Developments in Sedimentology 20." Elsevier, Amsterdam.

Walther, J. (1904). Die Fauna der Solnhofener Plattenkalke bionomisch betrachtet. *Festschr. Med.-Naturwissensch. Gesellsch. Jena* **11,** 133–214.

Williams, G. E. (1989). Late Precambrian tidal rhythms in South Australia and the history of the Earth's rotation. *J. Geol. Soc. London* **146,** 97–111.

Wilson, M. V. H., and Caldwell, M. W. (1993). New Silurian and Devonian fork-tailed 'thelodonts' are jawless vertebrates with stomachs and deep bodies. *Nature* **361,** 442–444.

Zorn, H. (1971). Paläontologische, stratigraphische und sedimentologische Untersuchungen des Salvatoredolomits (Mitteltrias) der Tessiner Kalkalpen. *Schweizer. Paläontol. Abhandl.* **91,** 1–90.

# INDEX


Abundance, estimation for intertidal fishes, 16
Acanthuridae, intertidal species, 321
Acid–base balance
  emersion effects, 70, 73–74
  ion transport, 83–85
  salt water concentration effects, 84
  urea effects, 87
Aerial respiration
  acid–base balance following emersion, 70, 73–74
  anatomical adaptations in intertidal fishes, 45, 47, 54–55
  blood oxygen affinity of air-emerging intertidal fishes, 73
  circulation, effects of air exposure, 68–69
  convergent evolution, 368–369
  emergence behavior, 56–57, 60
  esophagus, 61
  gills, 61–62
  respiratory gas exchange in air versus water, 63–64
  skin respiration, 61–63
  species of intertidal fishes, 58–60
*Alticus kirkii*
  aerial respiration, 45, 47, 63
  clinging and moving, 41–42, 44
Ammodytidae, intertidal species, 317
Ammonia
  blood levels, 91
  excretion, 85–91
  sources, 86
*Amphichthys cryptocentrus*, parental care, 166
Anablepidae, intertidal species, 305
Anaerobiosis, hypoxia in intertidal fishes, 63–64
Anguillidae, intertidal species, 298, 300
Anguilliformes, intertidal species, 298, 300–301
*Anoplarchus purpurescens*, parental care, 169
Antennariidae, intertidal species, 303–304
Aplodactylidae, intertidal species, 313
Apogonidae, intertidal species, 309–310
Ariidae, intertidal species, 302
*Artedius lateralis*, predation, 242
Atherinidae, intertidal species, 304
Aulopiformes, intertidal species, 302
Australia, biogeography of rocky intertidal fishes, 341

Balistidae, intertidal species, 323
Batrachoididae, intertidal species, 303
Behavior, laboratory experiments, 20–21
Belonidae, intertidal species, 304
Beryciformes, intertidal species, 305
Biogeography, rocky intertidal fishes
  Atlantic Ocean, 345–347
  data collection for survey, 332–334, 336–339
  Eastern Pacific, 343–345
  Indo-West Pacific, 340–343
  latitudinal patterns of species richness, 351–352
  overview of faunas, 334, 339
  similarity relationships among regional fishes, 347–349, 351
Blenniidae
  herbivory, 211–212
  intertidal species, 318–319
Blood affinity, oxygen of air-emerging intertidal fishes, 73
Body, shape and size
  classification of body plans, 362, 364, 366
  convergent evolution of features, 366–367
  effects of habitat, 127–128, 138–139
  histological and morphological features of fossils, 378–380
  minimum body size and gape, 249–250
  parental care correlations, 175–177
  size at recruitment, 188–189
Bothidae, intertidal species, 322
Bovichthyidae, intertidal species, 316
Brain, development comparison of intertidal and deep sea fishes, 136, 138
Bythitidae, intertidal species, 302

Callionymidae, intertidal species, 320
Capture, intertidal fishes
chemical methods
infrequently used chemicals, 12
precautions, 12
quinaldine, 11
rotenone, 10–11
manual methods, 9–10
netting
drop nets, 13–14
flumes, 14
Fyke nets, 14
gill nets, 14–15
lift nets, 14
seines, 12
stake nets, 14
trammel nets, 15
trawls and push nets, 12–13
weirs, 14
selection of method, 15–16
trapping, 15
Carangidae, intertidal species, 310
Carbon dioxide, respiratory gas exchange in water and air, 64
*Cebidichthys violaceus*
age structure of populations, 267
herbivory, 212, 214–215
prehistoric man food source, 3–4
Chaenopsidae, intertidal species, 318
Chaetodontidae, intertidal species, 311–312
Chandidae, intertidal species, 308
Cheilodactylidae, intertidal species, 313
Chemosensory anterior dorsal fin in rocklings
function, 136
solitary chemosensory cells, 134, 136
undulation, 134, 136
Chironemidae, intertidal species, 312
Circulation
cardiac responses
air exposure, 68–69
hyperoxia, 68
hypoxia, 68
temperature effects, 68
Cirrhitidae, intertidal species, 312
Clinging, coping with dynamic habitats, 41–43
Clinidae, intertidal species, 318
*Clinocottus analis*
interspecific interactions, 285–286
ontogenic changes in diet, 245–246, 249
*Clinocottus globiceps*
age structure of populations, 267
predation, 239
recruitment, 185, 187
*Clinus superciliosus*, intraspecific interactions, 287
Clupeidae, intertidal species, 301–302
Cold-Temperate Northeastern Pacific (CTNEP), biogeography of rocky intertidal fishes, 344
Color marking, intertidal fishes, 17, 20
Community, intertidal fishes
age structure of populations, 266–268
behavioral structure, 268, 270
convergence, *see* Convergent evolution
definition, 264
duration of occupancy, 265–266
interspecific interactions, 284–286
intraspecific interactions, 286–287
predation effects, 287–290
spatial variation
connectedness of geographically disjunct communities, 275–276
global patterns and latitudinal variation, 270–273, 289
individual tidepool variation, 278
locality studies, 276–278
regional variation, 271, 274–275
temporal variation
seasonal variation, 279–280
stability, resilience, and persistence, 281–284
tidal and diel cycles, 279
trophic structure, 268
Congiopodidae, intertidal species, 307
Congridae, intertidal species, 301
Conservation, intertidal fishes
ocean warming effects, 4
refuges, 4–5
Convergent evolution
community convergence, 370–371
community study sites
body plan classification, 362, 364, 366
taxonomic composition, 358–362
definition, 357
functional trait evidence
air-breathing, 368–369
desication tolerance, 367–368
parental care, 369
genetic isolation of communities, 356–357
structural trait evidence, 366–367
*Coryphoblennius galerita*
clinging and moving, 41
vertical range, 32–34
Cottidae
aerial respiration, 58
clinging and moving, 43
intertidal species, 307–308
Creedidae, intertidal species, 316
Cryptacanthodidae, intertidal species, 315
CTNEP, *see* Cold-Temperate Northeastern Pacific
Cyclopteridae, intertidal species, 308
Cynoglossidae, intertidal species, 322
Cyprinodontiformes, intertidal species, 305

Dactyloscopidae, intertidal species, 317
Demersal egg, *see* Egg

Desiccation
evaporative water loss rate, 80–82
resistance by intertidal fishes, 43–44
survival times of fishes in air, 80–81
tolerance, convergent evolution, 368–369
Diet, *see* Herbivory; Predation
Digestion
herbivore mechanisms, 209–211, 245
prey, 243
Diodontidae, intertidal species, 323
Diurnal changes
community structure effects, 279
spawning effects, 152
vertical distribution of intertidal fishes, 40
Diving, intertidal fish observation, 8
Drop net, capturing intertidal fishes, 13–14

Easter Island, biogeography of rocky intertidal fishes, 341
Egg, *see also* Spawning
accessory nest structures, 150
demersal eggs
coloration, 147–148
number, 147
size, 147
planktonic eggs, 147
protection, *see* Parental care
Eleotridae, intertidal species, 320
Embiotocidae, intertidal species, 313
Emergence behavior, *see* Aerial respiration
Engraulidae, intertidal species, 301
Esophagus, aerial respiration, 61
Estuary, movement of intertidal fishes, 106–107
Evolution, *see* Convergent evolution; Phylogenetic analysis
Eye
aerial vision adaptation, 44–45, 132
constraints on vision in small-sized fish, 133–134
integration centers for vision, 133, 138
intertidal versus deep sea fishes, 138
position in various species, 131–132
resolving power, 132–133
size and visual capacity, 132

Fin clipping, 17
Fistulariidae, intertidal species, 306
Flounder, *see Platichthys flesus*
Flume, capturing intertidal fishes, 14
Fossils, intertidal zone
criteria for recognizing fossil shores
histological and morphological features of fossils, 378–380
mixed assemblages and juvenile forms, 377
sediment structures, 373–374
stromatolites, 375
tracks and traces, 376
examples from different ages
Carboniferous, 384–385, 387–388
Cretaceous, 386
Devonian, 383–384, 387–388
Jurassic, 386
Ordovician, 381
Permian, 385, 387
Silurian, 381–383, 387
Tertiary, 386
Triassic, 385–386
freshwater versus seawater origin, 380–381
Fundulidae, intertidal species, 305
*Fundulus heteroclitus*
intertidal movements, 108–109
spawning, 155
Fyke net, capturing intertidal fishes, 14

Gadidae, intertidal species, 303
Gadiformes, intertidal species, 303
Gasterosteidae, intertidal species, 306
Gerreidae, intertidal species, 310
Gill
modifications for aerial respiration, 60–62
nitrogen excretion, 88
structure, 60
Gill net, capturing intertidal fishes, 14–15
Ginglymostomatidae, intertidal species, 298
Gobiesocidae
aerial respiration, 58
clinging and moving, 43
intertidal species, 319–320
*Gobiesox maeandricus*, parental care, 171
Gobiidae
aerial respiration, 58–59
intertidal species, 320–321

Haemulidae, intertidal species, 310
Hair cell senses, characteristic in intertidal fishes, 130–131
Hawaiian Islands, biogeography of rocky intertidal fishes, 341
Hearing, characteristic in intertidal fishes, 130–131
Heart, *see* Circulation
*Hemilepidotus hemilepidostus*, parental care, 173
Hemitripteridae, intertidal species, 308
Herbivory
case studies
Bleniidae, 211–212
Stichaeidae, 212, 214–215
digestive mechanisms, 209–211, 245
diversity of diets, 206
ecological impact, 215–216
family and species diversity, 198–201
feeding behavior
biting, 203

Herbivory (*continued*)
browsers, 203
grazers, 203
spin feeding, 203–204
food preferences and factors influencing, 206–207
gut length and transit times, 208–209
herbivore definition in fishes, 197
latitudinal diversity and hypotheses, 201–202
rocky intertidal habitats, 197–198
territorial defense, 204–205
tooth and jaw morphology, 205–206
Hexagrammidae, intertidal species, 307
*Hexagrammos decagrammus*, prehistoric man food source, 3
Holocentridae, intertidal species, 305
Homing
functions
feeding, 113
physiologically optimum environment searching, 117
predator and adverse condition avoidance, 113, 116–117
spawning, 118
resident intertidal fishes, examples, 112, 114–115
success, factors affecting, 112

Indian Ocean, biogeography of rocky intertidal fishes, 340
Intertidal zone
definition of fishes, 1
safety guidelines for fieldwork, 21
variations in space and time, 2

Kidney, nitrogen excretion, 88–89
Kuhliidae, intertidal species, 312
Kyphosidae, intertidal species, 312

Labridae, intertidal species, 314
Labrisomidae, intertidal species, 318
Lateral line, comparison between species, 45, 130
Latridae, intertidal species, 313
Leptoscopidae, intertidal species, 316
Lift net, capturing intertidal fishes, 14
Liparididae, intertidal species, 308
*Lipophrys pholis*
aerial respiration, 61
intraspecific interactions, 287
vertical range, 32–33, 40
*Lipophrys trigloides*
aerial respiration, 45
clinging and moving, 41
Lophiiformes, intertidal species, 303–304
Lutjanidae, intertidal species, 310

Mangal, dietary influences on intertidal fishes, 236, 238
Mangrove, movement of intertidal fishes, 110
Marking, intertidal fishes, 9, 16–20
*Menidia menidia*
intertidal movements, 109
spawning, 155–156
Microdesmidae, intertidal species, 321
Midshipman, parental care, 171–172
Monacanthidae, intertidal species, 323
Monodactylidae, intertidal species, 311
Moridae, intertidal species, 303
Moringuidae, intertidal species, 300
Moronidae, intertidal species, 309
Movement, intertidal fishes
coping with dynamic habitats, 41–43
directional cues, 118–119
estuaries, 106–107
functions
feeding, 113
physiologically optimum environment searching, 117
predator and adverse condition avoidance, 113, 116–117
spawning, 118
mangroves, 110
modeling, 119–120
mudflats, 107–108
rocky shores, 98–99, 101–103
salt marshes, 108–109
sandy beaches, 103–106
seagrass beds, 110, 112
spatial and temporal scales, 97–98
temporal cues, 118
Mudflat
dietary influences on intertidal fishes, 236
movement of intertidal fishes, 107–108
Mugilidae, intertidal species, 304
Mullidae, intertidal species, 311
Muraenidae, intertidal species, 301

Nemiptheridae, intertidal species, 311
New Zealand, biogeography of rocky intertidal fishes, 341–343
Nitrogen excretion
ammonia, 85–87, 89–91
emersion effects, 85–86, 89–91
gills, 88
kidney, 88–89
salinity effects, 86
skin, 89
urea, 85, 87–91
Northeastern Atlantic (TENEA), biogeography of rocky intertidal fishes, 345–346
Northwestern Atlantic (TENWA), biogeography of rocky intertidal fishes, 346

Notocheiridae, intertidal species, 304
Nototheniidae, intertidal species, 316

Observation, intertidal fishes in the field, 8–9
Odacidae, intertidal species, 314
Olfaction, characteristic in intertidal fishes, 130
*Oligocottus maculosus*
  age structure of populations, 266–267
  homing, 112
  interspecific interactions, 284–286
  intertidal movements, 98–99, 101, 118–119
  recruitment, 185, 187, 190
*Oligocottus snyderi*
  interspecific interactions, 284–286
  predation, 242
Ophichthidae, intertidal species, 301
Ophidiiformes, intertidal species, 302
Opistognathidae, intertidal species, 309
Orectolobiformes, intertidal species, 298
Ornithine–urea cycle (OUC), urea synthesis, 87
Osmeridae, intertidal species, 302
Osmoregulation
  evaporative water loss rate, 80–82
  ion transport, 79–80, 82
  salt secretion, 82
  seawater ingestion, 82
  sodium, plasma response to external concentration, 82–83
  survival times of fishes in air, 80–81
Ostraciidae, intertidal species, 323
OUC, *see* Ornithine–urea cycle
Oviparous reproduction, distribution among species, 144–145
Oxygen, *see also* Aerial respiration
  blood affinity of air-emerging intertidal fishes, 73
  respiratory environment of intertidal zone, 55–56
  respiratory gas exchange in water and air, 63–64
  seawater versus freshwater content, 55

*Parablennius sanguinolentus*, herbivory, 211–212
Paralichthyidae, intertidal species, 322
Parental care
  body size correlations, 175–177
  coiling, 169
  convergent evolution of traits, 369
  definition, 165
  duration, 174
  egg burying, 168
  egg scattering, 168–169
  evolution in vertebrates, 174–175
  exceptions in intertidal fishes, 173, 175
  forms, 166–174
  guarding, 165–167, 169–173
  oral brooding, 168
  overview, 165–167
  posthatching care, 166
  splashing, 168
  Stichaeoidea superfamily distribution, 175–177
  uniparental care, 166, 174, 177–178
  viviparity, 167–168
Pempheridae, intertidal species, 311
Perciformes, intertidal species, 308–322
Percophidae, intertidal species, 316
*Periophthalmus argentilineatus*, ventilation, 67–68
Pholididae, intertidal species, 315
*Pholis gunnellus*, parental care, 169–171
Phycidae, intertidal species, 303
Phylogenetic analysis, intertidal fishes
  diversity, 297, 299–300
  family distribution, 324–327
  Phylogenetic Analysis Using Parsimony program, 348–349
  similarity relationships among biogeographical regions, 347–349, 351
Pinguipedidae, intertidal species, 316
*Platichthys flesus*, intertidal movements, 106–107
Platycephalidae, intertidal species, 307
Plesiopidae, intertidal species, 309
Pleuronectidae, intertidal species, 322
Pleuronectiformes, intertidal species, 322
Plotosidae, intertidal species, 302
Poeciliidae, intertidal species, 305
*Pollachias virens*, prehistoric man food source, 3
Pomacanthidae, intertidal species, 312
Pomacentridae, intertidal species, 313–314
Predation
  capture of prey, 240–242
  cost/benefit considerations of nutrional gain, 251
  detection of prey, 251
  dietary studies, overview by region, 224–228
  digestion, 243
  ecological impact
    antipredator adaptations of intertidal organisms, 252
    community, 287–288
    direct and indirect costs on prey, 252
    experimental manipulations of predators or prey, 254
    overview, 251–252
    production versus consumption estimates of prey, 253–254
  functional interactions between fish and prey
    arthropods, 245
    crustaceans, 244
    gastropods, 244
    molluscs, 244
    shrimp, 243
    worms, 243–244
  habitat influence on diet
    mangals, 236, 238
    mudflats, 236
    overview, 228–229

Predation (*continued*)
rocky intertidal, 229–231
salt marsh, 239
sandy beaches, 233–235
seagrass beds, 236
ontogenic changes in diet
factors driving changes, 249–250
minimum body size and gape, 249–250
patterns of intraspecific and interspecific diet overlap, 247–249
summary of changes by prey type, 245–247
processing of prey, 242–243
residents versus transients, 223
Pseudochromidae, intertidal species, 309

Quinaldine, capturing intertidal fishes, 11

Recruitment
definition, 181
population consequences of recruitment patterns, 190–192
significance of study, 181–182
size at recruitment, 188–189
spatial and temporal variability in recruitment, 188, 190
success, factors affecting
habitat, 184–185
interactions with other fishes, 185, 187
timing, 182–184
Respiration, *see* Aerial respiration; Oxygen; Ventilation
Rock flat, fish species, 27–28
Rock pool, fish species, 28–30
Rocky intertidal
dietary influences on intertidal fishes, 229–231
herbivory, 197–198
movement of intertidal fishes, 98–99, 101–103
taxonomic structure of fish communities by location, 271–274
Rotenone, capturing intertidal fishes, 10–11

SAFR-A, *see* Temperate South Atlantic
Salt marsh
dietary influences on intertidal fishes, 239
movement of intertidal fishes, 108–109
Sandy beach
dietary influences on intertidal fishes, 233–235
movement of intertidal fishes, 103–106
Scaridae, intertidal species, 314
Scatophagidae, intertidal species, 321
SCCs, *see* Solitary chemosensory cells
Sciaenidae, intertidal species, 311
Scophthalmidae, intertidal species, 322
*Scorpaenichthys marmoratus*, prehistoric man food source, 3
Scorpaenidae, intertidal species, 306–307
Scorpaeniformes, intertidal species, 306–308
Scytalinidae, intertidal species, 315
Seagrass bed
dietary influences on intertidal fishes, 236
movement of intertidal fishes, 110, 112
Seasonal changes
community structure effects, 279–280
recruitment effects, 182–184
spawning effects, 152
vertical distribution of intertidal fishes, 39–40
*Sebastes*, prehistoric man food source, 3
Seine, capturing intertidal fishes, 12
Serranidae, intertidal species, 309
Siganidae, intertidal species, 321
Sillaginidae, intertidal species, 310
Siluformes, intertidal species, 302
Skin
aerial respiration, 61–63
chemosenses, 128, 130
nitrogen excretion, 89
Soleidae, intertidal species, 322
Solitary chemosensory cells (SCCs), *see* Chemosensory anterior dorsal fin in rocklings
Sparidae, intertidal species, 310–311
Spawning, *see also* Egg
adaptation versus preadaptation of traits, 157
case studies, 154–157
costs and benefits of intertidal spawning, 154, 158
demeral spawners, 148, 150
egg types, 147–148
influencing factors
diel, 152
endogenous rhythms, 153
horizontal habitat features, 151
lunar and tidal cycles, 152–153
season, 152
tidal stranding, 153–154
vertical zonation, 151–152
mixed subtidal and intertidal spawners, 157–158
oviparous versus viviparous reproduction, 144–145
resident versus migrant fish, 143, 158
taxonomic representations of spawning fishes, 144–145
Stake net, capturing intertidal fishes, 14
Stichaeidae
herbivory, 212, 214–215
intertidal species, 315
parental care distribution, 175–177
Stromatolite, fossil shore recognition, 375
Supratidal habitat, fish species, 30–32
*Symphodus melops*
prehistoric man food source, 3
spawning, 157–158
Syngnathidae, intertidal species, 306

*Syngnathus leptorhynchus*, parental care, 167
Synodontidae, intertidal species, 302

Tagging, intertidal fishes
  external tags, 17
  internal tags, 17
  ultrasonic transmitters, 9, 99
*Takifugu niphobles*, spawning, 156
Taste bud
  solitary chemosensory cell comparison, 134, 136
  species distribution, 128
Telemetry, intertidal fish observation, 9
Television, underwater observation, 9
Temperate South Atlantic (SAFR-A), biogeography of rocky intertidal fishes, 346–347
Temperate-Southeastern Pacific (TESEP), biogeography of rocky intertidal fishes, 344–345
TENEA, *see* Northeastern Atlantic
TENWA, *see* Northwestern Atlantic
TESEP, *see* Temperate-Southeastern Pacific
Tetraodontidae, intertidal species, 323
Tetraodontiformes, intertidal species, 323
Tide, rhythmic behavior of intertidal fishes, 2–3
Tidepool, volume measurement, 7–8
Trachichthyidae, intertidal species, 305
Trachinidae, intertidal species, 317
Trammel net, capturing intertidal fishes, 15
Transport, intertidal fishes, 20
Trawl, capturing intertidal fishes, 12–13
Tripterygiidae, intertidal species, 317
TRNWA, *see* Tropical Northwestern Atlantic
Tropical Northwestern Atlantic (TRNWA), biogeography of rocky intertidal fishes, 345

Uranoscopidae, intertidal species, 317
Urea
  excretion, 85, 88–91
  sources, 87

Ventilation
  activity on emersion, 67
  rate
    hyperoxia effects, 66
    hypoxia effects, 66–67
    normoxic water, 64–65
    temperature effects, 65–66
Vertical distribution, intertidal fishes
  anatomical adaptation, 26, 127–128
  body shape effects, 127–128, 138–139
  diurnal changes, 40
  ecological patterns
    rock flats, 27–28
    rock pools, 28–30
    supratidal habitats, 30–32
  intraspecific vertical zonation, 38–39
  Japan, 35–36
  North American Pacific coast, 36–38
  Northern Europe to Southern Africa, 32–35
  seasonal changes, 39–40
  spawning influence, 151–152
  terminology of fish types, 48
  transient versus resident species, 26–27
Vision, *see* Eye
Viviparous reproduction, distribution among species, 144–145, 167–168

Warm-Temperate and Subtropical Northeastern Pacific (WTSTNEP), biogeography of rocky intertidal fishes, 344
Weir, capturing intertidal fishes, 14
WTSTNEP, *see* Warm-Temperate and Subtropical Northeastern Pacific

Zeidae, intertidal species, 305
Zoarcidae, intertidal species, 315